Formation of structure in the universe

This advanced textbook provides an up-to-date and comprehensive introduction to the very active field of structure formation in cosmology. It is written by eleven world-leading authorities.

Written in a clear and pedagogical style appropriate for graduate students in astronomy and physics, this textbook introduces the reader to a wide range of exciting topics in contemporary cosmology – from recent advances in redshift surveys, to the latest models in gravitational lensing and cosmological simulations. The authors are all world-renowned experts both for their research and teaching skills.

In the fast-moving field of structure formation, this book provides advanced undergraduate and graduate students with a welcome textbook which unites the latest theory and observations.

Formation of structure in the universe

Edited by

Avishai Dekel
The Hebrew University of Jerusalem

Jeremiah P. Ostriker
Princeton University

CAMBRIDGE UNIVERSITY PRESS
Cambridge, New York, Melbourne, Madrid, Cape Town, Singapore,
São Paulo, Delhi, Dubai, Tokyo

Cambridge University Press
The Edinburgh Building, Cambridge CB2 8RU, UK

Published in the United States of America by Cambridge University Press, New York

www.cambridge.org
Information on this title: www.cambridge.org/9780521586320

© Cambridge University Press 1999

This publication is in copyright. Subject to statutory exception
and to the provisions of relevant collective licensing agreements,
no reproduction of any part may take place without the written
permission of Cambridge University Press.

First published 1999

A catalogue record for this publication is available from the British Library

Library of Congress Cataloguing in Publication data

Formation of structure in the universe / edited by Avishai Dekel,
Jeremiah P. Ostriker.
    p.   cm.
  Includes indexes.
  ISBN 0 521 58422 1 – ISBN 0 521 58632 1 (pb)
  1. Large scale structure (Astronomy)   2. Cosmology.
3. Astrophysics.   I. Dekel, Avishai, 1951– . II. Ostriker, J. P.
QB991.L37F67   1999
523.1–dc21   97-43008 CIP

ISBN 978-0-521-58422-7 Hardback
ISBN 978-0-521-58632-0 Paperback

Transferred to digital printing 2010

Cambridge University Press has no responsibility for the persistence or
accuracy of URLs for external or third-party internet websites referred to in
this publication, and does not guarantee that any content on such websites is,
or will remain, accurate or appropriate. Information regarding prices, travel
timetables and other factual information given in this work are correct at
the time of first printing but Cambridge University Press does not guarantee
the accuracy of such information thereafter.

To our teachers,
to whom we are forever grateful

Contents

List of contributors	page	xi
Preface		xiii

Part one: Introduction to structure formation   1

1 Dark matter and structure formation   3
J. R. Primack
1.1 Introduction   3
1.2 Cosmology basics   7
1.3 Age, expansion rate, and cosmological constant   12
1.4 Measuring Ω_0 23
1.5 Dark-matter particles   33
1.6 Origin of fluctuations: inflation and topological defects   42
1.7 Comparing DM models to observations: ΛCDM vs. CHDM   54
References   76

2 Gravitational instability   86
A. Yahil
2.1 Introduction   86
2.2 Linear theory and Zel'dovich approximation   88
2.3 Nonlinear methods and mixed boundary conditions   90
References   97

3 Microwave background and structure formation   98
J. Silk
3.1 Introduction   98
3.2 Gravitational instabilities: linear theory   99
3.3 Nonlinear evolution   105
3.4 Galaxy formation by reverse engineering   111
3.5 Cosmic microwave background anisotropies   118
3.6 Confrontation of theory and large-scale structure   123

viii Contents

	3.7 Future prospects	126
	References	131

Part two: Large-scale structure and flows — 133

4	Clusters and superclusters of galaxies	135
	N. A. Bahcall	
	4.1 Introduction	135
	4.2 Optical properties of galaxy clusters	136
	4.3 X-ray properties of galaxy clusters	143
	4.4 The baryon fraction in clusters	151
	4.5 Cluster masses	152
	4.6 Where is the dark matter?	155
	4.7 The mass function of clusters	155
	4.8 Quasar–cluster association	156
	4.9 Superclusters	158
	4.10 The cluster correlation function	160
	4.11 Peculiar motions of clusters	164
	4.12 Some unsolved problems	167
	References	169
5	Redshift surveys of the local universe	172
	M. A. Strauss	
	5.1 Introduction	172
	5.2 Varieties of redshift surveys	173
	5.3 The luminosity and selection function	178
	5.4 Clustering statistics	181
	5.5 Measurements of the power spectrum	184
	5.6 The relative distribution of galaxies and dark matter	195
	5.7 Surveys for the future	198
	5.8 Conclusions	206
	References	208
6	Measurement of galaxy distances	213
	J. A. Willick	
	6.1 Introduction	213
	6.2 Cepheid variables	216
	6.3 The Tully–Fisher relation for spiral galaxies	219
	6.4 Fundamental Plane relations for elliptical galaxies	225
	6.5 Surface brightness fluctuations	229
	6.6 Supernovae	233
	6.7 Brightest cluster galaxies	237

6.8	Redshift-distance catalogs	239
6.9	Malmquist and other biases	241
6.10	Summary	245
	References	247

7 Large-scale flows and cosmological implications 250
A. Dekel

7.1	Introduction	250
7.2	Reconstruction from peculiar velocities	253
7.3	Testing basic hypotheses	273
7.4	Statistics of mass-density fluctuations	280
7.5	Direct measurements of Ω from peculiar velocities	289
7.6	Measurements of β from galaxy density and velocities	293
7.7	Cosmological parameters	306
7.8	Conclusion	312
	References	315

Part three: Structure on galactic scales and lensing 319

8 Cosmological simulations 321
J. P. Ostriker

8.1	Introduction	321
8.2	Simulation methods	323
8.3	Results: comparison with observations	330
8.4	Conclusions	332
	References	335

9 Black holes in galaxy centers 337
S. M. Faber

9.1	Introduction	337
9.2	Surface-brightness and stellar distributions	339
9.3	Kinematic evidence for central massive black holes	345
	References	358

10 Gravitational lensing 360
R. Narayan & M. Bartelmann

10.1	Introduction	360
10.2	Lensing by point masses in the universe	364
10.3	Lensing by galaxies	384
10.4	Lensing by galaxy clusters and large-scale structure	406
	References	426

Part four: A conclusion 433

11 The mass of the universe 435
   P. J. E. Peebles
   11.1 Introduction 435
   11.2 Dynamical mass measurements 443
   11.3 The cosmological tests 455
   11.4 Scorecard and issues 460
   11.5 Concluding remarks 463
   References 464

   Index 466

Contributors

N. A. Bahcall
Department of Astrophysical Sciences, Peyton Hall, Princeton University, Princeton, NJ 08544, USA

M. Bartelmann
Max-Planck-Institut für Astrophysik, P. O. Box 1523, D-85740 Garching, Germany

A. Dekel
Racah Institute of Physics, The Hebrew University, Jerusalem 91904, Israel

S. M. Faber
UCO/Lick Observatories, University of California, Santa Cruz, CA 95064, USA

R. Narayan
Harvard-Smithsonian Center for Astrophysics, 60 Garden Street, Cambridge, MA 02138, USA

J. P. Ostriker
Department of Astrophysical Sciences, Peyton Hall, Princeton University, Princeton, NJ 08544, USA

P. J. E. Peebles
Joseph Henry Laboratories, Jadwin Hall, Princeton University, Princeton, NJ 08544, USA

J. R. Primack
Department of Physics, University of California, Santa Cruz, CA 95064, USA

J. Silk
Astronomy Department, Campbell Hall, University of California, Berkeley, CA 94720, USA

M. A. Strauss
Department of Astrophysical Sciences, Peyton Hall, Princeton University, Princeton, NJ 08544, USA

J. A. Willick
Department of Physics, Stanford University, Stanford, CA 94305, USA

A. Yahil
Department of Earth & Space Sciences, State University of New York, Stony Brook, NY 11794, USA

Preface

The study of structure formation in the universe has become a mature scientific field, where observations and theory confront one another in a quantitative way. This development has been driven by the rapid accumulation of quality data from advanced telescopes, on the ground or in space, covering the whole range of the radiation spectrum. The accompanying theoretical progress takes advantage of the rapid developments in desk-top computing and supercomputing, that now allow cosmological simulations with almost billion particles or fluid cells, and enable the comparison between the improved observations and the detailed predictions of a variety of theories. These developments make this field one of the most active areas of research in cosmology and astrophysics.

This book grew out of a winter school held in January 1996 at the Institute of Advanced Studies of the Hebrew University of Jerusalem. It contains reviews written by leading authorities, describing the current developments in their own fields of research, and is intended to serve as a supporting text for graduate or advanced undergraduate courses in astrophysics. The chapters can be used separately, or as a comprehensive collection of lectures with significant cross-talk among them. The content of the book covers most of the active topics in the research aimed at understanding the formation of structure in the universe on scales of galaxies and beyond.

We have organized the chapters into four parts. In Part one, *Introduction to structure formation*, the basics of cosmological theory are laid out. Primack opens with a comprehensive overview of the background cosmological model, the origin of fluctuations in the early universe and the dark-matter content of the universe. Then, Yahil provides a concise summary of the theory of gravitational instability, in the linear and the nonlinear regime. Silk concludes this part by addressing issues in galaxy formation, and by making the connection to the large-scale, small-amplitude fluctuations in the cosmic microwave background radiation. It was the direct observation of these fluctuations by the COBE satellite in 1991, on angular scales of order $10°$ corresponding to comoving lengths of $\sim 1000\,h^{-1}$ Mpc, which confirmed that our ideas for the growth of structure in the universe were basically correct.

In Part two, *Large-scale structure and flows*, we turn to observations of the local universe made at the largest scales measured to date (of order $100\,h^{-1}$ Mpc), where perturbations, having entered the nonlinear domain, produce great agglomerations of matter detectable in optical surveys of galaxies. The emphasis is on statistical properties of the galaxy distribution in real and velocity space. Bahcall provides a detailed review of the properties of galaxy clusters and superclusters both in the optical and X-ray bands. Strauss describes the analysis of redshift surveys and current developments towards redshift surveys of millions of galaxies. Willick provides an up-to-date account of the most useful extra-galactic distance indicators in the context of measuring the Hubble constant and peculiar velocities. Dekel concludes this part with a comprehensive discussion of the analysis of cosmic flows and their cosmological implications on the initial fluctuations and the universal mass density.

In Part three, *Structure on galactic scales and lensing*, we focus on phenomena on galactic scales, including Lyman-α clouds and black holes, for the insight that they give us into cosmological phenomena. Ostriker discusses cosmological simulations including gravity and gas processes, mostly in the context of clusters and absorption clouds, and addresses competing cosmological scenarios. Faber describes the effort to detect black holes in the centers of galaxies using the Hubble Space Telescope. Narayan & Bartelmann provide a comprehensive and pedagogical review of the rapidly developing subject of gravitational lensing, from stars, via galaxies, to clusters and large-scale structure.

Part four, *A conclusion*, is a summary by Peebles of the issues and tentative conclusions with regard to the overall background model and the nature of fluctuations. The stress is on the most important cosmological parameter, Ω, the mean mass density in the universe in units of the critical density. We highlight this contribution not because we all agree with each of the conclusions presented, but because Peebles' enormous contributions to physical cosmology warrant special prominence being given to his views.

There is certain overlap between the chapters, but the reader will notice that the discussion reflects different perspectives among the authors, where the same data may be interpreted in different ways and lead to somewhat different conclusions. For example, while Bahcall, Ostriker and Peebles conclude that the universal density parameter, Ω, must be significantly smaller than unity, which means that the universe is unbound and either open or dominated by a cosmological constant, Dekel and Primack stress the evidence for a somewhat larger value, of order 0.5 and perhaps even unity, consistent with a marginally-bound universe with no cosmological constant. This controversy reflects the honest division of opinion among experts in a rapidly developing field, and we have made no effort to impose ideological uniformity (or even consistency). However, the data expected in the early 2000s, the planned progress in simulation ability, and the quantitative scientific approach as reflected in this book, promise that we may be able to close in on the cosmological model and its parameters in the near future.

We believe that the book provides a useful summary of the principle questions being asked in worldwide cosmological investigations at the end of the 20th century. We are less sanguine that all the answers provided by our expert authors will stand the test of time, as the topics described are not only on the boundaries of the observable universe, but also at the cutting edge of our knowledge.

A. Dekel and J. P. Ostriker

Part one

Introduction to structure formation

1 Dark matter and structure formation

Joel R. Primack

Abstract

This chapter aims to present an introduction to current research on the nature of the cosmological dark matter and the origin of galaxies and large-scale structure within the standard theoretical framework: gravitational collapse of fluctuations as the origin of structure in the expanding universe. General relativistic cosmology is summarized, and the data on the basic cosmological parameters (t_0 and $H_0 \equiv 100h$ km s^{-1} Mpc^{-1}, Ω_0, Ω_Λ and Ω_b) are reviewed. The standard astrophysical classification of varieties of dark matter is used: *hot* and *cold*. Various particle physics candidates for hot, warm, and cold dark matter are briefly reviewed, together with current constraints and experiments that could detect or eliminate them. Also included is a very brief summary of the theory of cosmic defects, and a somewhat more extended exposition of the idea of cosmological inflation with a summary of some current models of inflation. The remainder is a discussion of observational constraints on cosmological model building, emphasizing models in which most of the dark matter is cold and the primordial fluctuations are the sort predicted by inflation. It is argued that the simplest models that have a hope of working are Cold Dark Matter with a cosmological constant (ΛCDM) if the Hubble parameter is high ($h \gtrsim 0.6$), and Cold + Hot Dark Matter (CHDM) if the Hubble parameter and age permit an $\Omega = 1$ cosmology; the most attractive variants of these models and the critical tests for each are discussed.

1.1 Introduction

The standard theory of cosmology is the Hot Big Bang, according to which the early universe was hot, dense, very nearly homogeneous, and expanding adiabatically according to the laws of general relativity (GR). This theory nicely accounts for the cosmic background radiation, and is at least roughly consistent with the abundances of the lightest nuclides. It is probably even true, as far as it goes; at least, I will assume so here. But as a fundamental theory of cosmology, the standard theory is seriously incomplete. One way of putting this is to say that it describes the middle of the story, but leaves us guessing about both the beginning and the end.

Galaxies and clusters of galaxies are the largest bound systems, and the filamentary or wall-like superclusters and the voids between them are the largest structures visible in the universe, but their origins are not yet entirely understood. Moreover, within the framework of the standard theory of gravity, there is compelling observational evidence that most of the mass detected gravitationally in galaxies and clusters, and especially on larger scales, is "dark" – that is, visible neither in absorption nor emission of any frequency of electromagnetic radiation. But we still do not know what this dark matter is.

Explaining the rich variety and correlations of galaxy and cluster morphology will require filling in much more of the history of the universe:

- *Beginnings,* in order to understand the origin of the fluctuations that eventually collapse gravitationally to form galaxies and large-scale structure. This is a mystery in the standard Hot Big Bang universe, because the matter that comprises a typical galaxy, for example, first came into causal contact about a year after the Big Bang. It is hard to see how galaxy-size fluctuations could have formed after that, but even harder to see how they could have formed earlier. The best solution to this problem yet discovered, and the one emphasized here, is cosmic inflation. The main alternative, discussed in less detail here, is cosmic topological defects.
- *Denouement,* since even given appropriate initial fluctuations, we are far from understanding the evolution of galaxies, clusters, and large-scale structure – or even the origins of stars and the stellar initial mass function.
- *Dark matter* is probably the key to unraveling the plot since it appears to be gravitationally dominant on all scales larger than the cores of galaxies. The dark matter is therefore crucial for understanding the evolution and present structure of galaxies, clusters, superclusters and voids.

The present chapter (updating Primack 1987, 1988, 1993–7) concentrates on the period *after* the first three minutes, during which the universe expands by a factor of $\sim 10^8$ to its present size, and all the observed structures form. This is now an area undergoing intense development in astrophysics, both observationally and theoretically. It is likely that the present decade will see the construction at last of a fundamental theory of cosmology, with perhaps profound implications for particle physics – and possibly even for broader areas of modern culture.

The current controversy over the amount of matter in the universe will be emphasized, discussing especially the two leading alternatives: a critical-density universe, i.e., with $\Omega_0 \equiv \bar{\rho}_0/\rho_c = 1$ (see Table 1.1), vs. a low-density universe having $\Omega_0 \approx 0.3$ with a positive cosmological constant $\Lambda > 0$ such that $\Omega_\Lambda \equiv \Lambda/(3H_0^2) = 1 - \Omega_0$ supplying the additional density required for the flatness predicted by the simplest inflationary models. (The significance of the cosmological parameters Ω_0, H_0, t_0, and Λ is discussed in §1.2. Note that the present-epoch matter density parameter Ω_0 is sometimes denoted Ω_m or simply Ω.) $\Omega = 1$ requires that the expansion rate of the universe, the Hubble parameter $H_0 \equiv 100h\,\mathrm{km\,s^{-1}\,Mpc^{-1}} \equiv 50h_{50}\,\mathrm{km\,s^{-1}\,Mpc^{-1}}$, be relatively low, $h \lesssim 0.5$, in order that the age of the universe t_0 be as large as the minimum estimate of the age of the stars in the oldest globular clusters. If the expansion rate turns out to be larger than this, we will see that GR then requires that $\Omega_0 < 1$, with a positive cosmological constant giving a larger age for any value of Ω_0.

Although this chapter will concentrate on the implications of Cold Dark Matter (CDM) and alternative theories of dark matter for the development of galaxies and large-scale structure in the relatively "recent" universe, one can hardly avoid recalling some of the earlier parts of the story. Inflation or cosmic defects will be important in this context for the nearly constant curvature (near-"Zel'dovich") spectrum of primordial fluctuations and as plausible solutions to the problem of generating these large-scale fluctuations without violating causality; and primordial nucleosynthesis will be important as a source of information on the amount of ordinary ("baryonic") matter in the universe. The fact that the observational lower bound on Ω_0 – namely $0.3 \lesssim \Omega_0$ – exceeds the most conservative upper limit on baryonic mass $\Omega_b \lesssim 0.03h^{-2}$ from Big Bang Nucleosynthesis (Copi, Schramm & Turner 1995; cf. Hata et al. 1995) is the main evidence that there must be such nonbaryonic dark-matter particles.

Of special concern will be evidence and arguments bearing on the astrophysical properties of the dark matter, which can also help to constrain possible particle physics candidates. The most popular of these are few-eV neutrinos (the "hot" dark matter candidate), heavy stable particles such as ~ 100 GeV photinos (or whatever neutralino is the lightest supersymmetric partner particle) or $10^{-6} - 10^{-3}$ eV invisible axions (these remain the favorite "cold" dark matter candidates), and various more exotic ideas such as keV gravitinos ("warm" dark matter) or primordial black holes (BH).

The usual astrophysical classification of the dark matter candidates is into *hot, warm,* or *cold,* depending on their thermal velocity in the early universe. Hot dark matter, such as few-eV neutrinos, is still relativistic when galaxy-size masses ($\sim 10^{12} M_\odot$) are first encompassed within the horizon. Warm dark matter is just becoming nonrelativistic then. Cold dark matter, such as axions or massive photinos, is nonrelativistic when even globular cluster masses ($\sim 10^6 M_\odot$) come within the horizon. As a consequence, fluctuations on galaxy scales are wiped out by the "free streaming" of the hot dark-matter particles which are moving at nearly the speed of light. But galaxy-size fluctuations are preserved with warm dark matter, and all cosmologically relevant fluctuations survive in a universe dominated by the sluggishly moving cold dark matter.

The first possibility for nonbaryonic dark matter that was examined in detail was massive neutrinos, assumed to have mass ~ 25 eV – both because that mass corresponds to closure density for $h \approx 0.5$, and because in the late 1970s the Moscow tritium β-decay experiment provided evidence (subsequently contradicted by other experiments) that the electron neutrino has that mass. Although this picture leads to superclusters

Table 1.1. *Physical constants for cosmology*

parsec	pc	$= 3.09 \times 10^{18}$ cm $= 3.26$ light years (lyr)
Newton's const.	G	$= 6.67 \times 10^{-8}$ dyne cm^2 g^{-2}
Hubble parameter	H_0	$= 100 h$ km s^{-1} Mpc^{-1}, $1/2 \lesssim h \lesssim 1$
Hubble time	H_0^{-1}	$= h^{-1}$ 9.78 Gyr
Hubble radius	R_H	$= cH^{-1} = 3.00 h^{-1}$ Gpc
critical density	ρ_c	$= 3H^2/8\pi G = 1.88 \times 10^{-29} h^2$ g cm^{-3}
		$= 10.5 h^2$ keV cm$^{-3} = 2.78 \times 10^{11} h^2$ M$_\odot$ Mpc^{-3}
speed of light	c	$= 3.00 \times 10^{10}$ cm s$^{-1} = 306$ Mpc Gyr^{-1}
solar mass	M$_\odot$	$= 1.99 \times 10^{33}$ g
solar luminosity	L$_\odot$	$= 3.85 \times 10^{33}$ erg s^{-1}
Planck's const.	$\hbar$	$= 1.05 \times 10^{-27}$ erg s $= 6.58 \times 10^{-16}$ eV s
Planck mass	m_{Pl}	$= (\hbar c/G)^{1/2} = 2.18 \times 10^{-5}$ g $= 1.22 \times 10^{19}$ GeV

and voids of roughly the size seen, superclusters are the first structures to collapse in this theory since smaller size fluctuations do not survive. The theory foundered on this point, however, since galaxies are almost certainly older than superclusters. The standard (adiabatic) form of this theory has recently been ruled out by the COsmic Background Explorer (COBE) data: if the amplitude of the fluctuation spectrum is small enough for consistency with the COBE fluctuations, superclusters would just be beginning to form at the present epoch, and hardly any smaller-scale structures, including galaxies, could have formed by the present epoch.

A currently popular possibility is that the dark matter is cold. After Peebles (1982), my co-workers and I were among those who first proposed and worked out the consequences of the CDM model (Primack & Blumenthal 1983, 1984; Blumenthal et al. 1984). Its virtues include an account of galaxy and cluster formation that at first sight appeared to be very attractive. Its defects took longer to uncover, partly because uncertainty about how to normalize the CDM fluctuation amplitude allowed for a certain amount of fudging, at least until COBE measured the fluctuation amplitude. The most serious problem with CDM is probably the mismatch between supercluster-scale and galaxy-scale structures and velocities, which suggests that the CDM fluctuation spectrum is not quite the right shape – which can perhaps be remedied if the dark matter content is a mixture of hot and cold, or if there is less than a critical density of cold dark matter.

The basic theoretical framework for cosmology is reviewed first (see a summary in §11.1.2), followed by a discussion of the current knowledge about the fundamental cosmological parameters.

Table 1.1 lists the values of the most important physical constants used in this chapter (cf. Barnett et al. 1996). The distance to distant galaxies is deduced from their redshifts; consequently, the parameter h appears in many formulas where the distance matters.

1.2 Cosmology basics

It is assumed here that Einstein's theory of general relativity accurately describes gravity. Although it is important to appreciate that there is no observational confirmation of this on scales larger than about 1 Mpc, the tests of GR on smaller scales are becoming increasingly precise, especially with pulsars in binary star systems (Will 1981, 1986, 1990; Taylor 1994). On galaxy and cluster scales, the general agreement between the mass estimated by velocity measurements and by gravitational lensing provides evidence supporting standard gravity. There are two other reasons most cosmologists believe in GR: it is conceptually so beautifully simple that it is hard to believe it could be wrong, and anyway it has no serious theoretical competition. Nevertheless, since a straightforward interpretation of the available data in the context of the standard theory of gravity leads to the disquieting conclusion that most of the matter in the universe is dark, there have been suggestions that perhaps our theory of gravity is inadequate on large scales. The suggested alternatives are mentioned briefly in §1.2.2.

The "Copernican" or "cosmological" principle is logically independent of our theory of gravity, so it is appropriate to state it before discussing GR further. First, some definitions are necessary.

- A *comoving observer* is at rest and unaccelerated with respect to nearby material (in practice, with respect to the center of mass of galaxies within, say, 100 h^{-1} Mpc).
- The universe is *homogeneous* if all comoving observers see identical properties.
- The universe is *isotropic* if all comoving observers see no preferred direction.

The *cosmological principle* asserts that the universe is homogeneous and isotropic on large scales. (Isotropy about at least three points actually implies homogeneity, but the counter-example of a cylinder shows that the reverse is not true.) In reality, the matter distribution in the universe is exceedingly inhomogeneous on small scales, and increasingly homogeneous on scales approaching the entire horizon. The cosmological principle is in practice the assumption that for cosmological purposes we can neglect this inhomogeneity, or treat it perturbatively. This has now been put on an improved basis, based on the observed isotropy of the cosmic background radiation and the (partially testable) Copernican assumption that other observers also see a nearly homogeneous CMB. The "COBE–Copernicus" Theorem (Stoeger, Maartens & Ellis 1995; Maartens Ellis & Stoeger 1995; reviewed by Ellis 1996) asserts that if all comoving observers measure the cosmic microwave background radiation to be almost isotropic in a region of the expanding universe, then the universe is locally almost spatially homogeneous and isotropic in that region.

The great advantage of assuming homogeneity is that our own cosmic neighborhood becomes representative of the whole universe, and the range of cosmological models to be considered is also enormously reduced. The cosmological principle also implies the existence of a universal cosmic time, since all observers see the same sequence of cosmic events with which to synchronize their clocks. (This assumption is sometimes explicitly included in the statement of the cosmological principle; e.g., Rindler (1977), p. 203.) In particular, they can all start their clocks with the Big Bang.

Astronomers observe that the redshift $z \equiv (\lambda - \lambda_0)/\lambda_0$ (where λ denotes wavelength) of distant galaxies is proportional to their distance. We assume, for lack of any viable alternative explanation, that this redshift is due to the expansion of the universe. Recent evidence for this includes higher CMB temperature at higher redshift (Songaila et al. 1994b) and time dilation of high-redshift Type Ia supernovae (Goldhaber et al. 1996). The cosmological principle then implies (see, for example, Rowan-Robinson 1981, §4.3) that the expansion is homogeneous: $r = a(t)r_0$, which immediately implies Hubble's law: $v = \dot{r} = \dot{a}a^{-1}r = H_0 r$. Here r_0 is the present distance of some distant galaxy (the subscript "0" in cosmology denotes the present era), r is its distance as a function of time, v is its velocity, and $a(t)$ is the scale factor of the expansion (scaled to be unity at the present: $a(t_0) = 1$). The scale factor is related to the redshift by $a = (1+z)^{-1}$. Hubble's "constant" $H(t)$ (constant in space, but a function of time except in an empty universe) is $H(t) = \dot{a}a^{-1}$.

Finally, it can be shown (see, e.g., Weinberg 1972, Rindler 1977) that the most general metric satisfying the cosmological principle is the Robertson–Walker metric

$$ds^2 = c^2 dt^2 - a(t)^2 \left[\frac{dr^2}{1-kr^2} + r^2(\sin^2\theta d\phi^2 + d\theta^2) \right], \tag{1.1}$$

where the curvature constant k, by a suitable choice of units for r, has the value 1, 0, or -1, depending on whether the universe is closed, flat, or open, respectively. For $k = 1$ the spatial universe can be regarded as the surface of a sphere of radius $a(t)$ in four-dimensional Euclidean space; and although for $k = 0$ or -1 no such simple geometric interpretation is possible, $a(t)$ still sets the scale of the geometry of space.

Formally, GR consists of the assumption of the Equivalence Principle (or the Principle of General Covariance) together with Einstein's field equations, labeled (E) in Table 1.2, where the key equations have been collected together. The Equivalence Principle implies that space-time is locally Minkowskian and globally (pseudo-)Riemannian, and the field equations specify precisely how space-time responds to its contents. The essential physical idea underlying GR is that space-time is not just an arena, but rather an active participant in the dynamics, as summarized by John Wheeler: "Matter tells space how to curve, curved space tells matter how to move."

Comoving coordinates are coordinates with respect to which comoving observers are at rest. A comoving coordinate system expands with the Hubble expansion. It is convenient to specify linear dimensions in comoving coordinates scaled to the present; for example, if we say that two objects were 1 Mpc apart in comoving coordinates at a redshift of $z = 9$, their actual distance then was 0.1 Mpc. In a non-empty universe with vanishing cosmological constant, the case first studied in detail by the Russian cosmologist Alexander Friedmann in 1922-4, gravitational attraction ensures that the expansion rate is always decreasing. As a result, the Hubble radius $R_H(t) \equiv cH(t)^{-1}$ is increasing. The Hubble radius of a non-empty Friedmann universe expands even in

Table 1.2. *Theoretical framework: GR cosmology*

GR: Matter tells space how to curve,   Curved space tells matter how to move.

(E) $R^{\mu\nu} - \frac{1}{2}Rg^{\mu\nu} = -8\pi G\, T^{\mu\nu} - \Lambda g^{\mu\nu}$

COBE–Copernicus Theorem: If all observers measure nearly isotropic CMB, then the universe is locally nearly homogeneous and isotropic – i.e., nearly FRW.

FRW E(00) $\dfrac{\dot{a}^2}{a^2} = \dfrac{8\pi}{3}G\rho - \dfrac{k}{a^2} + \dfrac{\Lambda}{3}$

FRW E(ii) $\dfrac{2\ddot{a}}{a} + \dfrac{\dot{a}^2}{a^2} = -8\pi G p - \dfrac{k}{a^2} + \Lambda$

$H_0 \equiv 100\, h\, \mathrm{km\, s^{-1}\, Mpc^{-1}}$
$\equiv 50\, h_{50}\, \mathrm{km\, s^{-1}\, Mpc^{-1}}$

$\dfrac{E(00)}{H_0^2} \Rightarrow 1 = \Omega_0 - \dfrac{k}{H_0^2} + \Omega_\Lambda$ with $H_0 \equiv \dfrac{\dot{a}_0}{a_0}$, $a_0 \equiv 1$, $\Omega_0 \equiv \dfrac{\rho_0}{\rho_c}, \Omega_\Lambda \equiv \dfrac{\Lambda}{3H_0^2}$,

$\rho_c \equiv \dfrac{3H_0^2}{8\pi G} = 0.70 \times 10^{11} h_{50}^2 M_\odot\, \mathrm{Mpc}^{-3}$

$E(ii) - E(00) \Rightarrow \dfrac{2\ddot{a}}{a} = -\dfrac{8\pi}{3}G\rho - 8\pi G p + \dfrac{2}{3}\Lambda$

Divide by $2E(00) \Rightarrow q_0 \equiv -\left(\dfrac{\ddot{a}}{a}\dfrac{a^2}{\dot{a}^2}\right)_0 = \dfrac{\Omega_0}{2} - \Omega_\Lambda$

$E(00) \Rightarrow t_0 = \int_0^1 \dfrac{da}{a}\left[\dfrac{8\pi}{3}G\rho - \dfrac{k}{a^2} + \dfrac{\Lambda}{3}\right]^{\frac{1}{2}} = H_0^{-1}\int_0^1 \dfrac{da}{a}\left[\dfrac{\Omega_0}{a^3} - \dfrac{k}{H_0^2 a^2} + \Omega_\Lambda\right]^{-\frac{1}{2}}$

$t_0 = H_0^{-1} f(\Omega_0, \Omega_\Lambda)$ $H_0^{-1} = 9.78 h^{-1}\mathrm{Gyr}$ $f(1,0) = \tfrac{2}{3}$
$f(0,0) = 1$
$f(0,1) = \infty$

$\dfrac{d}{dt}[E(00)a^3]'$ vs. $E(ii) \Rightarrow \dfrac{d}{da}(\rho a^3) = -3pa^2$ ("continuity")

Given eq. of state $p = p(\rho)$, integrate to determine $\rho(a)$,
                                 integrate $E(00)$ to determine $a(t)$.

Examples: $p = 0 \Rightarrow \rho = \rho_0 a^{-3}$ (assumed above in q_0, t_0 eqs.)

$p = \dfrac{\rho}{3}$, $k = 0 \Rightarrow \rho \propto a^{-4}$

---

comoving coordinates. Our backward lightcone encompasses more of the universe as time goes on.

1.2.1 *Friedmann–Robertson–Walker universes*

For a homogeneous and isotropic fluid of density ρ and pressure p in a homogeneous universe with curvature k and cosmological constant Λ, Einstein's system of partial differential equations reduces to the two ordinary differential equations labeled in Table 1.2 as FRW E(00) and E(ii), for the diagonal time and spatial components (see, e.g., Rindler 1977, §9.9). Dividing E(00) by H_0^2, and subtracting E(00) from E(ii) puts these equations into more familiar forms. Dividing the latter by 2E(00) and evaluating all expressions at the present epoch then gives the familiar expression for the deceleration parameter q_0 in terms of Ω_0 and Ω_Λ.

Multiplying E(00) by a^3, differentiating with respect to a, and comparing with E(ii) gives the equation of continuity. Given an equation of state $p = p(\rho)$, this equation can be integrated to determine $\rho(a)$; then E(00) can be integrated to determine $a(t)$.

Consider, for example, the case of vanishing pressure $p = 0$, which is presumably an excellent approximation for the present universe since the contribution of radiation and massless neutrinos (both having $p = \rho c^2/3$) to the mass-energy density is at the present epoch much less than that of nonrelativistic matter (for which p is negligible). The continuity equation reduces to $(4\pi/3)\rho a^3 = M = $ constant, and E(00) yields *Friedmann's equation*

$$\dot{a}^2 = \frac{2GM}{a} - kc^2 + \frac{\Lambda c^2 a^2}{3}. \qquad (1.2)$$

This gives an expression for the age of the universe t_0 which can be integrated in general in terms of elliptic functions, and for $\Lambda = 0$ or $k = 0$ in terms of elementary functions (cf. standard textbooks, e.g., Peebles 1993, §13, and Felton & Isaacman 1986).

Figure 1.1 (a) plots the evolution of the scale factor a for three interesting examples: $(\Omega_0, \Omega_\Lambda) = (1,0)$, $(0.3,0)$, and $(0.3,0.7)$. Figure 1.1 (b) shows how t/t_H depends on Ω_0 both for $\Lambda = 0$ (dashed) and $\Omega_\Lambda = 1 - \Omega_0$ (solid). Notice that for $\Lambda = 0$, t_0/t_H is somewhat greater for $\Omega_0 = 0.3$ ($t_0/t_H = 0.81$) than for $\Omega = 1$ ($t_0/t_H = 2/3$), while for $\Omega_0 = 1 - \Omega_\Lambda = 0.3$ it is substantially greater ($t_0/t_H = 0.96$). In the last case, the competition between the attraction of the matter and the repulsion of space by space represented by the cosmological constant results in a slowing of the expansion at $a \sim 0.5$; the cosmological constant subsequently dominates, resulting in an accelerated expansion (negative deceleration $q_0 = -0.55$ at the present epoch), corresponding to an inflationary universe. In addition to increasing t_0, this behavior has observational implications that we will explore in §1.3.3.

1.2.2 *Is the gravitational force* $\propto r^{-1}$ *at large r?*

Back to the question whether our conventional theory of gravity is trustworthy on large scales. The reason for raising this question is that interpreting modern observations within the context of the standard theory leads to the conclusion that at least 90% of the matter in the universe is dark. Moreover, there is no observational confirmation that the gravitational force falls as r^{-2} on large scales.

Tohline (1983) pointed out that a modified gravitational force law, with the gravitational acceleration given by $a' = (GM_{\text{lum}}/r^2)(1 + r/d)$, could be an alternative to dark matter galactic halos as an explanation of the constant-velocity rotation curves of

Dark matter and structure formation

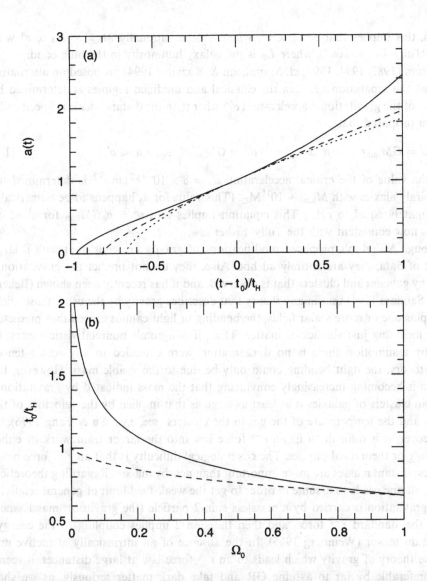

Figure 1.1. (a) Evolution of the scale factor $a(t)$ plotted vs. the time after the present $(t - t_0)$ in units of Hubble time $t_H \equiv H_0^{-1} = 9.78h^{-1}$ Gyr for three different cosmologies: Einstein–de Sitter ($\Omega_0 = 1, \Omega_\Lambda = 0$ dotted curve), negative curvature ($\Omega_0 = 0.3, \Omega_\Lambda = 0$: dashed curve), and low-Ω_0 flat ($\Omega_0 = 0.3$, $\Omega_\Lambda = 0.7$: solid curve). (b) Age of the universe today t_0 in units of Hubble time t_H as a function of Ω_0 for $\Lambda = 0$ (dashed curve) and flat $\Omega_0 + \Omega_\Lambda = 1$ (solid curve) cosmologies.

spiral galaxies. (The mass is written as M_{lum}, where "lum" is shorthand for 'luminous', to emphasize that there is not supposed to be any dark matter.) Indeed, this equation implies $v^2 = GM_{\text{lum}}/d = $ constant for $r \gg d$. However, with the distance scale d where the force shifts from r^{-2} to r^{-1} taken to be a physical constant, the same for all galaxies, this implies that $M_{\text{lum}} \propto v^2$, whereas observationally $M_{\text{lum}} \propto L_B \propto v^\alpha$ with $\alpha \sim 4$ ("Tully–Fisher law"), where L_B is the galaxy luminosity in the blue band.

Milgrom (1983, 1994, 1995; cf. Mannheim & Kazanas 1994) proposed an alternative idea, that the separation between the classical and modified regimes is determined by the value of the gravitational acceleration a' rather than the distance scale r. Specifically, Milgrom proposed that

$$a' = GM_{\text{lum}}r^{-2}, \quad a' \gg a'_0; \qquad a'^2 = GM_{\text{lum}}r^{-2}a'_0, \quad a \ll a'_0, \qquad (1.3)$$

where the value of the critical acceleration $a'_0 \approx 8 \times 10^{-8}h^2 \,\text{cm}\,\text{s}^{-2}$ is determined for large spiral galaxies with $M_{\text{lum}} \sim 10^{11} M_\odot$. (This value for a'_0 happens to be numerically approximately equal to cH_0.) This equation implies that $v^4 = a'_0 GM_{\text{lum}}$ for $a' \ll a'_0$, which is now consistent with the Tully–Fisher law.

Although Milgrom's proposed modifications of gravity are consistent with a large amount of data, they are entirely ad hoc. Also, they do not predict the gravitational lensing by galaxies and clusters that is observed, and it has recently been shown (Bekenstein & Sandars 1994) that in any theory that describes gravity by the usual tensor field of GR plus one or more scalar fields, the bending of light cannot exceed that predicted by GR including just the actual matter. Thus, if Milgrom's nonrelativistic theory, in which by assumption there is no dark matter, were extended to any scalar–tensor gravity theory, the light bending could only be due to the visible mass. However, the evidence is becoming increasingly convincing that the mass indicated by gravitational lensing in clusters of galaxies is at least as large as that implied by the velocities of the galaxies and the temperature of the gas in the clusters (see, e.g., Wu & Fang 1996).

Moreover, it is difficult to fit an r^{-1} force law into the larger framework of either cosmology or theoretical physics. The cosmological difficulty is that an r^{-1} force never saturates: distant masses are more important than nearby masses. Regarding theoretical physics, all one needs to assume in order to get the weak-field limit of general relativity is that gravitation is carried by a massless spin-2 particle (the graviton): masslessness implies the standard r^{-2} force, and then the spin 2 implies coupling to the energy–momentum tensor (Weinberg 1965). In the absence of an intrinsically attractive and plausible theory of gravity which leads to an r^{-1} force law at large distances, it seems to be preferable by far to assume GR and take dark matter seriously, as we shall do below. But until the nature of the dark matter is determined – e.g., by discovering dark-matter particles in laboratory experiments – it is good to remember that there may be alternative explanations for the data.

1.3 Age, expansion rate, and cosmological constant

1.3.1 Age of the universe t_0

The strongest lower limits for t_0 come from studies of the stellar populations of globular clusters (GCs). Standard estimates of the ages of the oldest GCs are $t_{\text{GC}} \approx 15\text{–}16$ Gyr

(Bolte & Hogan 1995; VandenBerg, Bolte & Stetson 1996; Chaboyer et al. 1996). A frequently quoted lower limit on the age of GCs is 12 Gyr (Chaboyer et al. 1996), which then places an even more conservative lower limit on $t_0 = t_{GC} + \Delta t_{GC}$, where $\Delta t_{GC} \gtrsim 0.5$ Gyr is the time from the Big Bang until GC formation. The main uncertainty in the GC age estimates comes from the uncertain distance to the GCs: a 0.25 magnitude error in the distance modulus translates to a 22% error in the derived cluster age (Chaboyer 1995). (We will come back to this in the next paragraph.) All the other obvious ways to lower the calculated t_{GC} have been considered and found to have limited effects, and many non-obvious ideas have also been explored (VandenBerg, Bolte & Stetson 1996). For example, stellar mass loss is a way of lowering t_{GC} (Willson, Bowen & Struck-Marcell 1987), but observations constrain the reduction in t_0 to be less than ~ 1 Gyr (Shi 1995, Swenson 1995). Helium sedimentation during the main sequence lifetime can reduce stellar ages by ~ 1 Gyr (Chaboyer & Kim 1995, D'Antona et al. 1997). Note that the higher primordial ^{4}He abundance implied by the new D/H (Deuterium/Hydrogen) values (Tytler, Burles & Kirkman 1996; Tytler, Fan & Burles 1996) lowers the central value of the GC ages by perhaps 0.5 Gyr (cf. Chaboyer 1995, Castellani et al. 1997). The usual conclusion has been that $t_0 \approx 12$ Gyr is probably the lowest plausible value for t_0, obtained by pushing many but not all the parameters to their limits.

However, in March and April 1997, preliminary reports (e.g., Watson 1997) of analyses of data from the Hipparcos astrometric satellite have indicated that the distances to GCs assumed in obtaining the ages just discussed were systematically underestimated. If this is true, it follows that their stars at the main sequence turn-off are brighter and therefore younger. Indeed, there are indications that this correction will be largest for the lowest-metallicity clusters that had the oldest ages according to the standard analysis, according to Reid (1997). His analysis, using a sample including 15 metal-poor stars with parallaxes determined to better than 12% accuracy to redefine the subdwarf main sequence, gives distance moduli ~ 0.3 magnitudes ($\sim 30\%$) brighter than current standard values for his four lowest-metallicity GCs (M13, M15, M30, and M92), and ages (*not* lower limits) of ~ 12 Gyr. The shapes of the theoretical isochrones (Bergbusch & VandenBerg 1992) used in previous GC age estimates (e.g., Bolte & Hogan 1995, Sandquist et al. 1996) are no longer acceptable fits to the subdwarf data with the revised distances, although the isochrones of D'Antona et al. (1997) give better fits to the local subdwarfs and to the GCs. Another analysis (Gratton et al. 1997) uses a sample including 11 low-metallicity non-binary subdwarf stars with Hipparcos parallaxes better than 10% and accurate metal abundances from high-resolution spectroscopy to determine the absolute location of the main sequence as a function of metallicity. They then derive ages for the old GCs (M13, M68, M92, NGC288, NGC6752, 47 Tuc) in their GC sample of $12.1^{+1.2}_{-3.6}$ Gyr. Their ages are lower both because of their 0.2 mag brighter distance moduli and because of their better metal determinations of cluster and field stars.

There are systematic effects that must be taken into account in the accurate determination of t_{GC}, including metallicity dependence and reddening corrections, and various physical phenomena such as stellar convection and helium sedimentation whose inclusion could lower ages still further and perhaps also bring theoretical isochrones into better agreement with the GC observations. Thus, after the full Hipparcos data becomes public in mid-1997, there is likely to be a period of perhaps a year or so during which additional data is sought and theoretical models are revised before a new consensus emerges regarding the GC ages. But it does appear that the older estimates $t_{GC} \approx$ 15–16 Gyr will be revised downward substantially.

Stellar age estimates are also relevant to another sort of argument for an old, low-density universe: observation of apparently old galaxies at moderately high redshift (Dunlop et al. 1996). In the most extreme example presented so far (Spinrad et al. 1997), a galaxy at redshift $z = 1.55$ has a rest-frame spectrum very similar to that of an F6 star, and a minimum age of 3.5 Gyr based on current stellar evolution models and reasonable (but perhaps not definitive) assumptions about stellar populations, reddening, etc.; the authors point out that for 3.5 Gyr to have elapsed at $z = 1.55$ requires $h < 0.45$ for $\Omega = 1$. (Note, however, that for 3.0 Gyr to have elapsed by $z = 1.55$ in an $\Omega = 1$ cosmology imposes the less restrictive requirement $h < 0.53$, for example.) Observations of old galaxies at high redshift will certainly constrain cosmological parameters, especially if the assumptions that go into the analysis can be independently verified.

Stellar age estimates are of course based on standard stellar evolution calculations. But the solar neutrino problem reminds us that we are not really sure that we understand how even our nearest star operates; and the sun plays an important role in calibrating stellar evolution, since it is the only star whose age we know independently (from radioactive dating of early solar system material). An important check on stellar ages can come from observations of white dwarfs in globular and open clusters (cf. Richer et al. 1995). And the two detached eclipsing binaries at the main sequence turn-off point recently discovered in Omega Centauri can be used both to measure the distance to this globular cluster accurately, and to determine their ages using the mass–luminosity relation (Paczynski 1996).

What if the GC age estimates are wrong for some unknown reason? The only other non-cosmological estimates of the age of the universe come from nuclear cosmochronometry – the chemical evolution of the Galaxy – and white dwarf cooling. Cosmochronometry age estimates are sensitive to a number of uncertain issues such as the formation history of the disk and its stars, and possible actinide destruction in stars (Malaney, Mathews & Dearborn 1989; Mathews & Schramm 1993). However, an independent cosmochronometry age estimate of 17 ± 4 Gyr has been obtained for a single ultra-low-metallicity star, based on the measured depletion of thorium (whose half-life is 14.2 Gyr) compared to stable heavy r-process elements (Cowan et al. 1997; cf. Bolte 1997, Sneden et al. 1996). This method will become very important if it is possible to obtain accurate measurements of r-process elements for a number of very-low-metallicity stars, and the resulting age estimates are consistent.

Independent age estimates come from the cooling of white dwarfs in the neighborhood of the sun. The key observation is that there is a lower limit to the luminosity,

Dark matter and structure formation

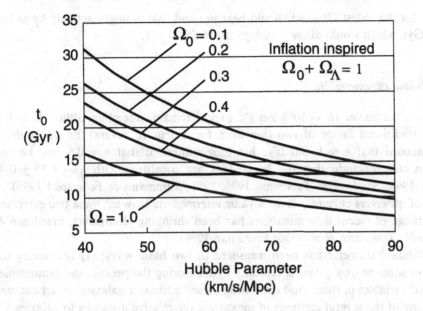

Figure 1.2. Age of the universe t_0 as a function of Hubble parameter H_0 in inflation-inspired models with $\Omega_0 + \Omega_\Lambda = 1$, for several values of the present-epoch cosmological density parameter Ω_0.

and therefore also the temperature, of nearby white dwarfs; although dimmer ones could have been seen, none have been found. The only plausible explanation is that the white dwarfs have not had sufficient time to cool to lower temperatures, which initially led to an estimate of 9.3 ± 2 Gyr for the age of the Galactic disk (Winget et al. 1987). Since there was evidence (based on the pre-Hipparcos GC distances) that the stellar disk of our Galaxy is about 2 Gyr younger than the oldest GCs (e.g., Stetson, VandenBerg & Bolte 1996), this, in turn, gave an estimate of the age of the universe of $t_0 \sim 11 \pm 2$ Gyr. More recent analyses (cf. Wood 1992, Hernanz et al. 1994) conclude that sensitivity to disk star formation history, and to effects on the white dwarf cooling rates due to C/O separation at crystallization and possible presence of trace elements such as ^{22}Ne, allow a rather wide range of ages for the disk of about 10 ± 4 Gyr. The latest determination of the white dwarf luminosity function, using white dwarfs in proper motion binaries, leads to a somewhat lower minimum luminosity and therefore a somewhat higher estimate of the age of the disk of $\sim 10.5^{+2.5}_{-1.5}$ Gyr (Oswalt et al. 1996).

Suppose that the old GC stellar age estimates that $t_0 \gtrsim 13$ Gyr are right, as we will assume in much of the rest of this chapter. Figure 1.2 shows that $t_0 > 13$ Gyr implies that $h \leq 0.50$ for $\Omega = 1$, and that $h \leq 0.73$ even for Ω_0 as small as 0.3 in flat cosmologies (i.e., with $\Omega_0 + \Omega_\Lambda = 1$). However, in view of the preliminary analyses using the new Hipparcos parallaxes and other new data that give strikingly lower age estimates for the oldest GCs, we should bear in mind that t_0 might actually be as low as ~ 11 Gyr, which would allow h as high as 0.6 for $\Omega = 1$.

1.3.2 Hubble parameter H_0

The Hubble parameter $H_0 \equiv 100h$ km s^{-1} Mpc^{-1} remains uncertain, although by less than the traditional factor of two (see Ch. 6 for a detailed review). de Vaucouleurs long contended that $h \approx 1$. Sandage has long contended that $h \approx 0.5$, and he and Tammann still conclude that the latest data are consistent with $h = 0.55 \pm 0.05$ (Sandage 1995; Sandage & Tammann 1995, 1996; Tammann & Federspiel 1996). A majority of observers currently favor a value intermediate between these two extremes, and the range of recent determinations has been shrinking (Kennicutt, Freedman & Mould 1995; Tammann et al. 1996; Freedman 1996).

The Hubble parameter has been measured in two basic ways: (1) Measuring the distance to some nearby galaxies, typically by measuring the periods and luminosities of Cepheid variables in them; and then using these "calibrator galaxies" to set the zero point in any of the several methods of measuring the relative distances to galaxies. (2) Using fundamental physics to measure the distance to some distant object(s) directly, thereby avoiding at least some of the uncertainties of the cosmic distance ladder (Rowan-Robinson 1985). The difficulty with method (1) was that there was only a handful of calibrator galaxies close enough for Cepheids to be resolved in them. However, the success of the Hubble Space Telescope (HST) Cepheid measurement of the distance to M100 (Freedman et al. 1994, Ferrarese et al. 1996) shows that the HST Key Project on the Extragalactic Distance Scale can significantly increase the set of calibrator galaxies – in fact, it already has done so. Adaptive optics from the ground may also be able to contribute to this effort, although the first published result of this approach (Pierce et al. 1994) is not entirely convincing. The difficulty with method (2) is that in every case studied so far, some aspect of the observed system or the underlying physics remains somewhat uncertain. It is nevertheless remarkable that the results of several different methods of type (2) are rather similar, and indeed not very far from those of method (1). This gives reason to hope for convergence.

1.3.2.1 Relative distance methods
One piece of good news is that the several methods of measuring the relative distances to galaxies now mostly seem to be consistent with each other (Jacoby et al. 1992; Fukugita, Hogan & Peebles 1993). These methods use either (a) "standard candles" or (b) empirical relations between two measurable properties of a galaxy, one distance-independent and the other distance-dependent.

(a) The old favorite standard candle is Type Ia supernovae (SNe); a new one is the apparent maximum luminosity of planetary nebulae (Jacoby et al. 1992). Sandage et al. (1996) and others (van den Bergh 1995, Branch et al. 1996, cf. Schaefer 1996) get low values of $h \approx 0.55$ from HST Cepheid distances to SN Ia host galaxies, including the seven SNe Ia with what Sandage *et al.* characterize as well-observed maxima that lie in six galaxies for which HST Cepheid distances are now available. But taking account of an empirical relationship between the SN Ia light curve shape and maximum luminosity leads to higher $h = 0.65 \pm 0.06$ (Riess, Press & Kirshner 1996) or $h = 0.63 \pm 0.03$ (Hamuy *et al.* 1996), although Tammann & Sandage (1995) disagree that the increase in h can be so large.

(b) The old favorite empirical relation used as a relative distance indicator is the Tully–Fisher relation between the rotation velocity and luminosity of spiral galaxies (and the related Faber–Jackson or D_n–σ relation). A newer one is based on the decrease in the fluctuations in elliptical galaxy surface brightness (SBF) on a given angular scale as comparable galaxies are seen at greater distances (Tonry 1991); a new SBF survey gives $h = 0.81 \pm 0.06$ (Tonry *et al.* 1997).

The "mid-term" value of the Hubble constant from the HST key project is $h = 0.73 \pm 0.10$ (Freedman 1996). This is based on the standard distance to the LMC of 50 kpc (corresponding to a distance modulus of 18.50), which is used to calibrate the Cepheid distances to more distant galaxies. But the preliminary results from the Hipparcos astrometric satellite suggest that the Cepheid distance scale must be recalibrated, and that this distance to the LMC is too low by about 10%, while taking into account its higher metallicity the distance to M31 is too low by about 17% (Feast & Catchpole 1997). An increase in the LMC distance of about 7% is also obtained using the preliminary Hipparcos recalibration of the zero point and metallicity dependence of the RR Lyrae distance scale (Gratton *et al.* 1997, Reid 1997; cf. Alcock *et al.* 1997b), thus removing a long-standing discrepancy between the RR Lyrae and Cepheid distances. If it is true that the standard distance to the LMC has been underestimated, the implication is that the Hubble parameter determined by Cepheid calibrators must be decreased, by perhaps 10% and possibly more. This applies to the HST key project, and it also applies to the SN Ia results for h, which are based on Cepheid distances; thus, for example, the Hamuy *et al.* (1996) value would decrease to about $h = 0.57$, with a corresponding $t_0 = 11.4$ Gyr for $\Omega = 1$.

1.3.2.2 Fundamental physics approaches

The fundamental physics approaches involve either Type Ia or Type II supernovae, the Sunyaev–Zel'dovich (S–Z) effect, or gravitational lensing. All are promising, but in each case the relevant physics remains somewhat uncertain.

The ^{56}Ni radioactivity method for determining H_0 using Type Ia supernovae avoids the uncertainties of the distance ladder by calculating the absolute luminosity of Type Ia supernovae from first principles using plausible but as yet unproved physical models. The first result obtained was that $h = 0.61 \pm 0.10$ (Arnet, Branch & Wheeler 1985; Branch 1992); however, another study (Leibundgut & Pinto 1992; cf. Vaughn *et al.* 1995) found that uncertainties in extinction (i.e., light absorption) toward each supernova increases the range of allowed h. Demanding that the ^{56}Ni radioactivity method agree with an expanding photosphere approach leads to $h = 0.60^{+0.14}_{-0.11}$ (Nugent *et al.* 1995). The expanding photosphere method compares the expansion rate of the SN envelope measured by redshift with its size increase inferred from its temperature and magnitude. This approach was first applied to Type II SN; the 1992 result $h = 0.6 \pm 0.1$ (Schmidt, Kirshner & Eastman 1992) was subsequently revised upward by the same authors (1994) to $h = 0.73 \pm 0.06 \pm 0.07$. However, there are various complications with the physics of the expanding envelope (Ruiz-Lapuente et al. 1995; Eastman, Schmidt & Kirshner 1996).

The S-Z effect is the Compton scattering of microwave background photons from the hot electrons in a foreground galaxy cluster. This can be used to measure H_0 since properties of the cluster gas measured via the S-Z effect and from X-ray observations have different dependences on H_0. The result from the first cluster for which sufficiently detailed data was available, A665 (at $z = 0.182$), was $h = (0.4 - 0.5) \pm 0.12$ (Birkinshaw, Hughes & Arnoud 1991); combining this with data on A2218 ($z = 0.171$) raised this somewhat to $h = 0.55 \pm 0.17$ (Birkinshaw & Hughes 1994). Early results from the ASCA X-ray satellite gave $h = 0.47 \pm 0.17$ for A665 and $h = 0.41^{+0.15}_{-0.12}$ for CL0016+16 ($z = 0.545$) (Yamashita 1994). A few S-Z results have been obtained using millimeter-wave observations (Wilbanks et al. 1994), and this method may allow more such measurements soon. New results for A2218 and A1413 ($z = 0.14$) using the Ryle radio telescope and ROSAT X-ray data gave $h = 0.38^{+0.17}_{-0.12}$ and $h = 0.47^{+0.18}_{-0.12}$, respectively (Lasenby 1997). Corrections for the near-relativistic electron motions (Rephaeli 1995) and for lensing by the cluster (Loeb & Refregier 1996) may raise these estimates for H_0 a little, but it seems clear that the S-Z results favor a smaller value than many optical astronomers obtain. However, since the S-Z measurement of H_0 is affected by the isothermality of the clusters (Roettiger, Stone & Mushotzky et al. 1997) and the unknown orientation of the cluster ellipticity with respect to the line of sight, and the errors in the derived values remain rather large, this lower S-Z H_0 can only become convincing with more detailed observations and analyses of a significant number of additional clusters. Perhaps this will be possible within the next several years.

Several quasars have been observed to have multiple images separated by a few arc seconds; this phenomenon is interpreted as arising from gravitational lensing of the source quasar by a galaxy along the line of sight. In the first such system discovered, QSO 0957+561 ($z = 1.41$), the time delay Δt between arrival at the earth of variations in the quasar's luminosity in the two images has been measured to be, e.g., 409 ± 23 days (Pelt et al. 1994), although other authors found a value of 540 ± 12 days (Press, Rybicki & Hewitt 1992). The shorter Δt has now been confirmed by the observation of a sharp drop in Image A of about 0.1 mag in late December 1994 (Kundic, Cohen & Blandford 1997) followed by a similar drop in Image B about 405-420 days later (Kundic et al. 1997). Since $\Delta t \approx \theta^2 H_0^{-1}$, this observation allows an estimate of the Hubble parameter, with the early results $h = 0.50 \pm 0.17$ (Rhee 1991), or $h = 0.63 \pm 0.21$ ($h = 0.42 \pm 0.14$) including (neglecting) dark matter in the lensing galaxy (Roberts et al. 1991), with additional uncertainties associated with possible microlensing and unknown matter distribution in the lensing galaxy and the cluster in which this is the first-ranked galaxy. Deep images allowed mapping of the gravitational potential of the cluster (at $z = 0.36$) using weak gravitational lensing, which led to the conclusion that

$h \leq 0.70(1.1\text{yr}/\Delta t)$ (Dahle, Maddox & Lilje 1994; Rhee et al. 1996, Fischer et al. 1997). Detailed study of the lensed QSO images (which include a jet) constrains the lensing and implies $h = 0.85(1 - \kappa)(1.1\text{yr}/\Delta t) < 0.85$, where the upper limit follows because the convergence due to the cluster $\kappa > 0$, or alternatively $h = 0.85(\sigma/322\,\text{km s}^{-1})^2(1.1\text{yr}/\Delta t)$ without uncertainty concerning the cluster if the one-dimensional velocity dispersion σ in the core of the giant elliptical galaxy responsible for the lensing can be measured (Grogin & Narayan 1996). The latest results for h from 0957+561, using all available data, are $h = 0.63 \pm 0.12$ (95% confidence level, C.L.) (Kundic et al. 1997), $h = 0.62 \pm 0.07$ (Falco et al. 1997, where the error does not include systematic errors in the assumed form of the mass distribution in the lens; uncertainties can also be reduced with new HST images of the system, allowing improved accuracy in the lens galaxy position).

The first quadruple-image quasar system discovered was PG1115+080. Using a recent series of observations (Schechter et al. 1997), the time delay between images B and C has been determined to be about 24 ± 3 days, or $25^{+3.3}_{-3.8}$ days by an alternative analysis (BarKana 1997). A simple model for the lensing galaxy and the nearby galaxies then leads to $h = 0.42 \pm 0.06$ (Schechter et al. 1997) or $h = 0.41 \pm 0.12$ (95% C.L.) (BarKana, private communication), although higher values for h are obtained by a more sophisticated analysis: $h = 0.60 \pm 0.17$ (Keeton & Kochanek 1997), $h = 0.62 \pm 17$ (Kundic, Cohen & Blandford 1997). The results depend on how the lensing galaxy and those in the compact group of which it is a part are modeled. Such models need to be constrained by new HST observations, especially of the light profile in the lensing galaxy, and spectroscopy to better determine the velocity dispersion of the lensing galaxy and of the group.

Although the most recent time-delay results for h from both lensed quasar systems are remarkably close, the uncertainty in the h determination by this method remains rather large. But it is reassuring that this completely independent method gives results consistent with the other determinations. The time-delay method is promising (Blandford & Kundic 1996), and when these systems are better understood and/or delays are reliably measured in several other multiple-image quasar systems, such as B1422+231 (Hammer, Rigaut & Angonin-Willaime 1995, Hjorth et al. 1996), or radio Einstein-ring systems, such as PKS 1830-211 (van Ommen et al. 1995) or B0218+357 (Corbett et al. 1996), this should lead to a more precise and reliable value for H_0.

1.3.2.3 Correcting for Virgocentric infall

What about the HST Cepheid measurement of H_0, giving $h = 0.80 \pm 0.17$ (Freedman et al. 1994), which received so much attention in the press? This calculated value is not based on either of the two methods (a) or (b) above, and it should not be regarded as being very reliable. Instead, this result is obtained by assuming that M100 is at the core of the Virgo cluster, and dividing the sum of the recession velocity of Virgo, about 1100 km s^{-1}, plus the calculated "infall velocity" of the local group toward Virgo, about 300 km s^{-1}, by the measured distance to M100 of 17.1 Mpc. (These recession and infall velocities are both a little on the high side, compared to other values one finds in the literature.) Adding the "infall velocity" is necessary in this method in order to bring the Virgo recession velocity to what it would be were it not for the gravitational attraction of Virgo for the Local Group of galaxies, but the problem with this is that the net motion of the Local Group with respect to Virgo is undoubtedly affected by much besides the Virgo cluster – e.g., the "Great Attractor." For example, in our CHDM supercomputer simulations (which appear to be a rather realistic match to observations), galaxies and groups at about 20 Mpc from a Virgo-sized cluster often have net outflowing rather than infalling velocities. Note that if the net "infall" of M100 were smaller, or if M100 were in the foreground of the Virgo cluster (in which case the actual distance to Virgo would be larger than 17.1 Mpc), then the indicated H_0 would be smaller.

Freedman et al. (1994) gave an alternative argument that avoids the "infall velocity" uncertainty: the relative galaxy luminosities indicate that the Coma cluster is about six times farther away than the Virgo cluster, and peculiar motions of the Local Group and the Coma cluster are relatively small corrections to the much larger recession velocity of Coma; dividing the recession velocity of the Coma cluster by six times the distance to M100 again gives $H_0 \approx 80$. However, this approach still assumes that M100 is in the core rather than the foreground of the Virgo cluster; and in deducing the relative distance of the Coma and Virgo clusters it assumes that the galaxy luminosity functions in each are comparable, which is uncertain in view of the very different environments. More general arguments by the same authors (Mould et al. 1995) lead them to conclude that $h = 0.73 \pm 0.11$ regardless of where M100 lies in the Virgo cluster. But Tammann et al. (1996), using all the available HST Cepheid distances and their own complete sample of Virgo spirals, conclude that $h \approx 0.54$.

To summarize, many observers, using mainly relative distance methods, favor a value $h \approx 0.6$–0.8 although Sandage's group and some others continue to get $h \approx 0.5$–0.6 and some of these values may need to be reduced by something like 10% if the full Hipparcos data set bears out the preliminary reports discussed above. Meanwhile the fundamental physics methods typically lead to $h \approx 0.4$–0.7. Among fundamental physics approaches, there has been important recent progress in measuring h via time delays between different images of gravitationally lensed quasars, with the latest analyses of both of the systems with measured time delays giving $h \approx 0.6 \pm 0.1$.

The fact that the fundamental physics measurements giving lower values for h (via time delays in gravitationally lensed quasars and the Sunyaev–Zel'dovich effect) are mostly of more distant objects has suggested to some authors (Turner, Cen & Ostriker 1992; Wu et al. 1995) that the local universe may actually be underdense and therefore be expanding faster than is typical. But in reasonable models where structure forms from Gaussian fluctuations via gravitational instability, it is extremely unlikely that a sufficiently large region has a density sufficiently smaller than average to make more than a rather small difference in the value of h measured locally (Suto, Suginohara & Inagaki 1995). Moreover, the small dispersion in the corrected maximum luminosity of distant Type Ia supernovae found by the LBL Supernova Cosmology Project (Kim et al. 1997) compared to nearby SNe Ia shows directly that the local and cosmological values of H_0 are approximately equal. The maximum deviation permitted is about 5%. Interestingly, preliminary results using 44 nearby Type Ia supernovae as yardsticks suggest that this maximum deviation is indeed realized, in the sense that in our local region of the universe, out to a radius of about $70\,h^{-1}$ Mpc (the distance of the Northern Great Wall), H_0 is about 5% larger than average (A. Dekel, private communication). The combined effect of this and the Hipparcos correction would, for example, reduce the 'mid-term' value $h \sim 0.73$ from the HST Key Project on the Extragalactic Distance Scale, to $h \sim 0.63$.

There has been recent observational progress in both relative distance and fundamental physics methods, and it is likely that the Hubble parameter will be known reliably to 10% within a few years. Most recent measurements are consistent with $h = 0.6 \pm 0.1$, corresponding to a range $t_0 = 6.52 h^{-1}$Gyr $= 9.3 - 13.0$ Gyr for $\Omega = 1$ – in good agreement with the preliminary estimates of the ages of the oldest globular clusters based on the new data from the Hipparcos astrometric satellite.

1.3.3 Cosmological constant Λ

Inflation is the only known solution to the horizon and flatness problems and the avoidance of too many Grand-Unified-Theory (GUT) monopoles. And inflation has the added bonus that at no extra charge (except the perhaps implausibly fine-tuned adjustment of the self-coupling of the inflaton field to be adequately small), simple inflationary models predict a near-Zel'dovich primordial spectrum (i.e., $P_p(k) \propto k^{n_p}$ with $n_p \approx 1$) of adiabatic Gaussian primordial fluctuations – which seems to be consistent with observations. All simple inflationary models predict that the curvature is vanishingly small, although inflationary models that are extremely contrived (at least, to my mind) can be constructed with negative curvature and therefore $\Omega_0 \lesssim 1$ without a cosmological constant (see §1.6.6 below). Thus most authors who consider inflationary models impose the condition $k = 0$, or $\Omega_0 + \Omega_\Lambda = 1$ where $\Omega_\Lambda \equiv \Lambda/(3H_0^2)$. This is what is assumed in ΛCDM models, and it is what was assumed in Fig. 1.2. (Note that Ω is used to refer only to the density of matter and energy, not including the cosmological constant, whose contribution in Ω units is Ω_Λ.)

The idea of a nonvanishing Λ is commonly considered unattractive. There is no known physical reason why Λ should be so small ($\Omega_\Lambda = 1$ corresponds to $\rho_\Lambda \sim 10^{-12}$ eV^4, which is small from the viewpoint of particle physics), though there is also no known reason why it should vanish (cf. Weinberg 1989, 1997). A very unattractive feature of $\Lambda \neq 0$ cosmologies is the fact that Λ must become important only at relatively low redshift – why not much earlier or much later? Also $\Omega_\Lambda \gtrsim \Omega_0$ implies that the universe has recently entered an inflationary epoch (with a de Sitter horizon comparable to the present horizon). The main motivations for $\Lambda > 0$ cosmologies are (1) reconciling inflation with observations that seem to imply $\Omega_0 < 1$, and (2) avoiding a contradiction between the lower limit $t_0 \gtrsim 13$ Gyr from globular clusters and $t_0 = (2/3)H_0^{-1} = 6.52h^{-1}$ Gyr for the standard $\Omega = 1$, $\Lambda = 0$ Einstein–de Sitter cosmology, if it is really true that $h > 0.5$.

The cosmological effects of a cosmological constant are not difficult to understand (Lahav et al. 1991; Carroll, Press & Turner 1992). In the early universe, the density of energy and matter is far more important than the Λ term on the right-hand side of the Friedmann equation. But the average matter density decreases as the universe expands, and at a rather low redshift ($z \sim 0.2$ for $\Omega_0 = 0.3$) the Λ term finally becomes dominant. If it has been adjusted just right, Λ can almost balance the attraction of the matter, and the expansion nearly stops: for a long time, the scale factor $a \equiv (1+z)^{-1}$ increases very slowly, although it ultimately starts increasing exponentially as the universe starts inflating under the influence of the increasingly dominant Λ term (see Fig. 1.1). The existence of a period during which expansion slows while the clock runs explains why t_0 can be greater than for $\Lambda = 0$, but this also shows that there is an increased likelihood of finding galaxies at the redshift interval when the expansion slowed, and a correspondingly increased opportunity for lensing of quasars (which mostly lie at higher redshift $z \gtrsim 2$) by these galaxies.

The frequency of such lensed quasars is about what would be expected in a standard $\Omega = 1$, $\Lambda = 0$ cosmology, so this data sets fairly stringent upper limits: $\Omega_\Lambda \leq 0.70$ at 90% C.L. (Maoz & Rix 1993, Kochanek 1993), with more recent data giving even tighter constraints: $\Omega_\Lambda < 0.66$ at 95% confidence (Kochanek 1996b). This limit could perhaps be weakened if there were (a) significant extinction by dust in the E/S0 galaxies responsible for the lensing or (b) rapid evolution of these galaxies, but there is much evidence that these galaxies have little dust and have evolved only passively for $z \lesssim 1$ (Steidel, Dickinson & Persson 1994; Lilly et al. 1995; Schade et al. 1996). (A recent paper – Malhotra, Rhodes & Turner 1997 – presents evidence for extinction of quasars by foreground galaxies and claims that this weakens the lensing bound to $\Omega_\Lambda < 0.9$, but there is no quantitative discussion in the paper to justify this claim.)

Yet another constraint comes from number counts of bright E/S0 galaxies in HST images (Driver et al. 1996), since as was just mentioned these galaxies appear to have evolved rather little since $z \sim 1$. The number counts are just as expected in the $\Omega = 1$, $\Lambda = 0$ Einstein–de Sitter cosmology. Even allowing for uncertainties due to evolution and merging of these galaxies, this data would allow Ω_Λ as large as 0.8 in flat cosmologies only in the unlikely event that half the Sa galaxies in the deep HST images were misclassified as E/S0. This number-count approach may be very promising for the future, as the available deep HST image data and our understanding of galaxy evolution both increase.

A model-dependent constraint comes from a detailed simulation of ΛCDM (Klypin, Primack & Holtzman 1996, hereafter KPH96): a COBE-normalized model with $\Omega_0 = 0.3$ and $h = 0.7$ has far too much power on small scales to be consistent with observations, unless there is unexpectedly strong scale-dependent anti-biasing of galaxies with respect to dark matter. (This is discussed in more detail in §1.7.4 below.) For ΛCDM models, the simplest solution appears to be raising Ω_0, lowering H_0, and tilting the spectrum ($n_p < 1$), though of course one could alternatively modify the primordial power spectrum in other ways.

Figure 1.2 shows that with $\Omega_\Lambda \leq 0.7$, the cosmological constant does not lead to a very large increase in t_0 compared to the Einstein–de Sitter case, although it may still be enough to be significant. For example, the constraint that $t_0 \geq 13$ Gyr requires $h \leq 0.5$ for $\Omega = 1$ and $\Lambda = 0$, but this becomes $h \leq 0.70$ for flat cosmologies with $\Omega_\Lambda \leq 0.66$.

1.4 Measuring Ω_0

1.4.1 Very-large-scale measurements

Although it would be desirable to measure Ω_0 and Λ through their effects on the large-scale geometry of space-time, this has proved difficult in practice since it requires comparing objects at higher and lower redshift, and it is hard to separate selection effects or the effects of the evolution of the objects from those of the evolution of the universe. For example, Kellermann (1993), using the angular-size vs. redshift relation for compact radio galaxies, obtained evidence favoring $\Omega \approx 1$; however, selection effects may invalidate this approach (Dabrowski, Lasenby & Saunders 1995). To cite another example, in "redshift-volume" tests (e.g., Loh & Spillar 1986) involving number counts of galaxies per redshift interval, how can we tell whether the galaxies at redshift $z \sim 1$ correspond to those at $z \sim 0$? Several galaxies at higher redshift might have merged, and galaxies might have formed or changed luminosity at lower redshift. Eventually, with extensive surveys of galaxy properties as a function of redshift using the largest telescopes such as Keck, it should be possible to perform classical cosmological tests at least on particular classes of galaxies–that is one of the goals of the Keck DEEP project.

At present, perhaps the most promising technique involves searching for Type Ia supernovae at high-redshift, since these are the brightest supernovae and the spread in their intrinsic brightness appears to be relatively small. Perlmutter et al. (1997a,b) have recently demonstrated the feasibility of finding significant numbers of such supernovae, but a dedicated campaign of follow-up observations of each one will be required in order to measure Ω_0 by determining how the apparent brightness of the supernovae depends on their redshift. This is therefore a demanding project. It initially appeared that ~ 100 high-redshift SNe Ia would be required to achieve a 10% measurement of $q_0 = \Omega_0/2 - \Omega_\Lambda$. However, using the correlation mentioned earlier between the absolute luminosity of a SN Ia and the shape of its light curve (slower decline correlates with higher peak luminosity), it now appears possible to reduce the number of SN Ia required. The Perlmutter group has now analyzed seven high-redshift SN Ia by this method, with the result for a flat universe that $\Omega_0 = 1 - \Omega_\Lambda = 0.94^{+0.34}_{-0.28}$, or equivalently $\Omega_\Lambda = 0.06^{+0.28}_{-0.34}$ (< 0.51 at the 95% confidence level) (Perlmutter et al. 1997b). In November 1995 they discovered an additional 11 high-redshift SN Ia, and they have subsequently discovered many more. Other groups, collaborations from ESO and MSSSO/CfA/CTIO, are also searching successfully for high-redshift supernovae to measure Ω_0 (Garnavich et al. 1996). There has also been recent progress understanding the physical origin of the SN Ia luminosity-light curve correlation. At the present rate of progress, a reliable answer may be available within perhaps a year or two if a consensus emerges from these efforts (see also §7.7).

1.4.2 Large-scale measurements

Ω_0 has been measured with some precision on a scale of about $\sim 50\,h^{-1}$ Mpc, using the data on peculiar velocities of galaxies, and on a somewhat larger scale using redshift surveys based on the IRAS galaxy catalog. Since the results of all such measurements to date have been reviewed in detail (see Ch. 7; also Dekel 1994; Strauss & Willick 1995), on only brief comments are provided here. The "POTENT" analysis tries to recover the scalar velocity potential from the galaxy peculiar velocities. It looks reliable, since it reproduces the observed large-scale distribution of galaxies – that is, many galaxies are found where the converging velocities indicate that there is a lot of matter, and there are voids in the galaxy distribution where the diverging velocities indicate that the density is lower than average. The comparison of the IRAS redshift surveys with POTENT and related analyses typically give fairly large values for the parameter $\beta_I \equiv \Omega_0^{0.6}/b_I$ (where b_I is the biasing parameter for IRAS galaxies), corresponding to $0.3 \lesssim \Omega_0 \lesssim 3$ (for an assumed $b_I = 1.15$). It is not clear whether it will be possible to reduce the spread in these values significantly in the near future – probably both additional data and a better understanding of systematic and statistical effects will be required.

A particularly simple way to deduce a lower limit on Ω_0 from the POTENT peculiar velocity data was proposed by Dekel & Rees (1994), based on the fact that high-velocity outflows from voids are not expected in low-Ω models. Data on just one void indicates that $\Omega_0 \geq 0.3$ at the 97% C.L. This argument is independent of assumptions about Λ or galaxy formation, but of course it does depend on the success of POTENT in recovering the peculiar velocities of galaxies.

However, for the particular cosmological models that are at the focus of this review – CHDM and ΛCDM – stronger constraints are available. This is because these models, in common with almost all CDM variants, assume that the probability distribution function (PDF) of the primordial fluctuations was Gaussian (the assumption of Gaussianity is also supported by observations, cf. §7.4.5). The PDF deduced by POTENT from observed velocities (i.e., the PDF of the mass, if the POTENT reconstruction is reliable) is far from Gaussian today, with a long positive-fluctuation tail and a sharp drop for negative $\delta\rho/\bar{\rho}$ (e.g., respecting the requirement that $\rho \geq 0$). It agrees with a Gaussian initial PDF if and only if Ω is about unity: $\Omega_0 \leq 0.3$ is ruled out at a high sigma level (§7.5.2; i.e., Nusser & Dekel 1993; Bernardeau et al. 1995).

1.4.3 Measurements on scales of a few Mpc

On smaller length scales, there are many measurements that are consistent with a smaller value of Ω_0 (e.g., Ch. 4; Ch. 11; Peebles 1993, Ch. 20). For example, the cosmic virial theorem gives $\Omega(\sim 1h^{-1}\,\text{Mpc}) \approx 0.15[\sigma(1h^{-1}\,\text{Mpc})/(300\,\text{km s}^{-1})]^2$, where $\sigma(1h^{-1}\,\text{Mpc})$ here represents the relative velocity dispersion of galaxy pairs at a separation of $1h^{-1}\,\text{Mpc}$. Although the classic paper (Davis & Peebles 1983) which first measured $\sigma(1h^{-1}\,\text{Mpc})$ using a large redshift survey (CfA1) obtained a value of 340 km s^{-1}, this result is now known to be in error since the entire core of the Virgo cluster was inadvertently omitted (Somerville, Davis & Primack 1996); if Virgo is included, the result is $\sim$ 500–600 km s^{-1} (cf. Mo, Jing & Börner 1993, Zurek et al. 1994), corresponding to $\Omega(\sim 1h^{-1}\,\text{Mpc}) \approx 0.4$–$0.6$. Various redshift surveys give a wide range of values for $\sigma(1h^{-1}\,\text{Mpc}) \sim 300 - 750$ km s^{-1}, with the most salient feature being the presence or absence of rich clusters of galaxies; for example, the IRAS galaxies, which are not found in clusters, have $\sigma(1h^{-1}\,\text{Mpc}) \approx 320$ km s^{-1} (Fisher et al. 1994), while the northern CfA2 sample, with several rich clusters, has much larger σ than the SSRS2 sample, with only a few relatively poor clusters (Marzke et al. 1995; Somerville, Primack & Nolthenius 1996). It is evident that the $\sigma(1h^{-1}\,\text{Mpc})$ statistic is not a very robust one. Moreover, the finite sizes of the dark-matter halos of galaxies and groups complicates the measurement of Ω using the cosmic virial theorem (CVT), generally resulting in a significant underestimate of the actual value (Bartlett & Blanchard 1996, Suto & Jing 1996).

A standard method for estimating Ω on scales of a few Mpc is based on applying virial estimates to groups and clusters of galaxies to try to deduce the total mass of the galaxies including their dark-matter halos from the velocities and radii of the groups; roughly, $GM \sim rv^2$. (What one actually does is to pretend that all galaxies have the same mass-to-light ratio M/L, given by the median M/L of the groups, and integrate over the luminosity function to get the mass density (Kirshner, Oemler & Schechter 1979; Huchra & Geller 1982; Ramella, Geller & Huchra 1989). The typical result is that $\Omega(\sim 1\,h^{-1}\,\text{Mpc}) \sim 0.1$–$0.2$. However, such estimates are at best lower limits, since they can only include the mass within the region where the galaxies in each group can act as test particles. It has been found in CHDM simulations (Nolthenius, Klypin & Primack 1997), that the effective radius of the dark-matter distribution associated with galaxy groups is typically 2-3 times larger than that of the galaxy distribution. Moreover, we find a velocity biasing (Carlberg & Couchman 1989) factor in CHDM groups $b_v^{\text{grp}} \equiv v_{\text{gal,rms}}/v_{\text{DM,rms}} \approx 0.75$, whose inverse squared enters in the Ω estimate. Finally, we find that groups and clusters are typically elongated, so only part of the mass is included in spherical estimators. These factors explain how it can be that our $\Omega = 1$ CHDM simulations produce group-velocity dispersions that are fully consistent with those of observed groups, even with statistical tests such as the median rms internal group velocity vs. the fraction of galaxies grouped (Nolthenius, Klypin & Primack 1994, 1997). This emphasizes the point that local estimates of Ω are at best lower limits on its true value.

Another approach to estimating Ω from information on relatively small scales has been pioneered by Peebles (1989, 1990, 1994). It is based on using the least action principle (LAP) to reconstruct the trajectories of the Local Group galaxies, and the assumption that the mass is concentrated around the galaxies. This is perhaps a reasonable assumption in a low-Ω universe, but it is not at all what must occur in an $\Omega = 1$ universe where most of the mass must lie between the galaxies. Although comparison with $\Omega = 1$ N-body simulations showed that the LAP often succeeds in qualitatively reconstructing the trajectories, the mass is systematically underestimated by a large factor by the LAP method (Branchini & Carlberg 1994). Surprisingly, a different study (Dunn & Laflamme 1995) found that the LAP method underestimates Ω by a factor of 4–5 even in an $\Omega_0 = 0.2$ simulation; the authors say that this discrepancy is due to the LAP neglecting the effect of "orphans" – dark-matter particles that are not members of any halo. Shaya, Peebles, and Tully (1995) have recently attempted to apply the LAP to galaxies in the local supercluster, again getting low Ω_0. The LAP approach should be more reliable on this larger scale, but the method still must be calibrated on N-body simulations of both high- and low-Ω_0 models before its biases can be quantified.

1.4.4 Estimates on galaxy halo scales

A classic paper by Little & Tremaine (1987) had argued that the available data on the Milky Way satellite galaxies required that the Galaxy's halo terminate at about 50 kpc, with a total mass of only about $2.5 \times 10^{11} M_\odot$. But by 1991, new data on local satellite galaxies, especially Leo I, became available, and the Little–Tremaine estimator increased to $1.25 \times 10^{12} M_\odot$. A recent, detailed study finds a mass inside 50 kpc of $(5.4 \pm 1.3) \times 10^{11} M_\odot$ (Kochanek 1996a). Work by Zaritsky *et al.* (1993) has shown that other spiral galaxies also have massive halos. They collected data on satellites of isolated spiral galaxies, and concluded that the fact that the relative velocities do not fall off out to a separation of at least 200 kpc shows that massive halos are the norm. The typical rotation velocity of ~ 200–250 km s^{-1} implies a mass within 200 kpc of $\sim 2 \times 10^{12} M_\odot$. A careful analysis taking into account selection effects and satellite orbit uncertainties concluded that the indicated value of Ω_0 exceeds 0.13 at 90% confidence (Zaritsky & White 1994), with preferred values exceeding 0.3. Newer data suggesting that relative velocities do not fall off out to a separation of ~ 400 kpc (Zaritsky *et al.* 1997) presumably would raise these Ω_0 estimates.

However, if galaxy dark-matter halos are really so extended and massive, that would imply that when such galaxies collide, the resulting tidal tails of debris cannot be flung very far. Therefore, the observed merging galaxies with extended tidal tails such as NGC4038/39 (the Antennae) and NGC7252 probably have halo:(disk + bulge) mass ratios less than 10:1 (Dubinski, Mihos & Hernquist 1996), unless the stellar tails are perhaps made during the collision process from gas that was initially far from the central galaxies (J. Ostriker, private communication, 1996); the latter possibility can be checked by determining the ages of the stars in these tails.

A direct way of measuring the mass and spatial extent of many galaxy dark-matter halos is to look for the small distortions of distant galaxy images due to gravitational lensing by foreground galaxies. This technique was pioneered by Tyson et al. (1984). Though the results were inconclusive (Kovner & Milgrom 1987), powerful constraints could perhaps be obtained from deep HST images or ground-based images with excellent seeing. Such fields would also be useful for measuring the correlated distortions of galaxy images from large-scale structure by weak gravitational lensing; although a pilot project (Mould et al. 1994) detected only a marginal signal, a reanalysis detected a significant signal suggesting that $\Omega_0 \sigma_8 \sim 1$ (Villumsen 1995). Several groups are planning major projects of this sort. The first results from an analysis of the Hubble Deep Field gave an average galaxy mass interior to $20h^{-1}$ kpc of $5.9^{+2.5}_{-2.7} \times 10^{11} h^{-1} M_\odot$ (Dell'Antonio & Tyson 1996).

1.4.5 Cluster baryons vs. Big Bang Nucleosynthesis

A recent review (Copi, Schramm & Turner 1995) of Big Bang Nucleosynthesis (BBN) and observations indicating primordial abundances of the light isotopes concludes that $0.009h^{-2} \leq \Omega_b \leq 0.02h^{-2}$ for concordance with all the abundances, and $0.006h^{-2} \leq \Omega_b \leq 0.03h^{-2}$ if only deuterium is used. For $h = 0.5$, the corresponding upper limits on Ω_b are 0.08 and 0.12, respectively. The recent observations (Songaila et al. 1994a, Carswell et al. 1994) of a possible deuterium line in a hydrogen cloud at redshift $z = 3.32$ in the spectrum of quasar 0014+813, indicating a deuterium abundance $D/H \sim 2 \times 10^{-4}$ (and therefore $\Omega_b \leq 0.006h^{-2}$), are inconsistent with D/H observations by Tytler and collaborators (Tytler, Burles & Kirkman 1996; Tytler, Fan & Burles 1996; Burles & Tytler 1996) in systems at $z = 3.57$ (toward Q1937-1009) and at $z = 2.504$, but with a deuterium abundance about ten times lower. These lower D/H values are consistent with solar system measurements of D and ^{3}He, and they imply $\Omega_b h^2 = 0.024 \pm 0.05$, or Ω_b in the range 0.08–0.11 for $h = 0.5$. If these represent the true D/H, then if the earlier observations were correct they were most probably of a Lyman α forest line. Rugers & Hogan (1996) argue that the width of the $z = 3.32$ absorption features is better fit by deuterium, although they admit that only a statistical sample of absorbers will settle the issue. There is a new possible detection of D at $z = 4.672$ in the absorption spectrum of QSO BR1202-0725 (Wampler et al. 1996) and at $z = 3.086$ toward Q0420-388 (Carswell et al. 1996), but they can only give upper limits on D/H. Wampler (1996) and Songaila, Wampler & Cowie (1997) claim that Tytler, Fan & Burles (1996) have overestimated the HI column density in their system, and therefore underestimated D/H. But Burles & Tytler (1997) argue that the two systems that they have analyzed are much more convincing as real detections of deuterium, that their HI column density measurement is reliable, and that the fact that they measure the same $D/H \sim 2.4 \times 10^{-5}$ in both systems makes it likely that this is the primordial value. Moreover, Tytler, Burles, & Kirkman (1996) have recently presented a higher-resolution spectrum of Q0014+813 in which "deuterium absorption is neither required nor suggested," which would of course completely undercut the argument of Hogan and collaborators for high D/H. Finally, the Tytler group has analyzed their new Keck LRIS spectra of the absorption system toward Q1937-1009, and they say that the lower HI column density advocated by Songaila et al. (1997) is ruled out (S. Burles and D. Tytler, private communications, 1997). Of course, one or two additional high-quality D/H measurements would be very helpful to really settle the issue.

There is an entirely different line of argument that also favors the higher Ω_b implied by the lower D/H of Tytler et al. This is the requirement that the high-redshift intergalactic medium contain enough neutral hydrogen to produce the observed Lyman α forest clouds given standard estimates of the ultraviolet ionizing flux from quasars. The minimum required $\Omega_b \gtrsim 0.05 h_{50}^{-2}$ (Gnedin & Hui 1996, Weinberg et al. 1997) is considerably higher than that advocated by higher D/H values, but consistent with that implied by the lower D/H measurements.

It thus seems that the lower D/H and correspondingly higher $\Omega_b \approx 0.1 h_{50}^{-2}$ are more likely to be correct, although it is worrisome that the relatively high value $Y_p \approx 0.25$ predicted by standard BBN for the primordial ^{4}He abundance does not appear to be favored by the data (Olive, Skillman & Steigman 1997).

White et al. (1993) have emphasized that X-ray observations of clusters, especially Coma, show that the abundance of baryons, mostly in the form of gas (which typically amounts to several times the mass of the cluster galaxies), is about 20% of the total cluster mass if h is as low as 0.5 (see also §4.4). For the Coma cluster they find that the baryon fraction within the Abell radius ($1.5h^{-1}$ Mpc) is

$$f_b \equiv \frac{M_b}{M_{tot}} \geq 0.009 + 0.050 h^{-3/2}, \tag{1.4}$$

where the first term comes from the galaxies and the second from gas. If clusters are a fair sample of both baryons and dark matter, as they are expected to be based on simulations (Evrard, Metzler & Navarro 1996), then this is 2-3 times the amount of baryonic mass expected on the basis of BBN in an $\Omega = 1$, $h \approx 0.5$ universe, though it is just what one would expect in a universe with $\Omega_0 \approx 0.3$ (Steigman & Felten 1995). The fair sample hypothesis implies that

$$\Omega_0 = \frac{\Omega_b}{f_b} = 0.3 \left(\frac{\Omega_b}{0.06}\right)\left(\frac{0.2}{f_b}\right). \tag{1.5}$$

A recent review of gas in a sample of clusters (White & Fabian 1995) finds that the baryon mass fraction within about 1 Mpc lies between 10 and 22% (for $h = 0.5$; the limits scale as $h^{-3/2}$), and argues that it is unlikely that: (a) the gas could be clumped enough to lead to significant overestimates of the total gas mass – the main escape route considered in White et al. (1993), (cf. Gunn & Thomas 1996). The gas mass would also be overestimated if large tangled magnetic fields provide a significant part of the pressure in the central regions of some clusters (Loeb & Mao 1994, but cf. Felten 1996); this can be checked by observation of Faraday rotation of sources behind clusters (Kronberg 1994). If $\Omega = 1$, the alternatives are then either: (b) that clusters have more mass than virial estimates based on the cluster galaxy velocities or estimates based on hydrostatic equilibrium (Balland & Blanchard 1997) of the gas at the measured X-ray temperature (which is surprising since they agree: Bahcall & Lubin 1994); (c) that the usual BBN estimate of Ω_b is wrong; or (d) that the fair sample hypothesis is wrong (for which there is even some observational evidence: Loewenstein & Mushotzky 1996). It is interesting that there are indications from weak lensing that at least some clusters (e.g., for A2218 see Squires *et al.* 1996; for this cluster the mass estimate from lensing becomes significantly higher than that from X-rays when the new ASCA satellite data, indicating that the temperature falls at large radii, is taken into account: Loewenstein 1996) may actually have extended halos of dark matter – something that is expected to a greater extent if the dark matter is a mixture of cold and hot components, since the hot component clusters less than the cold (Kofman *et al.* 1996). If so, the number density of clusters as a function of mass is higher than usually estimated, which has interesting cosmological implications (e.g., σ_8 is higher than usually estimated). It is of course possible that the solution is some combination of alternatives (a)–(d). If none of the alternatives is right, then the only conclusion left is that $\Omega_0 \approx 0.3$.

Notice that the rather high baryon density $\Omega_b \approx 0.1(0.5/h)^2$ implied by the recent Tytler *et al.* measurements of low D/H helps resolve the cluster baryon crisis for $\Omega = 1$ – it is escape route (c) above. With the higher Ω_b implied by the low D/H, there is now a "baryon cluster crisis" for low-Ω_0 models! Even with a baryon fraction at the high end of observations, $f_b \lesssim 0.2(h/0.5)^{-3/2}$, the fair sample hypothesis with this Ω_b implies $\Omega_0 \gtrsim 0.5(h/0.5)^{-1/2}$.

1.4.6 Cluster morphology and evolution

Cluster Morphology. Richstone, Loeb & Turner (1992) showed that clusters are expected to be evolved – i.e., rather spherical and featureless – in low-Ω cosmologies, in which structures form at relatively high redshift, and that clusters should be more irregular in $\Omega = 1$ cosmologies, where they have formed relatively recently and are still undergoing significant merger activity. There are few known clusters that seem to be highly evolved and relaxed, and many that are irregular – some of which are obviously undergoing mergers now or have recently done so (see e.g., Burns *et al.* 1994). This disfavors low-Ω models, but it remains to be seen just how low. Recent papers have addressed this. In one (Mohr *et al.* 1995), a total of 24 CDM simulations with $\Omega = 1$ or 0.2, the latter with $\Omega_\Lambda = 0$ or 0.8, were compared with data on a sample of 57 clusters. The conclusion was that clusters with the observed range of X-ray morphologies are very unlikely in the low-Ω cosmologies. However, these simulations have been criticized because the $\Omega_0 = 0.2$ ones included rather a large amount of ordinary matter: $\Omega_b = 0.1$. (This is unrealistic both because $h \approx 0.8$ provides the best fit for $\Omega_0 = 0.2$ CDM, but then the standard BBN upper limit is $\Omega_b < 0.02h^{-2} = 0.03$; and also because observed clusters have a gas fraction of $\sim 0.15(h/0.5)^{-3/2}$.) Another study (Jing et al. 1995) using dissipationless simulations and not comparing directly to observational data found that ΛCDM with $\Omega_0 = 0.3$ and $h = 0.75$ produced clusters with some substructure, perhaps enough to be observationally acceptable (cf. Buote & Xu 1997). Clearly, this important issue deserves study with higher-resolution hydrodynamic simulations, with a range of assumed Ω_b, and possibly including at least some of the additional physics associated with the galaxies which must produce the metallicity observed in clusters, and perhaps some of the heat as well (see Ch. 8). Better statistics for comparing simulations to data may also be useful (Buote & Tsai 1996).

Cluster Evolution. There is evidence for strong evolution of clusters at relatively low redshift, both in their X-ray properties (Henry et al. 1992, Castander et al. 1995, Ebeling et al. 1995) and in the properties of their galaxies. In particular, there is a strong increase in the fraction of blue galaxies with increasing redshift (the "Butcher–Oemler effect"), which may be difficult to explain in a low-density universe (Kauffmann 1995). Field galaxies do not appear to show such strong evolution; indeed, a recent study concludes that over the redshift range $0.2 \leq z \leq 1.0$ there is no significant evolution in the number density of "normal" galaxies (Steidel, Dickinson & Persson 1994). This is compatible with the predictions of various models, including CHDM with two neutrinos sharing a total mass of about 5 eV (see below), but the dependence of the number of clusters on redshift can be a useful constraint on theories (Jing & Fang 1994, Bryan et al. 1994, Walter & Klypin 1996, Eke, Cole & Frenk 1996).

1.4.7 Early structure formation

In linear theory, adiabatic density fluctuations grow linearly with the scale factor in an $\Omega = 1$ universe, but more slowly if $\Omega < 1$ with or without a cosmological constant. As a result, if fluctuations of a certain size in an $\Omega = 1$ and an $\Omega_0 = 0.3$ theory are equal in amplitude at the present epoch ($z = 0$), then at higher redshift the fluctuations in the low-Ω model had higher amplitude. Thus, structures typically form earlier in low-Ω models than in $\Omega = 1$ models.

Since quasars are seen at the highest redshifts, they have been used to try to constrain $\Omega = 1$ theories, especially CHDM which, because of the hot component, has additional suppression of small-scale fluctuations that are presumably required to make early structure (e.g., Haehnelt 1993). The difficulty is that dissipationless simulations predict the number density of halos of a given mass as a function of redshift, but not enough is known about the nature of quasars – for example, the mass of the host galaxy – to allow a simple prediction of the number of quasars as a function of redshift in any given cosmological model. A more recent study (Katz et al. 1994) concludes that very efficient cooling of the gas in early structures, and angular momentum transfer from it to the dark halo, allows for formation of *at least* the observed number of quasars even in models where most galaxy formation occurs late (cf. Eisenstein & Loeb 1995).

Observers are now beginning to see significant numbers of what may be the central regions of galaxies in an early stage of their formation at redshifts $z = 3$–3.5 (Steidel et al. 1996; Giavalisco, Steidel & Macchetto 1996) – although, as with quasars, a danger in using systems observed by emission is that they may not be typical. As additional observations (e.g., Lowenthal et al. 1997) clarify the nature of these objects, they can perhaps be used to constrain cosmological parameters and models. (This data is discussed in more detail in §1.7.5.)

Another sort of high-redshift object which may hold more promise for constraining theories is damped Lyman α systems (DLAS). DLAS are high column density clouds of neutral hydrogen, generally thought to be protogalactic disks, which are observed as wide absorption features in quasar spectra (Wolfe 1993). They are relatively common, seen in roughly a third of all quasar spectra, so statistical inferences about DLAS are possible. At the highest redshift for which data was published in 1995, $z = 3$–3.4, the density of neutral gas in such systems in units of critical density was reported to be $\Omega_{gas} \approx 0.006$, comparable to the total density of visible matter in the universe today (Lanzetta, Wolfe & Turnshek 1995). Several papers (Mo & Miralda-Escude 1994, Kauffmann & Charlot 1994, Ma & Bertschinger 1994) pointed out that the CHDM model with $\Omega_v = 0.3$ could not produce such a high Ω_{gas}. However, it has been shown that CHDM with $\Omega_v = 0.2$ could do so (Klypin et al. 1995, cf. Ma 1995). The power spectrum on small scales is a very sensitive function of the total neutrino mass in CHDM models. This theory makes two crucial predictions: Ω_{gas} must fall off at higher redshifts, and the DLAS at $z \gtrsim 3$ correspond to systems of internal rotation velocity or velocity dispersion less than about 100 km s^{-1} (this can be inferred from the Doppler widths of the metal line systems associated with the DLAS). Preliminary reports regarding the amount of neutral hydrogen in such systems deduced from the latest data at redshifts above 3.5 appear to be consistent with the first of these predictions (Storrie-Lombardi et al. 1996). But a possible problem for the second (Wolfe 1996) is the large velocity widths of the metal line systems associated with the highest-redshift DLAS (e.g., Lu et al. 1996, at $z = 4.4$); if these actually indicate that a massive disk galaxy is already formed at such a high redshift, and if discovery of other such systems shows that they are not rare, that would certainly disfavor CHDM and other theories with relatively little power on small scales. However, other interpretations of such data which would not cause such problems for theories like CHDM are perhaps more plausible (Haehnelt, Steinmetz & Rauch 1996). More data will help resolve this question, along with DLAS models including both dust absorption (Pei & Fall 1995) and lensing (Bartelmann & Loeb 1996).

One of the best ways of probing early structure formation would be to look at the main light output of the stars of the earliest galaxies, which is redshifted by the expansion of the universe to wavelengths beyond about 5 microns today. Unfortunately, it is not possible to make such observations with existing telescopes; since the atmosphere blocks almost all such infrared radiation, what is required is a large infrared telescope in space. The Space Infrared Telescope Facility (SIRTF) has long been a high priority, and it will be great to have access to the data such an instrument will produce when it is launched sometime in the next decade. In the meantime, the Near Infrared Camera/Multi-Object Spectrograph (NICMOS), installed on the Hubble Space Telescope since spring 1997, may help. An alternative method is to look for the starlight from the earliest stars as extragalactic background infrared light (EBL). Although it is difficult to see this background light directly because our Galaxy is so bright in the near infrared, it may be possible to detect it indirectly through its absorption of TeV gamma rays (via the process $\gamma\gamma \to e^+e^-$). Of the more than twenty active galactic nuclei (AGNs) that have been seen at ~ 10 GeV by the EGRET detector on the Compton Gamma Ray Observatory, only two of the nearest, Mk421 and Mk501, have also been clearly detected in TeV gamma rays by the Whipple Atmospheric Cerenkov Telescope (Quinn et al. 1996, Schubnell et al. 1996). Absorption of $\sim$ TeV gamma rays from AGNs at redshifts $z \sim 0.2$ has been shown to be a sensitive probe of the EBL and thus of the era of galaxy formation (MacMinn & Primack 1996).

1.4.8 Conclusions regarding Ω

The main issue that has been addressed so far is the value of the cosmological density parameter Ω. Arguments can be made for $\Omega_0 \approx 0.3$ (and models such as ΛCDM; Ch. 4; Ch. 8; Ch. 11) or for $\Omega = 1$ (§7; for which the best class of models is probably CHDM), but it is too early to tell which is right.

The evidence would favor a small $\Omega_0 \approx 0.3$ if: (1) the Hubble parameter actually has the high value $H_0 \approx 75$ favored by many observers, and the age of the universe $t_0 \geq 13$ Gyr; or (2) the baryonic fraction $f_b = M_b/M_{tot}$ in clusters is actually $\sim 15\%$, about 3 times larger than expected for standard Big Bang Nucleosynthesis in an $\Omega = 1$ universe. This assumes that standard BBN is actually right in predicting that the density of ordinary matter Ω_b lies in the range $0.009 \leq \Omega_b h^2 \leq 0.02$. High-resolution, high-redshift spectra are now providing important new data on primordial abundances of the light isotopes that should clarify the reliability of the BBN limits on Ω_b. If the systematic errors in the ^{4}He data are larger than currently estimated, then it may be wiser to use the deuterium upper limit $\Omega_b h^2 \leq 0.03$, which is also consistent with the value $\Omega_b h^2 \approx 0.024$ indicated by the only clear deuterium detection at high redshift, with the same D/H $\approx 2.4 \times 10^{-5}$ observed in two different low-metallicity quasar absorption systems (Tytler, Fan & Burles 1996); this considerably lessens the discrepancy between f_b and Ω_b. Another important constraint on Ω_b will come from the new data on small angle CMB anisotropies – in particular, the height of the first Doppler peak (Dodelson, Gates & Stebbins 1996; Jungman et al. 1996; Tegmark 1996), with the latest data consistent with low $h \approx 0.5$ and high $\Omega_b \approx 0.1$, or higher h and lower Ω_b. The location of the first Doppler peak at angular wavenumber $l \approx 220$ indicated by the presently available data (Netterfield et al. 1997, Scott et al. 1996) is evidence in favor of a flat universe; $\Omega_0 \sim 0.3$ with $\Lambda = 0$ is disfavored by this data.

The evidence would favor $\Omega = 1$ if: (1) the POTENT analysis of galaxy peculiar velocity data is right, in particular regarding outflows from voids or the inability to obtain the present-epoch non-Gaussian density distribution from Gaussian initial fluctuations in a low-Ω universe; or (2) the preliminary indication of high Ω_0 and low Ω_Λ from high-redshift Type Ia supernovae (Perlmutter et al. 1996) is confirmed.

The statistics of gravitational lensing of quasars is incompatible with a large cosmological constant Λ and low cosmological density Ω_0. Discrimination between models may improve as additional examples of lensed quasars are searched for in large surveys such as the Sloan Digital Sky Survey. The era of structure formation is another important discriminant between these alternatives, low Ω favoring earlier structure formation, and $\Omega = 1$ favoring later formation with many clusters and larger-scale structures still forming today. A particularly critical test for models like CHDM is the evolution as a function of redshift of Ω_{gas} in damped Lyman α systems. Reliable data on all of these issues is becoming available so rapidly today that there is reason to hope that a clear decision between these alternatives will be possible within the next few years.

What if the data ends up supporting what appear to be contradictory possibilities, e.g., large Ω_0 *and* large H_0? Exotic initial conditions (e.g., "designer" primordial fluctuation spectra, cf. Hodges *et al.* 1990) or exotic dark-matter particles beyond the simple "cold" vs. "hot" alternatives discussed in the next section (e.g., decaying 1–10 MeV tau neutrinos, Dodelson, Gyuk & Turner 1994; volatile dark matter, Pierpaoli *et al.* 1996) could increase the space of possible inflationary theories somewhat. But unless new observations, such as the new stellar parallaxes from the Hipparcos satellite, cause the estimates of H_0 and t_0 to be lowered, it may ultimately be necessary to consider going outside the framework of inflationary cosmological models and consider models with large-scale spatial curvature, with a fairly large Λ (or non-standard sorts of "matter" that violate the strong energy condition – cf. Visser 1997) as well as large Ω_0. This seems particularly unattractive, since in addition to implying that the universe is now entering a final inflationary period, it means that inflation probably did not happen at the beginning of the universe, when it would solve the flatness, horizon, monopole, and structure-generation problems. Moreover, aside from the H_0–t_0 problem, there is not a shred of reliable evidence in favor of $\Lambda > 0$, just increasingly stringent upper limits. Therefore, most cosmologists are rooting for the success of inflation-inspired cosmologies, with $\Omega_0 + \Omega_\Lambda = 1$. With the new upper limits on Λ from gravitational lensing of quasars, number counts of elliptical galaxies, and high-redshift Type Ia supernovae, this means that the cosmological constant is probably too small to lengthen the age of the universe significantly. So one hopes that when the dust finally settles, H_0 and t_0 will both turn out to be low enough to be consistent with General Relativistic cosmology. But of course the universe is under no obligation to live up to our expectations.

1.5 Dark-matter particles

1.5.1 *Hot, warm, and cold dark matter*

The current limits on the total and baryonic cosmological density parameters have been summarized, and it was argued in particular that $\Omega_0 \gtrsim 0.3$ while $\Omega_b \lesssim 0.1$. $\Omega_0 > \Omega_b$ implies that the majority of the matter in the universe is not made of atoms. If the dark matter is not baryonic, what *is* it? Summarized here are the physical and astrophysical implications of three classes of elementary-particle DM candidates, which are called

Table 1.3. *Dark-matter candidates*

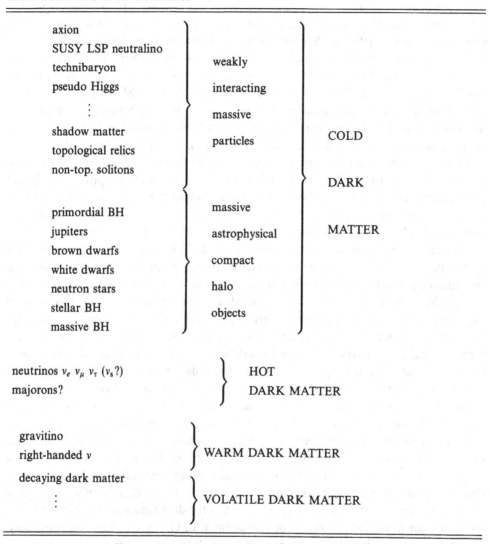

hot, warm, and cold.[1] Table 1.3 gives a list of dark-matter candidates, classified into these categories.

Hot DM (HDM) refers to low-mass neutral particles that were still in thermal equilibrium after the most recent phase transition in the hot early universe, the Quantum Chromo-Dynamics (QCD) confinement transition, which took place at $T_{QCD} \approx 10^2$

---

[1] Dick Bond suggested this terminology to me at the 1983 Moriond Conference, where I used it in my talk (Primack and Blumenthal 1983). George Blumenthal and I had thought of this classification independently, but we used a more complicated terminology.

MeV. Neutrinos are the standard example of hot dark matter, although other more exotic possibilities such as "majorons" have been discussed in the literature. Neutrinos have the virtue that v_e, v_μ, and v_τ are known to exist and, as summarized in §1.5.3, there is experimental evidence that at least some of these neutrino species have mass, though the evidence is not yet really convincing. Hot DM particles have a cosmological number density roughly comparable to that of the microwave background photons, which implies an upper bound to their mass of a few tens of eV: $m(v) = \Omega_v \rho_0 / n_v = \Omega_v 92 h^2$ eV. Having $\Omega_v \sim 1$ implies that free streaming destroys any adiabatic fluctuations smaller than supercluster size, $\sim 10^{15} M_\odot$ (Bond, Efstathiou & Silk 1980). With the COBE upper limit, HDM with adiabatic fluctuations would lead to hardly any structure formation at all, although Hot DM plus some sort of seeds, such as cosmic strings (see, e.g., Zanchin et al. 1996), might still be viable. Another promising possibility is Cold + Hot DM with $\Omega_v \sim 0.2$ (CHDM, discussed in some detail below).

Warm DM particles interact much more weakly than neutrinos. They decouple (i.e., their mean free path first exceeds the horizon size) at $T \gg T_{QCD}$, and they are not heated by the subsequent annihilation of hadronic species. Consequently their number density is expected to be roughly an order of magnitude lower, and their mass an order of magnitude higher, than hot DM particles. Fluctuations as small as large galaxy halos, $\gtrsim 10^{11} M_\odot$, could then survive free streaming. Pagels and Primack (1982) initially suggested that, in theories of local supersymmetry broken at $\sim 10^6$ GeV, gravitinos could be DM of the warm variety. Other candidates have also been proposed, for example light right-handed neutrinos (Olive & Turner 1982). Warm dark matter does not lead to structure formation in agreement with observations, since the mass of the warm particle must be chosen rather small in order to have the power spectrum shape appropriate to fit observations such as the cluster autocorrelation function, but then it is too much like standard hot dark matter and there is far too little small-scale structure (Colombi, Dodelson & Widrow 1996). (This, and also the possibly promising combination of hot and more massive warm dark matter, will be discussed in more detail in §1.7 below.)

Cold DM consists of particles for which free streaming is of no cosmological importance. Two different sorts of cold DM consisting of elementary particles have been proposed, a cold Bose condensate such as axions, and heavy remnants of annihilation or decay such as supersymmetric Weakly Interacting Massive Particles (WIMPs). As has been summarized above, a universe dominated by cold DM looks very much like the one astronomers actually observe, at least on galaxy to cluster scales.

1.5.2 Cold dark-matter candidates

The two sorts of particle candidates for cold dark matter that are best motivated remain supersymmetric WIMPs and the axion. They are both well motivated because both supersymmetry and Peccei–Quinn symmetry (associated with the axion) are key ideas in modern particle physics that were proposed independently of their implications for dark matter (for a review emphasizing direct and indirect methods of detecting both of these, see Primack, Seckel & Sadoulet 1988). There are many other dark-matter candidates whose motivations are more ad hoc (see Table 1.3) from the viewpoint of particle physics. But there is observational evidence that Massive Astrophysical Compact Halo Objects (MACHOs) may comprise a substantial part of the mass of the Milky Way's dark-matter halo.

1.5.2.1 Axions

Peccei–Quinn symmetry, with its associated particle the axion, remains the best solution known to the strong CP problem. A second-generation experiment is currently underway at Los Alamos National Lab. (LLNL) (Hagmann *et al.* 1996) with sufficient sensitivity to have a chance of detecting the axions that might make up part of the dark matter in the halo of our Galaxy, if the axion mass lies in the range 2–20 μeV. However, it now appears that most of the axions would have been emitted from axionic strings (Battye & Shellard 1994) and from the collapse of axionic domain walls (Nagasawa & Kawasaki 1994), rather than arising as an axion condensate as envisioned in the original cosmological axion scenario. This implies that if the axion is the cold dark-matter particle, the axion mass is probably $\sim$meV, above the range of the LLNL experiment. While current experiments looking for either axion or supersymmetric WIMP cold dark matter have a chance of making discoveries, neither type is yet sufficiently sensitive to cover the full parameter space and thereby definitively rule out either theory if they do not detect anything. But in both cases this may be feasible in principle with more advanced experiments that may be possible in a few years.

1.5.2.2 Supersymmetric WIMPs

From the 1930s through the early 1970s, much of the development of quantum physics was a search for ever bigger symmetries, from spin and isospin to the Poincaré group, and from electroweak symmetry to grand unified theories (GUTs). The larger the symmetry group, the wider the scope of the connections established between different elementary particles or other quantum states. The basic pattern of progress was to find the right Lie group and understand its role – $SU(2)$ as the group connecting different states in the cases of spin and isospin; $SU(3) \times SU(2) \times U(1)$ as the dynamical gauge symmetry group of the "Standard Model" of particle physics, connecting states without a gauge boson to states of the same particles including a gauge boson. Supersymmetry is a generalization of this idea of symmetry, since it mixes space-time symmetries, whose quantum numbers include the spin of elementary particles, with internal symmetries. It is based on a generalization of Lie algebra called graded Lie algebra, which involves anti-commutators as well as commutators of the operators that transform one particle state into another. Supersymmetry underlies almost all new ideas in particle physics since the mid-1970s, including superstrings. If valid, it is also bound to be relevant to cosmology. (Some reviews exist, see e.g. Collins, Martin & Squires 1989; de Boer 1994.)

The simplest version of supersymmetry, which should be manifest at the GUT scale ($\sim 10^{16}$ GeV) and below, has as its key prediction that for every kind of particle that

Table 1.4. *Supersymmetry*

A hypothetical symmetry between boson and fermion fields and interactions

Spin	Matter (fermions)	Forces (bosons)	Hypothetical superpartners	Spin
2		graviton	gravitino	3/2
1		photon, $W^\pm, Z^0$	photino, winos, zino,	1/2
		gluons	gluinos	
1/2	quarks u,d,...		squarks $\tilde{u}, \tilde{d}, ...$	0
	leptons $e, \nu_e, ...$		sleptons $\tilde{e}, \tilde{\nu}_e, ...$	
0		Higgs bosons	Higgsinos	1/2
		axion	axinos	

Note: Supersymmetric cold dark-matter candidate particles are underlined.

we have learned about at the relatively low energies which even our largest particle accelerators can reach, there should be an as-yet-undiscovered "supersymmetric partner particle" with the same quantum numbers and interactions except that the spin of this hypothetical partner particle differs from that of the known particle by half a unit. For example, the partner of the photon (spin 1) is the "photino" (spin 1/2), and the partner of the electron (spin 1/2) is the "selectron" (spin 0). Note that if a particle is a fermion (spin 1/2 or 3/2, obeying the Pauli exclusion principle), its partner particle is a boson (spin 0, 1, 2). The familiar elementary particles of matter (quarks and leptons) are all fermions, a fact that is responsible for the stability of matter, and the force particles are all bosons. Table 1.4 is a chart of the known families of elementary particles and their supersymmetric partners. It is these hypothetical partner particles among which we can search for the cold dark-matter particle. The most interesting candidates are underlined. (As have already been mentioned, the gravitino is a warm dark-matter particle candidate; this is discussed further below.)

Note the parallel with Dirac's linking of special relativity and quantum mechanics in his equation for spin-1/2 particles (Griest 1996). In modern language, the resulting CPT invariance (under the combination of charge-conjugation C, replacing each particle with its anti-particle; parity P, reversing the direction of each spatial coordinate; and time-reversal T) requires a doubling of the number of states: an anti-particle for every particle (except for particles, like the photon, which are their own anti-particles).

There are two other key features of supersymmetry that make it especially relevant to dark matter, R-parity and the connection between supersymmetry breaking and the electroweak scale. The R-parity of any particle is $R \equiv (-1)^{L+3B+2S}$, where L, B, and S are its lepton number, baryon number, and spin. Thus for an electron ($L = 1$, $B = 0$, $S = 1/2$) $R = 1$, and the same is true for a quark ($L = 0$, $B = 1/3$, $S = 1/2$) or a photon ($L = 0$, $B = 0$, $S = 1$). Indeed $R = 1$ for all the known particles. But for a selectron ($L = 1$, $B = 0$, $S = 1/2$) or a photino ($L = 0$, $B = 0$, $S = 1/2$), the R-parity is -1, or "odd". In most versions of supersymmetry, R-parity is exactly conserved. This has the powerful consequence that the lightest R-odd particle – often called the "lightest supersymmetric partner" (LSP) – must be stable, for there is no lighter R-odd particle for it to decay into. The LSP is thus a natural candidate to be the dark matter, as was first pointed out by Pagels & Primack (1982), although as mentioned above the LSP in the early form of supersymmetry that we considered would have been a gravitino weighing about a keV, which would now be classified as warm dark matter.

In the now-standard version of supersymmetry, there is an answer to the deep puzzle why there should be such a large difference in mass between the GUT scale $M_{GUT} \sim 10^{16}$ GeV and the electroweak scale $M_{EW} = 80$ GeV. Since both gauge symmetries are supposed to be broken by Higgs bosons which moreover must interact with each other, the natural expectation would be that $M_{GUT} \sim M_{EW}$. The supersymmetric answer to this "gauge hierarchy" problem is that the masses of the weak bosons $W^{\pm}$ and all other light particles are zero until supersymmetry itself breaks. Thus, there is a close relationship between the masses of the supersymmetric partner particles and the electroweak scale. Since the abundance of the LSP is determined by its annihilation in the early universe, and the corresponding cross-section involves exchanges of weak bosons or supersymmetric partner particles – all of which have electromagnetic-strength couplings and masses $\sim M_{EW}$ – the cross-sections will be $\sigma \sim e^2 s/M_{EW}^4$ (where s is the square of the center-of-mass energy) i.e., comparable to typical weak interactions. This in turn has the remarkable consequence that the resulting density of LSPs today corresponds to nearly critical density, i.e., $\Omega_{LSP} \sim 1$. The LSP is typically a spin-1/2 particle called a "neutralino" which is its own anti-particle – that is, it is a linear combination of the photino (supersymmetric partner of the photon), "zino" (partner of the Z^0 weak boson), "Higgsinos" (partners of the two Higgs bosons associated with electroweak symmetry breaking in supersymmetric theories), and "axinos" (partners of the axion, if it exists). In much of the parameter space, the neutralino χ is a "bino," a particular linear combination of the photino and zino. All of these neutralino LSPs are WIMPs. Because of their large masses, several tens to possibly hundreds of GeV, these supersymmetric WIMPs would be dark matter of the "cold" variety.

Having explained why supersymmetry is likely to be relevant to cold dark matter, one should also briefly summarize why supersymmetry is so popular with modern particle physicists. The reasons are that it is not only beautiful, it is even perhaps likely to be true. The supersymmetric pairing between bosons and fermions results in a cancellation of the high-energy (or "ultraviolet") divergences due to internal loops in Feynman diagrams. It is this cancellation that allows supersymmetry to solve the gauge hierarchy problem (how M_{GUT}/M_{EW} can be so big), and perhaps also unify gravity with the other forces ("superunification," "supergravity," "superstrings"). The one prediction of supersymmetry (Georgi, Quinn & Weinberg 1974) that has been verified so far is related to grand unification (Amaldi, de Boer & Furstenau 1991). The way this is usually phrased today is that the three gauge couplings associated with the three parts of the standard model – the $SU(3)$ "color" strong interactions, and the

$SU(2) \times U(1)$ electroweak interactions – do not unify at any higher-energy scale unless the effects of the supersymmetric partner particles are included in the calculation, and they do unify with the minimal set of partners (one partner for each of the known particles) as long as the partner particles all have masses not much higher than the electroweak scale M_{EW} (which, as explained above, is expected if electroweak symmetry breaking is related to supersymmetry breaking).

The expectations for the LSP neutralino, including prospects for their detection in laboratory experiments and via cosmic rays, have recently been exhaustively reviewed (Jungman, Kamionkowski & Griest 1996). Several ambitious laboratory search experiments for LSPs in the mass range of tens to hundreds of GeV are now in progress (e.g., Shutt et al. 1996), and within the next few years they will have adequate sensitivity to probe a significant amount of the supersymmetric model parameter space. There are also hints of supersymmetric effects from recent experiments, which suggest that supersymmetry may be definitively detected in the near future as collider energy is increased – and also hint that the LSP may be rather light (Kane & Wells 1996), possibly even favoring the gravitino as the LSP (Dimopoulos et al. 1996).

1.5.2.3 MACHOs

Meanwhile, the MACHO (Alcock et al. 1997a) and EROS (Ansari et al. 1996, Renault et al. 1997) experiments have detected microlensing of stars in the Large Magellanic Cloud (LMC). While the number of such microlensing events is small (six fairly convincing ones from two years of MACHO data discussed in their latest conference presentations, and one from three years of EROS observations), it is several times more than would be expected just from microlensing by the known stars. The MACHO data suggests that objects with a mass of $0.5^{+0.3}_{-0.2} M_\odot$ are probably responsible for this microlensing, with their total density equal to $\sim$ 20–50 percent of the mass of the Milky Way halo around $\sim$ 20 kpc radius (Gates, Gyuk & Turner 1996). Neither the EROS nor the MACHO groups have seen short duration microlensing events, which implies strong upper limits on the possible contribution to the halo of compact objects weighing less than about $0.05 M_\odot$. While the MACHO masses are in the range expected for white dwarfs, there are strong observational limits (Flynn, Gould & Bahcall 1996) and theoretical arguments (Adams & Laughlin 1996) against white dwarfs being a significant fraction of the dark halo of our Galaxy. Thus it remains mysterious what objects could be responsible for the observed microlensing toward the LMC. But the very large number of microlensing events observed toward the galactic bulge is probably explained by the presence of a bar aligned almost toward our position (Zhao, Rich & Spergel 1996; cf. Bissantz et al. 1997 for a dissenting view). Possibly the relatively small number of microlensing events toward the LMC represent lensing by a tidal tail of stars stretching toward us from the main body of the LMC (Zhao 1998); there is even some data on the colors and luminosities of stars toward the LMC suggesting that this may actually be true (D. Zaritsky, private communication 1997).

1.5.3 Hot dark matter: data on neutrino mass

The upper limit on the electron neutrino mass is roughly 10–15 eV; the current Particle Data Book (Barnett *et al.* 1996) notes that a more precise limit cannot be given since unexplained effects have resulted in significantly negative measurements of $m(v_e)^2$ in recent precise tritium beta decay experiments. The (90% C.L.) upper limit on an effective Majorana neutrino mass is 0.65 eV from the Heidelberg–Moscow ^{76}Ge neutrinoless double beta decay experiment (Balysh *et al.* 1995). The upper limits from accelerator experiments on the masses of the other neutrinos are $m(v_\mu) < 0.17$ MeV (90% C.L.) and $m(v_\tau) < 24$ MeV (95% C.L.). Since stable neutrinos with such large masses would certainly "overclose the universe" (i.e., prevent it from attaining its present age), the cosmological upper limits follow from the neutrino contribution to the cosmological density $\Omega_v = m(v)/(92h^2 \text{ eV}) < \Omega_0$. There is a small window for an unstable v_τ with mass $\sim 10 - 24$ MeV, which could have many astrophysical and cosmological consequences: relaxing the Big Bang Nucleosynthesis bound on Ω_b and N_v, allowing BBN to accommodate a low (less than 22%) primordial ^{4}He mass fraction or high deuterium abundance, improving significantly the agreement between the CDM theory of structure formation and observations, and helping to explain how Type II supernovae explode (Gyuk & Turner 1995).

But there is mounting astrophysical and laboratory data suggesting that neutrinos oscillate from one species to another, and therefore that they have non-zero mass. The implications if *all* these experimental results are taken at face value are summarized in Table 1.5. Of these experiments, the ones that are most relevant to neutrinos as hot dark matter are LSND (see below) and the higher-energy Kamiokande atmospheric (cosmic ray) neutrinos. But the experimental results that are probably most secure are those concerning solar neutrinos, suggesting that some of the electron neutrinos undergo MSW oscillations to another species of neutrino as they travel through the Sun (see, e.g., Hata & Langacker 1995, Bahcall 1996).

The recent observation of events that appear to represent $\bar{v}_\mu \to \bar{v}_e$ oscillations followed by $\bar{v}_e + p \to n + e^+$, $n + p \to D + \gamma$, with coincident detection of e^+ and the 2.2 MeV neutron-capture γ-ray in the Liquid Scintillator Neutrino Detector (LSND) experiment at Los Alamos suggests that $\Delta m_{e\mu}^2 \equiv |m(v_\mu)^2 - m(v_e)^2| > 0$ (Athanassopoulos *et al.* 1995, 1996). The analysis of the LSND data through 1995 strengthens the earlier LSND signal for $\bar{v}_\mu \to \bar{v}_e$ oscillations. Comparison with exclusion plots from other experiments implies a lower limit $\Delta m_{\mu e}^2 \equiv |m(v_\mu)^2 - m(v_e)^2| \gtrsim 0.2$ eV^2, implying in turn a lower limit $m_v \gtrsim 0.45$ eV, or $\Omega_v \gtrsim 0.02(0.5/h)^2$. This implies that the contribution of hot dark matter to the cosmological density is larger than that of all the visible stars ($\Omega_* \approx 0.004$, Peebles 1993, eq. 5.150). More data and analysis are needed from LSND's $v_\mu \to v_e$ channel before the initial hint (Caldwell 1995) that $\Delta m_{\mu e}^2 \approx 6$ eV^2 can be confirmed. Fortunately the KARMEN experiment has just added shielding to decrease its background so that it can probe the same region of $\Delta m_{\mu e}^2$ and mixing angle, with sensitivity as great as LSND's within about two years (Kleinfeller 1996). The Kamiokande data (Fukuda 1994) showing that the deficit of $E > 1.3$ GeV atmospheric

Table 1.5. *Data suggesting neutrino mass*

Solar ν_e deficit	$\Delta m^2_{ex} = 10^{-5}$ eV^2, $\sin^2 2\theta_{ex}$ small
Atm ν_μ deficit	$\Delta m^2_{\mu y} \simeq 10^{-2}$ eV^2, $\sin^2 2\theta_{\mu y} \sim 1$
	Kamiokande $E_\nu > 1.3$ GeV
Reactor ν_e	probably excludes $y = e$, so atm $\nu_\mu \to \nu_\tau$ or ν_s
BBN	excludes $\nu_\mu \to \nu_s$ with large mixing, so $y = \tau$
LSND	$\Delta m^2_{\mu e} \approx 1$–$10$ eV^2, $\sin^2 2\theta_{\mu e}$ small
	excludes $x = \mu$, so solar $\nu_e \to \nu_s$
Cold + Hot Dark Matter	$\Sigma m_\nu \approx 5 h^2_{50}$ eV

muon neutrinos increases with zenith angle suggests that $\nu_\mu \to \nu_\tau$ oscillations[2] occur with an oscillation length comparable to the height of the atmosphere, implying that $\Delta m^2_{\tau\mu} \sim 10^{-2}$ eV^2 – which in turn implies that if either ν_μ or ν_τ have large enough mass ($\gtrsim 1$ eV) to be a hot dark-matter particle, then they must be nearly degenerate in mass, i.e., the hot dark-matter mass is shared between these two neutrino species. The much larger Super-Kamiokande detector is now operating, and we should know by about the end of 1996 whether the Kamiokande atmospheric neutrino data that suggested $\nu_\mu \to \nu_\tau$ oscillations will be confirmed and extended. Starting in 1997 there will be a long-baseline neutrino oscillation disappearance experiment to look for $\nu_\mu \to \nu_\tau$ with a beam of ν_μ from the KEK accelerator directed at the Super-Kamiokande detector, with more powerful Fermilab-Soudan, KEK–Super-Kamiokande, and possibly CERN–Gran Sasso long-baseline experiments later.

Evidence for non-zero neutrino mass evidently favors CHDM, but it also disfavors low-Ω models. Because free streaming of the neutrinos damps small-scale fluctuations, even a little hot dark matter causes reduced fluctuation power on small scales and requires substantial cold dark matter to compensate; thus evidence for even 2 eV of neutrino mass favors large Ω and would be incompatible with a cold dark-matter density Ω_c as small as 0.3 (PHKC95). Allowing Ω_ν and the tilt to vary, CHDM can fit observations over a somewhat wider range of values of the Hubble parameter h than standard or tilted CDM (Pogosyan & Starobinsky 1995a, Liddle *et al.* 1996b). This is especially true if the neutrino mass is shared between two or three neutrino species (Holtzman 1989; Holtzman & Primack 1993; PHKC95; Pogosyan & Starobinsky 1995b; Babu, Schaefer & Shafi 1996), since then the lower neutrino mass results in a larger free-streaming scale over which the power is lowered compared to CDM. The

---

[2] The Kamiokande data is consistent with atmospheric ν_μ oscillating to any other neutrino species y with a large mixing angle $\theta_{\mu y}$. But as summarized in Table 1.5 (see further discussion and references in, e.g., Primack *et al.* 1995, hereafter PHKC95; Fuller, Primack & Qian 1995) ν_μ oscillating to ν_e with a large mixing angle is probably inconsistent with reactor and other data, and ν_μ oscillating to a sterile neutrino ν_s (i.e., one that does not interact via the usual weak interactions) with a large mixing angle is inconsistent with the usual Big Bang Nucleosynthesis constraints. Thus, by a process of elimination, if the Kamiokande data indicating atmospheric neutrino oscillations is right, the oscillation is $\nu_\mu \to \nu_\tau$.

result is that the cluster abundance predicted with $\Omega_\nu \approx 0.2$ and $h \approx 0.5$ and COBE normalization (corresponding to $\sigma_8 \approx 0.7$) is in reasonable agreement with observations without the need to tilt the model (Borgani, Masiero & Yamaguchi 1996) and thereby reduce the small-scale power further. (In CHDM with a given Ω_ν shared between $N_\nu = 2$ or 3 neutrino species, the linear power spectra are identical on large and small scales to the $N_\nu = 1$ case; the only difference is on the cluster scale, where the power is reduced by $\sim 20\%$ (Holtzman 1989, PHKC95, Pogosyan & Starobinsky 1995a, 1995b).

1.6 Origin of fluctuations: inflation and topological defects

1.6.1 Topological defects

A fundamental scalar field, the Higgs field, is invoked by particle theorists to account for the generation of mass; one of the main goals of the next generation of particle accelerators, including the Large Hadron Collider at CERN, will be to verify the Higgs theory for the generation of the mass of the weak vector bosons and all the lighter elementary particles. Another scalar field is required to produce the vacuum energy which may drive cosmic inflation (discussed in the next section). Scalar fields can also create topological defects that might be of great importance in cosmology. The basic idea is that some symmetry is broken wherever a given scalar field ϕ has a non-vanishing value, so the dimensionality of the corresponding topological defect depends on the number of components of the scalar field: for a single-component real scalar field, $\phi(\vec{r}) = 0$ defines a two-dimensional surface in three-dimensional space, a *domain wall*; for a complex scalar field, the real and imaginary parts of $\phi(\vec{r}) = 0$ define a one-dimensional locus, a *cosmic string*; for a three-component (e.g., isovector) field, $\phi_i(\vec{r}) = 0$ for $i = 1, 2, 3$ is satisfied at isolated points, *monopoles*; for more than three components, one gets *textures* that are not topologically stable but which can seed structure in the universe as they unwind.

To see how this works in more detail, consider a cosmic string. For the underlying field theory to permit cosmic strings, we need to couple a complex scalar field ϕ to a single-component (i.e., $U(1)$) gauge field A_α, like the electromagnetic field, in the usual way via the substitution $\partial_\alpha \to D_\alpha \equiv (\partial_\alpha - ieA_\alpha)$, so that the scalar field derivative term in the Lagrangian becomes $\mathscr{L}_{D\phi} = |D_\alpha \phi|^2$. Then if the scalar field ϕ gets a non-zero value by the usual Higgs "spontaneous symmetry breaking" mechanism, the gauge symmetry is broken because the field has a definite complex phase. But along a string where $\phi = 0$ the symmetry is restored. As one circles around the string at any point on it, the complex phase of $\phi(\vec{r})$ in general makes one, or possibly $n > 1$, complete circles $0 \to 2n\pi$. But since such a phase rotation can be removed at large distance from the string by a gauge transformation of ϕ and A_α, the energy density associated with this behavior of ϕ $\mathscr{L}_{D\phi} \to 0$ at large distances, and therefore the energy μ per unit length of string is finite. Since it would require an infinite amount of energy to unwind the phase of ϕ at infinity, however, the string is topologically stable. If the field theory describing the early universe includes a $U(1)$ gauge field and associated complex Higgs field ϕ, a rather high density of such cosmic strings will form when the string field ϕ acquires its non-zero value and breaks the $U(1)$ symmetry. This happens because there is no way for the phase of ϕ to be aligned in causally disconnected regions, and it is geometrically fairly likely that the phases will actually wrap around as required for a string to go through a given region (Kibble 1976). The string network will then evolve and can help cause formation of structure after the universe becomes matter dominated, as long as the string density is not diluted by a subsequent period of cosmic inflation (on the difficult problem of combining cosmic defects and inflation, see, e.g., Hodges & Primack 1991). A similar discussion can be given for domain walls and local (gauge) monopoles, but these objects are cosmologically pathological since they dominate the energy density and "overclose" the universe. But cosmic strings, a sufficiently low density of global (i.e., non-gauged monopoles), and global textures are potentially interesting for cosmology (recent reviews include Vilenkin & Shellard 1994, Hindmarsh & Kibble 1995, Shellard 1997). Cosmic defects are the most important class of models producing non-Gaussian fluctuations which could seed cosmic structure formation. Since they are geometrically extended objects, they correspond to non-local non-Gaussian fluctuations (Kofman et al. 1991).

The parameter μ, usually quoted in the dimensionless form $G\mu$ (where G is Newton's constant), is the key parameter of the theory of cosmic strings. The value required for the COBE normalization is $G\mu_6 \equiv G\mu \times 10^6 = 1-2$ (recent determinations include $G\mu_6 = 1.7 \pm 0.7$, Perivolaropoulos 1993; 2, Coulson et al. 1994; $(1.05^{+0.35}_{-0.20})$, Allen et al. 1996; 1.7, Allen et al. 1997). This is close enough to the value required for structure formation, $G\mu = (2.2 - 2.8)b_8^{-1} \times 10^{-6}$ (Albrecht & Stebbins 1992), with the smaller value for cosmic strings plus cold dark matter and the higher value for cosmic strings plus hot dark matter, so that the necessary value of the biasing factor b_8 is 1.3–3, which is high (probably leading to underproduction of clusters, and large-scale velocities that are low compared to observations – cf. Perivolaropoulos & Vachaspati 1994), but perhaps not completely crazy. (Here b_8 is the factor by which galaxies must be more clustered than dark matter, on a scale of $8h^{-1}$ Mpc.) Since generically $G\mu \sim (M/m_{\rm Pl})^2$, where M is the energy scale at which the string field ϕ acquires its non-zero value, the fact that $G\mu \sim 10^{-6}$, corresponding to M at roughly the Grand Unification scale, is usually regarded as a plus for the cosmic string scenario. (Even though there is no particular necessity for cosmic strings in GUT scenarios, GUT groups larger than the minimal SU(5) typically do contain the needed extra $U(1)$s.) Moreover, the required normalization is well below the upper limit obtained from the requirement that the gravitational radiation generated by the evolution of the string network not disrupt Big Bang Nucleosynthesis, $G\mu \lesssim 6 \times 10^{-6}$. However, there is currently controversy whether it is also below the upper limit from pulsar timing, which has been determined to be $G\mu \lesssim 6 \times 10^{-7}$ (Thorsett & Dewey 1996) vs. $G\mu \lesssim 5 \times 10^{-6}$ (McHugh et al. 1996; cf. Caldwell, Battye & Shellard 1996).

As for cosmic strings, the COBE normalization for global texture models also implies a high bias $b_8 \approx 3.4$ for $h = 0.7$ (Bennett & Rhie 1993), although the needed bias is somewhat lower for $\Omega \approx 0.3$ (Pen & Spergel 1995). The latest global defect simulations (Pen, Seljak & Turok, 1997) show that the matter power spectrum in all such models also has a shape very different than that suggested by the available data on galaxies and clusters.

But both cosmic string and global defect models have a problem which may be even more serious: they predict a small-angle CMB fluctuation spectrum in which the first peak is at rather high angular wavenumber $\ell \sim 400$ (Crittenden & Turok 1995, Durrer, Gangui & Sakellariadou 1996, Magueijo et al. 1996) and in any case is rather low in amplitude, partly because of incoherent addition of scalar, vector, and tensor modes, according to the latest simulations (strings: Allen et al. 1997; global defects: Pen, Seljak & Turok 1997). This is in conflict with the currently available small-angle CMB data (Netterfield et al. 1997, Scott et al. 1996), which shows a peak at $\ell \sim 200$ and a drop at $\ell \sim 400$, as predicted by flat ($\Omega_0 + \Omega_\Lambda = 1$) CDM-type models. Since the small-angle CMB data is still rather preliminary, it is premature to regard the cosmic defect models as being definitively ruled out. It will be interesting to see the nature of the predicted galaxy distribution and CMB anisotropies when more complete simulations of cosmic defect models are run. This is more difficult than simulating models with the usual inflationary fluctuations, both because it is necessary to evolve the defects, and also because the fact that these defects represent rare but high-amplitude fluctuations necessitates a careful treatment of their local effects on the ordinary and dark matter. It may be difficult to sustain the effort such calculations require, because the poor agreement between the latest defect simulations and current small-angle CMB data does not bode well for defect theories. Fortunately, there have been significant technical breakthroughs in calculational techniques (cf. Allen et al. 1997, Pen, Seljak & Turok 1997).

1.6.2 Cosmic inflation: introduction

The basic idea of inflation is that before the universe entered the present adiabatically expanding Friedmann era, it underwent a period of de Sitter exponential expansion of the scale factor, termed *inflation* (Guth 1981). Actually, inflation is never precisely de Sitter, and any superluminal (faster-than-light) expansion is now called inflation. Inflation was originally invented to solve the problem of too many GUT monopoles, which, as mentioned in the previous section, would otherwise be disastrous for cosmology.

The de Sitter cosmology corresponds to the solution of Friedmann's equation in an empty universe (i.e., with $\rho = 0$) with vanishing curvature ($k = 0$) and positive cosmological constant ($\Lambda > 0$). The solution is $a = a_0 e^{Ht}$, with constant Hubble parameter $H = (\Lambda/3)^{1/2}$. There are analogous solutions for $k = +1$ and $k = -1$ with $a \propto \cosh Ht$ and $a \propto \sinh Ht$ respectively. The scale factor expands exponentially because the positive cosmological constant corresponds effectively to a negative pressure. de Sitter space is discussed in textbooks on general relativity (for example, Rindler 1977, Hawking & Ellis 1973) mainly for its geometrical interest. Until cosmological

Table 1.6. *Inflation summary*

PROBLEM SOLVED

Horizon	Homogeneity, Isotropy, Uniform T
Flatness/Age	Expansion and gravity balance
"Dragons"	Monopoles, domain walls,... banished
Structure	Small fluctuations to evolve into galaxies, clusters, voids

Cosmological constant $\Lambda > 0 \Rightarrow$ space repels space, so the more space the more repulsion, $\Rightarrow$ de Sitter exponential expansion $a \propto e^{\sqrt{\Lambda/3}\,t}$.

Inflation is exponentially accelerating expansion caused by effective cosmological constant ("false vacuum" energy) associated with hypothetical scalar field ("inflaton").

FORCES OF NATURE

		Spin
Known	Gravity	2
	Strong, weak, and electromagnetic	1
Goal of LHC	Mass (Higgs boson)	0
Early universe	Inflation (inflaton)	0

Inflation lasting only $\sim 10^{-32}$ s suffices to solve all the problems listed above. Universe must then convert to ordinary expansion through conversion of false to true vacuum ("re-"heating).

---

inflation was considered, the chief significance of the de Sitter solution in cosmology was that it is a limit to which all indefinitely expanding models with $\Lambda > 0$ must tend, since as $a \to \infty$, the cosmological constant term ultimately dominates the right-hand side of the Friedmann equation.

As Guth (1981) emphasized, the de Sitter solution might also have been important in the very early universe because the vacuum energy that plays such an important role in spontaneously broken gauge theories also acts as an effective cosmological constant. A period of de Sitter inflation preceding ordinary radiation-dominated Friedmann expansion could explain several features of the observed universe that otherwise appear to require very special initial conditions: the horizon, flatness/age, monopole, and structure formation problems (see Table 1.6).

Let us illustrate how inflation can help with the horizon problem. At recombination $(p^+ + e^- \to H)$, which occurs at $a/a_0 \approx 10^{-3}$, the mass encompassed by the horizon was $M_h \approx 10^{18} M_\odot$, compared to $M_{h,0} \approx 10^{22} M_\odot$ today. Equivalently, the angular size today of the causally connected regions at recombination is only $\Delta\theta \sim 3°$. Yet the fluctuation in temperature of the cosmic background radiation from different regions is very small: $\Delta T/T \sim 10^{-5}$. How could regions far out of causal contact have come to temperatures that are so precisely equal? This is the "horizon problem". With inflation, it is no problem because the entire observable universe initially lay inside a single causally connected region that subsequently inflated to a gigantic scale. Similarly, inflation exponentially dilutes any preceding density of monopoles or other unwanted relics (a modern version of the "dragons" that decorated the unexplored borders of old maps).

In the first inflationary models, the dynamics of the very early universe was typically controlled by the self-energy of the Higgs field associated with the breaking of a Grand Unified Theory (GUT) into the standard 3-2-1 model: GUT$\rightarrow SU(3)_{color} \otimes [SU(2) \otimes U(1)]_{electroweak}$. This occurs when the cosmological temperature drops to the unification scale $T_{GUT} \sim 10^{14}$ GeV at about 10^{-35} s after the Big Bang. Guth (1981) initially considered a scheme in which inflation occurs while the universe is trapped in an unstable state (with the GUT unbroken) on the wrong side of a maximum in the Higgs potential. This turns out not to work: the transition from a de Sitter to a Friedmann universe never finishes (Guth & Weinberg 1981). The solution in the "new inflation" scheme (Linde 1982; Albrecht & Steinhardt 1982) is for inflation to occur *after* barrier penetration (if any). It is necessary that the potential of the scalar field controlling inflation ("*inflaton*") be nearly flat (i.e., decrease very slowly with increasing inflaton field) for the inflationary period to last long enough. This nearly flat part of the potential must then be followed by a very steep minimum, in order that the energy contained in the Higgs potential be rapidly shared with the other degrees of freedom ("reheating"). A more general approach, "chaotic" inflation, has been worked out by Linde (1983, 1990) and others; this works for a wide range of inflationary potentials, including simple power laws such as $\lambda\phi^4$. However, for the amplitude of the fluctuations to be small enough for consistency with observations, it is necessary that the inflaton self-coupling be very small, for example $\lambda \sim 10^{-14}$ for the ϕ^4 model. This requirement prevents a Higgs field from being the inflaton, since Higgs fields by definition have gauge couplings to the gauge field (which are expected to be of order unity), and these would generate self-couplings of similar magnitude even if none were present. Both the Higgs and inflaton are hypothetical fundamental (or possibly composite) scalar fields (see Table 1.6).

It turns out to be necessary to inflate by a factor $\gtrsim e^{66}$ in order to solve the flatness problem, i.e., that $\Omega_0 \sim 1$. (With $H^{-1} \sim 10^{-34}$ s during the de Sitter phase, this implies that the inflationary period needs to last for only a relatively small time $\tau \gtrsim 10^{-32}$ s.) The "flatness problem" is essentially the question why the universe did not become curvature dominated long ago. Neglecting the cosmological constant on the assumption that it is unimportant after the inflationary epoch, the Friedmann equation can be written

$$\left(\frac{\dot{a}}{a}\right)^2 = \frac{8\pi G}{3}\frac{\pi^2}{30}g(T)T^4 - \frac{kT^2}{(aT)^2}, \qquad (1.6)$$

where the first term on the right-hand side is the contribution of the energy density in relativistic particles and $g(T)$ is the effective number of degrees of freedom. The second term on the right-hand side is the curvature term. Since $aT \approx$ constant for adiabatic expansion, it is clear that as the temperature T drops, the curvature term becomes increasingly important. The quantity $K \equiv k/(aT)^2$ is a dimensionless measure of the curvature. Today, $|K| = |\Omega - 1| H_0^2/T_0^2 \leq 2 \times 10^{-58}$. Unless the curvature exactly vanishes, the most "natural" value for K is perhaps $K \sim 1$. Since inflation increases a by a tremendous factor $e^{H\tau}$ at essentially constant T (after reheating), it increases aT by the same tremendous factor and thereby decreases the curvature by that factor squared. Setting $e^{-2H\tau} \lesssim 2 \times 10^{-58}$ gives the needed amount of inflation: $H\tau \gtrsim 66$. This much inflation turns out to be enough to take care of the other cosmological problems mentioned above as well.

Of course, this is only the minimum amount of inflation needed; the actual inflation might have been much greater. Indeed, it is frequently argued that since the amount of inflation is a tremendously sensitive function of, e.g., the initial value of the inflaton field, it is extremely likely that there was much more inflation than the minimum necessary to account for the fact that Ω_0 is of order unity. It then follows that the curvature constant is probably vanishingly small after inflation, which implies (in the absence of a cosmological constant today) that $\Omega_0 = 1$ to a very high degree of accuracy. A way of evading this that has recently been worked out is discussed below.

1.6.3 Inflation and the origin of fluctuations

Thus far, it has been sketched how inflation stretches, flattens, and smoothes out the universe, thus greatly increasing the domain of initial conditions that could correspond to the universe that we observe today. But inflation also can explain the origin of the fluctuations necessary in the gravitational instability picture of galaxy and cluster formation. Recall that the very existence of these fluctuations is a problem in the standard Big Bang picture, since these fluctuations are much larger than the horizon at early times. How could they have arisen?

The answer in the inflationary universe scenario is that they arise from quantum fluctuations in the inflaton field ϕ whose vacuum energy drives inflation. The scalar fluctuations $\delta\phi$ during the de Sitter phase are of the order of the Hawking temperature $H/2\pi$. Because of these fluctuations, there is a time spread $\Delta t \approx \delta\phi/\dot\phi$ during which different regions of the same size complete the transition to the Friedmann phase. The result is that the density fluctuations when a region of a particular size re-enters the horizon are equal to $\delta_H \equiv (\delta\rho/\rho)_H \sim \Delta t/t_H = H\Delta t$ (Guth & Pi 1982; see Linde 1990 for alternative approaches). The time spread Δt can be estimated from the equation of motion of ϕ (the free Klein–Gordon equation in an expanding universe): $\ddot\phi + 3H\dot\phi = -(\partial V/\partial\phi)$. Neglecting the $\ddot\phi$ term, since the scalar potential V must be very flat in order for enough inflation to occur (this is called the "slow roll" approximation), $\dot\phi \approx -V'/(3H)$, so $\delta_H \sim H^3/V' \sim V^{3/2}/V'$. Unless there is a special feature in the potential $V(\phi)$ as ϕ rolls through the scales of importance in cosmology (producing such "designer inflation" features generally requires fine tuning – see e.g., Hodges *et al.* 1990), V and V' will hardly vary there and hence δ_H will be essentially constant. These are fluctuations of all the contents of the universe, so they are adiabatic fluctuations.

Thus *inflationary models typically predict a nearly constant curvature spectrum δ_H = constant of adiabatic fluctuations*. Some time ago Harrison (1970), Zel'dovich (1972), and others had emphasized that this is the only scale-invariant (i.e., power-law) fluctuation spectrum that avoids trouble at both large and small scales. If $\delta_H \propto M_H^{-\alpha}$, where M_H is the mass inside the horizon, then if $-\alpha$ is too large the universe will be less homogeneous on large than small scales, contrary to observation; and if α is too large, fluctuations on sufficiently small scales will enter the horizon with $\delta_H \gg 1$ and collapse to black holes (see e.g., Carr, Gilbert & Lidsey 1994, Bullock & Primack 1996); thus $\alpha \approx 0$. The $\alpha = 0$ case has come to be known as the Zel'dovich spectrum.

Inflation predicts more: it allows the calculation of the value of the constant δ_H in terms of the properties of the scalar potential $V(\phi)$. Indeed, this proved to be embarrassing, at least initially, since the Coleman–Weinberg potential, the first potential studied in the context of the new inflation scenario, results in $\delta_H \sim 10^2$ (Guth & Pi 1982) some six orders of magnitude too large. But this does not seem to be an insurmountable difficulty; as was mentioned above, chaotic inflation works, with a sufficiently small self-coupling. Thus inflation at present appears to be a plausible solution to the problem of providing reasonable cosmological initial conditions (although it sheds no light at all on the fundamental question why the cosmological constant is so small now). Many variations of the basic idea of inflation have been worked out, and the following sections will discuss two recent developments in a little more detail. Linde (1995) recently classified these inflationary models in an interesting and useful way: see Table 1.7.

1.6.4 Eternal inflation

Vilenkin (1983) and Linde (1986, 1990) pointed out that if one extrapolates inflation backward to try to imagine what might have preceded it, in many versions of inflation the answer is "eternal inflation": in most of the volume of the universe inflation is still happening, and our part of the expanding universe (a region encompassing far more than our entire cosmic horizon) arose from a tiny part of such a region. To see how eternal inflation works, consider the simple chaotic inflation model with $V(\phi) = (m^2/2)\phi^2$. During the de Sitter Hubble time H^{-1}, where, as usual, $H^2 = (8\pi G/3)V$, the slow rolling of ϕ down the potential will reduce it by

$$\Delta\phi = \dot\phi \Delta t = -\frac{V'}{3H}\Delta t = \frac{m_{\text{Pl}}^2}{4\pi\phi} \, . \tag{1.7}$$

Here m_{Pl} is the Planck mass (see Table 1.1). But there will also be quantum fluctuations that will change ϕ up or down by

Table 1.7. *Linde's classification of inflation models*

How Inflation Begins

Old Inflation	T_{in} high, $\phi_{in} \approx 0$ is false vacuum until phase transition
	Ends by bubble creation; Reheat by bubble collisions
New Inflation	Slow roll down $V(\phi)$, no phase transition
Chaotic Inflation	Similar to New Inflation, but ϕ_{in} essentially arbitrary:
	any region with $\frac{1}{2}\dot{\phi}^2 + \frac{1}{2}(\partial_i\phi)^2 \lesssim V(\phi)$ inflates
Extended Inflation	Like Old Inflation, but slower (e.g., power $a \propto t^p$),
	so phase transition can finish

Potential $V(\phi)$ During Inflation

Chaotic typically $V(\phi) = \Lambda\phi^n$, can also use $V = V_0 e^{\alpha\phi}$, etc.
$$\Rightarrow a \propto t^p, \; p = 16\pi/\alpha^2 \gg 1$$

How Inflation Ends

First-order phase transition – e.g., Old or Extended inflation
Faster rolling → oscillation – e.g., Chaotic $V(\phi)^2\Lambda\phi^n$
"Waterfall" – rapid roll of σ triggered by slow roll of ϕ

(Re)heating

Decay of inflatons
"Preheating" by parametric resonance, then decay

Before Inflation?

Eternal Inflation? Can be caused by:
- Quantum $\delta\phi \sim H/2\pi > $ rolling $\Delta\phi = \dot\phi\Delta t = \dot\phi H^{-1} \approx V'/V$
- Monopoles or other topological defects

$$\delta\phi = \frac{H}{2\pi} = \frac{m\phi}{\sqrt{3\pi}m_{Pl}} \,. \tag{1.8}$$

These will be equal for $\phi_* = m_{Pl}^{3/2}/2m^{1/2}$, $V(\phi_*) = (m/8m_{Pl})m_{Pl}^4$. If $\phi \gtrsim \phi_*$, *positive quantum fluctuations dominate* the evolution: after $\Delta t \sim H^{-1}$, an initial region becomes $\sim e^3$ regions of size $\sim H^{-1}$, in half of which ϕ increases to $\phi + \delta\phi$. Since $H \propto \phi$, this drives inflation faster in these regions. Various mechanisms probably cut this off as $\phi \to m_{Pl}^2/m$ and $V \to m_{Pl}^4$ – for further discussion and references, see Linde (1995). Thus, although ϕ at any given point is likely eventually to roll down the potential and end inflation, in most of the volume of the metauniverse $\phi > \phi_*$ and inflation is proceding at a very fast rate.

Eternal inflation is eternal in the sense that, once started, it never ends. But it remains uncertain whether or not it could have begun an infinite length of time ago. Assuming the "weak energy condition" $T_{\mu\nu}V^\mu V^\nu \geq 0$ for all timelike vectors V^μ, i.e., that any observer will measure a positive energy density, Borde & Vilenkin (1994) proved that a future-eternal inflationary model cannot be extended into the infinite past. However, Borde & Vilenkin (1997) have recently shown that the weak energy condition is quite likely to be violated in inflating space-times (except the open universe inflation models discussed below, §1.6.6), so a "steady-state" eternally inflating universe may be possible after all, with no beginning as well as no end.

1.6.5 A supersymmetric inflation model

We have already considered, in connection with cold dark-matter candidates, why supersymmetry is likely to be a feature of the fundamental theory of the particle interactions, of which the present "Standard Model" is presumably just a low-energy approximation. If the higher-energy regime within which cosmological inflation occurs is described by a supersymmetric theory, there are new cosmological problems that initially seemed insuperable. But recent work has suggested that these problems can plausibly be overcome, and that supersymmetric inflation might also avoid the fine-tuning otherwise required to explain the small inflaton coupling corresponding to the COBE fluctuation amplitude. Here the problems will be briefly summarized, and an explanation will be given of how one such model, due to Ross & Sarkar (1996; hereafter RS96) overcomes them. (An interesting alternative supersymmetric approach to inflation is sketched in Dine et al. 1996.)

When Pagels and I (1982) first suggested that the lightest supersymmetric partner particle (LSP), stable because of R-parity, might be the dark-matter particle, that particle was the gravitino in the early version of supersymmetry then in fashion. Weinberg (1982) immediately pointed out that if the gravitino were not the LSP, it could be a source or real trouble because of its long lifetime $\sim M_{Pl}^2/m_{3/2}^3 \sim (m_{3/2}/\text{TeV})^{-3} 10^3$ s, a consequence of its gravitational-strength coupling to other fields. Subsequently, it was realized that supersymmetric theories can naturally solve the gauge hierarchy problem, explaining why the electroweak scale $M_{EW} \sim 10^2$ GeV is so much smaller than the GUT or Planck scales. In this version of supersymmetry, which has now become the standard one, the gravitino mass will typically be $m_{3/2} \sim$ TeV; and the late decay of even a relatively small number of such massive particles can wreck BBN and/or the thermal spectrum of the CMB. The only way to prevent this is to make sure that the reheating temperature after inflation is sufficiently low: $T_{RH} \lesssim 2 \times 10^9$ GeV (for $m_{3/2} =$ TeV) (Ellis, Kim & Nanopoulos 1984, Ellis et al. 1992).

This can be realized in supergravity theories rather naturally (RS96). Define $M \equiv M_{Pl}/(8\pi)^{1/2} = 2.4 \times 10^{18}$ GeV. Break GUT by the Higgs field χ with vacuum expectation value (vev) $<\chi> \sim 10^{16}$ GeV. Break supersymmetry by a gaugino condensate $<\lambda\lambda> \sim (10^{13}\text{GeV})^3$; then the gravitino mass is $m_{3/2} \sim <\lambda\lambda>/M^2 \sim$ TeV. Inflation is expected to inhibit such breaking, so it must occur afterward. The inflaton superpotential has the form $I = \Delta^2 M f(\phi/M)$, with the corresponding potential

$$V(\phi) = e^{|\phi|^2/M^2}\left[\left(\frac{\partial I}{\partial \phi} + \frac{\phi I}{M^2}\right)^2 - \frac{3I^2}{M^2}\right], \tag{1.9}$$

with minimum at ϕ_0. Demanding that at this minimum the potential actually vanishes $V(\phi_0) = 0$, i.e., that the cosmological constant vanishes, implies that $I(\phi_0) = (\partial I/\partial \phi)_{\phi_0} = 0$. The simplest possibility is $I = \Delta^2(\phi - \phi_0)^2/M$. Requiring that $\partial V/\partial \phi|_0 = 0$ for a sufficiently flat potential, implies that $\phi_0 = M$ and that the second derivative also vanishes at the origin; thus

$$V(\phi) = \Delta^4\left[1 - 4\left(\frac{\phi}{M}\right)^3 + \frac{13}{2}\left(\frac{\phi}{M}\right)^4 - 8\left(\frac{\phi}{M}\right)^5 + ...\right] \tag{1.10}$$

(Holman, Raymond & Ross 1984). This particular inflaton potential is of the "new inflation" type, and corresponds to tilt $n_p = 0.92$ and a number of e-folds during inflation

$$N = \int_{\phi_{in}}^{\phi_{end}}\left(\frac{-V'}{V}\right)d\phi = \frac{M}{12\Delta}, \tag{1.11}$$

assuming that the starting value ϕ_{in} of the inflaton field is sufficiently close to the origin (which has relatively small but nonvanishing probability – the ϕ field presumably has a broad initial distribution). Matching the COBE fluctuation amplitude requires that $\Delta/M = 1.4 \times 10^{-4}$, which, in turn, implies that $N \sim 10^3$, $m_\phi \sim \Delta^2/M \sim 10^{11}$ GeV, $T_{RH} \sim 10^5$ GeV (parametric resonance reheating does not occur). Such a low reheat temperature insures that there will be no gravitino problem, and requires that the baryon asymmetry be generated by electroweak baryogenesis – which appears to be viable as long as the theory contains adequately large CP violation.

Note the following features of the above scenario: inflation occurs at an energy scale far below the GUT scale, so there is essentially no gravity wave contribution to the large-angle CMB fluctuations (i.e., $T/S \approx 0$) even though there is significant tilt ($n_p = 0.92$ for the particular potential above); there is a low reheat temperature, so electroweak baryogenesis is required; and the universe is predicted to be very flat since there are many more e-folds than required to solve the flatness problem.

1.6.6 Inflation with $\Omega_0 < 1$

Can inflation produce a region of negative curvature larger than our present horizon – for example, a region with $\Omega_0 < 1$ and $\Lambda = 0$? The old approach to this problem was to imagine that there might be just enough inflation to solve the horizon problem, but not quite enough to oversolve the flatness problem, e.g., $N \sim 60$ (Steinhardt 1990). This requires fine tuning, but the real problem with this approach is that the resulting region will not be smooth enough to agree with the small size of the quadrupole anisotropy Q measured by COBE. According to Grischuk & Zel'dovich (1978) (cf. Garcia-Bellido et al. 1995), $\delta \sim 1$ fluctuations on a super-horizon scale $L > H_0^{-1}$ imply $Q \sim (LH_0)^{-2}$.

COBE measured $Q_{rms} < 2 \times 10^{-5}$, which, in turn, implies that the region containing our horizon must be homogeneous on a scale $L \gtrsim 500 H_0^{-1}$, i.e., $N \gtrsim 70$, $|1 - \Omega_0| \lesssim 10^{-4}$.

A new approach was discovered, based on the fact that a bubble created from de Sitter space by quantum tunneling tends to be spherical and homogeneous if the tunneling is sufficiently improbable. The interior of such bubbles are quite empty, i.e., they are a region of negative curvature with $\Omega \to 0$. That was why, in "old inflation," the bubbles must collide to fill the universe with energy; and the fact that this does not happen (because the bubbles grow only at the speed of light while the space between them grows superluminally) was fatal for that approach to inflation (Guth & Weinberg 1983).[3] But now this defect is turned into a virtue by arranging to have a second burst of inflation inside the bubble, to drive the curvature back toward zero, i.e., $\Omega_0 \to 1$. By tuning the amount of this second period of inflation, it is possible to produce any desired value of Ω_0 (Sasaki, Tanaka & Yamamoto 1995; Bucher, Goldhaber & Turok 1995; Yamamoto, Sasaki & Tanaka 1995). The old problem of too much inhomogeneity beyond the horizon producing too large a value of the quadrupole anisotropy is presumably solved because the interior of the bubble produced in the first inflation is very homogeneous.

I personally regard this as an existence proof that inflationary models producing $\Omega_0 \sim 0.3$ (say) can be constructed which are not obviously wrong. But I do not regard such contrived models as being as theoretically attractive as the simpler models in which the universe after inflation is predicted to be flat. (Somewhat simpler two-inflaton models giving $\Omega_0 < 1$ have been constructed by Linde & Mezhlumian 1995.) Note also that if varying amounts of inflation are possible, much greater volume is occupied by the regions in which more inflation has occurred, i.e., where $\Omega_0 \approx 1$. But the significance of such arguments is uncertain, since no one knows whether volume is the appropriate measure to apply in calculating the probability of our horizon having any particular property.

The spectra of density fluctuations produced in inflationary models with $\Omega_0 < 1$ tend to have a lot of power on very large scales. However, when such spectra are normalized to the COBE CMB anisotropy observations, the spherical harmonics with angular wavenumber $\ell \approx 8$ have the most weight statistically, and all such models have similar normalization (Liddle *et al.* 1996a).

1.6.7 Inflation summary

The key features of all inflation scenarios are a period of superluminal expansion, followed by ("re-")heating which converts the energy stored in the inflaton field (for example) into the thermal energy of the hot big bang.

Inflation is *generic*: it fits into many versions of particle physics, and it can even be

---

[3] Although there have been attempts to revive Old Inflation within scenarios in which the inflation is slower so that the bubbles can collide, it remains to be seen whether any such Extended Inflation model can be sufficiently homogeneous to be entirely satisfactory.

made rather natural in modern supersymmetric theories as we have seen. The simplest models have inflated away all relics of any pre-inflationary era and result in a flat universe after inflation, i.e., $\Omega = 1$ (or more generally $\Omega_0 + \Omega_\Lambda = 1$). Inflation also produces scalar (density) fluctuations that have a primordial spectrum

$$\left(\frac{\delta\rho}{\rho}\right)^2 \sim \left(\frac{V^{3/2}}{m_{\rm Pl}^3 V'}\right)^2 \propto k^{n_{\rm p}}, \tag{1.12}$$

where V is the inflaton potential and $n_{\rm p}$ is the primordial spectral index, which is expected to be near unity (near-Zel'dovich spectrum). Inflation also produces tensor (gravity wave) fluctuations, with spectrum

$$P_{\rm t}(k) \sim \left(\frac{V}{m_{\rm Pl}}\right)^2 \propto k^{n_{\rm t}}, \tag{1.13}$$

where the tensor spectral index $n_{\rm t} \approx (1 - n_{\rm p})$ in many models.

The quantity $(1 - n_{\rm p})$ is often called the "tilt" of the spectrum; the larger the tilt, the more fluctuations on small spatial scales (corresponding to large k) are suppressed compared to those on larger scales. The scalar and tensor waves are generated by independent quantum fluctuations during inflation, and so their contributions to the CMB temperature fluctuations add in quadrature. The ratio of these contributions to the quadrupole anisotropy amplitude Q is often called $T/S \equiv Q_{\rm t}^2/Q_{\rm s}^2$; thus the primordial scalar fluctuation power is decreased by the ratio $1/(1 + T/S)$ for the same COBE normalization, compared to the situation with no gravity waves ($T = 0$). In power-law inflation, $T/S = 7(1 - n_{\rm p})$. This is an approximate equality in other popular inflation models such as chaotic inflation with $V(\phi) = m^2\phi^2$ or $\lambda\phi^4$. But note that the tensor wave amplitude is just the inflaton potential during inflation divided by the Planck mass, so the gravity wave contribution is negligible in theories like the supersymmetric model discussed above in which inflation occurs at an energy scale far below $m_{\rm Pl}$. Because gravity waves just redshift after they come inside the horizon, the tensor contributions to CMB anisotropies corresponding to angular wavenumbers $\ell \gg 20$, which came inside the horizon long ago, are strongly suppressed compared to those of scalar fluctuations. The indications from presently available data (Netterfield et al. 1997; cf. Tegmark 1996; Ch. 3) are that the CMB amplitude is rather high for $\ell \approx 200$, approximately in agreement with the predictions of standard CDM with $h \approx 0.5$, $\Omega_{\rm b} \approx 0.1$, and scalar spectral index $n_{\rm p} = 1$. This suggests that there is little room for gravity-wave contributions to the low-ℓ CMB anisotropies, i.e., that $T/S \ll 1$. Thus tests of inflation involving the gravity-wave spectrum will be very difficult. Fortunately, inflation can be tested with the data expected soon from the next generation of CMB experiments, since it makes very specific and discriminatory predictions regarding the relative locations of the acoustic peaks in the spectrum, for example the ratio of the first peak location to the spacing between the peaks $\ell_1/\Delta\ell \approx 0.7$–$0.9$ (Hu & White 1996, Hu, Spergel & White 1997).

On the other hand, inflation is also *Alice's restaurant* where, according to the Arlo Guthrie song, "... you can get anything you want ... excepting Alice". It's not even clear what "Alice" you can't get from inflation. It was initially believed that inflation predicts a flat universe. But now we know that you can

- get $\Omega_0 < 1$ (with $\Lambda = 0$), as discussed in the previous section.
- make models consistent with supergravity and the sort of four-dimensional physics expected from superstrings, in which case one may expect that inflation occurs at a relatively low energy scale, which implies $T/S \approx 0$, a low reheat temperature implying no production of topological defects and presumably requiring that baryosynthesis occur at the electroweak phase transition, and plenty of inflation implying that $\Omega_0 \approx 1$.
- alternatively get strings or other topological defects such as textures during or at the end of inflation (e.g., Hodges & Primack 1991) – which however probably requires tuning of the inflation and/or string model, for example to avoid a fractal pattern of structure-forming defects, which would conflict with the observed homogeneity of structure on very large scales.

And in many versions of inflation, the most reasonable answer to the question "what happened before inflation" appears to be eternal inflation, which implies that in most of the meta-universe, exponentially far beyond our horizon, inflation never stopped.

1.7 Comparing DM models to observations: ΛCDM vs. CHDM

1.7.1 Building a cosmology: overview

An effort has been made to summarize the main issues in cosmological model-building in Fig. 1.3. Here the choices of cosmological parameters, dark-matter composition, and initial fluctuations that specify the model are shown at the top of the chart, and the types of data that each cosmological model must properly predict are shown in the boxes with shaded borders in the lower part of the chart. Of course, the chart only shows a few of the possibilities. Models in which structure arises from gravitational collapse of adiabatic inflationary fluctuations and in which most of the dark matter is cold are very predictive. Since such models have also been studied in greatest detail, this class of models will be the center of attention here.

Perhaps the most decisive issue in model-building is the value of the cosmological expansion rate, the Hubble parameter h. If $h \approx 0.7$ as many observations suggest, and the age of the universe $t_0 \gtrsim 13$ Gyr, then only low-Ω_0 models can be consistent with general relativity.[4] Depending on just how large h and t_0 are, a positive cosmological

---

[4] It is important to appreciate that the possible t_0–H_0 (age–expansion rate) conflict goes to the heart of GR and does not depend on cosmological model-dependent issues like the growth rate of fluctuations. As explained in §1.2 besides GR itself the only other theoretical input needed is the cosmological principle: we do not live in the center of a spherical universe; any observer would see the same isotropy of the distant universe, as reflected in particular in the COBE observations. That is enough to imply the Friedmann–Robertson–Walker equations, which give the $t_0 - H_0$ connection. GR is not just a theory whose intrinsic beauty and great success in describing data on relatively small scales encourage us to extrapolate it to the scale of the entire observable universe. It is the only decent theory of gravity and cosmology that we have.

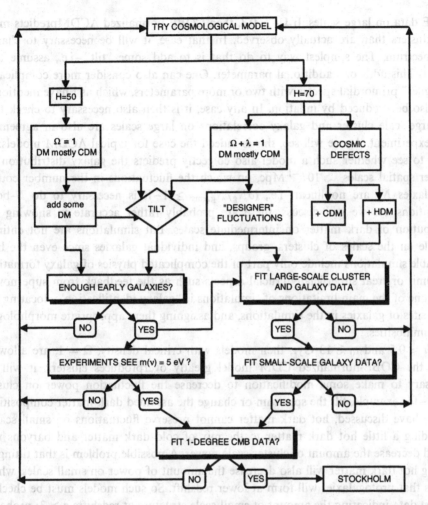

Figure 1.3. Building a cosmological model. (This figure was inspired by similar flowcharts on inventing dark-matter candidates, by David Weinberg and friends, and by Rocky Kolb.)

constant may be necessary for consistency with GR, since even in a universe with $\Omega \to 0$ the age $t_0 \to H_0^{-1} = 9.78 h^{-1}$ Gyr (see Fig. 1.1). Thus, with $\Lambda = 0$ and $\Omega_0 \to 0$, $h < 0.75(13\text{Gyr}/t_0)$. The upper limit on h is stronger, the larger Ω_0 is: with $\Lambda = 0$ and $\Omega_0 \geq 0.3$, $h < 0.61(13\text{Gyr}/t_0)$; with $\Lambda = 0$ and $\Omega_0 \geq 0.5$, $h < 0.57$. Since it has been argued above that the evidence strongly suggests that $\Omega_0 \geq 0.3$, especially if the initial fluctuations were Gaussian, if we assume a value of $h = 0.7$ we must include a positive cosmological constant. For definiteness, the specific choice shown is $\Omega_0 + \Omega_\Lambda = 1$, corresponding to the flat cosmology inspired by standard inflation.

In such a ΛCDM model, one might initially try the Zel'dovich primordial fluctuation spectrum, i.e., $P_s(k) = A_s k^{n_p}$ with $n_p = 1$. However, this might not predict the observed abundance of clusters when the amplitude of the spectrum is adjusted to agree with the

COBE data on large scales. If $\Omega_0 > 0.3$, then COBE-normalized ΛCDM predicts more rich clusters than are actually observed. In that case, it will be necessary to change the spectrum. The simplest way to do that is to add some "tilt" – i.e., assume that $n_p < 1$. This adds one additional parameter. One can also consider more complicated "designer" primordial spectra with two or more parameters, which as I have mentioned can also be produced by inflation. In any case, it is then also necessary to check that the large-scale cluster and galaxy correlations on large scales are also in agreement with experiment. As we will see, this is indeed the case for typical ΛCDM models. In order to see whether such a model also correctly predicts the galaxy distribution on smaller spatial scales $\lesssim 10\,h^{-1}$ Mpc, on which the fluctuations in the number counts of galaxies N_g are nonlinear – i.e., $(\delta N_g/N_g)_{\rm rms} \gtrsim 1$ – it is necessary to do N-body simulations. As we will discuss, these are probably rather accurate in showing the distribution of dark matter on intermediate scales. But simulations are not entirely reliable on the scales of clusters, groups, and individual galaxies since even the best available simulations include only part of the complicated physics of galaxy formation, and omit or treat superficially crucial aspects such as the feedback from supernovae. Thus one of the main limitations of simulations is "galaxy identification" – locating the likely sites of galaxies in the simulations, and assigning them appropriate morphologies and luminosities.

If $h \approx 0.5$ and $t_0 \lesssim 13$ Gyr, then models with critical density, $\Omega = 1$, are allowed. Since the COBE-normalized CDM model greatly overproduces clusters, it will be necessary to make some modification to decrease the fluctuation power on cluster scales – for example, tilt the spectrum or change the assumed dark-matter composition. As we have discussed, hot dark matter cannot preserve fluctuations on small scales, so adding a little hot dark matter to the mix of cold dark matter and baryons will indeed decrease the amount of cluster-scale power. A possible problem is that tilting or adding hot dark matter will also decrease the amount of power on small scales, which means that protogalaxies will form at lower redshift. So such models must be checked against data indicating the amount of small-scale structure at redshifts $z \geq 3$; probably the best data for this purpose is still the abundance of neutral hydrogen in damped Lyman α absorption systems in quasar spectra, since the precise nature of the several sorts of possible protogalaxies seen in emission at high redshift remains somewhat unclear. Such models must of course also fit the data on large- and small-scale galaxy distributions. As we will see, $\Omega = 1$ COBE-normalized models with a mixture of Cold and Hot Dark Matter (CHDM) can do this if the hot fraction $\Omega_\nu \approx 0.2$.

The ultimate test for all such cosmological models is whether they will agree with the CMB anisotropies on scales of a degree and below. Such data is just beginning to become available from ground-based and balloon-borne experiments, and continuing improvements in the techniques and instruments insure that the CMB data will become steadily more abundant and accurate. CMB maps of the whole sky must come from satellites, and it is great news for cosmology that NASA has approved the Microwave Anisotropy Probe (MAP) satellite which is expected to be ready for launch by 2001, and that the European Space Agency is planning an even more ambitious

COBRAS/SAMBA satellite, recently renamed Planck, to be launched a few years later (possibly in combination with the Far Infrared Space Telescope FIRST).

Both sorts of models that have been discussed, $\Omega = 1$ tilted CDM (TCDM) or CHDM, and $\Omega_0 + \Omega_\Lambda = 1$ ΛCDM, are simple, one-parameter modifications of the original standard CDM model. The astrophysics community has been encouraged by the great initial success of this theory in explaining the existence of galaxies and fitting galaxy and cluster data (Blumenthal et al. 1984; Davis et al. 1985), and the fact that biased CDM only missed predicting the COBE observations by a factor of about 2. The other reason why the CDM-variant models have been studied in much more detail than other cosmological models is that they are so predictive: they predict the entire dark-matter distribution in terms of only one or two model parameters (in addition to the usual cosmological parameters), unlike non-Gaussian models based on randomly located seeds, for example. Of course, despite the relatively good agreement between observations and the predictions of the best CDM variants, there is no guarantee that such models will ultimately be successful.

Although the cosmic defect models (cosmic strings, textures) are in principle specified in terms of only a small number of parameters (in the case of cosmic strings, the string tension parameter $G\mu$ plus perhaps a couple of parameters specifying aspects of the evolution of the string network), in practice it has not yet been possible for any group to work out the predicted galaxy distribution in such models. Most proponents of cosmic defect models have assumed an $\Omega = 1$, $H_0 \approx 50$ cosmology, but the chart refers instead to a cosmic defects option under $H_0 = 70$. This is done because it would be worthwhile to work out a low-Ω case as well, since in defect models there is less motivation to assume the inflation-inspired flat ($\Omega_0 + \Omega_\Lambda = 1$) cosmology.

1.7.2 Lessons from warm dark matter

As has been said, the chart in Fig. 1.3 only includes a few of the possibilities. But many possibilities that have been examined are not very promising. The problems with a pure Hot Dark Matter (HDM) adiabatic cosmology have already been mentioned. It will be instructive to look briefly at Warm Dark Matter (WDM), to see that some variants of CDM have less success than others in fitting cosmological observations, and also because there is renewed interest in WDM. Although CHDM and WDM are similar in the sense that both are intermediate models between CDM and HDM, CHDM and WDM are quite different in their implications. The success of some but not other modifications of the original CDM scenario shows that more is required than merely adding another parameter.

As explained above, WDM is a simple modification of HDM, obtained by changing the assumed average number density n of the particles. In the usual HDM, the dark-matter particles are neutrinos, each species of which has $n_\nu = 113$ cm^{-3}, with a corresponding mass of $m(\nu) = \Omega_\nu \rho_0 / n_\nu = \Omega_\nu 92 h^2$ eV. In WDM, there is another parameter, m/m_0, the ratio of the mass of the warm particle to the neutrino mass;

correspondingly, the number density of the warm particles is reduced by the inverse of this factor, so that their total contribution to the cosmological density is unchanged. It is true of both of the first WDM particle candidates, light gravitino and right-handed neutrino, that these particles interact much more weakly than neutrinos, decouple earlier from the Hot Big Bang, and thus have a diluted number density compared to neutrinos since they do not share in the entropy released by the subsequent annihilation of species such as quarks. This is analogous to the neutrinos themselves, which have lower number density today than photons because the neutrinos decouple before e^+e^- annihilation (and also because they are fermions).

In order to investigate the cosmological implications of any dark-matter candidate, it is necessary to work out the gravitational clustering of these particles, first in linear theory, and then after the amplitude of the fluctuations grows into the nonlinear regime. Colombi, Dodelson & Widrow (1996) did this for WDM, and Fig. 1.4 taken from their paper compares the square of the linear transfer functions $T(k)$ for WDM and CHDM. The power spectrum $P(k)$ of fluctuations is given by the quantity plotted times the assumed primordial power spectrum $P_p(k)$, $P(k) = P_p(k)T(k)^2$. The usual assumption regarding the primordial power spectrum is $P_p(k) = Ak^{n_p}$, where the "tilt" equals $1 - n_p$, and the untilted, or Zel'dovich, spectrum corresponds to $n_p = 1$.

One often can study large-scale structure just on the basis of such linear calculations, without the need to do computationally expensive simulations of the nonlinear gravitational clustering. Such studies have shown that matching the observed cluster and galaxy correlations on scales of about 20–30 h^{-1} Mpc in CDM-type theories requires that the "Excess Power" $EP \approx 1.3$, where

$$EP \equiv \frac{\sigma_{25}/\sigma_8}{(\sigma_{25}/\sigma_8)_{SCDM}}, \qquad (1.14)$$

and as usual σ_r is the rms fluctuation amplitude in randomly placed spheres of radius $r\,h^{-1}$ Mpc. The EP parameter was introduced by Wright et al. (1992), and Borgani et al. (1997) has shown that EP is related to the spectrum shape parameter Γ introduced by Efstathiou, Bond, and White (1992) (cf. Bardeen et al. 1986) by $\Gamma \approx 0.5(EP)^{-3.3}$. For CDM and the ΛCDM family of models, $\Gamma = \Omega h$; for CHDM and other models, the formula just quoted is a useful generalization of the spectrum shape parameter since the cluster correlations do seem to be a function of this generalized Γ, as shown in Fig. 1.5. As this figure shows, $\Gamma \approx 0.25$ to match cluster correlation data. Peacock & Dodds (1994) have shown that $\Gamma \approx 0.25$ also is required to match large-scale galaxy clustering data. This corresponds to $EP \approx 1.25$.

Since calculating $\sigma(r)$ is a simple matter of integrating the power spectrum times the top-hat window function,

$$\sigma^2(r) = \int_0^\infty P(k)W(kr)k^2 dk, \qquad (1.15)$$

the linear calculations shown in Fig. 1.4 immediately allow determination of EP for WDM and CHDM. The results are shown in Fig. 1.6, in which the lower horizontal

Dark matter and structure formation

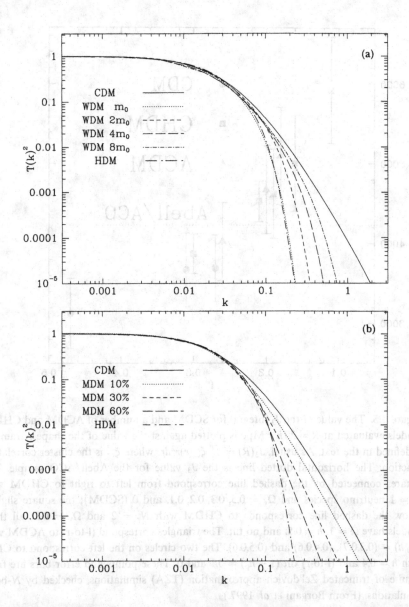

Figure 1.4. The square of the linear transfer function $T(k)$ vs. wavenumber $k = (2\pi)/\lambda$ (in units of h Mpc^{-1}) for (a) Warm Dark Matter (WDM), and (b) Mixed Dark Matter (MDM–CHDM with $N_\nu = 1$ neutrino species). (From Colombi, Dodelson & Widrow 1996, used by permission.)

axis represents the values of the WDM parameter m/m_0 (with $m/m_0 = 1$ representing the HDM limit), and the upper horizontal axis represents the values of the CHDM parameter Ω_ν. This figure shows that for WDM to give the required EP, the parameter value $m/m_0 \approx 1.5$–2, while for CHDM the required value of the CHDM parameter is $\Omega_\nu \approx 0.3$. But one can see from Fig. 1.4 that in WDM with $m/m_0 \gtrsim 2$, the spectrum lies

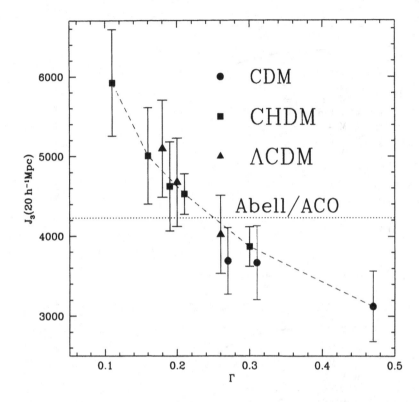

Figure 1.5. The value of the J_3 integral for SCDM and a number of ΛCDM and CHDM models evaluated at $R = 20\,h^{-1}$ Mpc is plotted against the value of the shape parameter Γ defined in the text. As usual, $J_3(R) = \int_0^R \xi_{cc}(r) r^2 dr$, where ξ_{cc} is the cluster correlation function. The horizontal dotted line is the J_3 value for the Abell/ACO sample. The squares connected by the dashed line correspond from left to right to CHDM with $n_\nu = 1$ neutrino species and $\Omega_\nu = 0.5, 0.3, 0.2, 0.1$, and 0 (SCDM); the square slightly below the dashed line corresponds to CHDM with $N_\nu = 2$ and $\Omega_\nu = 0.2$; all these models have $\Omega = 1$, $h = 0.5$. and no tilt. The triangles correspond (l-to-r) to ΛCDM with $(\Omega_0, h) = (0.3, 0.7), (0.4, 0.6)$ and $(0.5, 0.6)$. The two circles on the left correspond to CDM with $h = 0.4$ and (l-to-r) tilt $(1 - n_p) = 0.1$ and 0. These points and error bars are from a suite of truncated Zel'dovich approximation (TZA) simulations, checked by N-body simulations. (From Borgani et al. 1997.)

a lot lower than the CDM spectrum at $k \gtrsim 0.3h$ Mpc^{-1} (length scales $\lambda \lesssim 20\,h^{-1}$ Mpc). This in turn implies that formation of galaxies, corresponding to the gravitational collapse of material in a region of size ~ 1 Mpc, will be strongly suppressed compared to CDM. Thus WDM will not be able to accommodate simultaneously the distribution of clusters and galaxies. But CHDM will do much better – note how much lower $T(k)^2$ is at $k \gtrsim 0.3h$ Mpc^{-1} for WDM with $m/m_0 = 2$ than for CHDM with $\Omega_\nu = 0.3$. Actually, as we will discuss in more detail shortly, CHDM with $\Omega_\nu = 0.3$ turns out, on more careful examination, to have several defects – too many intermediate-size voids,

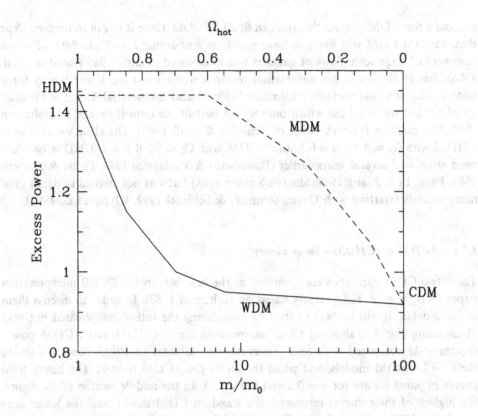

Figure 1.6. Excess power EP in the two models discussed that interpolate between CDM and HDM. Solid curve shows EP as a function of the WDM parameter m/m_0; note how quickly it becomes similar to CDM. Dashed curve shows how EP for Mixed Dark Matter (MDM–CHDM with $N_\nu = 1$ neutrino species) depends on Ω_ν. The observationally preferred value is $EP \approx 0.25$. (From Colombi, Dodelson & Widrow 1996, used by permission.)

too few early protogalaxies. Lowering Ω_ν to about 0.2, corresponding to a total neutrino mass of about $4.6(h/0.5)^{-2}$ eV, in a model in which $N_\nu = 2$ neutrino species share this mass, fits all this data (PHKC95).

Probably the only way to accommodate WDM in a viable cosmological model is as part of a mixture with hot dark matter (Malaney, Starkman & Widrow 1995), which might even arise naturally in a supersymmetric model (Borgani, Masiero & Yamaguchi 1996) of the sort in which the gravitino is the LSP (Dimopoulos et al. 1996). Cold plus "volatile" dark matter is a related possibility (Pierpaoli et al. 1996); in these models, the hot component arises from decay of a heavy unstable particle rather than decoupling of relativistic particles.

There are many more parameters needed to describe the presently available data on the distribution of galaxies and clusters and their formation history than the few parameters needed to specify a CDM-type model. Thus it should not be surprising that at most a few CDM-variant theories can fit all this data. Once it began to become clear that standard CDM was likely to have problems accounting for all the data, after the discovery of large-scale flows of galaxies was announced in early 1986 (Burstein et al. 1986), Jon Holtzman in his dissertation research worked out the linear theory for a wide variety of CDM variants (Holtzman 1989; cf. also Blumenthal, Dekel & Primack 1988) so that we could see which ones would best fit the data (Primack & Holtzman 1992, Holtzman & Primack 1993; cf. Schaefer & Shafi 1993). The clear winners were CHDM with $\Omega_\nu \approx 0.3$ if $h \approx 0.5$, and ΛCDM with $\Omega_0 \approx 0.2$ if $h \approx 1$. CHDM had first been advocated several years earlier (Bonometto & Valdarnini 1984, Dekel & Aarseth 1984, Fang, Li & Xiang 1984, Shafi & Stecker 1984) but was not studied in detail until more recently (starting with Davis, Summers & Schlegel 1992, Klypin et al. 1993).

1.7.3 ΛCDM vs. CHDM – linear theory

These two CDM variants were identified as the best bets in the COBE interpretation paper (Wright et al. 1992, largely based on Holtzman 1989). In order to discuss them in more detail, it will be best to start by considering the rather complicated but very illuminating Fig. 1.7, showing COBE-normalized linear CHDM and ΛCDM power spectra $P(k)$ compared with four observational estimates of $P(k)$.[5] Panel (a) shows the $\Omega = 1$ CHDM models, and panel (b) shows the ΛCDM models. The heavy solid curves in panel (a) are for $h = 0.5$ and $\Omega_b = 0.05$. In the middle section of the figure, the highest of these curves represents the standard CDM model, and the lower ones standard CHDM ($N_\nu = 1$) with $\Omega_\nu = 0.2$ (higher) and 0.3; the medium-weight solid curves represent the corresponding CHDM models with two neutrinos equally sharing the same total neutrino mass ($N_\nu = 2$). Note that the $N_\nu = 2$ CHDM power spectra are significantly smaller than those for $N_\nu = 1$ for $k \approx (0.04$–$0.4)h$ Mpc^{-1}; this arises because for $N_\nu = 2$ the neutrinos weigh half as much and correspondingly free stream over a longer distance. The result is that $N_\nu = 2$ COBE-normalized CHDM with $\Omega_\nu \approx 0.2$ can simultaneously fit the abundance and correlations of clusters (Primack et al. 1995, cf. Borgani, Masiero & Yamaguchi 1996). The light solid curve is CDM with $\Gamma = \Omega h = 0.2$.

The "bow" superimposed on these curves represents the approximate "pivot point" (cf. Gorski et al. 1994) for COBE-normalized "tilted" models (i.e., with $n_p \neq 1$), and the error bar there represents the 1σ COBE normalization uncertainty. The window functions for various spherical harmonic coefficients a_ℓ, bulk velocities V_R, and σ_8 are shown in the bottom part of this figure (see caption). The bow lies above the a_{11} window because the statistical weight of the COBE data is greatest for angular wavenumber $\ell \approx 11$ (cosmic variance is greater for lower ℓ, and the $\sim 7°$ resolution of the COBE DMR makes the uncertainty increase for higher ℓ).

---

[5] The normalization is actually according to the two-year COBE data, which is about 10% higher in amplitude than the final four-year COBE data (Gorski et al. 1996), but this relatively small difference will not be important for our present purposes.

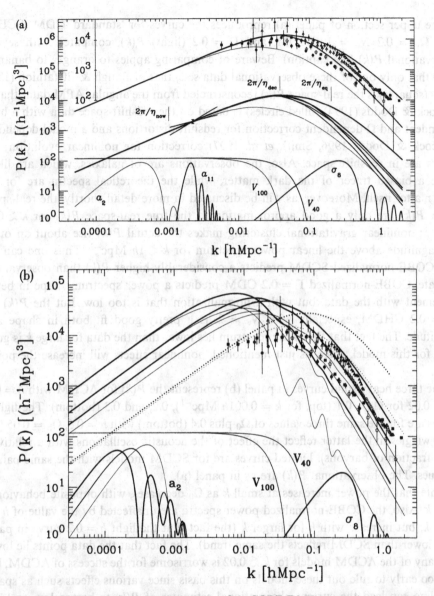

Figure 1.7. Fluctuation power spectra for COBE-DMR-normalized models: panel (a) $\Omega = 1$ CDM and CHDM models, panel (b) $\Omega_0 + \Omega_\Lambda = 1$ ΛCDM models. The theoretical spectra are discussed in the text. The data plotted for comparison is squares – real space $P(k)$ from angular APM data (Baugh & Efstathiou 1993), filled circles – estimate of real space $P(k)$ from redshift galaxy and cluster data (Peacock & Dodds 1994), pentagons – IRAS 1.2 Jy redshift space $P(k)$ (Fisher et al. 1993), open and filled triangles – CfA2 and SSRS2 redshift space $P(k)$ (da Costa et al. 1994). At the bottom of each panel are plotted window functions for CMB anisotropy expansion coefficients a_ℓ (panel (a): quadrupole a_2, and a_{11}; panel (b) left to right: a_2 for $\Omega_0 = 0.1$, 0.3, and 1), bulk flows V_R, and the rms mass fluctuation in a sphere of $8h^{-1}$ Mpc σ_8. (From Stompor, Gorski & Banday 1995, used by permission.)

The upper section of panel (a) reproduces the curves for 'standard' CDM (SCDM, top), $\Omega_v = 0.2$ $N_v = 1$ CHDM, and $\Gamma = 0.2$ (light) $P(k)$, compared with several observational $P(k)$ (see caption). Beware of comparing apples to oranges to bananas! Note that only one of these observational data sets, that of Baugh & Efstathiou (1993, 1994) (squares) is the real-space $P(k)$ reconstructed from the angular APM data; that of Peacock & Dodds (1994) (filled circles) is based on the redshift-space data with a bias-dependent and Ω-dependent correction for redshift distortions and a model-dependent (Peacock & Dodds 1996, Smith et al. 1997) correction for nonlinear evolution; the others are in redshift space. Also, the observations are of galaxies, which are likely to be a biased tracer of the dark matter, while the theoretical spectra are for the dark matter itself. Moreover, as will be discussed in more detail shortly, the real-space linear $P(k)$ are only a good approximation to the true real-space $P(k)$ for $k \lesssim 0.2h$ Mpc^{-1}; nonlinear gravitational clustering makes the actual $P(k)$ rise about an order of magnitude above the linear power spectrum for $k \gtrsim 1h$ Mpc^{-1}. Thus one can see that COBE-normalized SCDM predicts a considerably higher $P(k)$ than observations indicate. COBE-normalized $\Gamma = 0.2$ CDM predicts a power spectrum shape in better agreement with the data, but with a normalization that is too low. But the $P(k)$ for $\Omega_v = 0.2$ CHDM, especially with $N_v = 2$, is a pretty good fit both in shape and amplitude. The fact that the linear spectrum lies lower than the data for large k is good news for this model, since, as just mentioned, nonlinear effects will increase the power there.

The three heavy solid curves in panel (b) represent the $P(k)$ for ΛCDM with $h = 0.8$, $\Omega_b = 0.02$ for $\Omega_0 = 0.1$ (top, for $k = 0.001h$ Mpc^{-1}), 0.2, and 0.3 (bottom). The lighter curves are for the same three values of Ω_0 plus 0.4 (bottom) with $h = 0.5$, $\Omega_b = 0.05$ (the large wiggles in the latter reflect the effect of the acoustic oscillations with a relatively large fraction of baryons). Dotted curves are for SCDM models with the same pair of h values. The observational $P(k)$ are as in panel (a).

Note that the power increases at small k as Ω_0 decreases, with opposite behavior at large k. Also, the COBE-normalized power spectra are unaffected by the value of h for small k, but increase with h for larger k (the fact that the light $h = 0.2$ curve in panel (a) is lower than SCDM reflects the same trend). The fact that the data points lie lower than any of the ΛCDM models for $k \lesssim 0.02$ is worrisome for the success of ΛCDM, but it is too early to rule out these models on this basis since various effects such as sparse sampling can lead the current observational estimates of $P(k)$ to be too low on large scales (Efstathiou 1996). A better measurement of $P(k)$ on such large scales $k \lesssim 10^{-2}h$ Mpc^{-1} will be one of the most important early outputs of the next-generation very large redshift surveys: the 2° field (2DF) survey at the Anglo–Australian Telescope, and the Sloan Digital Sky Survey (SDSS) using a dedicated 2.5 m telescope at the Apache Point Observatory in New Mexico. $P(k)$ is much better determined for larger k by the presently available data, and the fact that the linear $\Omega_0 = 0.2$ and 0.3 curves lie higher than many of the data points for larger k means that these $h = 0.8$ models will lie far above the data when nonlinear effects are taken into account. This means that, unless some physical process causes the galaxies to be much less clustered than the dark matter ("anti-biasing"), such models could be acceptable only with a considerable amount of tilt – but that can make the shape of the spectrum fit more poorly.

1.7.4 Numerical simulations to probe smaller scales

"Standard" $\Omega = 1$ Cold Dark Matter (SCDM) with $h \approx 0.5$ and a near-Zel'dovich spectrum of primordial fluctuations (Blumenthal et al. 1984) until a few years ago seemed to many theorists to be the most attractive of all modern cosmological models. But although SCDMnormalized to COBE nicely fits the amplitude of the large-scale flows of galaxies measured with galaxy peculiar velocity data (Dekel 1994), it does not fit the data on smaller scales: it predicts far too many clusters (White, Efstathiou & Frenk 1993) and does not account for their large-scale correlations (e.g., Olivier et al. 1993, Borgani, Masiero & Yamaguchi 1996), and the shape of the power spectrum $P(k)$ is wrong (Baugh & Efstathiou 1994, Zaroubi et al. 1997). But as discussed above, variants of SCDM can do better. Here the focus is on CHDM and ΛCDM. The linear *matter* power spectra for these two models are compared in Fig. 1.8 to the real-space *galaxy* power spectrum obtained from the two-dimensional APM galaxy power spectrum (Baugh & Efstathiou 1994), which in view of the uncertainties is not in serious disagreement with either model for $10^{-2} \lesssim k \lesssim 1h$ Mpc^{-1}. The ΛCDM and CHDM models essentially bracket the range of power spectra in currently popular cosmological models that are variants of CDM.

The Void Probability Function (VPF) is the probability $P_0(r)$ of finding no bright galaxy in a randomly placed sphere of radius r. It has been shown that CHDM with $\Omega_v = 0.3$ predicts a VPF larger than observations indicate (Ghigna et al. 1994), but new results based on our $\Omega_v = 0.2$ simulations in which the neutrino mass is shared equally between $N_v = 2$ neutrino species (PHKC95) show that the VPF for this model is in excellent agreement with observations (Ghigna et al. 1997), as shown in Fig. 1.9. However, our simulations (Klypin, Primack & Holtzman 1996, hereafter KPH96) of COBE-normalized ΛCDM with $h = 0.7$ and $\Omega_0 = 0.3$ lead to a VPF that is too large to be compatible with a straightforward interpretation of the data. Acceptable ΛCDM models probably need to have $\Omega_0 > 0.3$ and $h < 0.7$, as discussed further below.

Another consequence of the reduced power in CHDM on small scales is that structure formation is more recent in CHDM than in ΛCDM. As discussed above (in §1.4.7), this may conflict with observations of damped Lyman α systems in quasar spectra, and other observations of protogalaxies at high redshift, although the available evidence does not yet permit a clear decision on this (see below). While the original $\Omega_v = 0.3$ CHDM model (Davis, Summers & Schlegel 1992, Klypin et al. 1993) certainly predicts far less neutral hydrogen in damped Lyman α systems (identified as protogalaxies with circular velocities $V_c \geq 50$ km s^{-1}) than is observed, as discussed already, lowering the hot fraction to $\Omega_v \approx 0.2$ dramatically improves this (Klypin et al. 1995). Also, the evidence from preliminary data of a falloff of the amount of neutral hydrogen in

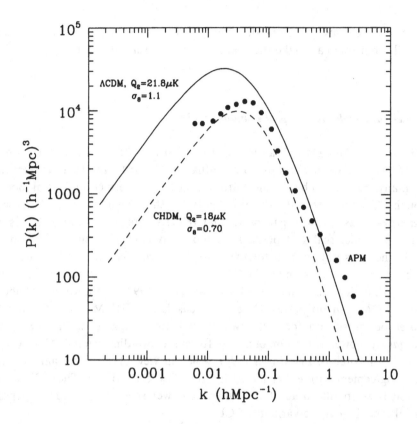

Figure 1.8. Power spectrum of dark matter for ΛCDM and CHDM models considered here, both normalized to COBE (2-year and 4-year data respectively, as in the simulations discussed in the text), compared to the APM galaxy real-space power spectrum. (ΛCDM: $\Omega_0 = 0.3$, $\Omega_\Lambda = 0.7$, $h = 0.7$, thus $t_0 = 13.4$ Gyr; CHDM: $\Omega = 1$, $\Omega_\nu = 0.2$ in $N_\nu = 2$ ν species, $h = 0.5$, thus $t_0 = 13$ Gyr; both models fit cluster abundance with no tilt, i.e., $n_p = 1$. (From Primack & Klypin 1996.)

damped Lyman α systems for $z \gtrsim 3$ (Storrie-Lombardi, McMahon & Irwin 1996) is in accord with predictions of CHDM (Klypin et al. 1995).

However, as for all $\Omega = 1$ models, $h \gtrsim 0.55$ implies $t_0 \lesssim 12$ Gyr, which conflicts with the pre-Hipparchos age estimates from globular clusters. The only way to accommodate both large h and large t_0 within the standard FRW framework of General Relativity, if in fact both $h \gtrsim 0.65$ and $t_0 \gtrsim 13$ Gyr, is to introduce a positive cosmological constant ($\Lambda > 0$). Low-Ω_0 models with $\Lambda = 0$ don't help much with t_0, and anyway are disfavored by the latest small-angle cosmic microwave anisotropy data (Netterfield et al. 1995, Scott et al. 1996; cf. Ganga, Ratra & Sugiyama 1996 for a contrary view).

ΛCDM flat cosmological models with $\Omega_0 = 1-\Omega_\Lambda \approx 0.3$, where $\Omega_\Lambda \equiv \Lambda/(3H_0^2)$, were discussed as an alternative to $\Omega = 1$ CDM since the beginning of CDM (Blumenthal et al. 1984, Peebles 1984a, 1984b, Davis et al. 1985). They have been advocated more

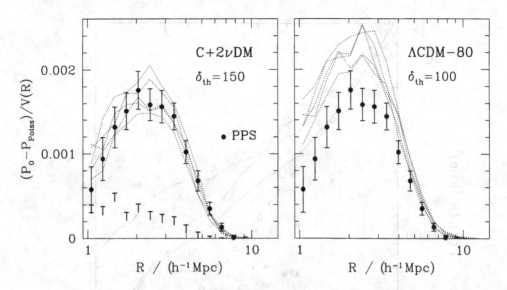

Figure 1.9. Void Probability Function $P_0(R)$ for (*left panel*) CHDM with $h = 0.5$ and $\Omega_\nu = 0.2$ in $N_\nu = 2$ species of neutrinos and (*right panel*) ΛCDM with $h = 0.7$ and $\Omega_0 = 0.3$. What is plotted here is difference between the actual VPF and that for a Poisson distribution, divided by $V(R) = 4\pi R^3/3$. Each plot shows also $P_0(R)$ for five typical different locations in the simulations (dotted lines) to give an indication of the sky variance. Data points are the VPF from the Perseus–Pisces Survey, with 3σ error bars; the VPF from the CfA2 survey is very similar. We have chosen the $\delta_{\rm th}$ for which the P_0 of each model best approaches the observational data. In the left panel, the T symbols at the bottom set the boundary of the region where the signal is indistinguishable from Poissonian. They are obtained from the 3σ scatters among measures for 50 different realizations of the Poissonian distribution in the same volume as our samples. (From Ghigna *et al.* 1997.)

recently (e.g., Efstathiou, Sutherland & Maddox 1990; Kofman, Gnedin & Bahcall 1993; Ostriker & Steinhardt 1995; Krauss & Turner 1995) both because they can solve the H_0–t_0 problem and because they predict a larger fraction of baryons in galaxy clusters than $\Omega = 1$ models (this is discussed in §1.4.5 above).

Early galaxy formation also is often considered to be a desirable feature of these models. But early galaxy formation implies that fluctuations on scales of a few Mpc spent more time in the nonlinear regime, as compared with CHDM models. As has been known for a long time, this results in excessive clustering on small scales. It has been found that a typical ΛCDM model with $h = 0.7$ and $\Omega_0 = 0.3$, normalized to COBE on large scales (this fixes $\sigma_8 \approx 1.1$ for this model), is compatible with the number-density of galaxy clusters (Borgani *et al.* 1997), but predicts a power spectrum of galaxy clustering in real space that is much too high for wavenumbers $k = (0.4 - 1)h/{\rm Mpc}$ (KPH96). This conclusion holds if we assume either that galaxies trace the dark matter, or just that a region with higher density produces more galaxies than a region with lower density. One can see immediately from Fig. 1.7 and Fig. 1.8

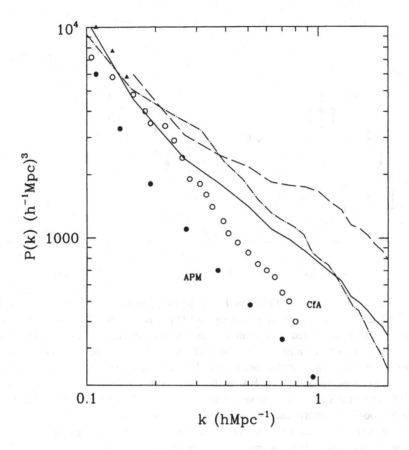

Figure 1.10. Comparison of the nonlinear power spectrum in the $\Omega_0 = 0.3$, $h = 0.7$ ΛCDM model with observational results. Filled circles are results for the APM galaxy survey. Results for the real-space power spectrum for the CfA survey are shown as open circles ($101h^{-1}$ Mpc sample) and filled triangles ($130h^{-1}$ Mpc sample). Formal error bars for each of the surveys are smaller than the difference between the open and filled points, which should probably be regarded as a more realistic estimate of the range of uncertainty. The full curve represents the power spectrum of the dark matter. Lower limits on the power spectrum of galaxies predicted by the ΛCDM model are shown as the dashed curve (higher-resolution ΛCDM$_f$ simulation in Klypin, Primack & Holtzman 1996) and the dot-dashed curve (lower-resolution ΛCDM$_c$ simulation).

that there will be a problem with this ΛCDM model, since the APM power spectrum is approximately equal to the linear power spectrum at wavenumber $k \approx 0.6h$ Mpc^{-1}, so there is no room for the extra power that nonlinear evolution certainly produces on this scale – illustrated in Fig. 1.10 for ΛCDM and in Fig. 1.11 for CHDM. The only way to reconcile the $\Omega_0 = 0.3$ ΛCDM model considered here with the observed power spectrum is to assume that some mechanism causes strong anti-biasing – i.e., that regions with high dark-matter density produce fewer galaxies than regions with low density.

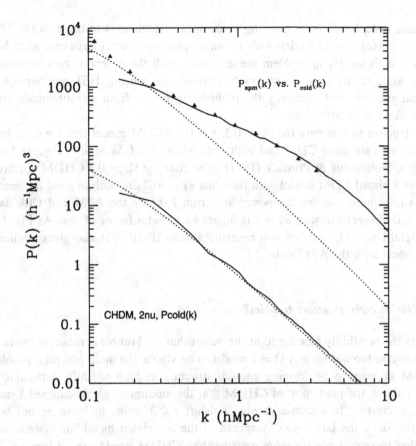

Figure 1.11. Comparison of APM galaxy power spectrum (triangles) with nonlinear cold particle power spectrum from the CHDM model (upper solid curve). The dotted curves are linear theory; upper curves are for $z = 0$, lower curves correspond to the higher redshift $z = 9.9$. (From Primack & Klypin 1996.)

While theoretically possible, this seems very unlikely; biasing rather than anti-biasing is expected, especially on small scales (e.g., Kauffmann, Nusser & Steinmetz 1997). Numerical hydro + N-body simulations that incorporate effects of UV radiation, star formation, and supernovae explosions (Yepes et al. 1997) do not show any anti-bias of luminous matter relative to the dark matter.

Our motivation to investigate this particular ΛCDM model was to have H_0 as large as might possibly be allowed in the ΛCDM class of models, which in turn forces Ω_0 to be rather small in order to have $t_0 \gtrsim 13$ Gyr. There is little room to lower the normalization of this ΛCDM model by tilting the primordial power spectrum $P_p(k) = Ak^{n_p}$ (i.e., assuming n_p significantly smaller than the "Zel'dovich" value $n_p = 1$), since then the fit to data on intermediate scales will be unacceptable – e.g., the number density of clusters will be too small (KPH96). Tilted ΛCDM models with higher Ω_0, and therefore lower H_0 for $t_0 \gtrsim 13$ Gyr, appear to have a better hope of fitting the available data, based on comparing quasi-linear calculations to the data (KPH96, Liddle et al. 1996c). But all models with a cosmological constant Λ large enough to help significantly with the H_0–t_0 problem are in trouble with the observations summarized above providing strong upper limits on Λ: gravitational lensing, HST number counts of elliptical galaxies, and especially the preliminary results from measurements using high-redshift Type Ia supernovae.

It is instructive to compare the $\Omega_0 = 0.3$, $h = 0.7$ ΛCDM model that we have been considering with standard CDM and with CHDM. At $k = 0.5h$ Mpc^{-1}, Figures 5 and 6 of Klypin, Nolthenius & Primack (1997) show that the $\Omega_\nu = 0.3$ CHDM spectrum and that of a biased CDM model with the same $\sigma_8 = 0.67$ are both in good agreement with the values indicated for the power spectrum $P(k)$ by the APM and CfA data, while the CDM spectrum with $\sigma_8 = 1$ is higher by about a factor of two. As Fig. 1.11 shows, CHDM with $\Omega_\nu = 0.2$ in two neutrino species (PHKC95) also gives nonlinear $P(k)$ consistent with the APM data.

1.7.5 CHDM: early structure troubles?

Aside from the possibility mentioned at the outset that the Hubble constant is too large and the universe too old for any $\Omega = 1$ model to be viable, the main potential problem for CHDM appears to be forming enough structure at high redshift. Although, as mentioned above, the prediction of CHDM that the amount of gas in damped Lyman α systems is starting to decrease at high redshift $z \gtrsim 3$ seems to be in accord with the available data, the large velocity spread of the associated metal-line systems *may* indicate that these systems are more massive than CHDM would predict (see e.g., Lu et al. 1996, Wolfe 1996). Also, results from a recent CDM hydrodynamic simulation (Katz et al. 1996) in which the amount of neutral hydrogen in protogalaxies seemed consistent with that observed in damped Lyman α systems (DLAS) led the authors to speculate that CHDM models would produce less than enough DLAS; however, since the regions identified as DLAS in these simulations were not actually resolved gravitationally, this will need to be addressed by higher-resolution simulations for all the models considered.

Finally, Steidel et al. (1996) have found objects by their emitted light at redshifts $z = 3$–3.5 apparently with relatively high velocity dispersions (indicated by the equivalent widths of absorption lines), which they tentatively identify as the progenitors of giant elliptical galaxies. *Assuming* that the indicated velocity dispersions are indeed gravitational velocities, Mo & Fukugita (1996, hereafter MF96) have argued that the abundance of these objects is higher than expected for the COBE-normalized $\Omega = 1$ CDM-type models that can fit the low-redshift data, including CHDM, but in accord with predictions of the ΛCDM model considered here. (In more detail, the MF96 analysis disfavors CHDM with $h = 0.5$ and $\Omega_\nu \gtrsim 0.2$ in a single species of neutrinos. They apparently would argue that this model is then in difficulty since it overproduces rich clusters – and if that problem were solved with a little tilt $n_p \approx 0.9$, the resulting decrease in fluctuation power on small scales would not lead to formation of enough early objects. However, if $\Omega_\nu \approx 0.2$ is shared between two species of neutrinos, the resulting model appears to be at least marginally consistent with both clusters and the Steidel objects even with the assumptions of MF96. The ΛCDM model with $h = 0.7$ consistent with the most restrictive MF96 assumptions has $\Omega_0 \gtrsim 0.5$, hence $t_0 \lesssim 12$ Gyr. ΛCDM models having tilt and lower h, and therefore more consistent with the small-scale power constraint discussed above, may also be in trouble with the MF96 analysis.) But in addition to uncertainties about the actual velocity dispersion and physical size of the Steidel et al. objects, the conclusions of the MF96 analysis can also be significantly weakened if the gravitational velocities of the observed baryons are systematically higher than the gravitational velocities in the surrounding dark-matter halos, as is perhaps the case at low redshift for large spiral galaxies (Navarro, Frenk & White 1996), and even more so for elliptical galaxies which are largely self-gravitating stellar systems in their central regions.

Given the irregular morphologies of the high-redshift objects seen in the Hubble Deep Field (van den Bergh et al. 1996) and other deep HST images, it seems more likely that they are mostly relatively low-mass objects undergoing starbursts, possibly triggered by mergers, rather than galactic protospheroids (Lowenthal et al. 1997). Since the number density of the brightest of such objects may be more a function of the probability and duration of such starbursts rather than the nature of the underlying cosmological model, it may be more useful to use the star formation or metal injection rates (Madau et al. 1996) indicated by the total observed rest-frame ultraviolet light to constrain models (Somerville, Faber & Primack 1997). The available data on the history of star formation (Gallego et al. 1996, Lilly et al. 1996, Madau et al. 1996) suggests that most of the stars and most of the metals observed formed relatively recently, after about redshift $z \sim 1$; and that the total star formation rate at $z \sim 3$ is perhaps a factor of 3 lower than at $z \sim 1$, with yet another factor of ~ 3 falloff to $z \sim 4$ (although the rates at $z \gtrsim 3$ could be higher if most of the star formation is in objects too faint to see). This is in accord with indications from damped Lyman α systems (Fall, Charlot & Pei 1996) and expectations for $\Omega = 1$ models such as CHDM, but perhaps not with the expectations for low-Ω_0 models which have less growth of fluctuations at recent epochs, and therefore must form structure earlier. But this must be investigated using more detailed modeling, including gas cooling and feedback from stars and supernovae (e.g., Kauffmann 1996, Somerville, Faber & Primack 1997), before strong conclusions can be drawn.

There is another sort of constraint from observed numbers of high-redshift protogalaxies that would appear to disfavor ΛCDM. The upper limit on the number of $z \gtrsim 4$ objects in the Hubble Deep Field (which presumably correspond to smaller-mass galaxies than most of the Steidel objects) is far lower than the expectations in low-Ω_0 models, especially with a positive cosmological constant, because of the large volume at high redshift in such cosmologies (Lanzetta, Yahil & Fernández-SotoLanzetta 1996). Thus evidence from high-redshift objects cuts both ways, and it is too early to tell whether high- or low-Ω_0 models will ultimately be favored by such data.

1.7.6 Advantages of mixed CHDM over pure CDM models

There are three basic reasons why a mixture of cold plus hot dark matter works better than pure CDM without any hot particles: **(1)** the power spectrum shape $P(k)$ is a better fit to observations, **(2)** there are indications from observations for a more weakly clustering component of dark matter, and **(3)** a hot component may help avoid the too-dense central dark-matter density in pure CDM dark-matter halos. Each will be discussed in turn.

(1) Spectrum shape. As explained in discussing WDM vs. CHDM above, the pure CDM spectrum $P(k)$ does not fall fast enough on the large-k side of its peak in order to fit indications from galaxy and cluster correlations and power spectra. The discussion there of "Excess Power" is a way of quantifying this. This is also related to the overproduction of clusters in pure CDM. The obvious way to prevent $\Omega = 1$ SCDM normalized to COBE from overproducing clusters is to tilt it a lot (the precise amount depending on how much of the COBE fluctuations are attributed to gravity waves, which can be increasingly important as the tilt is increased). But a constraint on CDM-type models that is likely to follow both from the high-z data just discussed and from the preliminary indications on cosmic microwave anisotropies at and beyond the first acoustic peak from the Saskatoon experiment (Netterfield et al. 1997) is that viable models cannot have much tilt, since that would reduce too much both their small-scale power and the amount of small-angle CMB anisotropy. As already explained, by reducing the fluctuation power on cluster scales and below, COBE-normalized CHDM naturally fits both the CMB data and the cluster abundance without requiring much tilt. The need for tilt is further reduced if a high baryon fraction $\Omega_b \gtrsim 0.1$ is assumed (White et al. 1996), and this also boosts the predicted height of the first acoustic peak. No tilt is necessary for $\Omega_\nu = 0.2$ shared between $N_\nu = 2$ neutrino species with $h = 0.5$ and $\Omega_b = 0.1$. Increasing the Hubble parameter in COBE-normalized models increases the amount of small-scale power, so that if we raise the Hubble parameter to $h = 0.6$ keeping $\Omega_\nu = 0.2$ and $\Omega_b = 0.1(0.5/h)^2 = 0.069$, then fitting the cluster abundance in this $N_\nu = 2$ model requires tilt $1 - n_p \approx 0.1$ with no gravity waves (i.e., $T/S = 0$; alternatively if $T/S = 7(1 - n_p)$ is assumed, about half as much tilt is needed, but the observational consequences are mostly very similar, with a little more small-scale power). The fit to the small-angle CMB data is still good, and the predicted Ω_{gas} in damped Lyman α systems is a little higher than for the $h = 0.5$ case. The only obvious problem with $h = 0.6$ applies to any $\Omega = 1$ model – the universe is rather young: $t_0 = 10.8$ Gyr. But the revision of the globular cluster ages with the new Hipparchos data may permit this.

(2) Need for a less-clustered component of dark matter. The fact that group and cluster mass estimates on scales of $\sim 1\ h^{-1}$ Mpc typically give values for Ω around 0.1–0.2 while larger-scale estimates give larger values around 0.3–1 (Dekel 1994) suggests that there is a component of dark matter that does not cluster on small scales as efficiently as cold dark matter is expected to do. In order to quantify this, the usual group M/L measurement of Ω_0 on small scales has been performed in "observed"

$\Omega = 1$ simulations of both CDM and CHDM (Nolthenius, Klypin & Primack 1997). We found that COBE-normalized $\Omega_\nu = 0.3$ CHDM gives $\Omega_{M/L} = 0.12$–0.18 compared to $\Omega_{M/L} = 0.15$ for the CfA1 catalog analyzed exactly the same way, while for CDM $\Omega_{M/L} = 0.34 - 0.37$, with the lower value corresponding to bias $b = 1.5$ and the higher value to $b = 1$ (still below the COBE normalization). Thus local measurements of the density in $\Omega = 1$ simulations can give low values, but it helps to have a hot component to get values as low as observations indicate. We found that there are three reasons why this virial estimate of the mass in groups misses so much of the matter in the simulations: (1) only the mass within the mean harmonic radius r_h is measured by the virial estimate, but the dark-matter halos of groups continue their roughly isothermal falloff to at least $2r_h$, increasing the total mass by about a factor of 3 in the CHDM simulations; (2) the velocities of the galaxies are biased by about 70% compared to the dark-matter particles, which means that the true mass is higher by about another factor of 2; and (3) the groups typically lie along filaments and are significantly elongated, so the spherical virial estimator misses perhaps 30% of the mass for this reason. Visualizations of these simulations (Brodbeck et al. 1998) show clearly how extended the hot dark-matter halos are. An analysis of clusters in CHDM found similar effects, and suggested that observations of the velocity distributions of galaxies around clusters might be able to discriminate between pure cold and mixed cold + hot models (Kofman et al. 1996). This is an area where more work needs to be done – but it will not be easy since it will probably be necessary to include stellar and supernova feedback in identifying galaxies in simulations, and to account properly for foreground and background galaxies in observations.

(3) **Preventing too dense centers of dark-matter halos.** Flores and Primack (1994) pointed out that dark-matter density profiles with $\rho(r) \propto r^{-1}$ near the origin from high-resolution dissipationless CDM simulations (Dubinski & Carlberg 1991; Warren et al. 1992; Crone, Evrard & Richstone 1994) are in serious conflict with data on dwarf spiral galaxies (cf. Moore 1994), and in possible conflict with data on larger spirals (Flores et al. 1993) and on clusters (cf. Miralda-Escudé 1995, Flores & Primack 1996). Navarro, Frenk & White (1996; cf. Cole & Lacey 1996) agree that rotation curves of small spiral galaxies such as DDO154 and DDO170 are strongly inconsistent with their universal dark-matter profile $\rho_{\text{NFW}}(r) \propto 1/[r(r+a)^2]$. Navarro, Eke & Frenk (1996) proposed a possible explanation for the discrepancy regarding dwarf spiral galaxies involving slow accretion followed by explosive ejection of baryonic matter from their cores, but it is implausible that such a process could be consistent with the observed regularities in dwarf spirals (Burkert 1995); in any case it will not work for low-surface-brightness galaxies. Work is in progress with Stephane Courteau, Sandra Faber, Ricardo Flores, and others to see whether the ρ_{NFW} universal profile is consistent with data from high- and low-surface-brightness galaxies with moderate to large circular velocities, and with Klypin, Kravtsov, and Bullock to see whether higher-resolution simulations for a wider variety of models continue to give ρ_{NFW}. The failure of simulations to form cores as observed in dwarf spiral galaxies either is a clue to a property of dark matter that we don't understand, or is telling us the simulations are inadequate. It is important to

discover whether this is a serious problem, and whether inclusion of hot dark matter or of dissipation in the baryonic component of galaxies can resolve it. It is clear that including hot dark matter will decrease the central density of dark-matter halos, both because the lower fluctuation power on small scales in such models will prevent the early collapse that produces the highest dark-matter densities, and also because the hot particles cannot reach high densities because of the phase space constraint (Tremaine & Gunn 1979, Kofman et al. 1996). But this may not be necessary, or alternatively it may not be enough.

1.7.7 Best bet CDM-type models

As said at the outset, the fact that the original CDM model did so well at predicting both the CMB anisotropies discovered by COBE and the distribution of galaxies makes it likely that a large fraction of the dark matter is cold – i.e., that one of the variants of the SCDM model might turn out to be right. Of these, CHDM is the best bet if Ω_0 turns out to be near unity and the Hubble parameter is not too large, while ΛCDM is the best bet if the Hubble parameter is too large to permit the universe to be older than its stars with $\Omega = 1$ (e.g., Chs. 4 and 8, §11.4).

Both theories do seem less "natural" than SCDM, in that they are both hybrid theories. But although SCDM won the beauty contest, it doesn't fit the data.

CHDM is just SCDM with some light neutrinos. After all, we know that neutrinos exist, and there is experimental evidence – admittedly not yet entirely convincing – that at least some of these neutrinos have mass, possibly in the few-eV range necessary for CHDM. Isn't it an unnatural coincidence to have three different sorts of matter – cold, hot, and baryonic – with contributions to the cosmological density that are within an order of magnitude of each other? Not necessarily. All of these varieties of matter may have acquired their mass from (super?)symmetry breaking associated with the electroweak phase transition, and when we understand the nature of the physics that determines the masses and charges that are just adjustable parameters in the Standard Model of particle physics, we may also understand why Ω_c, Ω_ν, and Ω_b are so close. In any case, CHDM is certainly not uglier than ΛCDM.

In the ΛCDM class of models, the problem of too much power on small scales that has been discussed here at some length for $\Omega_0 = 0.3$ and $h = 0.7$ ΛCDM implies either that there must be some physical mechanism that produces strong, scale-dependent anti-biasing of the galaxies with respect to the dark matter, or else that higher Ω_0 and lower h are preferred, with a significant amount of tilt to get the cluster abundance right and avoid too much small-scale power (KPH96). Higher $\Omega_0 \gtrsim 0.5$ also is more consistent with the evidence summarized above against large Ω_Λ and in favor of larger Ω_0, especially in models such as ΛCDM with Gaussian primordial fluctuations. But then $h \lesssim 0.63$ for $t_0 \gtrsim 13$ Gyr.

Among CHDM models, having $N_\nu = 2$ species share the neutrino mass gives a better fit to COBE, clusters, and small-scale data than $N_\nu = 1$, and moreover it appears to be favored by the available experimental data (PHKC95). But it remains to be seen whether CHDM models can fit the data on structure formation at high redshifts, and whether any models of the CDM type can fit all the data – the data on the values of the cosmological parameters, the data on the distribution and structure of galaxies at low and high redshifts, and the increasingly precise CMB anisotropy data. Reliable data is becoming available so rapidly now, thanks to the wonderful new ground- and space-based instruments, that the next few years will be decisive.

The fact that NASA and the European Space Agency plan to launch the COBE follow-up satellites MAP and COBRAS/SAMBA (renamed Planck) in the early years of the next decade, with ground- and balloon-based detectors promising to provide precise data on CMB anisotropies even earlier, means that we are bound to know much more soon about the two key questions of modern cosmology: the nature of the dark matter and of the initial fluctuations. Meanwhile, many astrophysicists, including my colleagues and I, will be trying to answer these questions using data on galaxy distribution, evolution, and structure, in addition to the CMB data. And there is a good chance that in the next few years important inputs will come from particle physics experiments on dark-matter candidate particles or the theories that lead to them, such as supersymmetry.

Acknowledgements

This work was partially supported by NASA and NSF grants at UCSC. JRP thanks his Santa Cruz colleagues and all his collaborators, especially Anatoly Klypin, for many helpful discussions of the material presented here. Special thanks to James Bullock, Avishai Dekel, Patrik Jonsson, and Tsafrir Kolatt for reading an earlier draft of this manuscript and for helpful suggestions for its improvement.

References

Adams, F. C. & Laughlin, G. 1996, *ApJ*, **468**, 586

Albrecht, A. & Steinhardt, P. J. 1982, *PRL*, **48**, 1220

Albrecht, A. & Stebbins, A. 1992, *PRL*, **68**, 2121; **69**, 2615

Alcock, C., et al. 1997a, *ApJ*, **486**, 697

Alcock, C., et al. 1997b, *ApJ*, **482**, 89

Allen, B., Caldwell, R. R., Shellard, E. P. S., Stebbins, A. & Veeraraghavan, S. 1996, *PRL*, **77**, 3061

Allen, B., Caldwell, R. R., Dodelson, S., Knox, L., Shellard, E. P. S. & Stebbins, A. 1997, *PRL*, **79**, 2624

Amaldi, U., de Boer, W. & Furstenau, H. 1991, *Phys. Lett. B.*, **260**, 447

Ansari, R., et al. 1996, *A&A*, 314, 94

Arnet, W. D., Branch, D. & Wheeler, J. C. 1985, *Nature*, **314**, 337

Athanassopoulos, C., et al. 1995, *PRL*, **75**, 2650

Athanassopoulos, C., et al. 1996, *PRL*, **77**, 3082; *Phys. Rev. C*, **54**, 2685

Babu, K. S., Schaefer, R. K. & Shafi, Q. 1996, *Phys. Rev. D*, **53**, 606

Bahcall, J. N. 1996, *ApJ*, **467**, 475

Bahcall, N. A. & Lubin, L. M. 1994, *ApJ*, **426**, 513

Balland, C. & Blanchard, A. 1997, *ApJ*, **487**, 33, and in *Birth of the Universe and Fundamental Physics*, ed. F. Occhionero (Berlin: Springer), p. 263

Balysh, A., et al. 1995, *Phys. Lett. B.*, **356**, 450

Bardeen, J. M., Bond, J. R., Kaiser, N. & Szalay, A. S. 1986, *ApJ*, **304**, 15

BarKana, R. 1997, *ApJ*, **489**, 21

Barnett, M., et al. 1996, Table of Astrophysical Constants from July 1996 Particle Physics Booklet available from pdg@lbl.gov; *Phys. Rev. D*, **54**, 1

Bartelmann, M. & Loeb, A. 1996, *ApJ*, **457**, 529

Bartlett, J. G. & Blanchard, A. 1996, *A&A*, **307**, 1

Battye, R. A. & Shellard, E. P. S. 1994, *PRL*, , **73**, 2954

Baugh, C. M. & Efstathiou, G. 1993, *MNRAS*, **265**, 145

Baugh, C. M. & Efstathiou, G. 1994, *MNRAS*, **267**, 32

Bekenstein, J. D. & Sandars, R. H. 1994, *ApJ*, **429**, 480

Bennett, D. P. & Rhie, S. H. 1993, *ApJ*, **406**, L7

Bergbusch, P. A. & VandenBerg, D. A. 1992, *ApJS*, **81**, 163

Bernardeau, F., Juskiewicz, R., Dekel, A. & Bouchet, F. R. 1995, *MNRAS*, **274**, 20

Birkinshaw, M. & Hughes, J. P. 1994, *ApJ*, **420**, 33

Birkinshaw, M., Hughes, J. P. & Arnoud, K. A. 1991, *ApJ*, **379**, 466

Bissantz, N., Englmaier, P., Binney, J. & Gerhard, O. 1997, *MNRAS*, **289**, 651

Blandford, R. D. & Kundic, T. 1996, astro-ph/9611229, to appear in *The Extragalactic Distance Scale*, eds. M. Livio, M.

Donahue & N. Panagia (Cambridge: Cambridge University Press)

Blumenthal, G. R., Dekel, A. & Primack, J. R. 1988, ApJ, **326**, 539

Blumenthal, G. R., Faber, S. M., Primack, J. R. & Rees, M. J. 1984, Nature, **311**, 517; erratum: **313**, 72

Bolte, M. & Hogan, C. J. 1995, Nature, **376**, 399

Bolte, M. 1997, Nature, **385**, 205

Bond, J. R., Efstathiou, G. & Silk, J. 1980, PRL, , **45**, 1980

Bonometto, S. A. & Valdarnini, R. 1984, Phys. Lett. A., **103**, 369

Borde, A. & Vilenkin, A. 1994, PRL, **72**, 3305

Borde, A. & Vilenkin, A. 1997, Phys. Rev. D., **56**, 717

Borgani, S., Masiero, A. & Yamaguchi, M. 1996, Phys. Lett. B., **386**, 189

Borgani, S., et al. 1997, New Astronomy, **1**, 299

Branch, D. 1992, ApJ, **392**, 35

Branch, D. & Khokhlov, A. M. 1995, Phys. Rep., **256**, 53

Branch, D., et al. 1996, ApJ, **470**, L7

Branchini, E. & Carlberg, R. G. 1994, ApJ, **434**, 37

Brodbeck, D., Hellinger, D., Nolthenius, R., Primack, J. R. & Klypin, A. 1998, ApJ, **495**, 1

Bryan, G., et al. 1994, ApJ, **437**, L5

Bucher, M., Goldhaber, A. S. & Turok, N. 1995, Phys. Rev. D., **52**, 3314

Bullock, J. & Primack, J. R. 1996, Phys. Rev. D., submitted

Buote, D. A. & Tsai, J. C. 1996, ApJ, **458**, 27

Buote, D. A. & Xu, G. 1997, MNRAS, **284**, 439

Burkert, A. 1995, ApJ, **447**, L25

Burles, S. & Tytler, D. 1997, in Proc. Origin of Matter and Evolution of Galaxies in the Universe, eds. T. Kajino, S. Kubono & Y. Yoshii (Singapore: World Scientific)

Burns, J. O., et al. 1994, ApJ, **427**, L87

Burstein, D., et al. 1986, in Galaxy Distances and Deviations from Universal Expansion, eds B. F. Madore & R. B. Tully (Dordrecht: Reidel)

Caldwell, D. O. 1995, Nucl. Phys. B, Proc. Suppl., **43**, 126

Caldwell, R. R., Battye, R. A. & Shellard, E. P. S. 1996, Phys. Rev. D., **54**, 7146

Carlberg, R. G. & Couchman, H. M. P. 1989, ApJ, **340**, 47

Carr, B. J., Gilbert, A. M. & Lidsey, J. E. 1994, Phys. Rev. D., **50**, 4853

Carroll, S. M., Press, W. H. & Turner, E. L. 1992, ARA&A, **30**, 499

Carswell, R. F., et al. 1994, MNRAS, **268**, L1

Carswell, R. F., et al. 1996, MNRAS, **278**, 506

Castander, F. J., et al. 1995, Nature, **377**, 39

Castellani, V., Ciacio, F., Degl'Innocenti, S. & Fiorentini, G. 1997, A&A, **322**, 801

Chaboyer, B. 1995, ApJ, **444**, L9

Chaboyer, B. & Kim, Y.-C. 1995, ApJ, **454**, 767

Chaboyer, B., Demarque, P., Kernan, P. J. & Krauss, L. M. 1996, Science, **271**, 957

Cole, S. & Lacey, C. 1996, MNRAS, **281**, 716

Collins, P. A. B., Martin, A. D. & Squires, E. J. 1989, Particle Physics and Cosmology (New York: Wiley)

Colombi, S., Dodelson, S. & Widrow, L. M. 1996, ApJ, **458**, 1

Copi, C. J., Schramm, D. N. & Turner, M. S. 1995, Science, **267**, 192; PRL, **75**, 3981

Corbett, E. A., Browne, I. W. A., Wilkinson, P. N. & Patnaik, A. R. 1996, in Astrophysical Applications of Gravitational Lensing, eds. C. S. Kochanek & J. N. Hewitt (Dordrecht: Kluwer), p. 37

Coulson, D. Ferreira, P., Graham, P. & Turok, N. 1994, Nature, **368**, 27

Cowan, J. J., McWilliam, A., Sneden, C. & Burris, D. L. 1997, ApJ, **480**, 246

Crittenden, R. G. & Turok, N. 1995, PRL, **75**, 2642

Crone, M., Evrard, A. & Richstone, D. 1994, ApJ, **434**, 402

da Costa, L. N., et al. 1994, ApJ, **437**, L1

Dabrowski, Y., Lasenby, A. & Saunders, R. 1995, *MNRAS*, , **277**, 753

Dahle, H., Maddox, S. J. & Lilje, P. B. 1994, *ApJ*, **435**, L79

D'Antona, F., Caloi, V. & Mazzitelli, I. 1997, *ApJ*, **477**, 519

Davis, M. & Peebles, P. J. E. 1983, *ApJ*, **267**, 465

Davis, M., Efstathiou, G., Frenk, C. S. & White, S. D. M. 1985, *ApJ*, **292**, 371

Davis, M., Summers, F. & Schlegel, D. 1992, *Nature*, **359**, 393

de Boer, W. 1994, *Prog. Particle Nucl. Phys.*, **33**, 201

Dekel, A. 1994, *ARA&A*, **32**, 371

Dekel, A. & Aarseth, S. J. 1984, *ApJ*, **283**, 1

Dekel, A. & Rees, M. J. 1994, *ApJ*, **422**, L1

Dell'Antonio, I. P. & Tyson, J. A. 1996, *ApJ*, **473**, L17

Dimopoulos, S., Dine, M., Raby, S. & Thomas, S. 1996, *PRL*, **76**, 3494

Dimopoulos, S., Thomas, S. & Wells, J. D. 1996, *Phys. Rev. D.*, **54**, 3283

Dimopoulos, S., Giudice, G. F. & Pomarol, A. 1996, *Phys. Lett. B.*, **389**, 37

Dine, M., Nelson, A., Nir, Y. & Shirman, Y. 1996, *Phys. Rev. D.*, **53**, 2658

Dodelson, S., Gyuk, G. & Turner, M. S. 1994, *PRL*, **72**, 3754

Dodelson, S., Gates, E. & Stebbins, A. 1996, *ApJ*, **467**, 10

Driver, S. P., Windhorst, R. A., Phillipps, S. & Bristow, P. D. 1996, *ApJ*, **461**, 525

Dubinski, J. & Carlberg, R. G. 1991, *ApJ*, **378**, 496

Dubinski, J., Mihos, J. C. & Hernquist L. 1996, *ApJ*, **462**, 576

Dunlop, J., et al. 1996, *Nature*, 381, 581

Dunn, A. M. & Laflamme R. 1995, *ApJ*, **443**, L1

Durrer, R., Gangui, A. & Sakellariadou, M. 1996, *PRL*, **76**, 579

Eastman, R. G., Schmidt, B. P. & Kirshner, R. 1996, *ApJ*, **466**, 911

Ebeling, H., et al. 1995, in *Roentgenstrahlung from the Universe*, Wurzberg, Sept. 1995.

Efstathiou, G. 1996, in *Critical Dialogues in Cosmology*, ed. N. Turok (Singapore: World Scientific)

Efstathiou, G., Bond, J. R. & White, S. D. M. 1992, *MNRAS*, **258**, 1p

Efstathiou, G., Sutherland, W. J. & Maddox, S. J. 1990, *Nature*, **348**, 705

Eisenstein, D. J. & Loeb, A. 1995, *ApJ*, **443**, 11

Eke, V. R., Cole, S. & Frenk, C. S. 1996, *ApJ*, **282**, 263

Ellis, G. F. R. 1996, in *Current Topics in Astrofundamental Physics*, ed. N. Sanchez (Singapore: World Scientific)

Ellis, J., et al. 1992, *Nucl. Phys. B.*, **373**, 399

Ellis, J., Kim, J. E. & Nanopoulos, D. V. 1984, *Phys. Lett. B.*, **145**, 181

Evrard, A. E., Metzler C. A. & Navarro J. F. 1996, *ApJ*, **469**, 494

Falco, E. E., Shapiro, I. I., Moustakas, L. A. & Davis, M. 1997, *ApJ*, **484**, 70

Fall, S. M., Charlot, S. & Pei, Y. C. 1996, *ApJ*, **464**, L43

Fang, L. Z., Li, S. X. & Xiang, S. P. 1984, *A&A*, **140**, 77

Feast, M. W. & Catchpole, R. W. 1997, *MNRAS*, 286, L1

Felten, J. E. 1996, in *Clusters, Lensing, and the Future of the Universe*, eds. V. Trimble & A. Reisenegger, ASP Conf. Series, 88, 271

Felton, J. E. & Isaacman, R. 1986, *Rev. Mod. Phys.*, **58**, 689

Ferrarese, L., et al. 1996, *ApJ*, **464**, 568

Fields, B. D. & Olive, K. 1996, *Phys. Lett. B.*, **368**, 103

Fischer, P., Bernstein, G., Rhee, G. & Tyson, J. A. 1997, *AJ*, **113**, 521

Fisher, A., et al. 1995, *ApJ*, **447**, L73

Fisher, K. B., et al. 1993, *ApJ*, **402**, 42

Fisher, K. B., et al. 1994, *MNRAS*, **267**, 927

Flores, R., Primack, J. R., Blumenthal, G. R. & Faber, S. M. 1993, *ApJ*, **412**, 443

Flores, R. & Primack, J. R. 1994, *ApJ*, **427**, L1

Flores, R. & Primack, J. R. 1996, *ApJ*, **457**, L5

Flynn, C., Gould, A. & Bahcall, J. N. 1996, *ApJ*, **466**, L55

Freedman, W. L., 1996, in *Critical Dialogues in Cosmology*, ed. N. Turok (Singapore: World Scientific)

Freedman, W. L., et al. 1994, *Nature*, **371**, 757

Fukuda, Y. 1994, *Phys. Lett. B.*, **335**, 237

Fukugita, M., Hogan, C. J. & Peebles, P. J. E. 1993, *Nature*, **366**, 309

Fuller, G. M., Primack, J. R. & Qian, Y.-Z. 1995, *Phys. Rev. D.*, **52**, 1288

Gallego, J., Zamorano, J., Aragon-Salamanca, A. & Rego, M. 1996, *ApJ*, **455**, L1

Ganga, K., Ratra, B. & Sugiyama, N. 1996, *ApJ*, **461**, L61

Garcia-Bellido, J., Liddle, A. R., Lyth, D. H. & Wands, D. 1995, *Phys. Rev. D.*, **52**, 6750

Garnavich, P., et al. 1996, IAU Circular no. 6358

Gates, E. I., Gyuk, G. & Turner, M. S. 1996, *Phys. Rev. D.*, **53**, 4138

Georgi, H., Quinn, H. R. & Weinberg, S. 1974, *PRL*, **33**, 451

Ghigna, S., et al. 1994, *ApJ*, **437**, L71

Ghigna, S., Borgani, S., Tucci, M., Bonometto, S., Klypin, A. & Primack, J.R. 1997, *ApJ*, **479**, 580

Giavalisco, M., Steidel, C. C. & Macchetto, F. D. 1996, *ApJ*, **470**, 189

Gnedin, N. Y. & Hui, L. 1996, *ApJ*, **472**, L73

Goldhaber, G., et al. 1996, in *Sources and Detection of Dark Matter in the Universe*, D. Cline & D. Sanders, eds., *Nucl. Phys. B, Proc. Suppl.*, **51**, 123

Goobar, A. & Perlmutter, S. 1995, *ApJ*, **450**, 14

Gorski, K. M., et al. 1994, *ApJ*, **430**, L89

Gorski, K. M., et al. 1996, *ApJ*, **464**, L11

Gratton, R. G., et al. 1997, *ApJ*, **491**, 749

Griest, K. 1996, astro-ph/9510089, in *International School of Physics "Enrico Fermi"*, Course CXXXII: Dark Matter in the Universe, Varenna 1995, eds. S. Bonometto,
J.R. Primack & A. Provenzale (Amsterdam: IOS Press), p. 243

Grischuk, L. P. & Zel'dovich, Ya. B. 1978, *Sov. Astron.*, **22**, 125

Grogin, N. A. & Narayan R. 1996, *ApJ*, **464**, 92; **473**, 570

Gunn, K. F. & Thomas, P. A. 1995, *MNRAS*, **281**, 1133

Guth, A. H. 1981, *Phys. Rev. D.*, **23**, 347

Guth, A. H. & Pi, S.-Y. 1982, *PRL*, **49**, 1110

Guth, A. H. & Weinberg, E. J. 1981, *Phys. Rev. D.*, **23**, 876

Guth, A. H. & Weinberg, E. J. 1983, *Nucl. Phys. B.*, **212**, 321

Gyuk, G. & Turner, M. S. 1995, *16th Internat. Conf. on Neutrino Physics and Astrophysics*, *Nucl. Phys. B, Proc. Suppl.*, **38**, 13

Haehnelt, M. G. 1993, *MNRAS*, **265**, 727

Haehnelt, M., Steinmetz, M. & Rauch, M. 1996, *ApJ*, **465**, L95

Hagmann, C., et al. 1996, in *Sources and Detection of Dark Matter in the Universe*, eds. D. Cline & D. Sanders, Nucl. Phys. B, Proc. Supp., **51**, 209

Hammer, F., Rigaut, F. & Angonin-Willaime, M.-C. 1995, *A&A*, **298**, 737

Hamuy, M., et al. 1996, *AJ*, **112**, 2398

Harrison, E. R. 1970, *Phys. Rev. D.*, **1**, 2726

Hata, N. & Langacker, P. 1995, *Phys. Rev. D.*, **52**, 420

Hata N., et al. 1995, *PRL*, **75**, 3977

Hawking, S. W. & Ellis, G. F. R. 1973, *The Large Scale Structure of Space-Time* (Cambridge: Cambridge University Press)

Henry, J. P., et al. 1992, *ApJ*, **386**, 408

Hernanz, M., et al. 1994, *ApJ*, **434**, 652

Hindmarsh, M. B. & Kibble, T. W. B. 1995, *Rep. Prog. Physics*, **58**, 477

Hjorth, J., et al. 1996, in *Astrophysical Applications of Gravitational Lensing*, eds. C. S. Kochanek & J. N. Hewitt (Dordrecht: Kluwer), p. 343

Hodges, H., Blumenthal, G. R., Kofman, L. A. & Primack, J. R. 1990, *Nucl. Phys. B.*, **335**, 197

Hodges, H. & Primack, J. R. 1991, *Phys. Rev. D.*, **43**, 3155

Holman, R., Ramond, P. & Ross, G. G. 1984, *Phys. Lett. B.*, **137**, 343

Holtzman, J. 1989, *ApJS*, **71**, 1

Holtzman, J. A. & Primack, J. R. 1993, *ApJ*, **405**, 428

Hu, W. & White, M. 1996, *PRL*, , **77**, 1687; *ApJ*, **471**, 30

Hu, W., Spergel, D. N. & White, M. 1997, *Phys. Rev. D.*, **55**, 3288

Huchra, J. P. & Geller, M. J. 1982, *ApJ*, **257**, 423

Jacoby, G. H., et al. 1992, *PASP*, **104**, 599

Jing, Y. P. & Fang, L. Z. 1994, *ApJ*, **432**, 438

Jing, Y. P., Mo, H. J., Börner, G. & Fang, L. Z. 1995, *MNRAS*, **276**, 417

Jungman, G., et al. 1996, *PRL*, 76, 1007; *Phys. Rev. D.*, **54**, 1332

Jungman, G., Kamionkowski, M. & Griest K. 1996, *Phys. Rep.*, **267**, 195

Kane, G. L. & Wells, J. D. 1996, *PRL*, **76**, 4458

Katz, N., Quinn, T., Bertschinger, E. & Gelb, J. M. 1994, *MNRAS*, **270**, L71

Katz, N., Weinberg, D. H., Hernquist, L. & Miralda-Escude, J. 1996, *ApJ*, **457**, L57

Kauffmann, G. 1995, *MNRAS*, **274**, 161

Kauffmann, G. 1996, *MNRAS*, **281**, 475

Kauffmann, G. & Charlot, S. 1994, *ApJ*, **430**, L97

Kauffmann, G., Nusser, A. & Steinmetz, M. 1997, *MNRAS*, **286**, 795

Keeton, C. R. & Kochanek, C. S. 1997, *ApJ*, **487**, 42

Kellermann, K. I. 1993, *Nature*, **361**, 134

Kennicutt, R. C., Freedman, W. L. & Mould, J. R. 1995, *AJ*, **110**, 1476

Kibble, T. W. B. 1976, *J. Phys.*, **A9**, 1387

Kibble, T. W. B. & Hindmarsh, M. B. 1995, *Rep. Prog. Phys.*, **58**, 477

Kim, A., et al. 1997, *ApJ*, **476**, L63

Kirschner, R., Oemler, A. & Schechter P. 1979, *ApJ*, **84**, 951

Kleinfeller, J. 1996, in *Proc. XVII Internat. Conf. in Neutrino Physics and Astrophysics*, Neutrino 96, Helsinki, Finland 13–19 June 1996, eds. K. Enqvist, K. Huitu & J. Maalampi (Singapore: World Scientific)

Klypin, A., Borgani, S., Holtzman, J. & Primack, J. R. 1995, *ApJ*, **444**, 1

Klypin, A., Holtzman, J., Primack, J. R. & Regős, E. 1993, *ApJ*, **416**, 1

Klypin, A., Nolthenius, R. & Primack, J. R. 1997, *ApJ*, **474**, 533

Klypin, A., Primack, J. R. & Holtzman, J. 1996, *ApJ*, **466**, 1 (KPH96)

Kochanek, C. S. 1993, *ApJ*, **419**, 12

Kochanek, C. S. 1996a, *ApJ*, **457**, 228

Kochanek, C. S. 1996b, *ApJ*, **466**, 638

Kofman, L. A., Blumenthal, G. R., Hodges, H. & Primack, J. R. 1991, in *Proceedings, Workshop on Large Scale Structure and Peculiar Motions in the Universe*, eds. D. W. Latham & L. N. da Costa (Astronomical Society of the Pacific), p. 251

Kofman, L. A., Gnedin, N. Y. & Bahcall, N. A. 1993, *ApJ*, **413**, 1

Kofman, L. A., Klypin, A., Pogosyan, D. & Henry, J. P. 1996, *ApJ*, **470**, 102

Kovner, I. & M. Milgrom 1987, *ApJ*, **321**, L113

Krauss, L. M. & Turner, M. S. 1995, *General Relativity & Gravitation*, **27**, 1137

Krauss, L. M. & Kernan, P. J. 1995, *Phys. Lett. B.*, **347**, 347

Kronberg, P. P. 1994, *Rep. Prog. Phys.*, **57**, 325

Kundic, T., et al. 1995, *ApJ*, **455**, L5

Kundic, T., et al. 1997, *ApJ*, **482**, 75

Kundic, T., Cohen, J. G. & Blandford, R. G. 1997, *AJ*, **114**, 507

Lahav, O., Lilje, P., Primack, J. R. & Rees, M.J. 1991, *MNRAS*, **251**, 128

Lanzetta, K. M., Wolfe, A. M. & Turnshek, D. A. 1995, *ApJ*, **440**, 435

Lanzetta, K. M., Yahil, A. & Fernández-Soto, A. 1996, *Nature*, **381**, 759, and work in prep.

Lasenby, A. 1997, in *Current Topics in Astrofundamental Physics*, ed. N. Sanchez (Singapore: World Scientific), p. 485

Leibundgut, B. & Pinto, P. A. 1992, *ApJ*, **401**, 49

Liddle, A. R., Lyth, D. H., Roberts, D. & Viana, P. T. P. 1996a, *MNRAS*, **278**, 644

Liddle, A. R., Lyth, D. H., Schaefer, R. K., Shafi, Q. & Viana, P. T. P. 1996b, *MNRAS*, **281**, 531

Liddle, A. R., Lyth, D. H., Viana, P. T. P. & White, M. 1996c, *MNRAS*, **282**, 281

Lilly, S., et al. 1995, *ApJ*, **455**, 108

Lilly, S. J., Le Fevre, O., Hammer, F. & Crampton, D. 1996, *ApJ*, **460**, L1

Linde, A. 1982, *Phys. Lett. B.*, **108**, 389

Linde, A. 1983, *Phys. Lett. B.*, **129**, 177

Linde, A. 1986, *Phys. Lett. B.*, **175**, 395

Linde, A. 1990, *Particle Physics and Inflationary Cosmology* (New York: Harwood)

Linde, A. 1995, hep-ph/9410082, in *Proc. Lake Louise Winter Institute on Particle Physics and Cosmology*, eds. A. Astbury, et al. (Singapore: World Scientific), p. 72

Linde, A. & Mezhlumian, A. 1995, *Phys. Rev. D.*, **52**, 6789

Little, B. & Tremaine, S. 1987, *ApJ*, **320**, 493

Loeb, A. & Mao S. 1994, *ApJ*, **435**, L109

Loeb, A. & Refregier, A. 1996, *ApJ*, **476**, L59

Loh, E. D. & Spillar, E. J. 1986, *ApJ*, **307**, L1

Loewenstein, M. & Mushotzky, R. F. 1996, *ApJ*, **471**, L83

Loewenstein, M. 1996, *Nucl. Phys. B, Proc. Suppl.*, **51**, 114

Lowenthal, J. D., et al. 1997, *ApJ*, **481**, 673

Lu, L., Sargent, W. L. W., Womble, D. S. & Barlow, T. A. 1996, *ApJ*, **457**, L1

Ma, C.-P. 1995, in *Dark Matter*, AIP Conference Proceedings 336, p. 420

Ma, C.-P. & Bertschinger, E. 1994, *ApJ*, **434**, L5

Maartens, R., Ellis, G. F. R. & Stoeger, W. R. 1995, *Phys. Rev. D.*, **51**, 5942

MacMinn, D. & Primack, J. R. 1996, *Space Sci. Rev.*, **75**, 413

Madau, P., et al. 1996, *MNRAS*, **283**, 1388

Magueijo, J., Albrecht, A., Ferreira, P. & Coulson, D. 1996, *Phys. Rev. D.*, **54**, 3727

Malaney, R. A., Mathews, G. J. & Dearborn, D. S. P. 1989, *ApJ*, **345**, 169

Malaney, R. A., Starkman, G. D. & Widrow, L. 1995, *Phys. Rev. D.*, **52**, 5480

Malhotra, S., Rhoads, J. E. & Turner, E. L. 1997, *MNRAS*, **288**, 138

Mannheim, P. D. & Kazanas, D. 1994, *General Relativity & Gravitation*, **26**, 337

Maoz, D. & Rix, H. W. 1993, *ApJ*, **416**, 425

Marzke, R. O., Geller M. J., da Costa L. N. & Huchra J. P. 1995, *AJ*, **110**, 477

Mathews, G. J. & Schramm, D. N. 1993, *ApJ*, **404**, 468

McHugh, M. P., Zalamamsky, G., Vermotte, F. & Lantz, R. 1996, *Phys. Rev. D.*, **54**, 5993

Milgrom, M. 1983, *ApJ*, **270**, 365

Milgrom, M. 1994, *Annals of Physics*, **229**, 384

Milgrom, M. 1995, *ApJ*, **455**, 439

Miralda-Escudé, J. 1995, *ApJ*, **438**, 514

Mo, H. J., Jing, Y. P. & Börner, G. 1993, *MNRAS*, **264**, 825

Mo, H. J. & Miralda-Escude, J. 1994, *ApJ*, **430**, L25

Mo, H. J. & Fukugita, M. 1996, *ApJ*, **467**, L9

Mohr, J. J., Evrard, A. E., Fabricant, D. G. & Geller, M. J. 1995, *ApJ*, **447**, 8

Moore, B. 1994, *Nature*, **370**, 629

Mould, J., et al. 1994, *MNRAS*, **271**, 31

Mould, J., et al. 1995, *ApJ*, **449**, 413

Nagasawa, M. & Kawasaki, M. 1994, *Phys. Rev. D.*, **50**, 4821

Navarro, J. F., Frenk, C. S. & White, S. D. M. 1996, *ApJ*, **462**, 563

Navarro, J. F., Eke, V. R. & Frenk, C. S. 1996, *MNRAS*, **283**, L72

Netterfield, C. B., Devlin, M. J., Jarosik, N., Page, L. & Wollack, E. J. 1997, *ApJ*, **474**, 47

Neumann, M. & Böhringer, H. 1996, *ApJ*, **461**, 572

Nolthenius, R., Klypin, A. & Primack, J. R. 1994, *ApJ*, **422**, L45

Nolthenius, R., Klypin, A. & Primack, J. R. 1997, *ApJ*, **480**, 43

Nugent, P., et al. 1995, *PRL*, **75**, 394; *Erratum* **75**, 1874

Nusser, A. & Dekel, A. 1993, *ApJ*, **405**, 437

Olive, K. A. & Turner, M. S. 1982, *Phys. Rev. D.*, **25**, 213

Olive, K. A., Skillman, E. & Steigman, G. 1997, *ApJ*, 489, 1006

Olivier, S., Primack, J. R., Blumenthal, G. R. & Dekel, A. 1993, *ApJ*, **408**, 17

Ostriker, J. P. & Steinhardt, P. J. 1995, *Nature*, 377, 600

Oswalt, T. D., Smith, J. A., Wood, M. A. & Hintzen, P. 1996, *Nature*, **382**, 692

Paczynzki, B. 1996 in *Astrophysical Applications of Gravitational Lensing*, IAU Symp. 173 (Dordrecht: Kluwer), p. 199

Pagels, H. & Primack, J. R. 1982, *PRL*, **48**, 223

Peacock, J. A. & Dodds, S. J. 1994, *MNRAS*, **267**, 1020

Peacock, J. A. & Dodds, S. J. 1996, *MNRAS*, **280**, L19

Peebles, P. J. E. 1982, *ApJ*, **258**, 413

Peebles, P. J. E. 1984a, *ApJ*, **263**, L1

Peebles, P. J. E. 1984b, *ApJ*, **284**, 439

Peebles, P. J. E. 1989, *ApJ*, **344**, 53

Peebles, P. J. E. 1990, *ApJ*, **362**, 1

Peebles, P. J. E. 1993, *Principles of Physical Cosmology* (Princeton: Princeton University Press)

Peebles, P. J. E. 1994, *ApJ*, **429**, 43

Peebles, P. J. E. 1996, in *Formation of Structure in the Universe*, eds. A. Dekel & J. P. Ostriker (Cambridge: Cambridge University Press)

Pei, Y. C. C. & Fall, S. M. 1995, *ApJ*, **454**, 69

Pelt, J., et al. 1994, *A&A*, **286**, 775

Pen, U. -L. & Spergel, D. N. 1995, *Phys. Rev. D.*, **51**, 4099

Pen, U. -L., Seljak, U. & Turok, N. 1997, *PRL*, **79**, 1611

Perivolaropoulos L. 1993, *Phys. Lett. B.*, **298**, 305

Perivolaropoulos L. & Vachaspati T. 1994, *ApJ*, **423**, L77

Perlmutter, S., et al. 1997a, in *Thermonuclear Supernovae* (NATO ASI), eds. R. Canal, P. Ruiz-LaPuente & J. Isern (Dordrecht: Kluwer Academic Publishers)

Perlmutter, S., et al. 1997b, *ApJ*, **483**, 65

Pierce, M. J., et al. 1994, *Nature*, **371**, 385

Pierpaoli, E., Coles, P., Bonometto, S. & Borgani, S. 1996, ApJ, **470**, 92

Pogosyan, D. & Starobinsky, A. A. 1995a, *ApJ*, **447**, 465

Pogosyan, D. & Starobinsky, A. A. 1995b, astro-ph/9502019, in *Internat. Workshop on Large Scale Structure in the Universe*, eds. P. Mucket, S. Gottlober & V. Muller (World Scientific)

Press, W. H., Rybicki, G. B. & Hewitt, J. N. 1992, *ApJ*, **385**, 416

Primack, J. R. 1987, in *Proc. Internat. School of Physics "Enrico Fermi"* 1984, XCII, ed. N. Cabibbo (Amsterdam: North-Holland), p. 137

Primack, J. R. 1988, in *Superstrings, Unified Theories, and Cosmology*, eds. G. Furlan et al. (Singapore: World Scientific), p. 618

Primack, J. R. 1993, in *Particle Physics and Cosmology at the Interface*, eds. J. Pati, P. Ghose, and J. Maharana (Singapore: World Scientific), p. 301

Primack, J. R. 1995, in *Particle and Nuclear Astrophysics and Cosmology in the Next Millenium*, eds. E. Kolb & R. Peccei (Singapore: World Scientific), p. 85

Primack, J. R. 1996a, astro-ph/9604184, in *International School of Physics "Enrico Fermi"*, Course CXXXII: Dark Matter in the Universe, Varenna, eds. S. Bonometto, J. R. Primack, & A. Provenzale (Amsterdam: IOS Press), p. 1

Primack, J. R. 1996b, astro-ph/9610078, in *Critical Dialogues in Cosmology*, ed. N. Turok (Singapore: World Scientific), p. 535

Primack, J. R. 1997, hep-ph/9610321, in *Proceedings of the XVII International Conference in Neutrino Physics and Astrophysics*, Neutrino 96, Helsinki, eds. K. Enqvist, K. Huitu & J. Maalampi (Singapore: World Scientific)

Primack, J. R. & Blumenthal, G. R. 1983, in *Formation and Evolution of Galaxies and Large Structures in the Universe*, eds. J. Audouze & J. Tran Thanh Van (Dordrecht: Reidel), p. 163

Primack, J. R. & Blumenthal, G. R. 1984, in *Clusters and Groups of Galaxies*, eds. F. Mardirossian, G. Giuricin, & M. Mezzetti (Dordrecht: Reidel), p. 435

Primack, J. R. & Holtzman, J. A. 1992, in *Gamma Ray–Neutrino Cosmology and Planck Scale Physics: Proc. 2nd UCLA Conf.*, ed. D. Cline (Singapore: World Scientific), p. 28

Primack, J. R. & Klypin, A. 1996, in *Sources and Detection of Dark Matter in the Universe*, eds. D. Cline & D. Sanders, *Nucl. Phys. B, Proc. Suppl.*, **51**, 30

Primack, J. R., Seckel, D. & Sadoulet, B. 1988, *Ann. Rev. Nucl. Part. Phys.* **38**, 751

Primack, J. R., Holtzman, J., Klypin, A. & Caldwell, D. O. 1995, *PRL*, **74**, 2160 (PHKC95)

Quinn, J., et al. 1996, *ApJ*, **456**, L83

Ramella, M., Geller, M. J. & Huchra, J. P. 1989, *ApJ*, **344**, 57

Renault, C., et al. 1997, *A&A*, **324**, 69

Rephaeli, Y. 1995, *ARA&A*, **33**, 541

Rhee, G., Bernstein, G., Tyson, A. & Fischer, P. 1996, in *Astrophysical Applications of Gravitational Lensing*, eds. C. S. Kochanek & J. N. Hewitt (Dordrecht: Kluwer), p. 49

Rhee, G. F. R. N. 1991, *Nature*, **350**, 211

Richer, H. B., et al. 1995, *ApJ*, **451**, L17

Richstone, D., Loeb, A. & Turner, E. L. 1992, *ApJ*, **393**, 477

Reid, I. N. 1997, *AJ*, **114**, 161

Riess, A. G., Press, W. H. & Kirshner, R. P. 1996, *ApJ*, **473**, 88

Rindler, W. 1977, *Essential Relativity: Special, General, and Cosmological* 2nd ed. (New York: Springer)

Roberts, D. H., et al. 1991, *Nature*, **352**, 43

Roettiger, K., Stone, J. M. & Mushotzky, R. 1997, *ApJ*, **482**, 588

Ross, G. G. & Sarkar, S. 1996, *Nucl. Phys. B*, **461**, 597

Roth, J. & Primack, J. R. 1996, *Sky & Telescope*, **91**(1), 20

Rowan-Robinson, M. 1981, *Cosmology* (Oxford: Clarendon Press)

Rowan-Robinson, M. 1985, *The Cosmological Distance Ladder* (New York: Freeman)

Rugers, M. & Hogan, C. J. 1996, *ApJ*, **459**, L1

Ruiz-Lapuente, P., et al. 1995, *ApJ*, **439**, 60

Sandage, A. 1995, in *Practical Cosmology: Inventing the Past*, 23rd Sass Fee lectures, ed. B. Binggeli & R. Buser (Berlin: Springer).

Sandage, A., et al. 1996, *ApJ*, **460**, L15

Sandage, A. & Tammann, G. A. 1995, in *Advances in Astrofundamental Physics, Proc. Int. School of Physics "D. Chalonge"*, eds. N. Sanchez & A. Zichichi (Dordrecht: Kluwer), p. 403

Sandage, A. & Tammann, G. A. 1996, in *Critical Dialogues in Cosmology*, ed. N. Turok (Singapore: World Scientific)

Sandquist, E. L., Bolte, M., Stetson, P. B. & Hesser, J. E. 1996, *ApJ*, **470**, 910

Sasaki, M., Tanaka, T. & Yamamoto, K. 1995, *Phys. Rev. D.*, **51**, 2979

Schade, D., et al. 1996, *ApJ*, **464**, L63

Schaefer, B. 1996, *ApJ*, **459**, 438

Schaefer, R. K. & Shafi, Q. 1993, *Phys. Rev. D.*, **47**, 1333

Schechter, P., et al. 1997, *ApJ*, **475**, L85

Schmidt, B. P., Kirschner, R. P. & Eastman, R. G. 1992, *ApJ*, **395**, 366

Schmidt, B. P., et al. 1994, *ApJ*, **432**, 42

Schubnell, M. S., et al. 1996, *ApJ*, **460**, 644

Scott, P. F., et al. 1996, *ApJ*, **461**, L1

Shafi, Q. & Stecker, F. W. 1984, *PRL*, **53**, 1292

Shaya, E. J., Peebles, P. J. E. & Tully, R. B. 1995, *ApJ*, **454**, 15

Shellard, E. P. S. 1997, in *Current Topics in Astrofundamental Physics*, ed. N. Sanchez (Singapore: World Scientific), p. 343

Shi, X. 1995, *ApJ*, **446**, 637

Shutt, T., et al. 1996, *Nucl. Instr. & Methods A*, **370**, 165

Smith, C. C., Klypin, A. A., Gross, M. A. K., Primack, J. R. & Holtzman, J. 1997, astro-ph/9702099, *MNRAS*, in press

Sneden, C., et al. 1996, *ApJ*, **467**, 819

Somerville, R., Davis, M. & Primack, J. R. 1996, *ApJ*, **479**, 616

Somerville, R., Primack, J. R. & Nolthenius, R. 1996, *ApJ*, **479**, 606

Somerville, R. S., Faber, S. M. & Primack, J. R. 1997, in *Proceedings of the Ringberg Workshop on Large Scale Structure*, ed. D. Hamilton (Kluwer Academic Publishers), and astro-ph/9806228

Songaila, A., et al. 1994a, *Nature*, **368**, 599

Songaila, A., et al. 1994b, *Nature*, **371**, 43

Songaila, A., Wampler, E. J. & Cowie, L. L. 1997, *Nature*, **385**, 137

Spinrad, H., et al. 1997, *ApJ*, **484**, 581

Squires, G., Kaiser, N., Babul, A., Fahlman, G. & Woods, D. 1996, *ApJ*, **461**, 572

Steidel, C. C., et al. 1996, *ApJ*, **462**, L17

Steidel, C. C., Dickinson, M. & Persson, S. E. 1994, *ApJ*, **437**, L75

Steigman, G. & Felten, J. E. 1995, *Space Science Reviews*, **74**, 245

Steinhardt, P. J. 1990, *Nature*, **345**, 47

Stetson, P. B., VandenBerg, D. A. & Bolte, M. 1996, *PASP*, **108**, 560

Stoeger, W. R., Maartens, R. & Ellis, G. F. R. 1995, *ApJ*, **443**, 1

Stompor, R., Gorski, K. M. & Banday, A. J. 1995, *MNRAS*, **277**, 1225

Storrie-Lombardi, L. J., McMahon, R. G. & Irwin, M. J. 1996, *MNRAS*, **283**, L79

Strauss, M. A. & Willick, J. A. 1995, *Phys. Rep.*, **261**, 271

Strickland, R. W. & Schramm, D. N. 1997, *ApJ*, **481**, 571

Suto, Y., Suginohara, T. & Inagaki, Y. 1995, *Prog. Theor. Phys.*, **93**, 839

Suto, Y. & Jing, Y.-P. 1996, *ApJS*, **110**, 167

Swenson, F. J. 1995, *ApJ*, **438**, L87

Tammann, G. A., et al. 1996, astro-ph/9603076, in *Science with the HST - II*, NASA

Tammann, G. A. & Sandage, A. 1995, *ApJ*, **452**, 16

Tammann, G. A. & Federspiel, M. 1996, astro-ph/961119, to appear in *The Extragalactic Distance Scale*, eds. M. Livio, M. Donahue & N. Panagia (Cambridge: Cambridge University Press)

Taylor, J. H. 1994, *Rev. Mod. Phys.*, **66**, 711

Tegmark, M. 1996, *ApJ*, **464**, L35

Thorsett, S. E. & Dewey, R. J. 1996, *Phys. Rev. D.*, **53**, 3468

Tohline, J. 1983, in *Internal Kinematics and Dynamics of Galaxies*, IAU Symp. 100, ed. E. Athanassoula (Dordrecht: Reidel), p. 205

Tonry, J. 1991, *ApJ*, **373**, L1

Tonry, J., Blakeslee, J. P., Ajhar, E. A. & Dressler, A. 1997, *ApJ*, **475**, 399

Tremaine, S. & Gunn, J. E. 1979, *PRL*, **42**, 407

Turner, E. L., Cen, R. & Ostriker, J. P. 1992, *AJ*, **103**, 1427

Tyson, J. A., et al. 1984, *ApJ*, **281**, L59

Tytler, D., Burles, S. & Kirkman, D. 1996, astro-ph/9612121

Tytler, D., Fan, X. & Burles, S. 1996, *Nature*, **381**, 207

Tytler, D. & Burles, S. 1997, in *Proc. Origin of Matter and Evolution of Galaxies in the Universe*, eds. T. Kajino, S. Kubono & Y. Yoshii (Singapore: World Scientific)

van den Bergh, S. 1995, *ApJ*, **453**, L55

van den Bergh, S., et al. 1996, *AJ*, **112**, 359

VandenBerg, D. A., Bolte, M. & Stetson, P. B. 1996, *ARA&A*, **34**, 461

van Ommen, T. D., Jones, D. L., Preston, R. A. & Jauncey, D. L. 1995 *ApJ*, **444**, 561

Vaughan, T. E., et al. 1995, *ApJ*, **439**, 558

Vilenkin, A. 1983, *Phys. Rev. D.*, **27**, 2848

Vilenkin, A. & Shellard, E. P. S. 1994, *Cosmic Strings and Other Topological Defects* (Cambridge: Cambridge University Press)

Villumsen, J. 1995, astro-ph/9507007

Visser, M. 1997, *Science*, **276**, 88

Walter, C. & Klypin, A. 1996, *ApJ*, **462**, 13

Wampler, E. J., et al. 1996, *A&A*, **316**, 33

Wampler, E. J. 1996, *Nature*, **383**, 308

Warren, M. S., Quinn, P. J., Salmon, J. K. & Zurek, W. H. 1992, *ApJ*, **399**, 405

Watson, A. 1997, *Science*, **275**, 1064

Weinberg, D. H., Miralda-Escude, J., Hernquist, L. & Katz, N. 1997, *ApJ*, **490**, 564

Weinberg, S. 1965, *Phys. Rev.*, **138**, 988

Weinberg, S. 1972, *Gravitation and Cosmology: Principles and Applications of the General Theory of Relativity* (New York: Wiley)

Weinberg, S. 1982, *PRL*, **48**, 1336

Weinberg, S. 1989, *Rev. Mod. Phys.*, **61**, 1

Weinberg, S. 1997, in *Critical Dialogues in Cosmology*, ed. N. Turok (Singapore: World Scientific), p. 195

White, D. A. & Fabian, A. C. 1995, *MNRAS*, **273**, 72

White, M., Viana, P. T. P., Liddle, A. R. & Scott, D. 1996, *MNRAS*, **283**, 107

White, S. D. M., Efstathiou, G. & Frenk, C. S. 1993, *MNRAS*, **262**, 1023

White, S. D. M., Navarro, J. F., Evrard, A. E. & Frenk, C. S. 1993, *Nature*, **366**, 429

Wilbanks, T. M., et al. 1994, *ApJ*, **427**, L75

Will, C. M. 1981, *Theory and Experiment in Gravitational Physics* (Cambridge: Cambridge University Press)

Will, C. M, 1986, *Was Einstein Right?* (New York: Basic Books)

Will, C. M. 1990, *Science*, **250**, 770

Willson, L. A., Bowen, G. H. & Struck-Marcell, C. 1987, *Comm. Astrophys.*, **12**, 17

Winget, D. E., et al. 1987, *ApJ*, **315**, L77

Wolfe, A. M. 1993, in *Relativistic Astrophysics and Particle Cosmology*, eds. C. W. Ackerlof, M. A. Srednicki (New York: New York Academy of Science), p. 281

Wolfe, A. M. 1996, in *Critical Dialogues in Cosmology*, ed. N. Turok (Singapore: World Scientific)

Wood, M. A. 1992, *ApJ*, **386**, 539

Wright, E. L., et al. 1992, *ApJ*, **396**, L13

Wu, X.-P., et al. 1995, *ApJ*, **448**, L65

Wu, X.-P. & Fang, L. -Z. 1996, *ApJ*, **467**, L45

Yamashita, K. 1994, in *New Horizon of X-ray Astronomy – First Results from ASCA*, eds. F. Makino & T. Ohashi (Tokyo: Universal Academy Press), p. 279

Yamamoto, K., Sasaki, M. & Tanaka, T. 1995, *ApJ*, **455**, 412

Yepes, G., Kates, R., Khokhlov, A. & Klypin, A. 1997, *MNRAS*, **284**, 235

Zanchin, V., Lima, J. A. S. & Brandenberger, R. 1996, *Phys. Rev. D.*, **54**, 7129

Zaritsky, D., Smith, R., Frenk, C. & White, S. D. M. 1993, *ApJ*, **405**, 464

Zaritsky, D., Smith, R., Frenk, C. & White, S. D. M. 1997, *ApJ*, **478**, 39

Zaritsky, D. & White, S. D. M. 1994, *ApJ*, **435**, 599

Zaroubi, S., Zehavi, I., Dekel, A., Hoffman, Y. & Kolatt, T. 1997, *ApJ*, **486**, 21

Zel'dovich, Ya. B. 1972, *MNRAS*, **160**, 1P

Zhao, H.S. 1998, *ApJ*, **294**, 139

Zhao, H. S, Rich, R. M. & Spergel, D. N. 1996, *MNRAS*, **282**, 175

Zurek, W., Quinn, P. J., Salmon, T. K. & Warren, M. S. 1994, *ApJ*, **431**, 559

2 Gravitational instability

Amos Yahil

Abstract

A brief summary is presented of the theory of nonrelativistic gravitational instability on galactic and larger scales in the expanding universe. The emphasis is on the growing mode, for which there is a dynamical relationship between position and velocity, a relationship which is exploited in comparisons of the large-scale density and velocity fields to extract the cosmological density parameter, Ω. Gravitational instability is traced from the linear regime of small perturbations, through the Zel'dovich approximation, to a useful implicit second-order relation between velocity and gravity, and on to fully nonlinear cases, which can be solved either by analytical techniques in special cases, or by using Hamilton's principle. This is followed by a discussion of computationally stable schemes for recovering the primordial perturbations from which present-epoch, nonlinear, structures grew.

2.1 Introduction

The linear theory of gravitational instability is well established. (For an excellent discussion and historical review see Peebles 1980.) An important result to emerge was that density perturbations grow at best in proportion to the expansion factor of the universe, $a(t)$, i.e., as power laws of time and not exponentially. Consequently, density perturbations could have grown by at most a factor ~ 1000 since recombination and should be detectable as small-scale anisotropies in the cosmic background radiation, as, indeed, they have been.

A second result of the linear instability theory has been the identification of two families of normal modes, the growing ones mentioned above and decaying ones. The decaying modes quickly become insignificant relative to the growing ones. Consequently, the peculiar velocities – the deviations of the velocity field from the homogeneous Hubble expansion – can be ascribed, at the present epoch, only to the growing modes. If these velocities, $\gtrsim 100 \text{ km s}^{-1}$, had been due to decaying modes, they would have been

Gravitational instability

of order c at recombination, causing small-scale anisotropies in the cosmic background radiation of order unity, which can clearly be ruled out.

The dominance of the growing modes results in a dynamical relationship between position and velocity, a relationship which is exploited in comparisons of the large-scale density and velocity fields to extract the cosmological density parameter, Ω.

The purpose of this chapter is to establish this relationship between the density and velocity perturbations. We begin with general comments on gravitational instability, followed by a discussion of linear perturbation theory, §2.2.1, and its extension to the Zel'dovich approximation, §2.2.2. We then consider nonlinear perturbations, beginning with a discussion of the mixed boundary conditions applicable to the growing modes, §2.3.1, followed by exact solutions for special cases, §2.3.2, a general second-order theory, §2.3.3, and the application of Hamilton's principle to more complicated cases, §2.3.4. We conclude with a discussion of "time machines": computationally stable methods to determine the initial fluctuations that have given rise to the configurations observed at the present epoch, §2.3.5.

Gravitational instability, like any other instability, is the growth of perturbations away from a known configuration, which need not be static, or even stationary. It is only required to be a zeroth-order solution of a given set of equations. The question is what happens if this solution is perturbed? Does the perturbation oscillate, ultimately to be damped by some inevitable dissipation, or does it begin to grow, and if so in what way? And if it does grow, what happens to it beyond the regime in which it might be considered to be "small".

Stated more precisely, if **y** are the variables in the problem and the equations which govern them can be written in operator form

$$\mathbf{L}(\mathbf{y}) = 0, \tag{2.1}$$

we seek to find the behavior of solutions around a known solution $\mathbf{y}_0$. By small perturbations we generally mean those which satisfy the linearized form of the equations

$$\mathbf{M}_0 \delta \mathbf{y} = 0, \tag{2.2}$$

where $\mathbf{M}_0$ is the Jacobian operator $\partial \mathbf{L}/\partial \mathbf{y}|_{\mathbf{y}_0}$. Clearly, nontrivial solutions exist if and only if $\mathbf{M}_0$ is singular. This requirement results in an eigenvalue problem, whose solutions, the normal modes, determine the stability of the zeroth-order configuration.

If the operator $\mathbf{M}_0$ does not depend explicitly on a spacetime variable, except that it may contain derivatives with respect to it, then the solution is exponential in that variable. Specifically, if the zeroth-order solution is time-independent, then $\delta \mathbf{y} \propto e^{i\omega t}$, and any differentiation with respect to time is equivalent to multiplication by $i\omega$. In this case, exponential instability occurs when ω has a negative imaginary part.

In the cosmological case, the zeroth-order configuration is the Friedmann-Robertson-Walker solution. Here we confine ourselves to nonrelativistic perturbations after recombination, when the universe is both matter-dominated and decoupled from the cosmic background radiation. For galactic-size and larger perturbations, pressure is also unimportant, and will be ignored.

It is convenient to use comoving coordinates

$$\mathbf{x} \equiv \mathbf{r}/a, \tag{2.3}$$

where $\mathbf{r}$ are the proper coordinates and $a(t)$ is the expansion factor. The peculiar velocity is then

$$\mathbf{v} = a\dot{\mathbf{x}}. \tag{2.4}$$

In these comoving coordinates the fully nonlinear equations for the perturbation are

$$\dot{\delta} + a^{-1}\nabla \cdot [(1 + \boxed{\delta})\mathbf{v}] = 0, \tag{2.5}$$

$$\dot{\mathbf{v}} + \boxed{a^{-1}(\mathbf{v} \cdot \nabla)\mathbf{v}} + H\mathbf{v} = -a^{-1}\nabla\phi, \tag{2.6}$$

$$a^{-2}\nabla^2\phi = 4\pi G\rho_b\delta, \tag{2.7}$$

where $H(t) = \dot{a}/a$ is the Hubble function, $\delta = \rho/\rho_b - 1$ is the relative density perturbation away from the background density, ρ_b, and ϕ is the perturbative gravitational field.

It is worth noting that there are only a couple of nonlinear terms in equations (2.5)–(2.7) which are emphasized by framing. The linear perturbation equations are obtained by neglecting these terms.

2.2 Linear theory and Zel'dovich approximation

2.2.1 Linear perturbation theory

Since the cosmological zeroth-order solution is spatially homogeneous, there is no spatial dependence in the perturbation equations, and the normal modes are waves, $\delta y \propto e^{i\mathbf{q}\cdot\mathbf{x}}$. There is, however, explicit time dependence, in spite of the stationary nature of the zeroth-order solution in comoving coordinates, because that reference frame is not inertial. The time dependence is therefore not exponential. Stated physically, since at any epoch the growth time is on the order of the expansion time, it changes within one e-folding time, leading to power-law, instead of exponential, behavior.

The linear form of equations (2.5)–(2.7), i.e., with the framed terms neglected, can be solved by eliminating $\mathbf{v}$ and ϕ to obtain

$$\ddot{\delta} + 2H(t)\dot{\delta} - 4\pi G\rho_b(t)\delta = 0. \tag{2.8}$$

This second-order, linear, equation has two solutions. One is decaying, $\delta \propto H(t)$, and is negligible by the present epoch. The important mode is the growing one, which is proportional to a universal growth function that can be written as (Heath 1977)

$$D(t) = H(t)\int_0^t \dot{a}(t')^{-2}\,dt'. \tag{2.9}$$

Gravitational instability

Turning next to the peculiar velocity, we note, by taking the curl of equation (2.6), that, in the linear regime, the growing mode is irrotational (potential flow). The Kelvin circulation theorem then guarantees that the flow remains irrotational as long as there is no orbit crossing, a condition satisfied well beyond the linear domain.

The divergence of the peculiar velocity in the linear approximation is obtained from the continuity equation

$$a^{-1}\nabla \cdot \mathbf{v} = -\dot{\delta} = -\frac{\dot{D}}{D}\delta , \qquad (2.10)$$

or, equivalently,

$$\dot{a}^{-1}\nabla \cdot \mathbf{v} = \nabla_{Hbor} \cdot \mathbf{v} = -f(\Omega, \lambda)\delta , \qquad (2.11)$$

where $\nabla_{Hr}\cdot$ is the divergence with respect to the Hubble expansion velocity (distance measured in km s^{-1}), and

$$f(\Omega, \lambda) = \frac{d\ln D}{d\ln a} \approx \Omega^{0.6} + \frac{\lambda}{70}\left(1+\frac{\Omega}{2}\right). \qquad (2.12)$$

The last equality is a phenomenological approximation (Lahav et al. 1991), from which it is seen that f depends mainly on the cosmological density parameter, Ω, and only very weakly on the dimensionless cosmological constant λ, which we ignore henceforth.

It is also seen from equation (2.12) that, in an open universe, density perturbations essentially cease to grow once Ω falls significantly below unity. All growth must therefore be accomplished at earlier epochs. (Alternative discussions of linear gravitational instability theory are provided in §§3.2 and 11.1.)

2.2.2 Zel'dovich approximation

The linear approximation breaks down when either of the nonlinear terms in equations (2.5)–(2.7), marked by the frames, becomes comparable to the other terms. As a perturbation grows, the continuity equation, equation (2.5), is the first to require a nonlinear term when $\delta \sim 1$. By contrast, the nonlinear term in the equation of motion, equation (2.6), becomes significant when the kinetic energy per unit mass in peculiar velocities is comparable to the perturbative potential, $v^2/2 \sim \phi$. This occurs when $\delta \approx 4$–5 (cf., spherical perturbations in §2.3.2).

This suggests that a stricter adherence to mass conservation would allow an approximation to be valid for $\delta \lesssim 4-5$, beyond the usual linear regime, even if the equation of motion is approximated. One way of guaranteeing mass conservation is to reject the Eulerian approach taken so far, in which the dynamical variables, such as density and velocity, are considered as functions of (comoving) position. Instead, adopt a Lagrangian view which explicitly follows the trajectories of mass elements, $\mathbf{x}(\mathbf{q}, t)$, as a function of time. The mass coordinate can be the initial comoving position, $\mathbf{q} \equiv \mathbf{x}(t=0)$. The continuity equation is then satisfied automatically, while the equation of motion takes the form

$$\frac{d(a\mathbf{v})}{dt} = -\nabla\phi. \tag{2.13}$$

In the linear regime, the time and mass dependencies of the displacement separate:

$$\mathbf{x} = \mathbf{q} + D(t)\mathbf{C}(\mathbf{q}). \tag{2.14}$$

This can be seen by evaluating the Eulerian density and velocity fields in the linear approximation. The peculiar velocity is given by

$$\mathbf{v} = a\dot{\mathbf{x}} = a\dot{D}\mathbf{C}, \tag{2.15}$$

while from mass conservation

$$1 + \delta = \det\left(\frac{\partial \mathbf{x}}{\partial \mathbf{q}}\right)^{-1} = \det\left(\mathbf{I} + D(t)\frac{\partial \mathbf{C}}{\partial \mathbf{q}}\right)^{-1} \approx 1 - D(t)\nabla_\mathbf{q}\cdot\mathbf{C}, \tag{2.16}$$

where the last approximation applies in the linear regime. The linear approximation, equation (2.10), then follows by eliminating $\mathbf{C}$ from equations (2.15)–(2.16).

Zel'dovich assumed that the displacement equation, equation (2.14) would be valid beyond the linear regime. The mathematical arguments for this approximation were given above. Physically, particles continue, through inertia, on the trajectories set up in the linear regime, even when convergent flow causes the density perturbation to exceed unity. N-body simulations show that the Zel'dovich prescription is, in fact, an excellent approximation, which remains valid until orbits begin to cross (Melott, Pellman & Shandarin 1994 and references therein; Melott, Buchert & Weiss 1994).

More recently, Nusser et al. (1991) demonstrated that, as long as the flow remains laminar, the density can be expressed equally well in terms of current spatial derivatives of the velocity field

$$1 + \delta = \det\left(\frac{\partial \mathbf{q}}{\partial \mathbf{x}}\right) = \det\left(\mathbf{I} - f(\Omega)^{-1}\frac{\partial \mathbf{v}}{\partial H\mathbf{r}}\right) \approx 1 - f(\Omega)^{-1}\nabla_{H\mathbf{r}}\cdot\mathbf{v}. \tag{2.17}$$

Equation (2.17) is very useful in determining the density structure corresponding to a given velocity field, but is difficult to invert, except in the linear limit.

2.3 Nonlinear methods and mixed boundary conditions

2.3.1 Mixed boundary conditions

The time evolution of a classical dynamical system is completely determined by the phase-space coordinates at a given epoch. For cosmological perturbations, some modes decay rapidly and are negligible by the present epoch, which means that the trajectories of all mass elements, as observed today, effectively satisfy an additional boundary condition in the early universe

$$\lim_{t\to 0} a^2\dot{\mathbf{x}} = 0. \tag{2.18}$$

As a result, not all phase-space coordinates can be specified independently at the present epoch, and there is a relationship between position and velocity.

It should be emphasized that, unlike initial-value problems, those with mixed boundary conditions do not, in general, have unique solutions. For example, M31 is 700 kpc away from the Galaxy, but this does not tell us if it is moving away from or towards us and whether it is on its first orbit, or has already gone around the Galaxy several times. The deduced mass of the Local Group depends, of course, strongly on how many orbits it has completed.

The linear solution is single-valued, equation (2.11); multiple solutions arise only in the nonlinear regime. We define the quasi-linear solutions as those which are "nearest" to the linear ones in the sense that they can be obtained from linear solutions by continuously varying parameters which define the perturbations. For example, if we seek to determine the velocity field associated with a density field $\delta(\mathbf{x})$, we could evolve a series of solutions for density fields $\epsilon\delta(\mathbf{x})$, with ϵ varying continuously from $\epsilon \ll 1$ in the linear regime to the desired nonlinear case $\epsilon = 1$. The result is a unique nonlinear solution which is termed quasi-linear.

Nonlinear solutions – even quasi-linear ones – are not simple to compute. Since the growing perturbations have zero measure in phase space, we can not find them by simple backward integration with an N-body code. The slightest numerical error introduces a decaying mode, which is amplified backward in time, violating the boundary condition, equation (2.18). The main need for approximation schemes to the growth of perturbations arises from the inability to use N-body codes to solve problems with mixed boundary conditions. (The numerous nonlinear approximations that have been devised for initial-value problems are rarely used in practice, since N-body codes are accurate and adequate tools.)

In the next few sections we consider a number of techniques to evaluate nonlinear problems with mixed boundary conditions: (1) analytical solutions for fully nonlinear perturbations in special cases, §2.3.2, (2) a general second-order solution for the quasi-linear regime, §2.3.3, and (3) solutions to nonlinear problems that need not be quasi-linear using Hamilton's principle, §2.3.4. We return to special methods for backward integration of perturbations in §2.3.5.

2.3.2 Exact analytical solutions for special cases

For planar symmetry the gravitational force is distance independent, as is readily deduced from Gauss's law. It follows that the Zel'dovich approximation is exact, as long as there is no shell crossing (Doroshkevich, Ryabenki & Shandarin 1973).

Non-crossing spherical shells behave like independent universes, each with its own Ω. In the linear regime, equation (2.11) gives

$$\mathbf{v} = -\frac{1}{3}f(\Omega)\langle\delta\rangle H_b \mathbf{r}, \qquad (2.19)$$

where $\langle\delta\rangle$ is the mean relative density perturbation enclosed by the shell, H_b is the background Hubble constant, and **r** is the proper radius of the shell. The peculiar velocity of a void ($\delta = -1$) in an Einstein–de Sitter universe, $\Omega_b = 1$, is therefore predicted to be

$$\mathbf{v} = \frac{1}{3} H_b \mathbf{r} , \qquad (2.20)$$

while for a bound perturbation at the point of maximum expansion – the point at which $\mathbf{v} = -H_b \mathbf{r}$ – the overdensity is

$$1 + \delta = 4 . \qquad (2.21)$$

The exact nonlinear solutions can be found by constructing for each shell the function

$$F(\Omega) = \frac{8\pi G}{3} \langle \rho \rangle t^2 = \Omega H^2 t^2 , \qquad (2.22)$$

where $\langle \rho \rangle$ is the mean density enclosed within the shell, and H and Ω are its "cosmological parameters". The common quantity to all the shells is the cosmological time, t. If it is eliminated by taking ratios of equation (2.22) for different shells, the results are relations between the dynamical variables. In particular, one can relate any shell to one evaluated far from the perturbation, where the mean density and the cosmological parameters are those of the underlying Friedmann–Robertson–Walker universe.

Returning to the above example of an Einstein–de Sitter universe, for which $H_b t = 2/3$, a void with $Ht=1$ has a peculiar velocity

$$\mathbf{v} = \frac{1}{2} H_b \mathbf{r} , \qquad (2.23)$$

while at maximum expansion $\Omega H^2 t^2 = \pi^2/4$, so the overdensity is

$$1 + \delta = \frac{9\pi^2}{16} \approx 5.55 . \qquad (2.24)$$

The exact nonlinear solutions are seen to differ from the linear approximation by factors ≈ 1.5 in both cases, which is characteristic. The purpose of the nonlinear approximations, therefore, is to improve on the linear limit when higher accuracy is required.

2.3.3 Second-order perturbations

To second order, the quasi-linear solution for the mixed-boundary-condition problem can be expressed in the form of an implicit equation which is easy to solve by iteration (Mancinelli & Yahil 1995). Here we outline the main points.

It is useful to change variables from t to D and to rescale δ, **v** and **g** to the linear solution, such that

$$\Delta \equiv \frac{\delta}{D} , \qquad (2.25)$$

Gravitational instability

$$V \equiv \frac{d\mathbf{x}}{dD} = \frac{\mathbf{v}}{a\dot{D}}, \qquad (2.26)$$

$$\mathbf{G} \equiv \frac{a}{\Omega \dot{a}^2 D}\mathbf{g}. \qquad (2.27)$$

The equation of motion, equation (2.13) can then be rewritten as

$$\frac{d\mathbf{V}}{dD} = \frac{\Omega}{f(\Omega)^2 D}\left(\mathbf{G} - \frac{3}{2}\mathbf{V}\right). \qquad (2.28)$$

In terms of these rescaled variables, the Poisson equation and continuity equation become, respectively,

$$\nabla \cdot \mathbf{G} = -\frac{3}{2}\Delta, \qquad (2.29)$$

and

$$D\frac{\partial \Delta}{\partial D} = -\Delta - \nabla \cdot (1 + D\Delta)\mathbf{V}. \qquad (2.30)$$

Equation (2.28), which is exact, is the starting point of the approximation. Noting that the growing mode is analytical in D at $D = 0$, we obtain the linear, Zel'dovich, approximation

$$\mathbf{V} - \frac{2}{3}\mathbf{G} = 0. \qquad (2.31)$$

Expansion of equation (2.28) to second order gives

$$\mathbf{V} = \frac{2}{3}\mathbf{G} - \frac{4f^2}{15\Omega}D\frac{d\mathbf{G}}{dD}. \qquad (2.32)$$

Substituting the Poisson and continuity equation then gives, after some algebra,

$$\nabla \times \mathbf{V} = \nabla \times \mathbf{G} = 0, \qquad (2.33)$$

and

$$\nabla \cdot \mathbf{V} - \frac{2}{3}\nabla \cdot \mathbf{G} = -\alpha D \left(G_{i,j} - G_{k,k}\delta_{ij}\right) V_{i,j}, \qquad (2.34)$$

where a subscript preceded by a comma indicates an Eulerian derivative with respect to that component, δ_{ij} is the Kronecker δ symbol, and the tensor summation convention is implied. The nonlinear coefficient is

$$\alpha = \frac{4}{6 + 15\Omega/f(\Omega)^2}. \qquad (2.35)$$

As a function of Ω, α ranges between 0.13 and 0.21 for $0.1 < \Omega < 2$, with a typical value of $4/21 \approx 0.19$ for $\Omega = 1$. The nonlinear correction is therefore also insensitive to Ω, a point often noted in previous work.

Equation (2.34) is easily used to solve for $\mathbf{V}$ in terms of $\mathbf{G}$, or vice-versa, by starting with the linear approximation, and then iteratively correcting the solution using the r.h.s. of the equation.

An alternative way to write equation (2.34) is

$$(1 + \alpha D \Delta)\nabla \cdot V + \alpha D \left(G_{i,j} - \frac{1}{3} G_{k,k} \delta_{ij} \right) V_{i,j} = -\Delta \,. \tag{2.36}$$

Without the second, shear, term on the l.h.s., equation (2.36) is identical to the phenomenological expression of Nusser et al. (1991), their eq. 38, and provides the first dynamical justification for it. Their nonlinear coefficient, estimated experimentally from N-body simulations to be 0.18, is close to the one derived here. The importance of the shear term, which Nusser et al. did not include, can be seen by considering laminar planar perturbations, for which linear theory is exact. The nonlinear term on the r.h.s. of equation (2.34) indeed vanishes in that case, but the scalar and shear terms in equation (2.36) separately are non-zero, and the omission of either one introduces an error.

2.3.4 Hamilton's principle

The second-order scheme outlined in §2.3.3 can, in principle, be extended to higher order. In practice, little is achieved, at a significant increase in algebraic effort. Mancinelli and Yahil (1995) explored third-order and fourth-order solutions for the spherical case, which can be compared with the analytical solution, and found little improvement over the second-order approximation. The perturbation seems to have a "radius of convergence" around maximum expansion, $\delta \sim 4$–5. Other attempts have been more phenomenological (cf., Dekel 1994), but they involve some fine-tuning, based on N-body simulations, and their general validity is less clear.

The only method to show greater promise for higher-density enhancements, or for nonlinear cases beyond the quasi-linear regime, has been an application of Hamilton's principle (§11.2.4; Peebles 1989, 1990, 1995; Giavalisco et al. 1993; Shaya, Peebles & Tully 1995). Hamilton's principle is ideally suited for problems with mixed boundary conditions, because the solutions are stationary variations of the action

$$\delta \int_{t_1}^{t_2} L \, dt = 0 \,, \tag{2.37}$$

with fixed boundaries at *both* ends

$$\delta \mathbf{x} = 0 \quad \text{or} \quad \mathbf{p} = 0 \,, \tag{2.38}$$

where **p** is the conjugate momentum of **x**.

For the cosmological problem at hand

$$L = \frac{1}{2} \sum m_i a^2 \dot{x}_i^2 - \sum m_i \phi_i(\mathbf{x}_i) \,, \tag{2.39}$$

and

$$\mathbf{p}_i = m_i a^2 \dot{\mathbf{x}}_i \,. \tag{2.40}$$

The boundary condition at $t=0$ is therefore satisfied automatically for the growing mode, equation (2.18). The boundary condition $\delta\mathbf{x}=0$ is also satisfied when considering given positions (a given density structure) at the present epoch.[1]

In principle, the solution should be found by considering all possible variations of the orbits with the above boundary conditions. In practice, the orbits are parameterized in some "sensible" manner and the stationary variations of the action are sought within that parameterization. Consider, by analogy, the determination of the ground energy level of a quantum-mechanical system using the variational method. The solution of the Schrödinger equation is a stationary variation of

$$L = \langle\psi|H|\psi\rangle - \lambda\langle\psi|\psi\rangle \,. \tag{2.41}$$

Even though the wave function, $|\psi\rangle$, is not known a priori, a very good estimate of the ground energy level is obtained by finding a minimum of L in equation (2.41) with a sensible parameterization of $|\psi\rangle$. But a poor parameterization can lead to a wrong answer. This is particularly problematic, for example, when seeking an excited energy level, which, while also a stationary variation, is not a minimum.

The second-order perturbation analysis suggests that a sensible parameterization of the cosmological trajectories is

$$\mathbf{x} = \mathbf{q} + \sum_n C_n(\mathbf{q})D^n \,, \tag{2.42}$$

or better yet (cf., Mancinelli & Yahil 1995)

$$\mathbf{x} = \mathbf{q} + \sum_n C_n(\mathbf{q})E^n \,, \tag{2.43}$$

where D and E are related by

$$\frac{dE}{dD} = \frac{\Omega E}{f(\Omega)^2 D} \,. \tag{2.44}$$

Giavalisco et al. (1993) demonstrated that the choice of expansion functions indeed makes a big difference to the accuracy of the determined trajectories, and presumably also to the ability to choose among the multiple solutions. A minimal requirement is for the first term in the expansion to reduce to the linear term.

An alternative explored recently is to parameterize the orbits directly by values at intermediate times (Peebles 1995), but this method also needs to find a way to discriminate between the multiple solutions.

---

[1] When seeking the density field given the velocity field, the roles of the coordinates and conjugate momenta are interchanged by a trivial canonical transformation, and the problem is solved in the same way. The more realistic case in which positions, or velocities, are given in redshift space is more complex, and Hamilton's principle can only be applied by iteration.

2.3.5 Time machines

Sooner or later, any investigation of cosmological structure today seeks to determine the initial perturbations from which they have grown. The goal may be statistical, to determine the initial probability distribution function of the primordial perturbations (§7.4.5; Nusser & Dekel 1993; Nusser, Dekel & Yahil 1995), or their power spectrum. Alternatively, one may seek to reproduce specific initial perturbations in order to compute mock models of the present universe for a variety of Monte Carlo tests (Kolatt et al. 1996). In any event, as noted earlier, simple backward integration with an N-body code is not possible, and other "time machines" are needed for such integrations.

Nusser & Dekel (1992) proposed to use the Zel'dovich approximation for the backward integration which, to use the language of §2.3.3, simply states that $\mathbf{V}$ of any mass element does not change with time. The problem with any Lagrangian scheme is that statistical noise, due to the finite number of mass elements used in the computation, is bound to overwhelm the small primordial density fluctuations.

Nusser and Dekel suggested that, instead, one recast the Zel'dovich approximation in Eulerian form

$$\frac{d\mathbf{V}}{dD} = \frac{\partial \mathbf{V}}{\partial D} + \mathbf{V} \cdot \nabla)\mathbf{V} = 0 , \qquad (2.45)$$

and integrate $\partial \mathbf{V}/\partial D$ back in time on the Eulerian grid. This poses no problem since $\mathbf{V}$ is finite and there are no singularities in the equation. The primordial density is then obtained from the linear continuity equation

$$\Delta = -\nabla \cdot \mathbf{V} . \qquad (2.46)$$

A second-order correction to equation (2.45) is obtained by using equation (2.28) and setting the r.h.s. to a time-independent approximation, e.g., that obtained from equation (2.34) evaluated at the present epoch. The scheme then proceeds as before, again with no singularities.

Time machines based on Hamilton's principle have so far been applied only to problems with a small number of bodies, in which the smooth initial density distribution is not an issue. Eulerian versions of Hamilton's principle are possible (Giavalisco et al. 1993) but they approximate mass conservation and consequently are not useful for $\delta \gtrsim 1$ (Susperregi & Binney 1994). A grid-based, mass-conserving, application of Hamilton's principle, which has the correct behavior as $D \to 0$ built in, remains to be formulated.

If only the initial probability distribution function is required, then it is not necessary to perform a grid-based integration, since only the initial density of each mass element is needed but not its location, and the density can be computed entirely from the velocity field at the present epoch. Nusser et al. (1991) provide the details for the Zel'dovich approximation; the generalization to second-order is quite straightforward.

References

Dekel, A. 1994, *ARA&A*, **32**, 371

Doroshkevich, A. G., Ryabenki, V. S. & Shandarin, S. F. 1973, *Astrofizica*, **9**, 257

Doroshkevich, A. G., Ryabenki, V. S. & Shandarin, S. F. 1975, *Astrophysics*, **9**, 144

Giavalisco, M., Mancinelli, B., Mancinelli, P. J. & Yahil, A. 1993, *ApJ*, **411**, 9

Heath, D. 1977, *MNRAS*, **179**, 351

Kolatt, T., Dekel, A., Ganon, G. & Willick, J. A. 1996, *ApJ*, **458**, 419

Lahav, O., Lilje, P. B., Primack, J. R. & Rees, M. J. 1991, *MNRAS*, **251**, 128

Mancinelli, P. J. & Yahil, A. 1995, *ApJ*, **452**, 75

Melott, A. L., Buchert, T. & Weiss, A. G. 1994, *A&A*, **294**, 345

Melott, A. L., Pellman, T. F. & Shandarin, S. F. 1994, *MNRAS*, **269**, 626

Nusser, A., Dekel, A., Bertschinger, E. & Blumenthal, G. R. 1991, *ApJ*, **379**, 6

Nusser, A., Dekel, A. & Yahil, A. 1995, *ApJ*, **449**, 439

Nusser, A. & Dekel, A. 1992, *ApJ*, **391**, 443

Nusser, A. & Dekel, A. 1993, *ApJ*, **405**, 437

Peebles, P. J. E. 1980, *The Large-Scale Structure of the Universe*, (Princeton: Princeton University Press)

Peebles, P. J. E. 1989, *ApJ*, **344**, L53

Peebles, P. J. E. 1990, *ApJ*, **362**, 1

Peebles, P. J. E. 1995, *ApJ*, **449**, 52

Shaya, E. J., Peebles, P. J. E. & Tully, R. B. 1995, *ApJ*, **454**, 15

Susperregi, M. & Binney, J. 1994, *MNRAS*, **271**, 719

3 Microwave background and structure formation

Joseph Silk

Abstract

An overview is presented of the formation of structure and gravitational instability in both linear and nonlinear regimes. The interaction between theory and observations is described, and ideas on how the luminous components of galaxies formed are presented.

Cosmic microwave background anisotropy studies have had an enormous impact on our understanding of the origin of the universe and of large-scale structure. Recent results are reviewed, and various aspects of the implications for cosmology are discussed.

3.1 Introduction

The theory of gravitational instability of small density fluctuations in the expanding universe has met with considerable success. Several important properties of large-scale structure and of galaxies can be explained, and the primordial fluctuations have been measured, at least on large scales. However, there are also some outstanding features that challenge the current theory, especially on the scales over which galaxies form.

The first segment of this chapter (§§3.2–3.4) presents an overview of the formation of structure and gravitational instability in both linear and nonlinear regimes (see also Chs. 1, 2 and 11). The interaction between theory and observations is described, and ideas on how the luminous components of galaxies formed are presented, with implications for the high-redshift universe.

The second part of the chapter (§§3.5–3.7) describes the ultimate direct probe of the structure of the early universe, the fluctuations in the cosmic microwave background radiation (CMB). These have had an enormous impact on our understanding both of the origin of the universe and of large-scale structure. Various aspects of the implications for cosmology of recent observational and theoretical results are reviewed, and the future prospects for advancement in this rapidly developing field are discussed.

3.2 Gravitational instabilities: linear theory

In a cold, static cloud, density fluctuations grow exponentially rapidly: $\delta\rho/\rho \propto \exp(t\sqrt{G\rho})$. However in the expanding universe the growth rate becomes a power law since the background density decreases with time (as t^{-2}). The Jeans length, $L_J \sim v_s(G\rho)^{-1/2}$ still demarcates the critical transition between stable and unstable modes.

Consider first the evolution of small density perturbations, superimposed on a Friedmann background. A qualitative discussion is followed by a formal Newtonian treatment (see also §§2.2 and 11.1).

3.2.1 Qualitative analysis

Denote mass fluctuations averaged over spheres containing mass M by $\delta M/M \propto M^{-(n+3)/6}$. While $n = 0$ yields a Poisson distribution, $n = 1$ is the requisite scale-invariant index. This follows from the definition of metric, or equivalently over small scales, gravitational potential fluctuations associated with density fluctuations δM:

$$|h_{ij}| \equiv |\delta\phi| = \left|\frac{G\delta M}{rc^2}\right| \qquad (3.1)$$

over a scale containing mass M. One has $\delta\rho/\rho = $ constant, if $\delta M/M \propto M^{-2/3}$, that is, if $n = 1$.

After setting $M = (4/3)\pi\rho r^3$, it is instructive to rewrite $|\delta\phi|$ as

$$|\delta\phi| \approx \frac{\delta M}{M}\left(\frac{r}{a}\right)^2 \left(\frac{a}{ct}\right)^2, \qquad (3.2)$$

where $\rho \sim 1/Gt^2$. Here r is the physical scale of a fluctuation, and $\lambda \equiv r/a(t)$ is the comoving scale.

One can now write, if $|\delta\phi|$ is constant,

$$\frac{\delta M}{M} \propto \lambda^{-2}t \propto M^{-2/3}t \qquad (3.3)$$

in the radiation-dominated regime ($a \propto t^{1/2}$), and

$$\frac{\delta M}{M} \propto \lambda^{-2}t^{2/3} \propto M^{-2/3}t^{2/3} \qquad (3.4)$$

in the ensuing matter-dominated epoch ($a \propto t^{2/3}$).

More generally,

$$|\delta\phi| \propto M_h^{(1-n)/6} \qquad (3.5)$$

where M_h is the mass contained within a horizon-scale density fluctuation. The scale-invariant spectrum is now seen to explicitly specify the same amplitude for density fluctuations at horizon crossing.

Hence constant metric fluctuations, $|h_{ij}| = $ constant, are equivalent to gravitational potential fluctuations, $\delta\phi$, evaluated at horizon crossing. Thus, for a comoving sphere of physical diameter $\lambda(1+z)^{-1}$ containing the invariant mass

$$M(\lambda) = 1.54 \times 10^{11} \Omega h^2 (\lambda/1 \text{ Mpc})^3 M_\odot ,\qquad(3.6)$$

one has

$$|\delta\phi| = \left|\frac{G\delta M}{\lambda c^2}\right|_h = \text{constant}.\qquad(3.7)$$

The resulting fluctuations δM over the sphere M are given by

$$\frac{\delta M}{M} \propto \left(\frac{ct}{\lambda}\right)^2 \propto M^{-2/3}\left(\frac{t}{1+z}\right)^2.\qquad(3.8)$$

In conclusion, a scale-invariant spectrum has mass fluctuations $\propto M^{-2/3}$ and, moreover, the fluctuations grow as either $t^{2/3}$ (matter-dominated) or t (radiation-dominated). These remarks apply provided pressure effects are neglected (a concern on sub-horizon scales). Their validity on super-horizon scales is evident once one realizes that arbitrary linear perturbations of a de Sitter model ($k = 0$) can be decomposed into a superposition of plane-wave curvature perturbations, which can in turn be regarded as a comparison of two Friedmann models with slightly different curvature.

3.2.2 Parametric analysis ($p = 0, k = 0$)

The $k = 0$ background with superimposed fluctuations may be examined by expanding the parametric solutions for the Friedmann model with arbitrary curvature:

$$a(t) = A(1 - \cos\theta) \approx \frac{A}{2}\theta^2\left(1 - \frac{\theta^2}{12}\right),\qquad(3.9)$$

$$t = B(\theta - \sin\theta) \approx B\frac{\theta^3}{6}.\qquad(3.10)$$

This immediately yields the growing mode

$$\frac{\delta\rho}{\rho} = -3\frac{\delta a}{a} \propto t^{2/3},\qquad(3.11)$$

in the matter-dominated regime. In general, there is also a decaying mode; one can write more generally

$$t - \tau = B(\theta - \sin\theta) \approx B\frac{\theta^3}{6},\qquad(3.12)$$

whence

$$\frac{\delta\rho}{\rho} \propto t^{2/3}\left(1 - \frac{2}{3}\frac{\tau}{t}\right).\qquad(3.13)$$

The decaying mode corresponds physically to non-simultaneous (on some initial hypersurface at $t \to 0$) initiation of fluctuations, where the growing mode describes curvature or gravitational potential fluctuations.

3.2.3 Newtonian analysis ($k = 0$)

Inclusion of pressure gradients is expected to stabilize gravitational instability, and is most simply described in the Newtonian approximation, valid over small scales ($\ell \ll ct$) and velocities ($v \ll c$). The following general equations are linearized about the Friedmann equation, including expansion in lowest order: the equation for mass conservation,

$$\frac{\partial \rho}{\partial t} + \nabla \cdot (\rho v) = 0 ; \tag{3.14}$$

the momentum equation,

$$\frac{dv}{dt} = \nabla \phi - \nabla p / \rho ; \tag{3.15}$$

and the generalized Poisson equation,

$$\nabla^2 \phi = 4\pi G (\rho + 3p/c^2) , \tag{3.16}$$

together with an appropriate equation of state, $p = p(\rho)$.

After linearizing these equations about the Friedmann background, one can make the following statements about the first-order perturbed quantities of interest. Decomposition of the perturbed velocity field into rotational $v_{\rm rot}$ and irrotational parts leads to a decaying mode,

$$\nabla \cdot \mathbf{v}_{\rm rot} = 0 , \tag{3.17}$$

and a compressible mode,

$$(\nabla \cdot \mathbf{v}_{\rm irrot})/a = -\dot{\delta} , \tag{3.18}$$

where a comoving coordinate $x = r/a$ has been introduced. Rotational velocities therefore decay as $d(a\omega)/dt = 0$ or $\omega \propto a^{-1}$, where $a^{-1}\nabla_x \times \mathbf{v}_{\rm rot} = \omega$. The irrotational component of velocity drives compressions and rarefactions of the density fluctuations, which are governed by

$$\frac{\partial^2}{\partial t^2}\delta + \frac{2\dot{a}}{a}\frac{\partial \delta}{\partial t} = 4\pi G \rho \delta + \frac{1}{a^2}\frac{dp}{d\rho}\nabla^2 \delta . \tag{3.19}$$

This equation explicitly reveals the damping effect of the expansion, the destabilizing role of self-gravity, and the stabilizing effect of pressure gradients. Definition of the physical wavelength $\lambda = 2\pi a/k$, with arbitrary perturbations described by

$$\delta(\mathbf{x}, t) = \sum_k \delta_k e^{i\mathbf{k} \cdot \mathbf{x}} , \tag{3.20}$$

leads to the generalized Jeans criterion for instability:

$$\lambda > \lambda_J \equiv \left(\frac{\pi v_s^2}{G\rho}\right)^{1/2}. \tag{3.21}$$

The "Jeans mass" is $M_J = (\pi/6)\lambda_J^3 \rho$. If pressure is negligible ($v_s^2 = dp/d\rho = 0$), we recover the zero-pressure solutions: $\delta \propto t^{2/3}, t^{-1}$. Mass scales above M_J are gravitationally unstable.

In the radiation era, $v_s = c/\sqrt{3}$, and $M_J \sim M_h \propto t$. Note that at phase transitions during the early universe, the sound speed can drop substantially for a brief period. This is of greatest significance during the quark–hadron transition. After matter–radiation equality at $z_{eq} = 4 \times 10^4 \Omega h^2$, the Jeans mass is approximately constant, since

$$v_s = \left[\frac{d(p_m + p_r)}{d(\rho_m + \rho_r)}\right]^{1/2} \approx \frac{c}{3^{1/2}}\left(1 + \frac{3}{4}\frac{\rho_m}{\rho_r}\right)^{-1/2} \propto (1+z), \tag{3.22}$$

where m and r refer to matter and radiation respectively. The maximum value of the Jeans mass is

$$M_J^{max} = 10^{16}(\Omega h^2)^{-2} M_\odot, \tag{3.23}$$

and drops abruptly after decoupling to $\sim 10^6 (\Omega h^2)^{-1} M_\odot$ (assuming for the temperature $T = T_r$).

3.2.4 Fluctuation modes

At late times, after the last scattering epoch, the growing mode of density fluctuations corresponds to a scalar metric perturbation. However, at much earlier times, other modes may also be present. The mode mix depends on the initial conditions at the epoch of fluctuation generation, associated with an early phase transition or with inflation. Vector perturbations are decaying modes and unimportant at late times. Tensor modes are gravitational waves, and do not involve any density compression. Hence they are unimportant for structure formation, but can contribute to microwave background fluctuations.

There are two types of scalar mode. The adiabatic mode corresponds to metric perturbations, and conserves entropy. The second mode involves no metric (or curvature) perturbations but consists of entropy fluctuations. In the adiabatic mode, one has

$$\frac{\delta \rho_r}{\rho_r} = \frac{4}{3}\frac{\delta \rho_m}{\rho_m}, \tag{3.24}$$

whereas in the isocurvature mode,

$$\delta \rho_r + \delta \rho_m = 0 \; ; \quad \delta s = \frac{\delta \rho_m}{\rho_m}\left(1 + \frac{4}{3}\frac{\rho_m}{\rho_r}\right). \tag{3.25}$$

Isocurvature perturbations are initially produced as entropy or baryon number fluctuations. Once the universe is matter-dominated, the isocurvature mode generates a pressure gradient at horizon crossing. This couples the radiation and matter to drive curvature fluctuations.

3.2.5 The role of dark matter

The sound speed controls the growth of baryon fluctuations. Prior to the epoch of last scattering, the sound speed of the coupled baryon–radiation fluid is high and baryon fluctuations behave like acoustic oscillations. They are described by

$$\frac{\delta\rho}{\rho} \propto \cos\left(\frac{kv_s t}{a}\right) ; \quad k = 2\pi a/\lambda , \qquad (3.26)$$

where a is the cosmological scale factor, λ is the comoving wavelength and v_s ($\approx c/\sqrt{3}$) is the sound speed. Only the cosine mode is involved, since as $k \to 0$ at the last scattering epoch, this gives the required initial condition component on super-horizon scales of primordial curvature, or energy density, fluctuations.

Cold dark-matter in the form of weakly interacting particles dominates the mass density. Nucleosynthesis of H, ^{2}H, and ^{7}Li constrains $\Omega_b \approx 0.02(\pm 0.01)h^{-2}$, whereas $\Omega_m \geq 0.2$. Adiabatic energy density perturbations involve ρ as well as ρ_b, and prior to decoupling, fluctuations in the cold dark-matter, unlike the baryon component, are Jeans unstable. Throughout matter domination, from matter–radiation equality at $z_{eq} \equiv 4 \times 10^4 \Omega_m h^2$, until the epoch of last scattering, at $z_{LS} \sim 1000$, the cold dark-matter fluctuations grow. Only after last scattering, when the matter and radiation decouple, does the baryon component of the fluctuations grow, by falling into the cold dark-matter potential wells that are growing by gravitational instability.

The cold dark matter prevents two disasters from occurring that vitiate a purely baryonic adiabatic model. Radiative diffusion smooths out baryon fluctuations, up to a scale of order $10^{15} M_\odot$. In the absence of cold dark-matter, this would result in a top-down model of structure formation, in contradiction with the observation that galaxy clusters are young, currently forming, objects. Moreover, such a model produces excessive microwave background temperature fluctuations. Cold dark-matter allows fluctuations to develop at all scales, so that galaxies can form first, and the growth between z_{eq} and z_{LS}, as well as logarithmic growth in the radiation era, reduces the amplitude of temperature fluctuations by an order of magnitude. The minimum fluctuation scale is determined by the ability of baryons to cool and condense in the cold dark-matter potential wells. This sets a limit of about $10^4 M_\odot$. A top-down formation sequence for structure development results if the primordial fluctuation spectrum is approximately scale-invariant. This is the actual prediction of inflationary models, as well as models that are seeded by topological defects such as cosmic strings.

It is instructive to compare three alternate representations of the density fluctuation spectrum. Define first the Fourier decomposition of the density fluctuation

$$\delta\rho(r)/\rho = \int \delta_k e^{i\mathbf{k}\cdot\mathbf{r}} d^3k . \tag{3.27}$$

The power spectrum $|\delta_k|^2$ specifies the variance in $\delta\rho/\rho$:

$$\langle(\delta\rho/\rho)^2\rangle = \int |\delta_k|^2 d^3k , \tag{3.28}$$

so that

$$\delta\rho/\rho \equiv \langle(\delta\rho/\rho)^2\rangle^{1/2} \approx \left(|\delta_k|^2 k^3\right)^{3/2} . \tag{3.29}$$

The gravitational potential fluctuations $\delta\phi$ may be defined by

$$\nabla^2 \delta\phi = 4\pi G \delta\rho , \tag{3.30}$$

so that

$$\delta\phi = \left(\int d^3k |\delta_k|^2 k^{-4}\right)^{1/2} . \tag{3.31}$$

The power spectrum itself may be expressed as a power law in k:

$$|\delta_k|^2 \propto k^n , \tag{3.32}$$

where $n = 1$ for a primordial scale-invariant spectrum. Note that $\delta\phi \propto k^{(n-1)/2}$, so that scale-invariance is equivalent to constant potential fluctuations, indeed, this properly defines scale-invariance. Also, since wavenumber $k = 2\pi/\lambda$, and is comoving, one can write

$$\delta\rho/\rho \propto k^{(n+3)/2} \propto M^{-(n+3)/6} \tag{3.33}$$

as a mass spectrum, where M represents the mass in a sphere of comoving diameter λ.

The primordial scale-invariant fluctuation spectrum develops a feature after the epoch of matter–radiation equality that corresponds to a peak in the power spectrum at the corresponding horizon scale, $\lambda_{eq} = 13(\Omega h^2)^{-1}$ Mpc. Subhorizon growth is suppressed for smaller-scale fluctuations in the radiation-dominated era: hence only larger-scale fluctuations retain the primordial shape. The suppression flattens the fluctuation spectrum on smaller scales by the fourth power of k in power. In terms of $\delta\rho/\rho$, this means that on large scales, a scale-invariant spectrum satisfies $n \approx 1$ and

$$\delta\rho/\rho \propto M^{-(n+3)/6} \propto M^{-2/3} , \tag{3.34}$$

whereas on scales below λ_{eq}, $n \approx -3$ and $\delta\rho/\rho \propto$ constant. In fact, $\delta\rho/\rho$ rises logarithmically towards smaller scales because $\delta\rho/\rho \propto \int k^{(n+3)/2} dk/k$. This suffices to result in a bottom-up sequence of structure formation. In fact, the effective value of n only approaches -3 on dwarf galaxy mass scales, so that larger scales are well separated in the clustering hierarchy as gravitational instability operates.

Cosmic microwave background fluctuations directly probe the linear regime of $\delta\rho$, at the epoch of last scattering. Attempts to reconstruct the primordial power spectrum have met with mixed success because of uncertain systematics due, in particular, to foregrounds. However there is unambiguous confirmation of a feature in the power spectrum that corresponds to λ_{eq} when measures of the local density fluctuations inferred from redshift surveys are compared with the density fluctuations that are reconstructed at last scattering, by inversion of the Boltzmann equation that couples matter and radiation (as discussed below). Important assumptions that enter into this comparison concern the cosmological model parameters, since one is comparing power at $z \sim 1000$ with power at $z \sim 0$, and the biasing factor, since the galaxy surveys generally probe only the luminous component of matter. Nevertheless, one can see that while the standard COBE-normalized CDM power spectrum results in excessive small-scale power, a modest tinkering of parameters, such as introducing a slight spectral tilt, yields satisfactory agreement.

3.3 Nonlinear evolution

3.3.1 The galaxy mass function

The linear theory of gravitational instability in the expanding universe provides an adequate description of fluctuation growth until self-gravity becomes significant. One can describe the nonlinear evolution by numerical simulations, but a simple analytic description captures the salient features.

Consider a simple spherical top-hat model for the nonlinear growth of a cold dark-matter fluctuation. It is described by the Friedmann equation for a bound spherical shell taken to be embedded in, for simplicity, a $k = 0$ background. The shell is described parametrically by

$$a_{sh} = a_{max}(1 - \cos\theta) ; \tag{3.35}$$

$$t/t_{max} = (\theta - \sin\theta)/\pi , \tag{3.36}$$

while the background satisfies

$$a = a_{max}(\theta - \sin\theta)^{2/3}\, 3^{2/3}\, 2^{-4/3} . \tag{3.37}$$

The density contrast within the shell relative to the background is

$$\rho_{sh}(t)/\rho(t) = (a/a_{sh})^3 , \tag{3.38}$$

which may be evaluated at the time t_{max} of maximum shell radius to be $9\pi^2/16$. A dissipationless shell of matter collapses by a factor of 2 in radius, when it rapidly reaches equilibrium. The collapse time from maximum radius is t_{max}. We can identify epoch $2t_{max}$ with the virialization epoch for the shell. At this epoch, the overdensity is $18\pi^2$. Linear theory would give an overdensity at this epoch that is given by

$$\delta(t) = -3\frac{\delta a}{a} = \frac{3}{20}\theta^2 = \frac{3}{20}\left(6\pi\frac{t}{t_{max}}\right)^{2/3} \tag{3.39}$$

evaluated at $2\,t_{max}$, or $\delta_c = (3/20)(12\pi)^{2/3} = 1.686$.

One now has the machinery to be able to calculate the collapsed mass fraction. Take the matter density to be described by a random Gaussian field. For cold dark matter, with power on arbitrarily small scales, one needs to smooth the density field in order to be able to compute the mean density and its variance. Smoothing is effected by introducing a filter function, the simplest form of which is a spherical top-hat filter of comoving radius R. Implementing the filter, we can compute the rms density fluctuations on scale R:

$$\sigma^2(R, t) = \left\langle \left(\frac{\delta\rho}{\rho}(x)\right)^2 \right\rangle \equiv D^2(t)\sigma_0^2(R), \qquad (3.40)$$

where, for $\Omega = 1$, the linear theory growth factor $D(t) \propto t^{2/3}$ and $\sigma_0(R) = (R_{nl}/R)^{(n+3)/2}$ for a power-law spectrum. The fraction of mass in spheres of radius R with overdensity $\delta > \delta_c$, the linear overdensity at virialization, is

$$F(R, t) = \int_{\delta_c}^{\infty} \frac{d\delta}{\sqrt{2\pi}\,\sigma} e^{-\delta^2/2\sigma^2}. \qquad (3.41)$$

The normalization of $\sigma_0(R)$ is such that if mass traces light, $R_{nl} = 8h^{-1}$ Mpc, the scale over which galaxy count fluctuations have unit variance. In terms of mass,

$$\sigma(M) = (M_{nl}/M)^{(n+3)/6}, \qquad (3.42)$$

where $M_{nl} = 4 \times 10^{13}(1+z)^{-6/(3+n)}\,M_\odot$.

To proceed further, one has to confront the following issue: counting spheres of radius aR is not necessarily equivalent to counting lumps of mass $M = 4\pi\bar\rho a^3 R^3/3$, because mass lumps can contain substructure in the form of smaller spheres that are no longer distinct entities. This uncertainty has to be addressed with numerical simulations. Assuming that one can identify the fraction of mass in virialized spheres with the fraction of mass in lumps of the equivalent scale, one can then infer the mass function of newly virialized lumps,

$$\frac{dN}{dM}(M, t) = -2\frac{\bar\rho}{M}\frac{\partial F}{\partial R}\frac{dR}{dM} \qquad (3.43)$$

$$= -\sqrt{\frac{2}{\pi}}\frac{\bar\rho}{M^2}\frac{\delta_c}{M}\frac{d\ln\sigma(M)}{d\ln M}e^{-M/2\sigma(M,t)}. \qquad (3.44)$$

A factor of 2 has been added to account for the fact that there is an equal amount of matter in underdense as in overdense regions, relative to the background, and the underdense matter is presumably accreted by the mass lumps. Remarkably, despite these assumptions, the expression for the mass function of newly formed objects is found to agree with numerical simulations of structure formation in the expanding universe.

One can also deduce the spherically-averaged properties of the galaxies and clusters, or more precisely, of their dark halos, that form by hierarchical clustering. From linear theory

$$\delta\rho/\rho \propto M^{-(n+3)/6}\,t^{2/3}, \qquad (3.45)$$

and "formation" occurs at $\delta\rho/\rho = 1.67$. Hence the formation time t_f scales as $t_f \propto M^{(n+3)/4}$. The redshift at which an object of present mass M has on average acquired half its mass is (Lacey & Cole 1993)

$$z_f = \left(2^{(n+3)/3} - 1\right)^{1/2} (M/M_{\rm nl})^{-(n+3)/6}. \tag{3.46}$$

Applying the condition of virial equilibrium, one then infers, since the mean density is approximately $180\bar\rho$, where $\bar\rho = (1/6\pi)Gt_f^2$, that the velocity dispersion is $V^2 \propto M^{(1-n)/6}$, the surface density is $\epsilon \propto M^{-(n+2)/3}$ and density is $\rho \propto V^{-3(n+3)/(1-n)}$. The dispersion in mean properties is large, of order unity, and the scaling relations are valid provided that the effective spectral index lies in the range $-3 < n < 1$. This is always satisfied for primordial spectra that are already scale-invariant. For example, the effective index on galaxy scales for a scale-invariant initial spectrum is approximately -2.

3.3.2 Comparison with observations

To the extent that luminosity tracks mass, luminous galaxies should be associated with massive dark halos, and dwarfs with smaller halos. The luminosity function of galaxies is well described by the Schechter function

$$\frac{dN}{dL} = \frac{\phi_*}{L_*}\left(\frac{L_*}{L}\right)^\alpha e^{-L/L_*}, \tag{3.47}$$

where $\phi_* = 0.01\,h^3\,{\rm Mpc}^{-3}$, $L_* = 10^{10}\,h^{-2}\,{\rm L}_\odot$, and $\alpha \approx 1.1$.

Comparison with the Press–Schechter mass function immediately raises two questions. The general form is similar. However the current epoch scale at which luminous mass structures are becoming nonlinear, as inferred from the rms fluctuations in galaxy counts, is $\sim 5h^{-1}$ Mpc, equivalent to a present epoch mass scale $M_{\rm nl} \sim 4 \times 10^{13}\Omega h^{-1}\,{\rm M}_\odot$. Moreover, the slope of the predicted mass function at the low-mass end is steeper by about one power in mass than the equivalent slope of the galaxy luminosity function for luminosities below L_*. Reconciliation of characteristic mass with luminosity and mass function slope with luminosity function slope requires additional physics that incorporates the effects of dissipative matter (baryons) and of star formation. Another distinctive feature of galaxies is rotation: disks are rotationally supported, but spheroids are not.

3.3.3 The characteristic luminosity L_*

Since L_* refers to the stellar mass, one can immediately ask what baryonic mass associated with $M_{\rm nl}$ can have cooled to have formed stars. This presumably is a necessary condition to form the luminous mass of a galaxy. Moreover, the baryonic matter dissipation must have occurred at an epoch corresponding to the redshift of galaxy formation. Observations of damped Lyman α absorption line systems towards quasars, considered to be disk precursors, and of high-redshift galaxies suggest that the bulk of galaxy formation occurred at z ~ 2–3. Consider a $\nu\sigma$ fluctuation, where σ is the rms density fluctuation so that a galaxy precursor satisfies $\delta\rho/\rho = \nu\sigma_0(1+z)^{-1}$ if $\Omega = 1$. Since $M_{nl} \propto [\nu/(1+z)]^{6/(n+3)}$, I infer that typical 2σ galaxy scale fluctuations undergoing collapse at $z \approx 2.5$, say, have mass $M = (10^{12} - 10^{13})\Omega h^{-1} M_\odot$, since $-n_{\text{eff}} = 1.5$–2. It is encouraging that this mass scale, that of dark halos, lies in the expected range. The associated baryonic mass is

$$M_b = (2 \times 10^{10} - 2 \times 10^{11})(\Omega_b h^2/0.02) h^{-3} f_b \, M_\odot , \qquad (3.48)$$

where f_b allows for a possible baryon enhancement on galactic scales over the primordial value that was scaled as $\Omega_b h^2 = 0.02$: this has an uncertainty of about a factor of 2.

The baryonic mass represents an upper limit on the luminous mass since there is no guarantee that all of the baryons have condensed into galactic stars. One may expect cooling to be a necessary precursor and, in particular, one could argue that cooling with a dynamical time-scale is required in order for star formation to occur efficiently within a pregalactic structure. However, the mass of cooled gas that condenses within a dark halo is found to increase without limit as the potential well depth increases. Cooling does not therefore account for L_*, there being an effective upper limit to the potential well depth of luminous galaxies that corresponds to a central velocity dispersion for an L_* elliptical of about 270 km s^{-1}.

To limit the mass in cooled gas to $M_* \sim 10^{11} h^{-1} M_\odot$ as expected for the stellar mass in L_* galaxies, one has to appeal to feedback from star formation and death. Supernovae provide an attractive means of feedback from luminous protogalaxies, because of the inference from intracluster gas abundance studies that significant ejection of iron and other heavy elements occurred early in the history of the early-type galaxies that dominate rich clusters. For a nominal supernova rate of one per 250 M_{250} solar masses per year that form stars, corresponding to a solar neighborhood initial mass function for which $M_{250} \approx 1$, one finds that the protogalactic gas can radiate away the injected supernova remnant kinetic energy provided that the protogalactic potential well satisfies

$$\sigma < 270 \left(\epsilon_{0.2} E_{51} M_{250}^{-1}\right)^{1/2} \text{ km s}^{-1} , \qquad (3.49)$$

where $E_{SN} \equiv 10^{51} E_{51}$ ergs is the initial injected energy of a supernova and $\epsilon \equiv 0.2\epsilon_{0.2}$ denotes the fraction of gas turned into stars per protogalactic dynamical time. This demonstrates that despite the uncertain efficiency of star formation and uncertainty in the early initial-mass function (IMF), supernova feedback more than suffices to constrain M_* within the observed bound, given that the protogalactic scale length (or central surface density) complies with the empirical (M, σ) scaling that is found in the Faber–Jackson or Tully–Fisher relations.

3.3.4 Surface brightness

Central surface density is presumably determined by rotational support for disks and by dynamical relaxation for spheroids. Central surface brightness peaks for L_* galaxies and declines both towards high and low luminosities. We lack an understanding of the central surface brightness of galaxies. The problem is primarily that of understanding disks, since spheroid formation can be satisfactorily simulated by mergers of disks.

Analytic collapse calculations appear to explain the scale of disks, via accounting for the origin of disk angular momentum. Tidal torques between neighboring protogalaxies generate an initial amount of angular momentum, expressible in terms of a dimensionless parameter $\lambda \equiv |E|^{1/2} J / GM^{1/2}$, where E is the potential energy and J is the angular momentum of a halo of mass M, that spans the range $0.01 \leq \lambda \leq 1$ but has a median initial value

$$\lambda \approx 0.05 \Omega^{0.1}, \tag{3.50}$$

at turn-around.

In rotationally supported disks with rotational velocity $V_{\rm rot}$ and halo velocity dispersion σ, one can express λ in the form

$$\lambda \approx 0.4 V_{\rm rot}/\sigma \approx 0.4. \tag{3.51}$$

Self-gravitating non-dissipative collapse fails to bridge the gap between initial and current values of λ in disks, since in this case $\lambda \propto R^{-1/2}$. Disks typically have $R_{\rm disk} \approx J/MV_{\rm rot} \approx 5\,{\rm kpc}$, requiring collapse from 500 kpc, an absurdly large critical disk radius. Dissipative baryonic collapse within a dark non-dissipative halo effectively transfers angular momentum via tidal torquing against the halo dark matter. Since specific angular momentum is conserved ($V_{\rm rot} R \approx$ constant) within an isothermal halo ($\sigma \approx$ constant), one now obtains $\lambda \propto R^{-1}$, which implies an initial radius of 50–100 kpc at maximum extent of the protogalaxy.

An initial gas extent of order 50 kpc is consistent with the interpretation of damped Lyman α absorption clouds, seen in absorption towards high-redshift quasars, as being protodisks. Gas collapse within a dark halo can also explain the shapes of galaxy rotation curves, observed to be approximately flat outside a disk scale length, provided that disk self-gravity plays a role in helping account for the flattening of the inner rotation curve. The dominant dark-matter distribution produces a nearly flat rotation curve at large radii. Such a conspiracy between baryonic and dark-matter components is a natural outcome of simulations of disk galaxy formation.

However the simulations have revealed a serious problem. In hierarchical clustering, the initial halos are clumpy. The substructure results in efficient dynamical friction of infalling lumps, and the resulting baryonic disk is found to be far too small. The analytic prediction $\sigma_{\rm disk} \approx \lambda \sigma_{\rm halo}$ for uniform spherical collapse overpredicts the disk size by approximately a factor of 5. For spheroids, such efficient collapse is exactly what is required to account for the observed centrally concentrated light profiles, provided that the final structure is not rotationally supported. This is more or less a natural outcome of disk mergers. Feedback from star formation may heat the gas sufficiently during disk formation to avert this catastrophe, but there is as yet no detailed modelling of collapse with energy feedback.

3.3.5 Successes and failures of the hierarchical model

Bottom-up structure formation has been extensively simulated, usually in the context of a cold dark matter-dominated universe at critical density. The theory is well-formulated for dark matter, and modelled via N-body simulations, and has been extended to include the baryonic component, with inclusion of smoothed particle hydrodynamics. Gas cooling has been included on galaxy formation scales, but the theory lacks any fundamental prescription for star formation, and is consequently even more seriously deficient in the ability to include effects of feedback from star formation.

The successes of hierarchical structure formation are numerous. One can account for galaxy clustering: simulations of large-scale structure are indistinguishable from actual surveys. On the largest scales where effective comparison is made, $(10-100)h^{-1}$ Mpc, and structure is in the linear regime, one can measure the shape of the power spectrum of luminous matter. This is expressible as $P(k) \propto k^{-1.4}$, and corresponds to a CDM model with $\Omega h \approx 0.2$–0.3, as naturally occurs either in a flat, vacuum-dominated or open cosmological model if the primordial index is scale-invariant ($\lambda = 1$) at least over these scales. Bulk flows can, in principle, measure the dark-matter power spectrum over similar scales, although the results do not discriminate between rival cosmological models. The amplitude of the bulk flows is determined by the parameter $\sigma_8 \Omega^{0.6}$, where σ_8 is the value of $\sigma_0(M)$ at $8h^{-1}$ Mpc, the scale at which the galaxy number counts in spheres have unit variance. The observed value of this parameter (§7.4; Kolatt & Dekel 1997) (~ 0.8, but with considerable uncertainty) provides the strongest evidence that supports a high value of Ω.

The abundance of galaxy clusters effectively probes the power spectrum shape at a scale of $\sim 10 h^{-1}$ Mpc, the comoving scale from which rich clusters condensed. A best-fit value $\sigma_8 \approx 0.6(\pm 0.1)$ gives a reasonably robust measure of the degree of biasing, or the ratio of dark to luminous matter, on the largest nonlinear scales that have been usefully probed to date. The present epoch abundance of luminous galaxies is a further natural outcome of a model with primordial index $\Omega \approx 1$, although at high redshift ($z = 3 - 4$), recent observations are beginning to discriminate between cosmological models. A significant abundance of luminous star-forming galaxies has been found in this redshift range, as have damped Lyman α clouds that appear to have rotational velocities of 200 km s^{-1} or more. Models with diminished power on sub-galactic scales, due to a component of hot or warm dark matter, are most strongly constrained by such observations. In general, however, galaxy formation at redshifts $z \approx 1$–5 is a natural outcome of hierarchical models.

Mergers are also a natural outcome of hierarchical models. Major mergers, between nearly equal mass systems are rare today, but more common at high redshift. Major mergers lead to formation of spherical systems that are not rotationally supported, assuming high star formation efficiency as is needed to account for the old stellar populations of spheroids. Minor mergers incorporate substructures in which the gas fraction is easily disrupted by stellar feedback, and so are expected to form stars inefficiently. Such systems are the logical precursors of disk galaxies, for which the predicted rotation curves, appropriate to self-gravitating massive gas disks embedded within dark halos, constitute another success of the bottom-up theory.

Evolutionary and morphological studies of high-redshift galaxies fit well into a hierarchical formation scheme. Deep redshift surveys reveal that blue, star-forming galaxies have evolved by about 0.5 magnitude to $z \sim 1$. HST observations reveal many of these to be disks. However, red galaxies, identified as Es and S0s, show little evidence of any evolution in luminosity. Field studies show a significantly increasing population of blue, irregular galaxies towards fainter magnitudes, these galaxies dominating the very faint galaxy counts. All of this is consistent with early formation via mergers and strong tidal interactions, and may be considered to be a qualitative success of the hierarchical galaxy formation theory.

The quantitative failures of the hierarchical formation theory are a consequence of its failure to provide a unique recipe for star formation and associated feedback. The excess of faint galaxies at low luminosities, the upper limit on galaxy luminosity characterized by L_*, and the upper limit on disk galaxy surface brightness specified by the Freeman law for spiral galaxies, are all issues whose explanation seemingly demands incorporation of feedback effects.

Bottom-up formation predicts that massive galaxies form more recently. This produces the following dilemma: luminous ellipticals have deeper potential wells, and hence more massive halos, on the average, than do luminous spirals. Yet ellipticals have predominantly old stellar populations. Resolution of this difficulty has been achieved in a somewhat ad hoc manner, by assuming that the major mergers, elliptical precursors, form stars efficiently over a dynamical time-scale, leaving little gas behind for late star formation, whereas the minor mergers and slow infall accumulation of gas into disks, in low density environments, result in continued gas-rich star formation at low efficiency for a Hubble time. Such a scheme may be said to "work," in the sense that ellipticals are red and spirals are blue, but clearly leaves something to be desired in the sense of extracting more from a model than one inputs into the model.

3.4 Galaxy formation by reverse engineering

3.4.1 Spiral galaxies

Numerical simulations of structure formation adopt what may be called the "forwards" approach to galaxy formation. One assumes initial conditions that have a plausible cosmological ancestry, adopts an ad hoc prescription for star formation, and runs a numerical simulation. Now star formation in our local patch of the universe, let alone at remote locations and epochs, depends on many parameters, including the gas density, molecular abundances, dust opacity, ionization, magnetic field strength, turbulence, protostellar outflows, and feedback from dying stars. One can more readily aspire to make long-term predictions of the weather from first principles than develop a predictive theory of star formation. As with meteorology, only a highly phenomenological approach that incorporates as much local data as is available is likely to have even a remote chance of success. For the study of galaxy formation, incorporation of star formation knowledge acquired locally results in a "backwards" approach.

One may readily outline the ingredients of the backwards approach to galaxy formation. Commence with a semi-phenomenological theory for the instability of cold, self-gravitating disks, and apply that to the Milky Way galaxy. The phenomenological ingredients with which one begins are that the star formation rate has been approximately constant over the past 10 Gyr, to within a factor of 2, and that the star formation rate surface density (per unit disk area) is proportional to the total gas surface density above a threshold value. The threshold is determined by either of the following arguments. Within the disk co-rotation radius, clouds passing through the spiral density wave acquire a non-circular component of velocity proportional to $\Omega(R) - \Omega_p$, where Ω_p is the density wave pattern angular velocity and $\Omega(R)$ is the disk rotation rate at radius R.

Alternatively, one can argue that the disk is unstable to non-local gravitational instabilities as well as being locally Jeans unstable if the Toomre parameter $Q \le 1$, where Q is defined to be given by

$$Q \equiv f \kappa \sigma_{gas} / \pi G \mu_{gas} \equiv \mu_{cr} / \mu_{gas} \tag{3.52}$$

for a gas disk, and σ_{gas} is the gas velocity dispersion, μ_{gas} is the gas surface density (κ is the epicycle frequency ($\approx \sqrt{2}\Omega(R)$) for a flat rotation curve), and f is a correction factor of order unity, that allows for the contribution of the stellar component to the self-gravity of the gas. One can show that the linear instability growth rate is approximately equal to $\kappa(1 - Q^2)^{1/2}/Q$ if $Q < 1$, and equal to zero if $Q > 1$.

One can now write the star formation rate in the following physically motivated forms (Wyse & Silk 1989; Wang & Silk 1993):

$$\text{SFR} = \epsilon \Sigma_{gas} \left[\Omega(R) - \Omega_p\right] \tag{3.53}$$

or

$$\text{SFR} = \epsilon \Sigma_{gas} \kappa (1 - Q^2)^{1/2}/Q . \tag{3.54}$$

In either case, one has, for a flat rotation curve, in the star-forming region of the disk that

$$\text{SFR} \propto \Sigma_{gas}/R . \tag{3.55}$$

Hence the models predict that disks form inside out. This is a generic feature of disk models (Prantzos & Aubert 1995). Figure 3.1 shows the predicted surface brightness

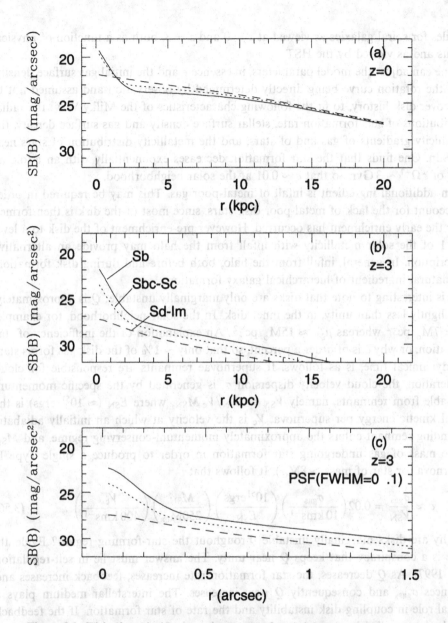

Figure 3.1. Disks form inside-out: apparent surface brightness profiles as a function of physical radius for Sb (solid line), Sbc-Sc (dotted line) and Sd-Im (dashed line) model galaxies with infall at (a) $z = 0$ and (b) $z = 3$, for $\Omega = 1$, $H_0 = 50$ km s^{-1}Mpc^{-1}, and a formation redshift $z_f = 10$. (c) same as (b) but as a function of angular radius and after convolving the model profile with a PSF of FWHM= 0.1 arcsec. From (Cayon, Silk & Charlot 1996).

profiles for spiral galaxies as viewed at $z = 0$ and $z = 3$, both as a function of physical radius and as viewed by the HST.

One can adjust the model parameters, in essence ϵ and the initial gas surface density, with the rotation curve being directly determined by observation and assumed not to vary over disk history, to fit the following characteristics of the Milky Way: the radial distributions of star formation rate, stellar surface density and gas surface density, the metallicity gradients of gas and of stars, and the metallicity distribution of stars near the Sun. One finds that the star formation decreases exponentially with an e-folding time of $\epsilon \Omega^{-1} \sim 3 \, \text{Gyr}$, so that $\epsilon \sim 0.01$ at the solar neighborhood.

An additional ingredient is infall of metal-poor gas. This may be required in order to account for the lack of metal-poor disk stars, since most of the disk is then formed after the early enrichment has occurred. However pre-enrichment of the disk to a level of 0.1 of the solar metallicity with infall from the halo may provide an alternative prescription. In general, infall from the halo, both before and during disk formation, is a natural ingredient of hierarchical galaxy formation.

It is interesting to note that disks are only marginally unstable: Q is approximately, but slightly less than unity, in the inner disk. In the solar neighborhood, for example, $\mu_{cr} \approx 7 M_\odot \, \text{pc}^{-2}$ whereas $\mu_{gas} \approx 15 M_\odot \, \text{pc}^{-2}$. An explanation of the inefficiency of star formation, or why ϵ is of order a percent, so that only $\sim 1\%$ of the disk gas forms stars per dynamical time, is as follows. If supernovae remnants are responsible for cloud acceleration, the cloud velocity dispersion σ, is generated by the specific momentum available from remnants, namely $V_{SN} \equiv E_{SN}/V_c M_{SN}$, where E_{SN} ($\approx 10^{51}$ ergs) is the initial kinetic energy per supernova, V_c is the velocity at which an initially adiabatic expanding remnant enters the approximately momentum-conserving regime, and M_{SN} is the mass of gas undergoing star formation in order to produce a single Type II supernova (or star of mass $> 8 M_\odot$). It follows that

$$\epsilon \approx \frac{\sigma_{gas}}{V_{SN}} = 0.02 \left(\frac{\sigma_{gas}}{10 \, \text{kms}^{-1}}\right) \left(\frac{10^{51} \text{ergs}}{E_{SN}}\right) \left(\frac{M_{SN}}{250 M_\odot}\right) \left(\frac{V_c}{400 \, \text{kms}^{-1}}\right). \quad (3.56)$$

Why are disks marginally unstable throughout the star-forming region? Evidently there is a conspiracy that keeps Q near unity. The answer must lie in self-regulation (Silk 1997). As Q decreases, the star formation rate increases, feedback increases and enhances σ_{gas}, and consequently Q now increases. The interstellar medium plays a crucial role in coupling disk instability and the rate of star formation. If the feedback is indeed controlled by supernovae remnants, long believed to be the primary source of interstellar cloud turbulence and heating, then a relevant parameter is the porosity P of the interstellar medium to the hot, $\sim 10^6$ K, gas associated with the interiors of supernova remnants. The porosity is defined to be the product of supernova rate and supernova 4-volume (age multiplied by volume at maximum expansion). The fraction of volume of the interstellar medium filled by the hot phase (gas at $\sim 10^6$ K) is $1 - e^{-P}$. Effective feedback requires $P \sim 1$, and $P \gg 1$ results in a supernova-driven galactic wind. In fact, $P \sim 0.1$–1 is observed for the local interstellar medium. Even if $P \sim 1$, one expects occasional hot bubbles to break out of the cold interstellar gas in the disk, since V_{SN} exceeds the disk escape velocity. This phenomenon manifests itself in the form of the chimneys that are observed in the local interstellar medium.

The self-regulation is expected to operate as follows. As Q decreases, stars form and die at an enhanced rate, leading to more supernovae, and P increases. At fixed P, the cold gas fraction is inversely proportional to Q ($\equiv \mu_{cr}/\mu_{gas}$) but at fixed Q, the cold gas fraction decreases as P increases. Hence one can show that the cold gas fraction remains constant while the cold, unstable gas layer is compressed, occupying a volume proportional to e^{-P}, as P increases and Q decreases, as long as $P \lesssim 1$. The star formation rate, which depends primarily on the surface density of the cold gas, should be unchanged. Perhaps the competition between P and Q accounts for the universality and low dispersion in the Tully–Fisher relation.

3.4.2 Elliptical galaxies

A semi-phenomenological theory fares well for forming disk galaxies. One has the basic ingredients of the gravitational instability of disks to non-axisymmetric perturbations well under control, and it is possible to embed this framework into a scheme that incorporates the rich data available in the Milky Way and other nearby disk galaxies.

Elliptical and, more generally, spheroid formation presents rather more of a challenge. There is no theory of star formation in dynamically hot systems. A purely phenomenological aproach is essential in the absence of any robust rules. Formation of spheroids is approached as follows, using the philosophy of "backwards" evolution. One can apply population synthesis techniques, adopting a universal form for the initial stellar mass function, to model the spectral energy distribution of a spheroid. One finds that most of the stars formed within the first 1–2 Gyr of the birth of a stellar population that is now at least as old as the oldest globular star clusters. Such a starburst is a purely empirical model for the current epoch spectrum of a typical elliptical galaxy. There is evidence for some intermediate age stars in nearby ellipticals, and this can be modelled in terms of a starburst as recently as 4–5 Gyr ago.

Starbursts are naturally explained in a hierarchical model of galaxy formation. In a galaxy merger, any gas is rapidly concentrated via inelastic cloud encounters into the center of the resulting potential well. The massive concentration of gas should provide a fertile environment for a starburst. In a sufficiently deep potential well, with escape velocity $\lesssim 100$ km s^{-1}, the debris from supernova explosions should be trapped within the cloud, so that star formation should be capable of efficiently consuming the available gas. The contrast between protodisk and protoelliptical star formation is likely to be (Silk & Wyse 1998) that the star-forming units in a disk galaxy have masses comparable to giant molecular clouds ($\sim 10^5 - 10^6$ M$_\odot$) with escape velocities of a few km s^{-1}, whereas stars form efficiently in the massive substructures that characterize protoellipticals. In the smaller, shallow potential well, substructures, star formation is inevitably a highly inefficient process.

The dynamics of major mergers are consistent with the hypothesis of spheroidal formation. Relaxation is rapid, with ongoing mergers already having developed, when azimuthally averaged, a de Vaucouleurs light profile. The high central surface density of a spheroid is attained via baryonic dissipation and settling, and the associated star formation and enrichment is capable of generating the observed gradients in metallicity. In the nearby universe, major starbursts which are extremely rare at the present epoch, are shrouded by dust, and most of the star formation luminosity is emitted at far-infrared wavelengths. This is consistent with the merger hypothesis, which predicts a strongly decreasing merger rate with cosmic epoch.

3.4.3 Galaxies at high redshift

Armed with a theoretical description of nearby galaxies, one can attempt to project the models back in time. For disks, the model predictions are dramatic. Only modest evolution, by a magnitude or two, is found to redshift unity, as is seen in the deep redshift surveys. However disk angular sizes are greatly reduced, because of the inside-out evolution. One implication is that galaxy angular size cannot be used as a cosmological probe of geometry because of the dominant role of evolution.

Protodisks constitute a more speculative area for model predictions. Damped Lyman α clouds, seen in absorption towards quasars especially at high redshift, have been long conjectured to be protodisks on the basis of column density, HI dominance, and spatial extent. The number density evolves strongly with redshift, and translates directly into the HI gas fraction (Ω_{HI}) in such systems, which is found to peak at a redshift of about 3. The observed ratio of Ω_{HI} at this redshift corresponds to the total mass in stellar disks at present, and has been inferred to therefore be a representative measure of protodisks in the early universe. The substantial decrease of Ω_{HI} between $z = 3$ and the present epoch can only be reconciled with the chemical evolution of disk galaxies, inferred to be relatively modest over the past 10 Gyr of disk evolution, if (Fall & Pei 1993) considerable amounts of dust are present that result in undercounting of quasars and therefore underestimates of Ω_{HI} at redshifts between 1 and 3.

The kinematics of the damped Lyman α clouds, as probed by high-resolution optical spectrometry, substantiate the hypothesis that these clouds are massive protospiral galaxies, rotational velocities of 100–300 km s^{-1} being measured (Wolfe 1996). However their identification as protodisks remains elusive, for several reasons. The kinematic rotation signature of line distribution asymmetry can only be produced by rotating systems with a substantial scale length, of at least several kpc. Moreover the abundances are low, although with a large dispersion especially at low redshift (Lu et al. 1998). At high redshift, the abundances are generally around a percent of the solar value, about an order of magnitude below the oldest disk star abundances. The abundance ratios are found to reveal the pattern of nucleosynthetic yields associated with Type II supernovae. The abundances of damped Lyman α clouds resemble those of the galactic halo rather than the old disk stars. A halo origin may be more appropriate for these clouds, although it is also possible that in the outer disk, where stellar abundance ratio determinations are unavailable, the scale height is large, and the inner disk metallicity gradient could continue, may serve as an evolutionary endpoint for the damped Lyman α systems.

The lack of luminosity and spectral evolution for ellipticals found in deep galaxy redshift surveys favors an origin for ellipticals at high redshift. Theory predicts that the elliptical formation phase should be luminous. One has to assemble $\sim 10^{11} M_\odot$ of stars on a dynamical time-scale (~ 1 Gyr) in order to reproduce the morphological and kinematic characteristics of an elliptical, and this assembly must involve a substantial amount of star formation as well as of dynamical merging. Population synthesis confirms the time-scale for the star formation, but cannot distinguish between a series of small bursts and a single luminous starburst. The well-known difficulty has been that optical and near-infrared surveys for luminous protogalaxies at high redshift have been unsuccessful.

There are two possibilities for elliptical formation. Assembly by many mergers, with an associated sequence of many ministarbursts, or a coherent collapse in which the luminous starburst is shrouded by dust and hence only visible in the far infrared. The former hypothesis is consistent with some interpretations of the faint blue galaxy counts, and the latter with detections of ultra-luminous starbursts detected by the IRAS survey that may represent relatively low-redshift examples of a large population of such systems. Either hypothesis is consistent with the discovery of vast amounts of intracluster iron and other heavy elements. The intracluster iron mass has an abundance of about 1/3 of the solar value, and requires a nucleosynthetic yield that is about 5 times that in the solar neighborhood. The abundance ratios are consistent with those of Type II supernova ejecta. Explanations of the enrichment of the intracluster gas appeal to galactic winds from a population of dwarfs, now mostly disrupted, or to early protogalactic winds from luminous E and S0 galaxies. Some support for the dwarf hypothesis may come from evidence for a large population of low-surface-brightness dwarfs in clusters. However, these dwarfs are metal-poor, and a more logical origin for the enrichment is from metal-rich galaxies. Several observations lend credence to this interpretation: the intracluster iron abundance as measured in groups and clusters is proportional to the luminosity in early-types, as opposed to all galaxies, one sees a similar abundance pattern for Mg/Fe ratios in the central regions of ellipticals, and the stellar Mg abundance is found to be proportional to the local escape velocity, suggestive of regulation by an early wind (Zepf & Silk 1996).

A resolution of the puzzle of elliptical formation may come with submillimeter measurements. There already are indications via the tentative discovery of a diffuse background at submillimeter wavelengths (Puget et al. 1996) that the integrated emission from dust-shrouded starbursts at high redshifts may have been detected, and protoellipticals provide an attractive interpretation of the diffuse flux. Confirmation will come with mapping of blank fields at high latitudes by ISO (out to 200 μ) and by submillimeter arrays (at 400, 800 μ). If elliptical starbursts are responsible for the diffuse background, deep imaging should reveal their presence. Alternatively, if ellipticals are assembled by mergers of many smaller star-forming units, studies of deep fields in the optical and near–infrared with HST should reveal the conclusive signature that is inevitable in a merging model.

3.5 Cosmic microwave background anisotropies

3.5.1 Introduction

Cosmic microwave background anisotropy studies have had an enormous impact on our understanding of the origin of large-scale structure in the universe. This has come about for two reasons. The spectrum of the CMB shows no discernible deviation from a black-body spectrum (Fixsen et al. 1998). The black-body temperature is 2.728($\pm$0.004, 95%C.L.). Fluctuations have been measured at a level of about $\delta T/T \sim 10^{-5}$, consistent with the amplitudes of the primordial fluctuations inferred to seed the growth of large-scale structure.

Consider first the spectral distortion limits. The Comptonisation y-parameter describes the spectral distortion due to Compton scattering of CMB photons by hot intergalactic gas. y is defined to be the average fractional photon energy change per scattering $\times$ mean number of scatterings back to epoch t, or

$$y = \int_t^{t_0} \frac{k}{m_e c^2} (T_e - T) n_e \sigma_T c \, dt , \qquad (3.57)$$

where t_0 denotes the present epoch, T_e is the electron temperature, n_e is the intergalactic plasma density and σ_T is the Thomson cross-section. The current observational limit on a possible y distortion of the cosmic black-body radiation is $y < 1.5 \times 10^{-5}$. The y limit constrains the temperature of the uniform intergalactic medium to satisfy

$$T < 2 \times 10^6 \, (1+z)^{-3/2}(\Omega_b h^2)^{-1/2} . \qquad (3.58)$$

This eliminates the possibility of an intergalactic medium that is hot enough ($kT \approx 40$ keV) to account for the 2–100 keV diffuse X-ray background radiation.

The chemical potential (μ) distortion describes the relic spectral distortion at early epochs, well before last scattering. At $T \gtrsim 10^4$ K, the rate of photon diffusion towards low frequencies of photons near the black-body peak via bremsstrahlung exceeds that of Comptonisation heating of the photons. Blackbody radiation has $\mu = 0$; $\mu < 0$ corresponds to a Bose distribution in which the number of photons is conserved, and μ/kT measures, for a near-black-body distortion, the fractional energy distortion. The current observational limit on μ is $|\mu| < 9 \times 10^{-5}$.

The discovery (Smoot et al. 1992) of the large-scale angular anisotropies in the CMB by the COBE DMR experiment, announced in 1992, has enabled us to normalize model predictions of primordial density fluctuations to observational data. Cosmology has come of age. Simulations are no longer random shots in the dark, with a token appeal to the ubiquitous biasing parameter, namely the ratio of luminous matter to dark-matter density fluctuations, but have developed precise predictions, at least in the inflationary context.

The intermediate angular scale observations over the last four years have filled the gap in fluctuation scale coverage with regard to the fluctuations measured in large-scale structure studies. A coherent picture of the fluctuation spectrum has emerged which challenges theoretical models.

3.5.2 Motivation

Inflationary cosmology provides a unique predictor of initial conditions for the later evolution of large-scale structure. The CMB photons stream freely to us from redshift $z \approx 1000$, when the last scatterings occurred for the bulk of the photons. The radiation carries information about the fluctuations that subsequently gave rise to the formation of large-scale structure by gravitational instability.

Acausal angular scales carry information directly to us from the inflationary epoch. The causal horizon at last scattering is the distance that photons have travelled by $z \approx 1000$. This distance subtends an angle as measured by a local observer,

$$\theta_{LS} \approx 2\Omega_0^{1/2} \text{ degrees} \tag{3.59}$$

for $0.3 \leq \Omega_0 \leq 1$. On larger angular scales, one primarily measures fluctuations attributable to inflationary physics. On smaller scales, the microphysics of the last scattering epoch interactions modifies, and adds, to the inflationary signal.

If the universe failed to recombine on schedule, but remained ionized because of some source of energy input, last scattering occurred by $z \sim 50$. The maximum angular scale up to which causal interactions, such as diffusion damping, can modify primordial temperature fluctuations increases to at most 10 degrees. The thickness of the last scattering surface is approximately $\sim 0.1 ct_{LS}$, and angular scales smaller than ~ 10 arcmin suffer strong damping in a flat universe that recombined on schedule. Any early reionization is strongly constrained by the lack of spectral distortions in the CMB. The Comptonization limit restricts reionization to occur below $z \sim 100$.

Cosmic microwave background fluctuation predictions from inflation may be summarized as follows. Quantum fluctuations both in the scalar field that drives inflation and in the metric are amplified to macroscopic scales and possibly to large amplitudes. Long-lived modes are either scalar density fluctuations or tensor gravitational waves. The amplitudes are not predictable a priori, but are constrained by observations both of the remarkable isotropy of the CMB and of large-scale structure. Only the scalar mode generated density fluctuations that seeded large-scale structure and survived to late times. The gravitational wave modes are only sustained in the radiation-dominated phases and subsequently phase mix away on sub-horizon scales, namely on angular scales greater than a degree or two. The concordance between temperature and density fluctuations in terms of a common underlying power spectrum directly tells us that the tensor mode contribution is small, although not necessarily negligible.

One makes this argument quantitative by considering the spectral slope of the fluctuations. The underlying density fluctuations are defined by a power spectrum

$$P(k) = |\delta_k|^2 \propto k^n,\tag{3.60}$$

where the Fourier component at spatial wavenumber k is given by

$$|\delta_k|^2 = \frac{1}{(2\pi)^3}\int (\delta\rho/\rho)^2 e^{i\mathbf{k}\cdot\mathbf{r}}d^3r.\tag{3.61}$$

If the amplitude ratio of tensor to scalar modes is T/S, then inflation generically predicts, to within small correction terms,

$$n \approx 1 - \frac{1}{7}\frac{T}{S}.\tag{3.62}$$

Now in order of magnitude

$$\delta\rho/\rho \sim [k^3 P(k)]^{1/2} \propto k^{(n+3)/2},\tag{3.63}$$

and the associated spatial curvature or gravitational potential fluctuations are

$$\delta\phi \propto k^{-2}\delta\rho/\rho \propto k^{(n-1)/2}.\tag{3.64}$$

Thus one finds that the inflationary models predict nearly scale-invariant fluctuations ($n \approx 1$) in the gravitational potential. These generate temperature fluctuations on super-horizon scales at the last scattering epoch that provide a direct measurement of the inflation-generated power spectrum, albeit over a very limited range of inflation scales.

3.5.3 Detection

To compare with experiments, it is customary to expand the temperature fluctuations $\Delta T/T$ on the sky in spherical harmonics

$$\frac{\Delta T}{T}(\theta,\phi) = \Sigma a_{lm} Y_l^m(\theta,\phi),\tag{3.65}$$

assume random phases and average over all points on the sky and observers to define

$$C_l = \langle |a_{lm}|^2 \rangle.\tag{3.66}$$

Introducing an experimental filter function F_l, defined by scanning or switching at low l and beam width at high l, one has

$$\langle \frac{\Delta T}{T}^2 \rangle = \frac{1}{4\pi}\int_{l\geq 2}^{\infty}(2l+1)l\, C_l F_l\, dlnl.\tag{3.67}$$

The COBE DMR detections of fluctuations provide overwhelming evidence that the fluctuations, on pre-last scattering super-horizon scales, are nearly scale-invariant:

$$n = 1.2 \pm 0.3\ (95\%\text{C.L.});\quad l \leq 20,\tag{3.68}$$

where the angular smoothing scale or beamwidth $\theta \approx 135 l^{-1}$ degrees, l being the corresponding spherical harmonic wavenumber. Allowing for a possible tilt ($n \neq 1$) in the range $0.7 < n < 1.2$ and in the absence of any gravitational wave component of the primordial fluctuations, the inferred value of the density fluctuation amplitude at horizon crossing in a universe at critical density is (Bunn & White 1997)

$$\delta\rho/\rho = 1.9 \times 10^{-5} e^{1-n} \pm 7\% \pm 4\%, \qquad (3.69)$$

where the first (95%C.L.) uncertainty represents the COBE 4-year data error range and the second the effects of varying h and Ω_b.

There are some 10 detections of temperature fluctuations to $l \approx 400$, and upper limits to $l \approx 4000$. Broad-band power, averaged over the filter function appropriate to a particular experiment, can be expressed as either $(2l + 1)C_l W_l / 4\pi$, based on the preceding definition of $\langle(\delta T/T)^2\rangle$, or by assuming a flat spectrum, a reasonably accurate assumption for any given experiment. One can define a quadrupole-normalized fluctuation amplitude by writing $l(l + 1)C_l = (24/5)\pi Q_{\text{flat}}^2$. Results for the various experiments are shown in Fig. 3.2.

There clearly is evidence for the first acoustic peak. Equally clearly, one can conclude that early reionization did not occur, so that the optical depth of the primary last scattering surface must be less than unity. Comparison of the grid of models compiled by Sugiyama (1995) with the data shows that there are indications of a surprisingly high acoustic peak, especially from consideration of the latest Saskatoon data, even when the quoted 14% calibration uncertainty is included. One can fit the data points in a standard CDM model with high Ω_b, or in a flat Λ model with $\Omega \approx 0.3$. A more exotic fit is $h \approx 0.3$ in standard CDM (Tegmark 1998). Note that an open model with $\Omega \approx 0.3$ is also possible, but is beginning to be constrained by the CAT experiment.

3.5.4 Theory: from acoustic peaks to cosmological parameters

Prior to last scattering, the sound speed is maintained near $c_s/\sqrt{3}$ by the tight coupling between matter and radiation. The comoving horizon scale at last scattering is approximately 100 Mpc, which is also comparable to the sound horizon. One can therefore depict primordial fluctuations, containing masses up to $\sim 10^{18} M_\odot$, as gigantic sound waves, the amplitudes of which are fixed by the initial conditions set down at the inflationary epoch as a consequence of quantum fluctuations. What is more remarkable, perhaps, is that the fluctuation phases are fixed by inflationary models on super-horizon scales. This is because only the growing mode has survived. The absence of a decaying mode means that one can write $\delta\rho/\rho \propto \cos(kv_s t)$. Effective growth of fluctuations only occurs once the universe is matter-dominated, and baryon fluctuations only begin to grow once matter and radiation decouple at the epoch of last scattering of matter and radiation. Evaluating $\delta\rho/\rho$ at the last scattering epoch, one sees that the emerging density fluctuations are a series of alternating peaks and troughs, commencing at wavelength $\sim \pi v_s t_{\text{LS}}$. This was first pointed out by Sunyaev and Zel'dovich, and the series of peaks are known as Sakharov oscillations. In principle, such peaks are observable in the autocorrelation function of density fluctuations at present, if the universe is sufficiently baryon-dominated.

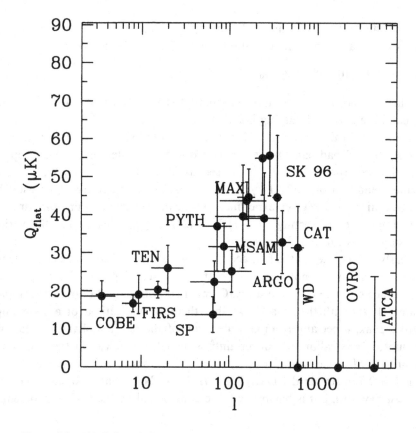

Figure 3.2. "ΔT" fluctuations are plotted for each experiment, as a function of scale (multipole $\ell \sim \theta^{-1}$). Q_{flat} is the best-fitting amplitude of a flat power spectrum, quoted at the quadrupole. The vertical error bars are $\pm 1\sigma$, while the horizontal lines represent the half-power ranges of the window functions (Scott, Silk & White 1995). There are also three smaller-scale upper limits plotted at the 95%C.L. The general rise in the area around $\ell \simeq 200$ can be interpreted as evidence for an adiabatic peak in the radiation power spectrum.

However, the CMB fluctuations carry a far more dramatic signature of primordial acoustic oscillations. Temperature fluctuations are due primarily to the intrinsic component $\delta\rho/3\rho$ (Silk 1968) and the gravitational potential fluctuations (Sachs & Wolfe 1968) $\delta\phi/3c^2$ where $\delta\phi = k^{-2}4\pi G(\delta\rho/\rho)$ with the 1/3 factor arising from a partial cancellation between curvature and intrinsic fluctuations. There is also a small, out-of-phase contribution from Doppler scattering on the last scattering surface, $v_r/c = -k_r^{-1}\partial(\delta\rho/\rho)/\partial t$. With the phase at horizon crossing specified by the inflationary requirement for primordial adiabatic fluctuations, one finds a series of modified sound wave peaks and compressions that are damped on sufficiently small scales, about 0.1 of the last scattering surface scale.

Consider an expansion of the CMB in spherical harmonics with coefficient $a_{\ell,m}$.

Averaging over the azimuthal wavenumber m, and assuming random phases, one can define $C_\ell = \langle |a_{\ell,m}|\rangle^2$ to be the radiation power spectral amplitude coefficient, as viewed in projection on the last scattering surface. To understand the physics of the resulting C_ℓ peaks, one can visualize the radiation fluctuations as arising from a forced harmonic oscillator acting on the coupled radiation–baryon plasma. Baryon inertia enhances the peaks relative to the rarefactions, and peak height increases with baryon densitiy $\Omega_b h^2$. The peaks (rarefactions in quadrature are viewed as peaks) occur at wavelengths $n^{-1}\pi v_s t_{LS}$ for $n = 1, 2, 3, ...$, before damping exponentially suppresses the temperature fluctuations at $n > 3$–4. The peak amplitude is also enhanced by an increase in the effective sound speed. In a flat universe, reducing the matter density raises the peak height. Hence as the redshift of the matter–radiation equality epoch, proportional to the matter density $\Omega_0 h^2$, is reduced, the peaks are also enhanced in amplitude. An additional effect is that peak locations are shifted as the curvature of the universe is varied, due to geodesic curvature, approximately as $\Omega_{curv}^{1/2}$.

3.6 Confrontation of theory and large-scale structure

The COBE-DMR observations of CMB fluctuations have revolutionized cosmological model building. The COBE data, effectively centered at $\ell \approx 10$, have provided the normalization to the predicted fluctuation spectrum that enables quantitative comparison to be made with large-scale structure data. All acceptable models are now "COBE-normalized." Indeed, it was already apparent in 1992 that even the first-year DMR data were within a factor of 2 in fluctuation amplitude, or 4 in power, of the simplest model predictions arising from the large-scale structure data of the IRAS galaxy redshift survey. The simplest model was cold dark matter with $\Omega = 1$, a scale-invariant fluctuation spectrum ($n = 1$), and the canonical baryon abundance $\Omega_b h^2 = 0.0125$ (± 0.005).

While with the 4-year DMR data, the normalization has come down by about 15%, the large-scale structure data base has been expanded and consolidated to include a three-dimensional power spectrum from the APM survey, as well as several other redshift surveys that yield the redshift-space power spectrum. From the modelling perspective, it is now realized that inflation can generate $\Omega < 1$ and $n \approx 1$ (but not necessarily equal to 1) and that $\Omega_b h^2$ may be increased by about a factor of 2 if some of the recent high-redshift D/H measurements are correct. The major development, however, has been with the intermediate angular scale microwave background measurements, ranging from scales of 10 arcminutes to 5 degrees. While these detections are subject to uncertain systematic corrections arising primarily from cosmic foregrounds, one can now attempt detailed comparison of the primordial fluctuation power spectrum models with data over comoving scales that span the present horizon, $9000h^{-1}$ Mpc, to the scale of present-epoch nonlinearity, $10h^{-1}$ Mpc.

A straightforward procedure for effecting an experiment-by-experiment comparison is as follows. Assume a set of cosmological parameters, but leave the primordial spectral index unspecified. One may readily compute the radiation transfer function, including

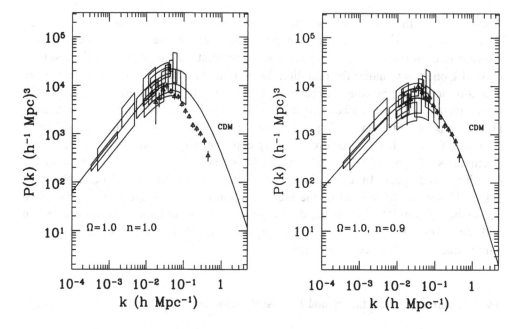

Figure 3.3. The matter power spectrum, $P(k)$, as reconstructed from large-scale structure and CMB data (updated from Scott, Silk & White 1995). Boxes are $\pm 1\sigma$ values of $P(k)$ inferred from CMB measurements, assuming CDM ($\Omega_0 = 1$, $h = 0.5$, $\Omega_b = 0.03$) with $n = 1$ and $n = 0.9$. The overall amplitude of the LSS data (triangles), taken from a compilation by Peacock and Dodds (1994), is uncertain to $\sim 20\%$.

sources of temperature fluctuations both on the last scattering surface and integrated along the line of sight, for a specified matter power spectrum. Any specific experiment is sensitive to a window in ℓ-space, which is the projection of a less well-defined window in k-space. For each experiment, one can take a broad-band average over the window function. The transfer function can be now inverted to obtain the matter power spectrum over a specified range in k. The error bars on $P(k)$ are determined by the errors on the measured C_ℓ's.

The CMB anisotropy experiments effectively probe a limited range in ℓ. In principle, this inversion approach can be used to do a piecewise reconstruction of the matter power spectrum over the observed ℓ-range. Hitherto a comparison has only been performed for specific power spectra of the reconstructed amplitudes with the large-scale structure data. Some results from such comparisons are shown in Fig. 3.3 for variants of the cold dark-matter model.

The implications for cosmology are summarized in Fig. 3.4, in the form of a constraint diagram for Ω_0 and n in open inflationary models (White & Silk 1998). The parameter space consists of $\Omega_0 < 1$, n, h and $\Omega_b h^2$. There are a variety of observational tests which any model of structure formation must pass: the most constraining tests, based on the well-defined linear theory predictions, are the shape of the CDM power

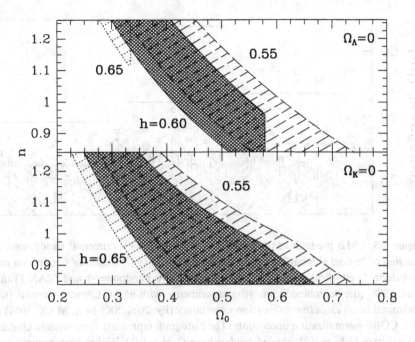

Figure 3.4. Allowed regions of parameter space (White & Silk 1998) for Ω_0 and n, at fixed h, for which CDM models do not violate any large-scale structure constraints at 95%C.L. Upper limits on Ω_0 come from requiring $t_0 \geq 12\,\mathrm{Gyr}$, assuming $\Omega_b h^2 = 0.015$ and minimal large-angle CMB anisotropies.

spectrum (Peacock & Dodds 1994) and the cluster abundance (Viana & Liddle 1996). The latter constraint can be expressed as a limit on

$$\sigma_8^2 = \int \frac{dk}{k}\, \Delta^2(k)\, W^2(kR)\,, \tag{3.70}$$

where $R = 8h^{-1}$ Mpc, that probes scales $k \sim 0.2h$ Mpc^{-1}. The amount of small-scale power required to fit the cluster abundance in these models is,

$$\sigma_8 = 0.6\, \Omega_0^{-0.36-0.31\Omega_0+0.28\Omega_0^2}. \tag{3.71}$$

The abundances of high-redshift massive objects (e.g., quasars and damped Lyman α systems) are compatible with the observations for all of these models, which are required to be at least 12 Gyr old.

A large region of parameter space is allowed after applying these constraints for a variety of slices with h =constant. The baryon content of the universe is set to $\Omega_b h^2 = 0.015$, although the conclusions do not depend strongly on this assumption. Lowering $\Omega_b h^2$ allows slightly higher Ω_0 for fixed h while increasing it allows slightly lower Ω_0. Flat models require a tilted "red" primordial fluctuation spectrum. In open models, very low-Ω_0 is allowed only if the primordial fluctuation spectrum is "blue" ($n > 1$). Such blue spectra arise mostly in hybrid or multiple scalar field models where

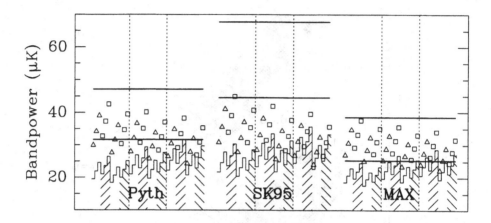

Figure 3.5. CMB predictions (White & Silk 1998) of the experimental bandpower, as a function of "model number", for the Python (Platt et al. 1997) (highest ℓ bandpower only), Saskatoon (Netterfield et al. 1997) (the third of five bandpowers) and MAX (Tanaka et al. 1998) (the "combined" analysis) experiments, with 95%C.L. measurements (solid, horizontal lines) *excluding* calibration uncertainty (Py: 20%, SK: 14%, MAX: 10%) and 10% COBE-normalization uncertainty. The histogram represents open models. Unshaded models have $\Omega_b h^2 = 0.01$, shaded models have $\Omega_b h^2 = 0.02$. Within each group $n = 0.8$, 1, 1.2. Shading running bottom-left to top-right indicates $h = 0.6$, top-left to bottom-right $h = 0.7$, with the unshaded predictions having the same h as the shaded prediction to their right. Vertical dotted lines separate $\Omega_0 = 0.3$, 0.4 and 0.5 (left to right). The open triangles and squares are the predictions for $\Omega_\Lambda = 1 - \Omega_0$ for comparison.

inflation ends not by a field rolling down an ever steepening potential but through an instability. No models of open inflation have been constructed with this property.

Another constraint on open models comes from small angular scale measurements of the CMB anisotropy. Experiments which probe angular scales around 0.5° will observe more power in a model without spatial curvature than in a model with with negative spatial curvature. Figure 3.5 quantifies this comparison for three representative experiments which probe scales near the first peak in a flat model (White & Silk 1998). The predicted bandpower, or level of fluctuation, is shown for 32 models with $\Omega_0 = 0.3$, 0.4, 0.5, $n = 0.8$, 1.0, 1.2, $h = 0.6$, 0.7 and $\Omega_b h^2 = 0.01$, 0.02. Late reionization is ignored: this could reduce the temperature fluctuations by a factor $\sim 1\%$ to 30% (e.g., Tegmark & Silk 1995).

3.7 Future prospects

The new generation of CMB anisotropy experiments will feature the making of maps as an important goal. Hitherto, the pre-COBE experiments have largely been one-dimensional, and have generated radiation power spectra. There is much additional information to be found in the phase distribution, which need not necessarily be random even at early epochs in the universe. Sources of non-Gaussianity in the CMB include initial conditions, if structure is seeded by topological defects, nonlinearity, lensing by large-scale structure and reionization by discrete sources, as well as the integrated Sachs–Wolfe and Rees–Sciama effects.

It is not feasible to make map predictions given so much uncertainty in the astrophysics of the origin of non-Gaussian phases. However an empirical approach can be developed towards generalizing maps. Take a three-dimensional galaxy survey that spans scales that probe the regime of linear density fluctuations, say 10–100 Mpc. One can reconstruct a map of the primordial fluctuation distribution over these scales.

To proceed further, if one assumes that mass traces light, one can go from a two-dimensional map of the galaxy distribution, inverting Limber's equation, to get a three-dimensional map. By application of the linear (or quasi-linear) theory of gravitational instability to fluctuation growth, one can extrapolate this map back to epochs well before last scattering. Now apply the Boltzmann equation, mode by mode, and calculate the resulting CMB anisotropy map (Fig. 3.6). With surveys such as 2DF or especially SDSS, one will have thousands of independent cubes 100 Mpc on the side either in redshift or in real space that one can compare with the comparable numbers of CMB maps of similar resolution, namely over 0.1–1 degree, that will be produced by future experiments, both long duration balloon and satellite.

What can we learn from such a comparison? The CMB fluctuation maps are unbiased, whereas the galaxy surveys yield biased maps. Moreover, non-Gaussianity may be present in the CMB maps that is not apparent in the large-scale structure maps since the latter are a highly processed version of the former. By cross-correlating the two types of maps, one should be able to study biasing and the possible development of non-Gaussianity over 10–100 Mpc scales.

One can answer other questions by studying nearby large-scale structure. How typical is our local patch of the universe? And what is the curvature of the universe? An ideal example is the Mark III catalog of galaxy peculiar velocities, generated by A. Dekel and collaborators, from which an unbiased map of the mass distribution over $(12-80)h^{-1}$ Mpc has been produced. The corresponding CMB map prediction of the appearance of the precursor of our local region to an observer located on some appropriate space-like hypersurface is shown in Fig. 3.6. The combination of intrinsic potential and out-of-phase Doppler components results in a differential filter that varies the map structure as a function of geodesic curvature. Identical features are dramatically reduced in scale for an open model as Ω_m is lowered, and even, because of the angular-size redshift relation for a flat low Ω_m model. To reconcile the amplitude and scale sizes with that seen in the CMB, one ideally needs a large sample of CMB maps with which to compare the local large-scale predictions. A statistical comparison would enable one to make a quantative determination of whether our local universe is, for example, characterized by unusually large or small bulk flows.

Low Ω_m models provide an outstanding description of the universe on scales below ~ 10 Mpc, but fail on larger scales primarily because of the low amplitudes predicted for

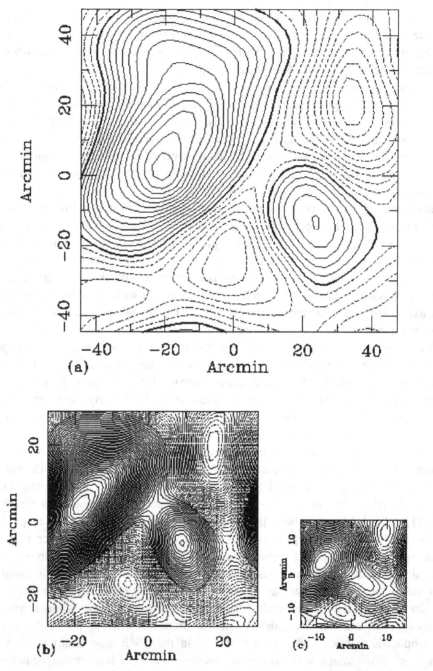

Figure 3.6. Contour map of $\Delta T/T$ produced from our local patch of the universe (Zaroubi et al. 1997), with the supergalactic plane assumed to be the last scattering surface seen by a distant observer. The contour spacing is 5×10^{-6} in $\Delta T/T$. The maps assume $h = 0.5$, $\Omega_b = 0.0125h^{-2}$, for flat universes (a) $\Omega = 1$, (b) $\Omega = 0.3$, $\lambda = 0.7$, and an open universe (c) with $\Omega = 0.3$.

bulk flows. The CMB provides a unique prospect for verifying the currently acceptable menagerie of cosmological models. This is an exciting time to be studying the CMB fluctuations. Two major satellite experiments have been approved, and there are also projects underway using ground-based interferometers and balloon-borne bolometer arrays that are capable of being competitive on a shorter time-scale. At the same time, our sampling of large-scale structure via deep redshift surveys will increase by an order of magnitude.

Complementary efforts are underway to map the CMB. Long duration balloon-borne arrays utilizing both bolometers and HEMTs are planned that will map several percent of the sky at a resolution of 10–30 arcmin and a sensitivity of 30 µK per pixel. Interferometers are being constructed that will extend the resolution range to several arcmin over small patches of sky.

The ultimate maps will be generated by two planned satellite experiments. NASA's MAP satellite will carry HEMTs that perform at 22–90 GHz with 35 µK per pixel sensitivity at a resolution of 12 arcmin. MAP is scheduled to be launched in 2000. ESA is planning the Planck (COBRAS/SAMBA) satellite for launch in 2007. This platform will include both bolometers and HEMTs that operate over 31–860 GHz at a sensitivity of up to 3 µK per pixel and a resolution of up to 7 arcmin for the CMB.

The all-sky coverage provided by the satellite experiments offers the prospect of measuring the C_ℓ's to a precision of 1 or 2% (Jungman et al. 1994). The optimal experiment requires only modest signal-to-noise per pixel, but extensive sky coverage. Specifically

$$\frac{\delta C_\ell}{C_\ell} = \left(\frac{2}{2l+1}\right)^{1/2} \left(\frac{4\pi}{\Omega}\right)^{1/2} \left(1 + \frac{1}{S/N}\right), \qquad (3.72)$$

where the first factor on the right of this equation describes the cosmic variance and the second factor is the sample variance. With $N \gtrsim 10^6$ pixels expected in a future satellite experiment, one should be able to measure C_ℓ's at $\ell \sim N^{1/2}$ to the desired accuracy.

One will be able to utilize clean areas of sky and model out foregrounds with adequate frequency coverage and angular resolution. The prospects are excellent for resolving all issues in classical cosmology to do with evaluating the standard model parameters. The Hubble constant, baryon fraction, Ω_0, the cosmological constant, and the primordial fluctuation spectral index can all be evaluated to a precision of a few percent. Even such subtle ingredients as a small admixture of hot dark matter carry a uniquely decipherable signature. It will prove more difficult to evaluate possible tensor or vector contributions: polarization will help in this regard. A decade from now, it may be timely for cosmologists to focus on understanding the pressing issue of galaxy formation, one of the few outstanding problems that CMB maps will not resolve.

Acknowledgments

JS is indebted to various collaborators for many discussions of topics covered in these lectures given at the Jerusalem Winter School. This chapter is based in part on talks given at the Strasbourg NATO ASI on *The Cosmic Microwave Background* (Kluwer, in press, 1997), at the XXXIst Rencontre de Moriond on Microwave Background Anisotropies (Editions Frontières, in press, 1997), and at *Critical Dialogues in Cosmology*, Princeton 250th Anniversary Conference (World Scientific, in press, 1997). The research has been supported in part by a grant from NASA.

References

Bunn, E., & White, M. 1997, *ApJ*, **480**, 6

Cayon, L., Silk, J., & Charlot, L. 1996, *ApJ*, **467**, L53

Fall, S. M., & Pei, Y. C. 1993, *ApJ*, **402**, 479

Fixsen, D. J., *et al.* 1998, *ApJ*, in press

Jungman, G., Kamionkowski, M., Kosowsky, A. & Spergel, D. N. 1994, *PRL*, **77**, 1007

Kolatt, T. & Dekel, A. 1997, *ApJ*, **479**, 592

Lacey, C. & Cole, S. 1993, *MNRAS*, **262**, 627

Lu, L., Sargent, W. L. W., Barlow, T. A., Churchill, C. W. & Vogt, S. 1998, *ApJS*, in press

Netterfield, S. B., Devlin, M. J., Jarosik, N., Page, L. & Wollack, E. L. 1997, *ApJ*, **474**, 47

Peacock, J. A. & Dodds, D. D. 1994, *MNRAS*, **267**, 1020

Platt, S. R., Kovac, J., Dragovan, M., Peterson, J. B. & Ruhl, J. E. 1997, *ApJ*, **475**, L1

Prantzos, N. & Aubert, O. 1995, *A&A*, **302**, 69

Puget, J.-L., *et al.* 1996, *A&A*, **308**, L5

Sachs, R. K. & Wolfe, A. M. 1968, *ApJ*, **147**, 73

Scott, D., Silk, J. & White, M. 1995, *Science*, **268**, 8

Silk, J. 1968, *ApJ*, **151**, 459

Silk, J. 1997, *ApJ*, **481**, 703

Silk, J. & Wyse, R. F. G. 1998, in preparation

Smoot, G., *et al.* 1992, *ApJ*, **396**, L1

Sugiyama, N. 1995, *ApJS*, **100**, 281

Tanaka, S., *et al.* 1998, in preparation

Tegmark, M. 1998, *ApJ*, in press

Tegmark, M., & Silk, J. 1995, *ApJ*, **441**, 458

Viana, P. T. P., & Liddle, A. R. 1996, *MNRAS*, **281**, 323

Wang, B., & Silk, J. 1993, *ApJ*, **427**, 759

White, M., & Silk, J. 1998, *PRL*, in press

Wolfe, A. M. 1996, in *QSO Absorption Lines*, ed. G. Meylan (Heidelberg: Springer), p. 13

Wyse, R. F. G. & Silk, J. 1989, *ApJ*, **339**, 700

Zaroubi, S., Sugiyama, N., Silk, J., Hoffman, Y. & Dekel, A. 1997, *ApJ*, **490**, 473

Zepf, S. & Silk, J. 1996, *ApJ*, **466**, 114

Part two

Large-scale structure and flows

4 Clusters and superclusters of galaxies

Neta A. Bahcall

Abstract

Rich clusters of galaxies are the most massive virialized systems known. Even though they contain only a small fraction of all galaxies, rich clusters provide a powerful tool for the study of galaxy formation, dark matter, large-scale structure, and cosmology. Superclusters, the largest known systems of galaxies, extend to $\sim 100h^{-1}$ Mpc in size and highlight the large-scale structure of the universe. This large-scale structure reflects initial conditions in the early universe and places strong constraints on models of galaxy formation and on cosmology.

Some of the questions that can be addressed with clusters and superclusters of galaxies include: How did galaxies and larger structures form and evolve? What is the amount, composition, and distribution of matter in clusters and larger structures? How does the cluster mass density relate to the matter density in the universe? What constraints can the cluster and supercluster data place on cosmology? This chapter discusses the important properties of clusters and superclusters that can be used to investigate these topics.

4.1 Introduction

Clusters and superclusters of galaxies have been studied extensively both for their intrinsic properties and to investigate the dark matter in the universe, the baryon content of the universe, large-scale structure, evolution, and cosmology. For previous reviews see Zwicky (1957), Bahcall (1977, 1988, 1996), Oort (1983), Dressler (1984), Rood (1988), and Peebles (1993).

This chapter discusses the following topics and their implications for structure formation and cosmology: (§4.2) optical properties of galaxy clusters, (§4.3) X-ray properties of galaxy clusters, (§4.4) the baryon fraction in clusters, (§4.5) cluster masses,

(§4.6) where is the dark matter?, (§4.7) the mass function of clusters, (§4.8) quasar–cluster association, (§4.9) superclusters, (§4.10) the cluster correlation function, (§4.11) peculiar motions of clusters, and (§4.12) some unsolved problems. A Hubble constant of $H_0 = 100h$ km s^{-1} Mpc^{-1} is used throughout.

4.2 Optical properties of galaxy clusters

4.2.1 Typical properties of clusters and groups

Clusters of galaxies are bound, virialized, high-overdensity systems of galaxies, held together by the cluster's self gravity. Rich clusters contain, by traditional definition (Abell 1958), at least 30 galaxies brighter than $m_3 + 2$ (where m_3 is the magnitude of the third-brightest cluster member) within a radius of $R \simeq 1.5h^{-1}$ Mpc of the cluster center. This galaxy count is generally defined as the richness of the cluster. The galaxies in rich clusters move with random peculiar velocities of typically ~ 750 km s^{-1} (median line-of-sight velocity dispersion). This motion corresponds to a typical rich cluster mass (within $1.5h^{-1}$ Mpc) of $\sim 5 \times 10^{14} h^{-1}$ M$_\odot$. In addition to galaxies, all rich clusters contain an intracluster medium of hot plasma, extending as far as the main galaxy concentration ($R \sim 1.5h^{-1}$ Mpc). The typical temperature of the hot intracluster gas is ~ 5 keV, with a range from ~ 2 to 14 keV; the central gas density is $\sim 10^{-3}$ electrons cm^{-3}. The hot plasma is detected through the luminous X-ray emission it produces by thermal bremsstrahlung radiation, with $L_x \sim 10^{44}$ erg s^{-1}.

Like mountain peaks on earth, the high-density rich clusters are relatively "rare" objects; they exhibit a spatial number density of $\sim 10^{-5}$ clusters Mpc^{-3}, compared with $\sim 10^{-2}$ galaxies Mpc^{-3} for the density of bright galaxies.

The main properties of clusters and groups of galaxies are summarized in Table 4.1 (taken from Bahcall 1996). The table lists the typical range and/or median value of each observed property. Groups and poor clusters, whose properties are also listed, provide a natural and continuous extension to lower richness, mass, size, and luminosity from the rich and rare clusters.

The following subsections discuss in more detail some of these intrinsic cluster properties.

4.2.2 Distribution of clusters with richness and distance

An illustration of the distribution of rich clusters with richness and distance is summarized in Table 4.2. The table refers to the statistical sample of the Abell (1958) cluster catalog [i.e., richness class $R \geq 1$, corresponding to a richness threshold count of $N_R \geq 50$ galaxies within $1.5h^{-1}$ Mpc radius and magnitude $m \leq m_3 + 2$ (see Table 4.1); redshift range $z \simeq 0.02$ to 0.2; and sky coverage $\delta > -27°$ and $|b| \gtrsim 30°$]. The Abell catalog covers $\sim 1/3$ of the entire sky to $z \lesssim 0.2$. Recent smaller automated surveys

Table 4.1. *Typical properties of clusters and groups*

Property[a]	Rich clusters	Groups and poor clusters
Richness[b]	30–300 galaxies	3–30 galaxies
Radius[c]	$(1-2)h^{-1}$ Mpc	$(0.1-1)h^{-1}$ Mpc
Radial velocity dispersion[d]	400–1400 km s^{-1}	100–500 km s^{-1}
Radial velocity dispersion[d] (median)	~ 750 km s^{-1}	~ 250 km s^{-1}
Mass ($r \leq 1.5h^{-1}$ Mpc)[e]	$(10^{14}-2\times 10^{15})h^{-1}$ $M_\odot$	$(10^{12.5}-10^{14})h^{-1}$ $M_\odot$
Luminosity (B)[f] ($r \leq 1.5h^{-1}$ Mpc)	$(6\times 10^{11}-6\times 10^{12})h^{-2}$ $L_\odot$	$(10^{10.5}-10^{12})h^{-2}$ $L_\odot$
$\langle M/L_B \rangle$[g]	$\sim 300h$ $M_\odot/L_\odot$	$\sim 200h$ $M_\odot/L_\odot$
X-ray temperature[h]	2–14 keV	$\lesssim 2$ keV
X-ray luminosity[h]	$(10^{42.5}-10^{45})h^{-2}$ erg s^{-1}	$\lesssim 10^{43}h^{-2}$ erg s^{-1}
Cluster number density[i]	$(10^{-5}-10^{-6})h^3$ Mpc^{-3}	$(10^{-3}-10^{-5})h^3$ Mpc^{-3}
Cluster correlation scale[j]	$(22\pm 4)h^{-1}$ Mpc $(R \geq 1)$	$(13\pm 2)h^{-1}$ Mpc
Fraction of galaxies in clusters or groups[k]	$\sim 5\%$	$\sim 55\%$

[a] In most entries, the typical range in the listed property or the median value is given. Groups and poor clusters are a natural and continuous extension to lower richness, mass, size, and luminosity from the rich and rare clusters.
[b] Cluster richness: the number of cluster galaxies brighter than $m_3 + 2^m$ (where m_3 is the magnitude of the third-brightest cluster galaxy), and located within a $1.5h^{-1}$ Mpc radius of the cluster center (§4.2.2).
[c] The radius of the main concentration of galaxies (where, typically, the galaxy surface density drops to $\sim 1\%$ of the central core density). Many clusters and groups are embedded in larger-scale structures (to tens of Mpc).
[d] Typical observed range and median value for the radial (line-of-sight) velocity dispersion in groups and clusters (§4.2.9).
[e] Dynamical mass range of clusters within $1.5h^{-1}$ Mpc radius (§4.2.10).
[f] Luminosity range (blue) of clusters within $1.5h^{-1}$ Mpc radius (§4.2.10).
[g] Typical mass-to-light ratio of clusters and groups (median value) (§4.2.10).
[h] Typical observed ranges of the X-ray temperature and 2–10-keV X-ray luminosity of the hot intracluster gas (§4.3).
[i] The number density of clusters decreases sharply with cluster richness (§4.7).
[j] The cluster correlation scale for rich ($R \geq 1$, $N_R \geq 50$, $n_{cl} = 0.6\times 10^{-5}h^3$ Mpc^{-3}) and poor ($N_R \gtrsim 20$, $n_{cl} = 2.4\times 10^{-5}h^3$ Mpc^{-3}) clusters (§4.2.3, §4.7).
[k] The fraction of bright galaxies ($\gtrsim L_*$) in clusters and groups within $1.5h^{-1}$ Mpc.

Table 4.2. *Distribution of Abell clusters with distance and richness*[a]

Distance distribution			Richness distribution		
D	$\langle z_{est} \rangle$	$N_{cl}(R \geq 1)$	R	N_R	N_{cl}
1	0.0283	9	(0)[b]	(30–49)	($\sim 10^3$)
2	0.0400	2	1	50–79	1224
3	0.0577	33	2	80–129	383
4	0.0787	60	3	130–199	68
5	0.131	657	4	200–299	6
6	0.198	921	5	≥ 300	1
	Total	1682		Total ($R \geq 1$)	1682

	Nearby redshift sample[c,d] $D \leq 4$	Distant projected sample[d] $D = 5+6$
N_{cl}(total)	104	1574
$N_{cl}(b \geq 30°)$	71	984
$N_{cl}(b \leq -30°)$	33	563
$N_{cl}(R = 1)$	82	1125
$N_{cl}(R \geq 2)$	22	422

[a] Statistical sample. $|b|$ boundaries as given in Table 1 of Abell (1958). Notation: D = distance group (defined by the estimated redshifts of the clusters); $\langle z_{est} \rangle$ = average estimated redshift from the magnitude of the tenth-brightest galaxy; N_{cl} = number of clusters; R = richness class; N_R = number of galaxies brighter than $m_3 + 2^m$ within the Abell radius $R_A = 1.5h^{-1}$ Mpc (richness count).
[b] $R = 0$ clusters are not part of the statistical sample.
[c] Redshifts by Hoessel, Gunn & Thuan (1980).
[d] This sample is limited to $|b| \geq 30°$ in addition to the $|b|$ boundaries of the statistical sample.

from digitized plates are reported by the Edinburgh–Durham catalog (EDCC; Lumsden et al. 1992), and the APM survey (Dalton et al. 1992). A large automated cluster catalog will be available in the near future from the Sloan Digital Sky Survey (SDSS; see §5.7). This and other planned surveys will allow a more accurate determination of the distribution of rich clusters with richness and distance and other statistical studies of clusters.

4.2.3 Number density of clusters

The number density of clusters is a strong function of cluster richness. Integrated cluster densities, $n_{cl}(> N_R)$, representing the number density of clusters above a given

Clusters and superclusters of galaxies 139

Table 4.3. *Number density of clusters*

R	N_R	n_{cl} $(> N_R)h^3$ (Mpc^{-3})^a	d $(> N_R)h^{-1}$ (Mpc)
≥ 0	≥ 30	13.5×10^{-6}	42
≥ 1	≥ 50	6.0×10^{-6}	55
≥ 2	≥ 80	1.2×10^{-6}	94
≥ 3	≥ 130	1.5×10^{-7}	188

^{a} Approximate uncertainties are $10^{\pm 0.2}$ for the $R \geq 0, 1, 2$ densities and $10^{\pm 0.3}$ for $R \geq 3$.

richness threshold, and the associated mean cluster separation, d ($\equiv n_{cl}^{-1/3}$), are listed in Table 4.3 (Bahcall & Cen 1993).

4.2.4 Fraction of galaxies in clusters

The fraction of galaxies in rich $R \gtrsim 0$ clusters is $\sim 5\%$ (within the Abell radius $R_A = 1.5h^{-1}$ Mpc). The fraction of all galaxies that belong in clusters increases with increasing radius R_A and with decreasing cluster richness threshold.

The average number of galaxies per cluster for $R \geq 0$ clusters within $1.5h^{-1}$ Mpc radius and $m \leq m_3 + 2$ is $\langle N_R \rangle_{\text{median}} \simeq 50$, or $\langle N_R \rangle_{\text{mean}} \simeq 56$. For $R \geq 1$ clusters the average is $\langle N_R \rangle_{\text{median}} \simeq 60$, or $\langle N_R \rangle_{\text{mean}} \simeq 75$. The number of galaxies increases to fainter luminosities following the Schechter (1976) luminosity function.

4.2.5 Galaxy overdensity in rich clusters

The average number density of bright ($\gtrsim L_*$) galaxies in $R \gtrsim 0$ clusters (within $R_A = 1.5h^{-1}$ Mpc) is

$$n_g(\text{cluster}) \sim 3h^3 \text{ galaxies Mpc}^{-3}. \tag{4.1}$$

The average overall (field) number density of bright ($\gtrsim L_*$) galaxies is

$$n_g(\text{field}) \sim 1.5 \times 10^{-2} h^3 \text{ galaxies Mpc}^{-3}. \tag{4.2}$$

The average galaxy overdensity in rich ($R \geq 0$) clusters (within $1.5h^{-1}$ Mpc radius) is thus

$$n_g(\text{cluster})/n_g(\text{field}) \sim 200. \tag{4.3}$$

The typical threshold galaxy overdensity in clusters (within $1.5h^{-1}$ Mpc radius) is

$$R \geq 0 \text{ clusters}: \quad n_g(\text{cluster})/n_g(\text{field}) \gtrsim 100, \tag{4.4}$$

$$R \geq 1 \text{ clusters}: \quad n_g(\text{cluster})/n_g(\text{field}) \gtrsim 200. \tag{4.5}$$

The galaxy overdensity increases at smaller radii from the cluster center. The galaxy overdensity in the cores of typical compact rich clusters is approximately

$$n_g^0(\text{cluster core})/n_g(\text{field}) \sim 10^4 - 10^5 . \tag{4.6}$$

4.2.6 Density profile

The radial density distribution of galaxies in a rich cluster can be approximated by a bounded Emden isothermal profile (Zwicky 1957; Bahcall 1977), or by its King approximation (King 1972) in the *central* regions.

In the central regions, the King approximation for the galaxy distribution is

$$n_g(r) = n_g^0 (1 + r^2/R_c^2)^{-3/2}, \quad \text{spatial profile} \tag{4.7}$$

$$S_g(r) = S_g^0 (1 + r^2/R_c^2)^{-1}, \quad \text{projected profile.} \tag{4.8}$$

$n_g(r)$ and $S_g(r)$ are, respectively, the space and projected profiles (of the number density of galaxies, or brightness), n_g^0 and S_g^0 are the respective central densities, and R_c is the cluster core radius (where $S(R_c) = S^0/2$). Typical central densities and core radii of clusters are listed below. The projected and space central densities relate as

$$S_g^0 = 2 R_c n_g^0 . \tag{4.9}$$

A bounded Emden isothermal profile of galaxies in clusters yields a profile slope that varies approximately as (Bahcall 1977)

$$S_g(r \lesssim R_c/3) \sim \text{constant} , \tag{4.10}$$

$$S_g(R_c \lesssim r \lesssim 10 R_c) \propto r^{-1.6} ; \tag{4.11}$$

therefore

$$n_g(R_c \lesssim r \lesssim 10 R_c) \propto r^{-2.6} . \tag{4.12}$$

The galaxy–cluster cross-correlation function (Peebles 1980; Lilje & Efstathiou 1988) also represents the average radial density distribution of galaxies around clusters. For $R \geq 1$ clusters, and r in h^{-1} Mpc, these references suggest, respectively

$$\xi_{gc}(r) \simeq 130 r^{-2.5} + 70 r^{-1.7} \tag{4.13}$$

or

$$\xi_{gc}(r) \simeq 120 r^{-2.2} . \tag{4.14}$$

The average density distribution profile of galaxies in clusters thus follows, approximately,

$$n_g(r) \propto r^{-2.4 \pm 0.2} \quad (\text{spatial}), \quad r > R_c \tag{4.15}$$

$$S_g(r) \propto r^{-1.4 \pm 0.2} \quad (\text{projected}), \quad r > R_c . \tag{4.16}$$

Some substructure (subclumping) in the distribution of galaxies exists in a significant fraction of rich clusters ($\sim 40\%$) (Geller 1990).

4.2.7 Central density and core size

Central number density of galaxies in rich compact clusters (Bahcall 1975, 1977; Dressler 1978) is (for galaxies in the brightest 3 magnitude range)

$$n_g^0(\Delta m \simeq 3^m) \sim 10^3 h^3 \text{ galaxies Mpc}^{-3}. \tag{4.17}$$

The central density reaches $\sim 10^4 h^3$ galaxies Mpc^{-3} for the richest compact clusters. The typical central *mass* density in rich compact clusters, determined from cluster dynamics is

$$\rho_0(\text{mass}) \simeq 9\sigma_{rc}^2/4\pi G R_c^2$$
$$\sim 4 \times 10^{15} M_\odot \text{ Mpc}^{-3} [(\sigma_{rc}/10^3 \text{ km s}^{-1})/(R_c/0.2 \text{ Mpc})]^2 h^2 \tag{4.18}$$

where σ_{rc} is the radial (line-of-sight) central cluster velocity dispersion (in km s^{-1}) and R_c is the cluster core radius (in Mpc).

Core radii of typical rich compact clusters, determined from the galaxy distribution (Bahcall 1975; Dressler 1978; Sarazin 1986) are in the range

$$R_c \simeq (0.1-0.25)h^{-1} \text{ Mpc}. \tag{4.19}$$

Core radii of the X-ray emitting intracluster gas (§4.3) are

$$R_c(\text{X-rays}) \simeq (0.1-0.3)h^{-1} \text{ Mpc}. \tag{4.20}$$

The core radius of the mass distribution determined from gravitational lensing observations of some clusters may be smaller, $R_c \lesssim 50$ kpc, than determined by the galaxy and gas distribution.

The typical central density of the hot intracluster gas in rich clusters (§4.3) is

$$n_e \sim 10^{-3} \text{ electrons cm}^{-3}. \tag{4.21}$$

4.2.8 Galactic content in rich clusters

The fraction of elliptical (E), S0, and spiral (Sp) galaxies in rich clusters (Table 4.4) differs from that in the field, and depends on the classification type, or density, of the cluster (see §4.2.11) (Bahcall 1977; Oemler 1974; Dressler 1980).

The fraction of elliptical and S0 galaxies increases and the fraction of spirals decreases toward the central cores of rich compact clusters. The fraction of spiral galaxies in the dense cores of some rich clusters (e.g., the Coma cluster) may be close to zero.

The galactic content of clusters as represented in Table 4.4 is part of the general density–morphology relation of galaxies (Dressler 1980; Postman & Geller 1984); as the local density of galaxies increases, the fraction of E and S0 galaxies increases and the fraction of spirals decreases. For local galaxy densities $n_g \lesssim 5$ galaxies Mpc^{-3}, the fractions remain approximately constant at the average "field" fractions listed above.

Table 4.4. *Typical galactic content of clusters* ($r \lesssim 1.5h^{-1}$ Mpc)

Cluster type	E	S0	Sp	(E+S0)/Sp
Regular clusters (cD)	35%	45%	20%	4.0
Intermediate clusters (spiral-poor)	20%	50%	30%	2.3
Irregular clusters (spiral-rich)	15%	35%	50%	1.0
Field	10%	20%	70%	0.5

4.2.9 Velocity dispersion

The typical radial (line-of-sight) velocity dispersion of galaxies in rich clusters (median value) is

$$\sigma_r \sim 750 \text{ km s}^{-1} . \tag{4.22}$$

The typical range of radial velocity dispersion in rich clusters (Struble & Rood 1991) is

$$\sigma_r \sim 400\text{--}1400 \text{ km s}^{-1} . \tag{4.23}$$

A weak correlation between σ_r and richness exists; richer clusters exhibit, on average, larger velocity dispersion (Bahcall 1981). The observed velocity dispersion of galaxies in rich clusters is generally consistent with the velocity implied by the X-ray temperature of the hot intracluster gas (§4.3.5), as well as with the cluster velocity dispersion implied from observations of gravitational lensing by clusters (except possibly in the central core). Velocity dispersion and temperature profiles as a function of distance from the cluster center have been measured only for a small number of clusters so far. The profiles are typically isothermal [$\sigma_r^2(r) \sim T_x(r) \sim$ constant] for $r \lesssim (0.5\text{--}1)h^{-1}$ Mpc, and drop somewhat at larger distances.

4.2.10 Mass, luminosity, and mass-to-luminosity ratio

The typical dynamical mass of rich clusters within a $1.5h^{-1}$ Mpc radius sphere (determined from the virial theorem for an isothermal distribution) is

$$M_{cl}(\leq 1.5) \simeq \frac{2\sigma_r^2(1.5h^{-1} \text{ Mpc})}{G} \simeq 0.7 \times 10^{15} \left(\frac{\sigma_r}{1000}\right)^2$$
$$\simeq 0.4 \times 10^{15} h^{-1} M_\odot \text{ (for } \sigma_r \sim 750 \text{ km s}^{-1}\text{)} . \tag{4.24}$$

The approximate range of masses for $R \gtrsim 0$ clusters (within $1.5h^{-1}$ Mpc) is

$$M_{cl}(\leq 1.5) \sim (0.1\text{--}2) \times 10^{15} h^{-1} M_\odot . \tag{4.25}$$

Comparable masses are obtained using the X-ray temperature and distribution of the hot intracluster gas (Hughes 1989; Bahcall & Cen 1993; Lubin & Bahcall 1993; §4.3).

The typical (median) blue luminosity of rich clusters (within $1.5h^{-1}$ Mpc) is

$$L_{cl}(\leq 1.5) \sim 10^{12} h^{-2} \, L_\odot \; . \tag{4.26}$$

The approximate range of rich cluster blue luminosities is

$$L_{cl}(\leq 1.5) \sim (0.6\text{--}6) \times 10^{12} h^{-2} L_\odot \; . \tag{4.27}$$

The typical mass-to-luminosity ratio of rich clusters is thus

$$(M/L_B)_{cl} \sim 300h \; (M_\odot/L_\odot) \; . \tag{4.28}$$

The inferred mass-density in the universe based on cluster dynamics is

$$\Omega_{dyn} \sim 0.2 \tag{4.29}$$

(if mass follows light, $M \propto L$, on scales $\gtrsim 1h^{-1}$ Mpc). $\Omega = 1$ corresponds to the critical mass-density needed for a closed universe and $M/L_B \, (\Omega = 1) \simeq 1500h$.

4.2.11 Cluster classification

Rich clusters are classified in a sequence ranging from early- to late-type clusters, or equivalently, from regular to irregular clusters. Many cluster properties (shape, concentration, dominance of brightest galaxy, galactic content, density profile, and radio and X-ray emission) are correlated with position in this sequence. A summary of the sequence and its related properties is given in Table 4.5. Some specific classification systems include the Bautz–Morgan (BM) System (Bautz & Morgan 1970), which classifies clusters based on the relative contrast (dominance in extent and brightness) of the brightest galaxy to the other galaxies in the cluster, ranging from type I to III in decreasing order of dominance; and the Rood–Sastry (RS) system (Rood & Sastry 1971) which classifies clusters based on the distribution of the ten brightest members (from cD, to binary (B), core (C), line(L), flat (F), and irregular (I)).

4.3 X-ray properties of galaxy clusters

4.3.1 X-ray emission from clusters

All rich clusters of galaxies produce extended X-ray emission due to thermal bremsstrahlung radiation from a hot intracluster gas (Jones & Forman 1984; Sarazin 1986; David et al. 1993; Edge et al. 1990; Edge & Stewart 1991; Henry & Arnaud 1991; Henry et al. 1992; Burg et al. 1994). The cluster X-ray luminosity emitted in the photon energy band E_1 to E_2 by thermal bremsstrahlung from a hot (T_x degrees) intracluster gas of uniform electron density n_e and a radius R_x is

Table 4.5. *Cluster classification and related characteristics*

Property	Regular (early) type clusters	Intermediate clusters	Irregular (late) type clusters
Zwicky type	Compact	Medium-compact	Open
BM type	I, I–II, II	(II), II–III	(II–III), III
RS type	cD,B,(L,C)	(L),(F),(C)	(F),I
Shape symmetry	Symmetrical	Intermediate	Irregular shape
Central concentration	High	Moderate	Low
Galactic content	Elliptical-rich	Spiral-poor	Spiral-rich
E fraction	35%	20%	15%
S0 fraction	45%	50%	35%
Sp fraction	20%	30%	50%
E:S0:Sp	3:4:2	2:5:3	1:2:3
Radio emission	~ 50% detection	~ 50% detection	~ 25% detection
X-ray luminosity	High	Intermediate	Low
Fraction of clusters	~ 1/3	~ 1/3	~ 1/3
Examples	A401, Coma	A194	Virgo, A1228

$$L_x \propto n_e^2 R_x^3 T_x^{0.5} g(e^{-E_1/kT_x} - e^{-E_2/kT_x}) \,. \tag{4.30}$$

The Gaunt factor correction g (of order unity) is a slowly varying function of temperature and energy. The bolometric thermal bremsstrahlung luminosity of a cluster core can be approximated by

$$L_x(\text{core}) \simeq 1.4 \times 10^{42} n_e(\text{cm}^{-3})^2 R_c(\text{kpc})^3 kT_x(\text{keV})^{0.5} h^{-2} \text{ erg s}^{-1} \,. \tag{4.31}$$

Some of the main properties of the hot intracluster gas are summarized below.

4.3.2 X-ray properties of clusters

Some of the main properties of the X-ray emission from rich clusters of galaxies are summarized in Table 4.6.

4.3.3 The intracluster gas: some relevant questions

Some of the fundamental questions about the intracluster gas relate to its origin, evolution, metal enrichment, hydrodynamical state in the cluster, and its relation to the distribution of galaxies and mass in the cluster. Listed below are some of the relevant questions and a selection of these topics are discussed in the following subsections. Some of the questions posed do not yet have sufficient observational constraints to suggest a possible solution.

- What is the hydrodynamical state of the hot gas in clusters? Is it in approximate hydrostatic equilibrium with the cluster potential?

Table 4.6. *X-ray properties of rich clusters*

Property	Typical value or range	Notes
L_x (2–10 keV)	$\sim (10^{42.5}\text{–}10^{45})h^{-2}\,\mathrm{erg\,s^{-1}}$	a
$I_x(r)$	$I_x(r) \propto [1 + (r/R_c)^2]^{-3\beta+1/2}$	b
$\langle \beta \rangle$	~ 0.7	c
$\rho_{\mathrm{gas}}(r)$	$\rho_{\mathrm{gas}}(r) \propto [1 + (r/R_c)^2]^{-3\beta/2}$ $\propto [1 + (r/R_c)^2]^{-1}$	d
kT_x	$\sim 2\text{–}14$ keV	e
T_x	$\sim 2 \times 10^7\text{–}10^8$ K	e
$\beta_{\mathrm{spec}} = \dfrac{\sigma_r^2}{kT_x/\mu m_p}$	~ 1	f
$R_c(x)$	$\sim (0.1\text{–}0.3)h^{-1}$ Mpc	g
n_e	$\sim 3 \times 10^{-3} h^{1/2}$ cm^{-3}	h
M_{gas} ($\lesssim 1.5h^{-1}$ Mpc)	$\sim 10^{13.5} M_\odot$ [range: $(10^{13}\text{–}10^{14})M_\odot\,h^{-2.5}$]	i
$M_{\mathrm{gas}}/M_{\mathrm{cl}}$ ($\lesssim 1.5h^{-1}$ Mpc)	~ 0.07 (range: $0.03\text{–}0.15\,h^{-1.5}$)	i
Iron abundance	~ 0.3 solar (range: 0.2–0.5)	j

[a] The X-ray luminosity of clusters [(2-10) keV band]. $\langle L_x \rangle$ increases with cluster richness and with cluster type (toward compact, elliptical-rich clusters) (Bahcall 1977a,b; Sarazin 1986; Edge et al. 1990; Jones & Forman 1992; David et al. 1993; Burg et al. 1994).

[b] X-ray surface brightness distribution, $I_x(r)$; R_c is the cluster core radius.

[c] Mean $\langle \beta \rangle$ from observations of X-ray brightness profiles (Jones & Forman 1984; Sarazin 1986).

[d] Implied spatial density profile of the hot gas in the cluster [from $I_x(r)$; isothermal].

[e] Range of observed X-ray gas temperature in rich clusters (Edge et al. 1990; Henry & Arnaud 1991; Arnaud et al. 1992).

[f] β_{spec} is the ratio of galaxy to gas velocity dispersion: μ is mean molecular weight in amu ($\mu \simeq 0.6$), m_p is mass of the proton, σ_r is radial velocity dispersion of galaxies in the cluster, and T_x is the X-ray temperature of the gas (Lubin & Bahcall 1993).

[g] Cluster core radius determined from the X-ray distribution in the cluster (Jones & Forman 1992).

[h] Typical intracluster gas density in rich cluster cores (Jones & Forman 1992).

[i] Typical mass (and range of masses) of hot gas in rich clusters and its fraction of the total (virial) cluster mass ($M_{\mathrm{gas}}/M_{\mathrm{cl}}$) within $r \lesssim 1.5h^{-1}$ Mpc of the cluster center (Edge & Stewart 1991; Jones & Forman 1992; White & Fabian 1995; Lubin et al. 1996).

[j] Typical iron abundance (and range) of the intracluster gas (in solar units) (Edge & Stewart 1991; Jones & Forman 1992).

- What is the relation of the intracluster gas to the galaxies and mass in the cluster? For example (the subscripts below refer to gas, galaxies, and mass, respectively):

  — Density profiles: $\rho_{\mathrm{gas}}(r)$ vs. $\rho_g(r)$ vs. $\rho_m(r)$?

- Temperature–velocity relation: T vs. σ_g vs. σ_m?
- Mass: M_{gas} vs. M_g vs. M_{cl}?
- Profiles of the above properties:
  $T(r)$ vs. $\sigma_g(r)$ vs. $\sigma_m(r)$?
  $M_{gas}(r)$ vs. $M_g(r)$ vs. $M_{cl}(r)$?
- Core radii: $R_c(gas)$ vs. $R_c(g)$ vs. $R_c(m)$?
- Luminosity versus galaxy-type relation: L_x vs. spiral fraction?

- What is the origin of the hot intracluster gas?
- What is the origin of the metal enrichment of the gas?
- What is the evolution of the intracluster gas?

4.3.4 The intracluster gas: hydrostatic equilibrium?

The standard model of clusters assumes that both the gas and the galaxies are in approximate hydrostatic equilibrium with the binding cluster potential (Bahcall & Sarazin 1977; Jones & Forman 1984; Sarazin 1986; Evrard 1990; Bahcall & Lubin 1994). In this model the gas distribution obeys

$$\frac{dP_{gas}}{dr} = -\frac{GM_{cl}(\leq r)\rho_{gas}}{r^2} \tag{4.32}$$

where P_{gas} and ρ_{gas} are the gas pressure and density, and $M_{cl}(\leq r)$ is the total cluster binding mass within a radius r. The cluster mass can thus be represented as

$$M_{cl}(\leq r) = -\frac{kT}{\mu m_p G}\left(\frac{d\ln\rho_{gas}(r)}{d\ln r} + \frac{d\ln T}{d\ln r}\right)r, \tag{4.33}$$

where T is the gas temperature and μm_p is the mean particle mass of the gas.

The galaxies in the cluster respond to the same gravitational field, and they satisfy

$$M_{cl}(\leq r) = -\frac{\sigma_r^2}{G}\left(\frac{d\ln\rho_g(r)}{d\ln r} + \frac{d\ln\sigma_r^2}{d\ln r} + 2A\right)r, \tag{4.34}$$

where σ_r is the radial velocity dispersion of galaxies in the cluster, $\rho_g(r)$ is the galaxy density profile, and A represents a possible anisotropy in the galaxy velocity distribution [$A = 1-(\sigma_t/\sigma_r)^2$, where t and r represent the tangential and radial velocity components].

The above two relations yield

$$\beta_{spec} \equiv \frac{\sigma_r^2}{kT/\mu m_p} = \frac{d\ln\rho_{gas}(r)/d\ln r + d\ln T/d\ln r}{d\ln\rho_g(r)/d\ln r + d\ln\sigma_r^2/d\ln r + 2A}, \tag{4.35}$$

where the β_{spec} parameter, defined by the left-hand side of the above relation, can be determined directly from observations of cluster velocity dispersions and gas temperatures. The β_{spec} parameter represents the ratio of energy per unit mass in the galaxies to that in the gas. Observations of a large sample of clusters yield a mean best-fit value of $\beta_{spec} \simeq 1 \pm 0.1$ (Lubin & Bahcall 1993; see also §4.3.5). This suggests that, on average, the gas and galaxies follow each other with comparable energies ($\sigma_r^2 \simeq kT/\mu m_p$). The

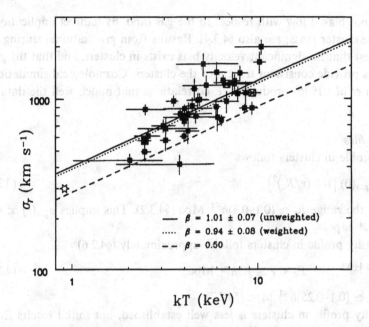

Figure 4.1. Cluster radial velocity dispersion (σ_r) vs. gas temperature (kT) for 41 clusters (Lubin & Bahcall 1993). The best-fit $\beta \equiv \sigma_r^2/(kT/\mu m_p)$ lines are shown by the solid and dotted curves, with $\beta \simeq 1$. The $\beta \simeq 0.5$ line previously proposed for a velocity bias in clusters is shown by the dashed curve; the velocity bias is inconsistent with the data.

observed mean value $\beta_{\text{spec}} \simeq 1 \pm 0.1$ is consistent with the value of β determined from the right-hand side of the β relation (referred to as β_{fit}, and determined from the gas and galaxy density profile fits). Using $\rho_{\text{gas}}(r) \propto r^{-2}$ (§4.3.2) and $\rho_g(r) \propto r^{-2.4 \pm 0.2}$ (§4.2.6), one finds $\beta_{\text{fit}} \simeq 0.85 \pm 0.1$ (for an isothermal distribution) (Bahcall & Lubin 1994). The above consistency supports the assumption that the gas is in an approximate hydrostatic equilibrium with the cluster potential, and suggests that the galaxies and gas approximately trace each other in the clusters.

4.3.5 The relation between gas and galaxies

The hot intracluster gas in rich clusters appears to trace reasonably well the galaxies in the clusters, and – with larger uncertainty – also the cluster mass.

4.3.5.1 Velocity–temperature relation

The galaxy velocity dispersion in clusters is well correlated with the temperature of the intracluster gas; it is observed (Fig. 4.1) that $\sigma_r^2 \simeq kT/\mu m_p$ (Lubin & Bahcall 1993). The best-fit σ–T relation is listed in §4.3.7. The observed correlation indicates that, on average, the energy per unit mass in the gas and in the galaxies is the same. Figure 4.1 shows that, unlike previous expectations, the galaxy velocities (and therefore the implied cluster mass) are not biased low with respect to the gas (and, by indirect implications, with respect to the cluster mass; see also §4.3.4). Results from gravitational lensing by clusters also suggest that no significant velocity bias exists in clusters, and that the gas, galaxies, and mass provide consistent tracers of the clusters. Cosmological simulations of clusters (Lubin et al. 1996) produce σ–T correlations that match well the data in Fig. 4.1.

4.3.5.2 Density profiles
The gas density profile in clusters follows

$$\rho_{gas}(r) \simeq \rho_{gas}(0) \left[1 + (r/R_c)^2\right]^{-1}, \tag{4.36}$$

with core radii in the range $R_c \simeq (0.1$–$0.3)h^{-1}$ Mpc (§4.3.2). This implies $\rho_{gas}(r) \propto r^{-2}$ for $R_c < r \lesssim 1.5h^{-1}$ Mpc.

The galaxy density profile in clusters follows approximately (§4.2.6)

$$\rho_g(r) \propto r^{-2.4\pm0.2} \qquad R_c < r \lesssim 1.5h^{-1} \text{ Mpc} \tag{4.37}$$

with core radii $R_c \simeq (0.1$–$0.25)h^{-1}$ Mpc (§4.2.7).

The mass density profile in clusters is less well established, but initial results from gravitational lensing distortions of background galaxies by foreground clusters suggest that the mass profile is consistent with the galaxy density profile (Fischer & Tyson 1997; Fischer et al. 1997). In the small central core regions of some clusters ($r \lesssim 100$ kpc), the mass distribution may be more compact than the gas or galaxies, with a small mass core radius of $R_c(m) \lesssim 50h^{-1}$ kpc. The results for the overall cluster, however, suggest that the distributions of gas, galaxies, and mass are similar (with the gas distribution possibly somewhat more extended than the galaxies, as seen by the mean density slopes above).

4.3.5.3 Beta discrepancy
The mean $\beta_{spec} \equiv \sigma_r^2/kT/\mu m_p \simeq 1$ result discussed above, combined with the similarity of the gas and galaxy density profile slopes (that yields $\beta_{fit} \simeq 0.85\pm0.1$; §4.3.3) show that the long-claimed β-discrepancy for clusters (where $B_{spec} > \beta_{fit}$ was claimed) has been resolved (Bahcall & Lubin 1994). The gas and galaxies trace each other both in their spatial density distribution and in their energies, as expected for a quasi-hydrostatic equilibrium.

4.3.5.4 Gas mass fraction
The ratio of the mass of gas in clusters to the total virial cluster mass (within $\sim 1.5h^{-1}$ Mpc) is observed to be in the range

$$\frac{M_{gas}}{M_{cl}} \simeq (0.03 - 0.15)h^{-1.5}, \tag{4.38}$$

with a median value of

$$\langle \frac{M_{gas}}{M_{cl}}(\lesssim 1.5h^{-1} \text{ Mpc})\rangle_{median} \simeq 0.07h^{-1.5} \tag{4.39}$$

(Jones & Forman 1992; White *et al.* 1993; White & Fabian 1995; Lubin *et al.* 1996). The implications of this result, which shows a high fraction of baryons in clusters, are discussed in §4.4.

The total gas mass in clusters, $\sim (10^{13} - 10^{14})h^{-2.5}M_\odot$, is generally larger than the total mass of the luminous parts of the galaxies (especially for low values of h). With so much gas mass, it is most likely that a large fraction of the intracluster gas is of cosmological origin (rather than all the cluster gas being stripped out of galaxies). Additional optical-X-ray correlations of clusters are summarized in §4.3.7.

4.3.6 Metal abundance in intracluster gas

The iron-abundance in the intracluster gas is observed to be ~ 0.3 solar, with only small variations ($\pm \sim 0.1$) from cluster to cluster (e.g., Jones & Forman 1992). A strong correlation between the total iron mass in clusters and the total luminosity of the E + S0 cluster galaxies is observed (Jones & Forman 1992). The metal enrichment of the intracluster gas is likely caused by gas stripped out of the elliptical galaxy members.

The iron abundance profile as a function of radius from the cluster center is generally flat, i.e., a constant abundance at all radii (except for some poor, low-mass clusters dominated by a single massive galaxy).

No evolution is observed in the overall iron abundance of clusters to $z \sim 0.4$ (Mushotzky & Loewenstein 1997).

Recent results using ASCA observations (Mushotzky & Loewenstein 1997) of different element abundances in nearby clusters (O, Ne, Mg, Si, S, Ar, Ca) suggest a SN II origin for the metals (resulting from early massive stars) rather than the expected SN Ia. These new results will be expanded in the near future as additional accurate X-ray data become available and will provide further clues regarding the origin of the metallicity of the intracluster gas.

4.3.7 X-ray–optical correlations of cluster properties

Some observed correlations between X-ray and optical properties are listed in Table 4.7 (Bahcall 1977a,b; Edge & Stewart 1991; David *et al.* 1993; Lubin & Bahcall 1993).

4.3.8 The X-ray luminosity function of clusters

The X-ray luminosity function of clusters (the number density of clusters with X-ray luminosity L_x to $L_x + dL_x$) is approximately (Edge *et al.* 1990)

$$\Phi_x(L_x)dL_x \simeq 2.7 \times 10^{-7}(L_x/10^{44})^{-1.65} \exp(-L_x/8.1 \times 10^{44})(dL_x/10^{44}) \text{ Mpc}^{-3} \quad (4.40)$$

for $h = 0.5$, where L_x is the 2–10-keV X-ray luminosity in units of erg s^{-1} (for $h = 0.5$). The luminosity function can also be approximated as a power law (Edge *et al.* 1990)

Table 4.7. *Correlations between X-ray and optical properties*[a]

Properties	Correlation
σ_r–T	σ_r (km s^{-1}) $\simeq (332 \pm 52)[kT \text{ (keV)}]^{0.6\pm0.1}$
T–$N_{0.5}$	kT (keV) $\simeq 0.3 N_{0.5}^{0.95\pm0.18}$
L_x–$N_{0.5}$	L_x(bol) $\sim 1.4 \times 10^{40} N_{0.5}^{3.16\pm0.15} h^{-2}$
L_x–f_{sp}	L_x(bol) $\simeq 0.6 \times 10^{43} f_{sp}^{-2.16\pm0.11} h^{-2}$
f_{sp}–T	$f_{sp} \simeq 1.2[kT \text{ (keV)}]^{-0.94\pm0.38}$
T–L_x	kT (keV) $\simeq 0.3[L_x(\text{bol})h^2/10^{40}]^{0.297\pm0.004}$

[a] σ_r is the galaxy line-of-sight velocity dispersion in the cluster (km s^{-1}). T is the temperature of the intracluster gas [kT (keV)]. $N_{0.5}$ is the central galaxy density in the cluster [number of galaxies brighter than $m_3 + 2^m$, within $r \leq 0.5 h_{50}^{-1} = 0.25 h^{-1}$ of the cluster center (Bahcall 1977a; Edge and Stewart 1991)]. L_x(bol) is the bolometric X-ray luminosity of the cluster (erg s^{-1}). f_{sp} is the fraction of spiral galaxies in the cluster ($\lesssim 1.5 h^{-1}$ Mpc) (Bahcall 1977b; Edge and Stewart 1991). Typical uncertainties of the coefficients are $\sim 50\%$ (see references).

$$\Phi_x(L_x)dL_x \simeq 2.2 \times 10^{-7}(L_x/10^{44})^{-2.17}(dL_x/10^{44}) \text{ Mpc}^{-3} \quad (h=0.5). \tag{4.41}$$

The number of X-ray clusters with X-ray luminosity brighter than L_x is approximately

$$n_{cl}(>L_x) \simeq 2 \times 10^{-7}(L_x/10^{44})^{-1.17} \text{ Mpc}^{-3} \quad (h=0.5). \tag{4.42}$$

The observed evolution of the X-ray cluster luminosity function suggests fewer high-luminosity clusters in the past ($z \gtrsim 0.5$) (Edge et al. 1990, Henry et al. 1992). Additional data is required, however, to confirm and better assess the cluster evolution.

4.3.9 Cooling flows in clusters

Cooling flows are inferred to be common at the dense cores of rich clusters; X-ray images and spectra of $\sim 50\%$ of clusters suggest that the gas is cooling rapidly at their centers (Sarazin 1986; Fabian 1992). Typical inferred cooling rates are $\sim 100 M_\odot$/yr. The gas cools within $r \lesssim 100 h^{-1}$ kpc of the cluster center (generally centered on the brightest galaxy). The cooling flows often show evidence for optical line emission, blue stars, and in some cases evidence for colder material in HI or CO emission, or X-ray absorption.

4.3.10 The Sunyaev–Zel'dovich effect in clusters

The Sunyaev–Zel'dovich effect (Sunyaev & Zel'dovich 1972) is a perturbation to the spectrum of the cosmic microwave background radiation (CMB) as it passes through the hot dense intracluster gas. It is caused by inverse Compton scattering of the radiation by the electrons in the cluster gas.

At the long-wavelength side of the CMB spectrum, the hot gas lowers the brightness temperature seen through the cluster center by the fractional decrement

$$\frac{\delta T}{T} = -2\tau_0 \frac{kT_x}{m_e c^2}, \tag{4.43}$$

where $T = 2.73$ K is the microwave radiation temperature, τ_0 is the Thomson scattering optical depth through the cluster ($\tau_0 = \sigma_T \int n_e dl$, where σ_T is the Thomson scattering cross-section and dl is the distance along the line of sight), T_x is the intracluster gas temperature, and m_e is the electron mass.

For typical observed rich cluster parameters of $L_x \sim 10^{44} h^{-2}$ erg s^{-1}, $R_c \sim 0.2 h^{-1}$ Mpc, and $kT_x \simeq 4$ keV, the bremsstrahlung relation ($L_x \propto n_e^2 R_c^3 T_x^{0.5}$, §4.3.1) implies a central gas density of $n_e \simeq 3 \times 10^{-3} h^{1/2}$ electrons cm^{-3}, thus yielding $\tau_0 \simeq 3 \times 10^{-3} h^{-1/2}$ [$\tau_0 = 0.0064 n_e (\text{cm}^{-3}) R_c(\text{kpc})$]. Therefore

$$\frac{\delta T}{T} \sim -6 \times 10^{-5} h^{-1/2}. \tag{4.44}$$

This temperature decrement remains constant over the cluster core diameter

$$\theta_c \simeq \frac{2H_0 R_c}{cz} \simeq \frac{0.5}{z} \text{ arcmin} \tag{4.45}$$

and decreases at larger separations.

The effect has been detected in observations of some rich, X-ray luminous clusters (e.g., Coma, A665, A2163, A2218, Cl0016+16) (Birkinshaw, Gull & Hordebeck 1984; Jones et al. 1993; Wilbanks et al. 1994; Herbig et al. 1995).

4.4 The baryon fraction in clusters

Rich clusters of galaxies provide the best laboratory for studying the baryon fraction (i.e., the ratio of the mass in baryons to the total mass of the system) on relatively large scales of $\sim$ Mpc. The mass of baryons in clusters is composed of *at least* two components: the hot intracluster gas (§4.3) and the luminous parts of the galaxies.

The baryon fraction in clusters is therefore

$$\frac{\Omega_b}{\Omega_m} \gtrsim \frac{M_{\text{gas}} + M_{\text{stars}}}{M_{\text{cl}}} \simeq 0.07 h^{-1.5} + 0.05, \tag{4.46}$$

where the first term on the right-hand side represents the gas mass ratio (§4.3.5) and the second term corresponds to the stellar (luminous) contribution.

The baryon density required by big-bang nucleosynthesis is (Walker et al. 1991)

$$\Omega_b(\text{BBN}) \simeq 0.015 h^{-2}. \tag{4.47}$$

Comparison of the above relations indicates that if $\Omega_m = 1$ then there are many more baryons observed in clusters than allowed by nucleosynthesis. In fact, combining the two relations yields an Ω_m value for the mass-density of

$$\Omega_m \lesssim \frac{\Omega_b}{0.07h^{-1.5} + 0.05} \simeq \frac{0.015h^{-2}}{0.07h^{-1.5} + 0.05} \sim 0.2 \tag{4.48}$$

for the observed range of $h \sim 0.5$–0.8. Therefore, the baryon density given by nucleosynthesis and the high baryon content observed in clusters (mainly in the hot intracluster gas) suggest that $\Omega_m \simeq 0.2$. This assumes, as expected on this large scale (and as seen in simulations), that the baryons are not segregated relative to the dark matter in clusters (White et al. 1993).

Figure 4.2 compares the observed gas mass fraction in clusters ($\sim$ baryon fraction) with expectations from cosmological simulations of $\Omega_m = 1$ and $\Omega_m = 0.45$ CDM flat models (Lubin et al. 1996). The results show, as expected, that the $\Omega_m = 1$ model predicts a much lower gas mass fraction than observed, by a factor of ~ 3. A low-density CDM model with $\Omega_m \sim 0.2$–0.3 (in mass) best matches the data, as expected from the general analysis discussed above (see White et al. 1993; White & Fabian 1995; Lubin et al. 1996).

In summary, the high baryon fraction observed in clusters suggests, independent of any specific model, that the mass density of the universe is low, $\Omega_m \sim 0.2$–0.3. This provides a powerful constraint on high-density ($\Omega_m = 1$) models; if $\Omega_m = 1$, a resolution of this baryon problem needs to be found (compare §1.4.5, §7.7, Ch. 11).

4.5 Cluster masses

The masses of clusters of galaxies within a given radius, $M(\leq r)$, can be determined from three independent methods:

a) Optical: galaxy velocity dispersion (and distribution) assuming hydrostatic equilibrium (§4.3.3);

b) X-rays: hot gas temperature (and distribution) assuming hydrostatic equilibrium (§4.3.3);

c) Lensing: gravitational distortions of background galaxies (see §10.4; Tyson, Valdes & Wenk 1990; Fischer & Tyson 1997; Fischer et al. 1997; Kaiser & Squires 1993). This method determines directly the surface mass overdensity.

The first two methods were discussed in §§4.2 and 4.3; the last (and newest) method is discussed in Ch. 10. (See also Tyson et al. 1990, 1996; Kaiser & Squires 1993; Smail et al. 1995, and references therein.)

The galaxy and hot gas methods yield cluster masses that are consistent with each other as discussed in §4.3 (with $\sigma_r^2 \simeq kT/\mu m_p$). Gravitational lensing results, which provide direct information about cluster masses, are available only for a small number of clusters so far (with data rapidly increasing). For all but one of the clusters the masses determined from lensing are consistent, within the uncertainties, with the

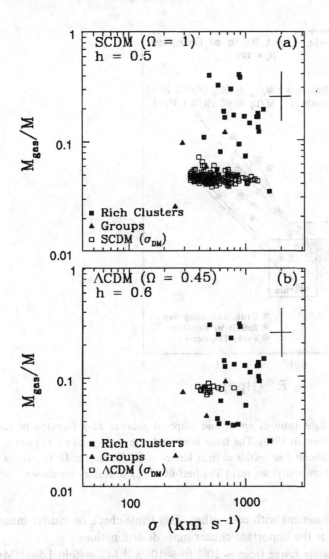

Figure 4.2. Observed and simulated gas mass fraction (M_{gas}/M) vs. line-of-sight velocity dispersion for rich clusters (Lubin et al. 1996). A typical 1σ uncertainty is shown. The simulated results (open squares) present the dark-matter velocity dispersion; the galaxy velocity dispersion is lower by $b_v \sim 0.8$ for SCDM and by $b_v \sim 0.9$ for ΛCDM. (a) SCDM ($\Omega_m = 1, h = 0.5$); (b) ΛCDM ($\Omega_m = 0.45, h = 0.6$).

masses determined from the galaxies and hot gas methods (see Bahcall 1995 for a summary). Some differences in masses between the different methods are expected for individual clusters due to anisotropic velocities, cluster orientation (yielding larger lensing surface mass densities for clusters elongated in the line-of-sight, and vice versa), and substructure in clusters. On average, however, all three independent methods yield

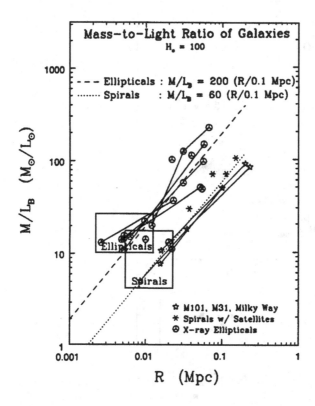

Figure 4.3. Mass-to-light ratio of spiral and elliptical galaxies as a function of scale (Bahcall, Lubin & Dorman 1995). The large boxes indicate the typical ($\sim 1\sigma$) range of M/L_B for bright ellipticals and spirals at their luminous (Holmberg) radii. (L_B refers to *total* corrected blue luminosity; see text.) The best-fit $M/L_B \propto R$ lines are shown.

cluster masses that are consistent with each other. This triple check on cluster masses provides strong support for the important cluster mass determinations.

The masses of rich clusters range from $\sim 10^{14}$ to $\sim 10^{15} h^{-1}$ $M_\odot$ within $1.5 h^{-1}$ Mpc radius of the cluster center. When normalized by the cluster luminosity, a robust mass-to-light ratio is determined for nearby clusters, with only small variations from cluster to cluster (§4.2.10)

$$\frac{M}{L_B}(\text{clusters}) \simeq (300 \pm 100) h (M_\odot/L_\odot) \,. \tag{4.49}$$

This result is similar to the one obtained from the baryon fraction in §4.4.

If, as desired by theoretical arguments, the universe has critical mass density $\Omega_m = 1$, then most of the mass in the universe *cannot* be associated with galaxies, groups, and clusters; the mass distribution in this case would be strongly biased (i.e., mass does not follow light, with the mass distributed considerably more diffusely than the light).

4.6 Where is the dark matter?

A recent analysis of the mass-to-light ratio of galaxies, groups and clusters (Bahcall, Lubin & Dorman 1995) suggests that while the M/L ratio of galaxies increases with scale up to radii of $R \sim (0.1–0.2)h^{-1}$ Mpc, due to the large dark halos around galaxies (see Fig. 4.3; also Ostriker, Peebles & Yahil 1974), this ratio appears to flatten and remain approximately constant for groups and rich clusters, up to scales of ~ 1.5 Mpc, and possibly even up to the larger scales of superclusters (Fig. 4.4). The flattening occurs at $M/L_B \simeq (200–300)h$, corresponding to $\Omega_m \sim 0.2$. This observation suggests that most of the dark matter is associated with the dark halos of galaxies. Unlike previous expectations, this result implies that clusters do *not* contain a substantial amount of *additional* dark matter, other than that associated with (or torn-off from) the galaxy halos, and the hot intracluster medium. Bahcall, Lubin & Dorman (1995) suggest that the relatively large M/L_B ratio of clusters ($\sim 300h$) results mainly from a high M/L_B ratio of elliptical/S0 galaxies. They show (Fig. 4.3) that ellipticals have an M/L_B ratio that is approximately 3 to 4 times larger than typical spirals at the same radius [$(M/L_B)_s \sim 100h$ and $(M/L_B)_e \sim 400h$ within $r \lesssim 200h^{-1}$ kpc]. Since clusters are dominated by elliptical and S0 galaxies, a high M/L_B ratio results.

Unless the distribution of matter is very different from the distribution of light, with large amounts of dark matter in the "voids" or on very large scales, the above results suggest that the mass density in the universe may be low, $\Omega_m \sim 0.2$ [or $\Omega_m \sim 0.3$ for a small biasing of $b \sim 1.5$, where the biasing factor b relates the overdensity in galaxies to the overdensity in mass: $b \equiv (\Delta\rho/\rho)_g/(\Delta\rho/\rho)_m$]. (Compare §§1.4, 7.5, 7.6, and Ch. 8)

4.7 The mass function of clusters

The observed mass function (MF), $n(> M)$, of clusters of galaxies, which describes the number density of clusters above a threshold mass M, can be used as a critical test of theories of structure formation in the universe. The richest, most massive clusters are thought to form from rare high peaks in the initial mass-density fluctuations; poorer clusters and groups form from smaller, more common fluctuations. Bahcall & Cen (1993) determined the MF of clusters of galaxies using both optical and X-ray observations of clusters. Their MF is presented in Fig. 4.5. The function is well fit by the analytic expression

$$n(> M) = 4 \times 10^{-5}(M/M_*)^{-1} \exp(-M/M_*) h^3 \text{ Mpc}^{-3}, \qquad (4.50)$$

with $M_* = (1.8 \pm 0.3) \times 10^{14} h^{-1}$ M$_\odot$, (where the mass M represents the cluster mass within $1.5h^{-1}$ Mpc radius).

The observed cluster mass function is compared in Fig. 4.5 with expectations from different CDM cosmologies using large-scale simulations (Bahcall & Cen 1992). The comparison shows that the cluster MF is indeed a powerful discriminant among models. The standard CDM model ($\Omega_m = 1$) cannot reproduce the observed MF for any biasing

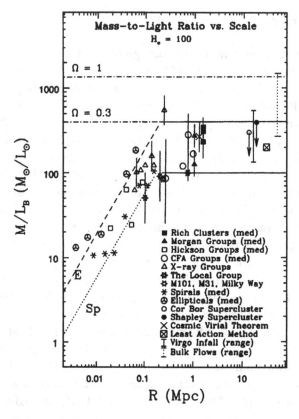

Figure 4.4. Composite mass-to-light ratio of different systems–galaxies, groups, clusters, and superclusters–as a function of scale (Bahcall, Lubin & Dorman 1995). The best-fit $M/L_B \propto R$ lines for spirals and ellipticals (from Fig. 4.3) are shown. We present median values at different scales for the large samples of galaxies, groups and clusters, as well as specific values for some individual galaxies, X-ray groups, and superclusters. Typical 1σ uncertainties and 1σ scatter around median values are shown. Also presented, for comparison, are the M/L_B (or equivalently Ω) determinations from the cosmic virial theorem, the least action method, and the *range* of various reported results from the Virgocentric infall and large-scale bulk flows (assuming mass traces light). The M/L_B expected for $\Omega = 1$ and $\Omega = 0.3$ are indicated.

parameter; when normalized to the COBE microwave background fluctuations on large scales, this model produces too many massive clusters, unseen by the observations. A low-density CDM model on the other hand, with $\Omega_m \sim 0.2$–0.3 (with or without a cosmological constant), appears to fit well the observed cluster MF (Fig. 4.5).

4.8   Quasar–cluster association

Imaging and spectroscopic data (Yee & Green 1987; Ellingson, Yee & Green 1991; Yee 1992) indicate that quasars are found in environments significantly richer than those

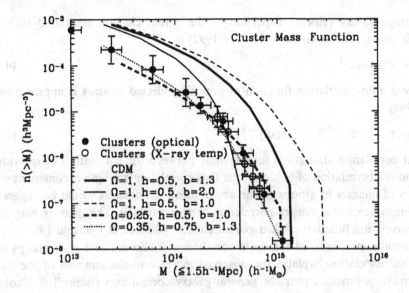

Figure 4.5. Cluster mass functions from observations and from CDM simulations (Bahcall & Cen 1992).

of average galaxies. The data show a positive association of quasars with neighboring galaxies.

Optically selected quasars to $z \lesssim 0.7$ exhibit a quasar–galaxy cross-correlation function amplitude, A_{qg}, that is approximately 2.3 times stronger than the galaxy–galaxy correlation amplitude (to separations $r \lesssim 0.25h^{-1}$ Mpc):

$$\langle A_{qg} \rangle \simeq 2.3 \langle A_{gg} \rangle \simeq 46 \,. \tag{4.51}$$

This excess correlation suggests that the quasars are typically located in groups of galaxies with a *mean* richness

$$\langle N_R \rangle = n_g \int_0^{1.5} A_{qg} r^{-1.8} 4\pi r^2 dr \simeq 12 \text{ galaxies} \tag{4.52}$$

(where $n_g \simeq 0.015$ Mpc^{-3} is the mean density of galaxies). The range of individual group richnesses is, however, wide.

Radio-loud quasars at $z \lesssim 0.5$ are found in similar environments to those of the optical quasars above. At $0.5 \lesssim z \lesssim 0.7$, the radio quasars appear to be located in richer environments, with

$$\langle A_{qg} \rangle \simeq 8 \langle A_{gg} \rangle \simeq 160 \qquad \text{(radio quasars, } 0.5 \lesssim z \lesssim 0.7\text{)}\,. \tag{4.53}$$

This cross-correlation amplitude corresponds to a *mean* environment of rich clusters ($R \sim 0$, $N_R \sim 40$). Radio quasars at these redshifts are thus typically found in rich clusters.

The average galaxy velocity dispersion of the parent clusters associated with the quasars (Ellingson, Yee & Green 1991; Yee 1992) is

$$\sigma_r \sim 500 \text{ km s}^{-1}. \tag{4.54}$$

The observed auto-correlation function of optically selected quasars is approximately (Shaver 1988)

$$\xi_{qq}(r, z \sim 0) \simeq 10^{2\pm0.2}[r(\text{Mpc})]^{-1.8}. \tag{4.55}$$

The quasar correlation strength is intermediate between the correlation of individual galaxies and the correlation of rich clusters. This correlation strength is consistent with the location of quasars in groups of the above mean richness, as would be suggested by the richness-dependent cluster correlation function (§4.10). The quasars may thus trace the correlation function of their parent clusters (Bahcall & Chokshi 1991).

Similar results are observed for the association of radio galaxies with groups and clusters. This association explains the observed increase in the strength of the radio galaxy correlation function over the general galaxy correlations (Bahcall & Chokshi 1992).

4.9 Superclusters

Rich clusters of galaxies are powerful tracers of the large-scale structure of the universe (Bahcall 1988, Peebles 1993). The use of clusters in tracing the large-scale structure is summarized below, including superclusters (§4.9), statistics of the cluster correlation function (§4.10), and peculiar motions on large scales (§4.11).

4.9.1 Supercluster properties

Redshift surveys of galaxies reveal that superclusters are very large, high-density systems of galaxies that are flattened or filamentary in shape, extending to tens of Mpc. The superclusters appear to surround large underdense regions ("voids") of comparable sizes creating a "cellular-like" morphology of the universe on large scales (Gregory and Thompson 1978; Gregory, Thompson & Tifft 1981; Chincarini, Rood & Thompson 1981; Giovanelli, Haynes & Chincarini 1986; de-Lapparent, Geller & Huchra 1986; da Costa et al. 1988; Rood 1988; Schectman et al. 1996; Landy et al. 1996).

Large-scale superclusters have been identified very effectively by rich clusters of galaxies (Bahcall & Soneira 1984), like high mountain peaks tracing mountain chains. Superclusters are generally defined as clusters of rich clusters of galaxies above a given spatial density enhancement f. Here $f \equiv n_{cl}(SC)/n_{cl}$, where $n_{cl}(SC)$ is the number density of clusters in a supercluster and n_{cl} is the mean number density of clusters. The observed superclusters are large flattened systems, extending to $\sim 150h^{-1}$ Mpc in size. The superclusters typically contain several clusters. The high-density superclusters generally surround low-density regions.

Table 4.8. *Global properties of Bahcall–Soneira superclusters*

Property	$f = 20$ superclusters
Number density of SCs	$\sim 10^{-6} h^3$ Mpc^{-3}
Number of clusters per SC	2–15 clusters
Fraction of clusters in SCs	54%
Size of largest SC	$\sim 150 h^{-1}$ Mpc
SC shape	Flattened
Volume of space occupied by SCs	$\sim 3\%$

A complete catalog of superclusters was constructed by Bahcall & Soneira (1984) to $z \lesssim 0.08$. The catalog identifies all superclusters that have a spatial density enhancement $f \geq 20$. The mean number density of the Bahcall–Soneira superclusters is $\sim 10^{-6} h^3$ Mpc^{-3}, with a mean separation between superclusters of $\sim 100 h^{-1}$ Mpc. A summary of the main properties of the superclusters is presented in Table 4.8.

The superclusters trace well the structure observed in the more detailed, but smaller, galaxy redshift surveys.

4.9.2 Superclusters and pencil-beam surveys

Observations of the redshift distribution of galaxies in narrow (~ 40 arcmin) pencil-beam surveys to $z \lesssim 0.3$ (Broadhurst *et al.* 1990; hereafter BEKS) reveal a highly clumped and apparently periodic distribution of galaxies. The distribution features peaks of galaxy counts with an apparently regular separation of 128 Mpc, with few galaxies between the peaks. What is the origin of this clumpy, periodic distribution of galaxies? What does it imply for the nature of the large-scale structure and the properties discussed above? Bahcall (1991) investigated these questions observationally, by comparing the specific galaxy distribution with the distribution of known superclusters.

Bahcall showed that the observed galaxy clumps originate from the tails of large superclusters (§4.9.1). When the narrow beams intersect these superclusters, which have a mean separation of ~ 100 Mpc, the BEKS galaxy distribution is reproduced.

The redshift distribution of the superclusters is essentially identical to the galaxy redshift distribution, i.e., it reproduces the observed peaks in the BEKS survey, for $z \lesssim 0.1$. This indicates that the galaxy clumps observed in the pencil-beam survey originate from these superclusters as the beam crosses the superclusters' surfaces. The main superclusters that contribute to the clumps were identified. For example, the first northern clump originates from the Coma–Hercules supercluster (= the Great Wall); the second northern clump is mostly due to the large Corona Borealis supercluster.

The narrow-beam survey of BEKS is directed toward the north and south galactic poles. Some of the Bahcall–Soneira superclusters coincident with the BEKS peaks are located at projected distances of up to $\sim (50\text{--}100)$ Mpc from the poles. This suggests that the high-density supercluster regions are embedded in still larger halo surfaces, ~100 Mpc in size, and that these large structures surround large underdense regions. The observed number of clumps and their mean separation are consistent with the number density of superclusters and their average extent (§4.9.1).

The narrow widths of the BEKS peaks are consistent with, and imply, flat superclusters. From simulations of superclusters and pencil-beams, Bahcall, Miller & Udomprasert (1998) find that the observed peak-widths distribution is consistent with that expected of randomly placed superclusters with $\lesssim 15$ Mpc width and ~ 150 Mpc extent.

The Bahcall–Soneira superclusters may exhibit weak positive correlations on scales $\sim(100$–$150)$ Mpc (Bahcall & Burgett 1986). This implies that the superclusters, and thus their related galaxy clumps, are not randomly distributed but are located in some weakly correlated network of superclusters and voids, with typical mean separation of ~ 100 Mpc. This picture is consistent with statistical analyses of the BEKS distribution as well as with the observational data of large-scale structure. The apparent periodicity in the galaxy distribution suggested by BEKS is expected to be greatly reduced when pencil-beams in various directions are combined; the scale reflects the typical *mean* separation between large superclusters, $\sim(100$–$150)h^{-1}$ Mpc, but with large variations at different locations.

4.10 The cluster correlation function

The correlation function of clusters of galaxies efficiently quantifies the large-scale structure of the universe. Clusters are correlated in space more strongly than are individual galaxies, by an order of magnitude, and their correlation extends to considerably larger scales ($\sim 50h^{-1}$ Mpc). The cluster correlation strength increases with richness ($\propto$ luminosity or mass) of the system from single galaxies to the richest clusters (Bahcall & Soneira 1983; Bahcall & West 1992). The correlation strength also increases with the mean spatial separation of the clusters (Szalay & Schramm 1985; Bahcall & Burgett 1986; Bahcall & West 1992). This dependence results in a "universal" dimensionless cluster correlation function; the cluster dimensionless correlation scale is constant for all clusters when normalized by the mean cluster separation.

Empirically, two general relations have been found (Bahcall & West 1992) for the correlation function of clusters of galaxies, $\xi_i = A_i r^{-1.8}$:

$$A_i \propto N_i , \tag{4.56}$$

$$A_i \simeq (0.4 d_i)^{1.8} , \tag{4.57}$$

where A_i is the amplitude of the cluster correlation function, N_i is the richness of the galaxy clusters of type i (§4.2.2), and d_i is the mean separation of the clusters. Here $d_i = n_i^{-1/3}$, where n_i is the mean spatial number density of clusters of richness N_i (§4.2.3) in a volume-limited, richness-limited complete sample. The first relation, equation (4.56), states that the amplitude of the cluster correlation function increases with cluster richness, i.e., rich clusters are more strongly correlated than poorer clusters.

The second relation, equation (4.57), states that the amplitude of the cluster correlation function depends on the mean separation of clusters (or, equivalently, on their number density); the rarer, large mean separation richer clusters are more strongly correlated than the more numerous poorer clusters. Equations (4.56) and (4.57) relate to each other through the richness function of clusters, i.e., the number density of clusters as a function of their richness. Equation (4.57) describes a universal scale-invariant (dimensionless) correlation function with a correlation scale $r_{0i} = A_i^{1/1.8} \simeq 0.4 d_i$ (for $30 \lesssim d_i \lesssim 90 h^{-1}$ Mpc).

There are some conflicting statements in the literature about the precise values of the correlation amplitude, A_i. Nearly all these contradictions are caused by not taking account of equation (4.56). When apples are compared to oranges, or the clustering of rich clusters is compared to the clustering of poorer clusters, differences are expected and observed. Figure 4.6 clarifies the observational situation. The $A_i(d_i)$ relation for groups and clusters of various richnesses is presented in the figure. The recent automated cluster surveys of APM (Dalton et al. 1992) and EDCC (Nichol et al. 1992) are consistent with the predictions of equations (4.56) and (4.57), as is the correlation function of X-ray selected ROSAT clusters of galaxies (Romer et al. 1994). Bahcall & Cen (1994) show that a flux-limited sample of X-ray selected clusters will exhibit a correlation scale that is smaller than that of a volume-limited, richness-limited sample of comparable apparent spatial density since the flux-limited sample contains poor groups nearby and only the richest clusters farther away. Using the richness-dependent cluster correlations of equations (4.56) and (4.57), Bahcall & Cen (1994) find excellent agreement with the observed flux-limited X-ray cluster correlations of Romer et al. (1994).

The strong correlation amplitude of galaxy clusters, and the large scales to which clusters are observed to be positively correlated [$\sim (50-100) h^{-1}$ Mpc], complement and quantify the superclustering of galaxy clusters discussed in §4.9. Clusters of galaxies are strongly clustered in superclusters of large scales (§4.9), consistent with the strong cluster correlations to these scales (§4.10).

This fundamental observed property of clusters of galaxies–the cluster correlation function–can be used to place strong constraints on cosmological models and the density parameter Ω_m by comparison with model expectations. Bahcall & Cen (1992) contrasted these observations with standard and non-standard CDM models using large N-body simulations ($400 h^{-1}$ box, $10^{7.2}$ particles). They find that none of the standard $\Omega_m = 1$ CDM models can fit consistently the strong cluster correlations. A low-density ($\Omega_m \sim 0.2-0.3$) CDM-type model (with or without a cosmological constant), however, provides a good fit to the cluster correlations (see Figs. 4.7–4.9) as well as to the observed cluster mass function (§4.7, Fig. 4.5). This is the first CDM-type model that is consistent with the high amplitude and large extent of the correlation function of the Abell, APM, and EDCC clusters. Such low-density models are also consistent with other observables as discussed in this paper. The Ω_m constraints of these cluster results are model-dependent; a mixed hot + cold dark-matter model, for example, with $\Omega_m = 1$, is also consistent with these cluster data (see Ch. 1).

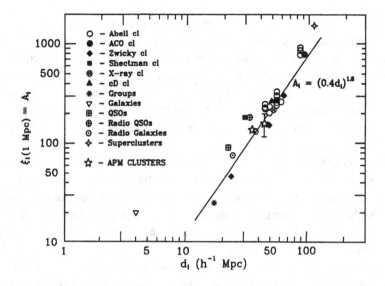

Figure 4.6. The *universal* dimensionless cluster correlations: the dependence of correlation amplitude on mean separation (Bahcall & West 1992). Data points include different samples and catalogs of clusters and groups, as well as X-ray selected and cD clusters. Quasars and radio galaxies, as represented by their parent groups, are also included. The APM results are presented; they are consistent with the expected relation.

The CDM results for clusters corresponding to the rich Abell clusters (richness class $R \geq 1$) with $d = 55h^{-1}$ Mpc are presented in Fig. 4.7 together with the observed correlations (Bahcall & Soneira 1983; Peacock & West 1992). The results indicate that the standard $\Omega_m = 1$ CDM models are inconsistent with the observations; they cannot provide either the strong amplitude or the large scales ($\gtrsim 50h^{-1}$ Mpc) to which the cluster correlations are observed. Similar results are found for the APM and EDCC clusters.

The low-density, low-bias model is consistent with the data; it reproduces both the strong amplitude and the large scale to which the cluster correlations are detected.

The dependence of the observed cluster correlation on d was also tested in the simulations. The results are shown in Fig. 4.8 for the low-density model. The dependence of correlation amplitude on mean separation is clearly seen in the simulations. To compare this result directly with observations, Fig. 4.9 shows the dependence of the correlation scale, r_0, on d for both the simulations and the observations. The low-density model agrees well with the observations, yielding $r_0 \approx 0.4d$, as observed. The $\Omega_m = 1$ model, while also showing an increase of r_0 with d, yields considerably smaller correlation scales and a much slower increase of $r_0(d)$.

What causes the observed dependence on cluster richness [equations (4.56)–(4.57)]? The dependence, seen both in the observations and in the simulations, is most likely caused by the statistics of rare peak events, which Kaiser (1984) suggested as an

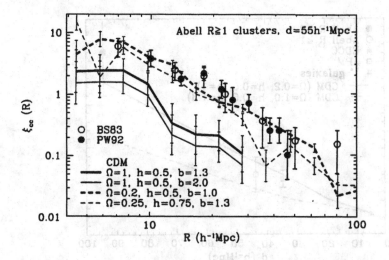

Figure 4.7. Two-point correlation function of Abell $R \geq 1$ clusters, with mean separation $55h^{-1}$ Mpc, from observations and the CDM simulations (Bahcall & Cen 1992).

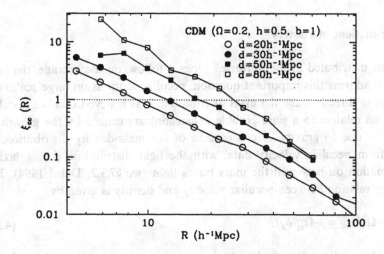

Figure 4.8. Model dependence of the cluster correlation function on mean separation d (CDM simulation: $\Omega = 0.2, h = 0.5, b = 1$) (from Bahcall & Cen 1992).

explanation of the observed strong increase of correlation amplitude from galaxies to rich clusters. The correlation function of rare peaks in a Gaussian field increases with their selection threshold. Since more massive clusters correspond to a higher threshold, implying rarer events and thus larger mean separation, equation (4.57) results. A fractal distribution of galaxies and clusters would also produce equation (4.57) (e.g., Szalay & Schramm 1985).

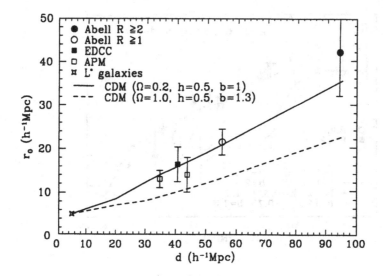

Figure 4.9. Correlation length as a function of cluster separation, from both observations and simulations (Bahcall & Cen 1992).

4.11 Peculiar motions of clusters

How is the mass distributed in the universe? Does it follow, on the average, the light distribution? To address this important question, peculiar motions on large scales are studied in order to directly trace the mass distribution. It is believed that the peculiar motions (motions relative to a pure Hubble expansion) are caused by the growth of cosmic structures due to gravity. A comparison of the mass-density distribution, as reconstructed from peculiar velocity data, with the light distribution (i.e., galaxies) provides information on how well the mass traces light (see §7.3.2; Dekel 1994). The basic underlying relation between peculiar velocity and density is given by

$$\vec{\nabla} \cdot \vec{v} = -\Omega_m^{0.6} \delta_m = -\Omega_m^{0.6} \delta_g / b \tag{4.58}$$

where $\delta_m \equiv (\Delta\rho/\rho)_m$ is the mass overdensity, δ_g is the galaxy overdensity, and $b \equiv \delta_g/\delta_m$ is the biasing parameter discussed in §4.6. A formal analysis yields a measure of the parameter $\beta \equiv \Omega_m^{0.6}/b$. Other methods that place constraints on β include the anisotropy in the galaxy distribution in the redshift direction due to peculiar motions (see Strauss & Willick 1995 for a review).

Measuring peculiar motions is difficult. The motions are usually inferred with the aid of measured distances to galaxies or clusters that are obtained using some (moderately reliable) distance indicators (such as the Tully–Fisher or D_n–σ relations), and the measured galaxy redshift. The peculiar velocity v_p is then determined from the difference between the measured redshift velocity, cz, and the measured Hubble velocity, v_H, of the system (the latter obtained from the distance–indicator): $v_p = cz - v_H$.

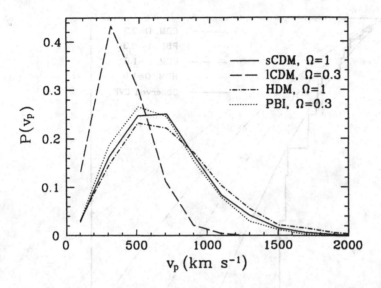

Figure 4.10. Differential three-dimensional peculiar velocity distribution of rich clusters of galaxies for four cosmological models (Bahcall, Gramann & Cen 1994).

A summary of all measurements of β made so far is presented in Table 1 of §7.6, and in Strauss & Willick (1995). The dispersion in the current measurements of β is very large; the various determinations range from $\beta \sim 0.4$ to ~ 1, implying, for $b \simeq 1, \Omega_m \sim 0.2$ to ~ 1. No strong conclusion can therefore be reached at present regarding the values of β or Ω_m. The larger and more accurate surveys currently underway, including high-precision velocity measurements, will likely lead to the determination of β and possibly its decomposition into Ω_m and b (e.g., Cole, Fisher & Weinberg 1994).

Clusters of galaxies can also serve as efficient tracers of the large-scale peculiar velocity field in the universe (Bahcall, Gramann & Cen 1994). Measurements of cluster peculiar velocities are likely to be more accurate than measurements of individual galaxies, since cluster distances can be determined by averaging a large number of cluster members as well as by using different distance indicators. Using large-scale cosmological simulations, Bahcall, Gramann & Cen (1994) find that clusters move reasonably fast in all the cosmological models studied, tracing well the underlying matter velocity field on large scales. The clusters exhibit a Maxwellian distribution of peculiar velocities as expected from Gaussian initial density fluctuations. The model cluster 3-D velocity distribution, presented in Fig. 4.10, typically peaks at $v \sim 600$ km s^{-1} and extends to high cluster velocities of ~ 2000 km s^{-1}. The low-density CDM model exhibits lower velocities (Fig. 4.10). Approximately 10% of all model rich clusters (1% for low-density CDM) move with $v \gtrsim 10^3$ km s^{-1}. A comparison of model expectation with recent, well-calibrated cluster velocity data (Giovanelli, Haynes & Chincarini 1996) is presented in Fig. 4.11 (Bahcall & Oh

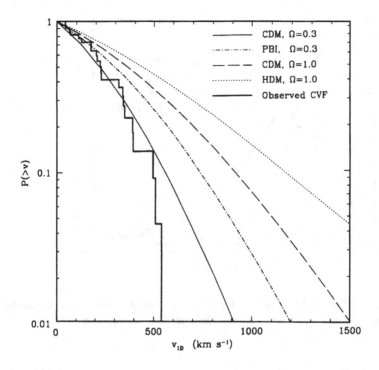

Figure 4.11. Observed vs. model cluster peculiar velocity functions (from Bahcall & Oh 1996). The Giovanelli et al. (1997) data are compared with model expectations convolved with the observational errors. Note the absence of a high velocity tail in the observed cluster velocity function.

1996). The comparison between models and observations suggests that the cluster velocity data is consistent with a low-density CDM model, and is inconsistent with a standard $\Omega_m = 1$ CDM model, since no high velocity clusters are observed.

Cen, Bahcall & Gramann (1994) determined the expected velocity correlation function of clusters in different cosmologies. They find that close cluster pairs, with separations $r \lesssim 10h^{-1}$ Mpc, exhibit strong attractive motions; the pairwise velocities depend sensitively on the model. The mean pairwise attractive cluster velocities on $5h^{-1}$ Mpc scale ranges from ~ 1700 km s^{-1} for $\Omega_m = 1$ CDM to ~ 700 km s^{-1} for $\Omega_m = 0.3$ CDM. The cluster velocity correlation function, presented in Fig. 4.12, is negative on small scales – indicating large attractive velocities, and is positive on large scales, to $\sim 200h^{-1}$ Mpc – indicating significant bulk motions in the models. None of the models reproduce the very large bulk flow of clusters on $150h^{-1}$ Mpc scale, $v \simeq 689 \pm 178$ km s^{-1}, recently reported by Lauer & Postman (1994). The bulk flow expected on this large scale is generally $\lesssim 200$ km s^{-1} for all the models studied ($\Omega_m = 1$ and $\Omega_m \simeq 0.3$ CDM, and PBI).

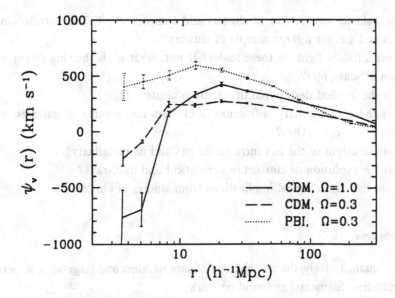

Figure 4.12. Velocity correlation function of rich ($R \geq 1$) clusters of galaxies for three models. Error bars indicate the 1σ statistical uncertainties (from Cen, Bahcall & Gramann 1994).

4.12 Some unsolved problems

Considerable progress has been made over the last two decades in the study of clusters and superclusters of galaxies, as described above. However, many problems still remain open. Some of the unsolved problems in this field that are likely to be solved in the coming decade are highlighted here. Currently planned large redshift surveys of galaxies and clusters such as the Sloan Digital Sky Survey and the 2dF survey, deep optical and X-ray surveys using HST, Keck, ROSAT, ASCA, and AXAF, among other, should allow a considerable increase in our understanding of the nature and evolution of these fundamental systems. At the same time, state of the art cosmological simulations to be available in the next decade (e.g., Ch. 8) should greatly enhance our ability to compare the observations with detailed expectations from various cosmologies and hopefully narrow down the correct cosmological model of our universe.

Here is a partial list of some of the interesting unsolved problems in the field of clusters and superclusters of galaxies.

4.12.1 Clusters of galaxies

- What is the mass distribution and its extent in clusters of galaxies? Using gravitational lensing distortions, one can determine the mass density profile, $\rho_m(r)$,

and total cluster mass, $M(r)$, of clusters and compare it with the distribution of galaxies and gas for a large sample of clusters.
- Does mass follow light on these scales? If not, what is the biasing factor as a function of scale, $b(r)$?
- What is the implied density parameter from clusters, $\Omega_m(r)$?
- What is the accurate baryon-fraction in clusters and groups of galaxies, as a function of scale, $\Omega_b/\Omega(r)$?
- What is the origin of the hot intracluster gas and its metallicity?
- What is the evolution of clusters in the optical and in X-rays?
- What are the cosmological implications from studies of clusters?

4.12.2 Superclusters

- What is, quantitatively, the morphology of superclusters and large-scale structure (superclusters, filaments, and void network)?
- What is the dependence of the superclustering properties on galaxy luminosity, surface brightness, type (E, S), and system (galaxies versus clusters)?
- What are the peculiar motions in superclusters and on large scales?
- What is the mass, and mass distribution, in superclusters and on large scales? Does mass follow light?
- What is in the "voids"?
- What is Ω_m on large scales?
- What is the baryon fraction in superclusters?
- What is the time evolution of superclusters?
- What are the constraints made by the observed superclusters and large-scale structure on cosmology and galaxy formation models?

Many of these questions will be addressed and possibly solved in the coming decade.

Acknowledgments

NAB thanks the organizers of the Jerusalem Winter School 1995, J. P. Ostriker and A. Dekel, for an outstanding, productive, and fun school. The work by NAB and collaborators is supported by NSF grant AST93-15368.

References

Abell, G. O. 1958, *ApJS*, **3**, 211

Arnaud, M., et al. 1992, *ApJ*, **390**, 345

Bahcall, J. N. & Sarazin, C. 1977, *ApJ*, **213**, L99

Bahcall, N. A. 1975, *ApJ*, **198**, 249

Bahcall, N. A. 1977, *ARA&A*, **15**, 505

Bahcall, N. A. 1977a, *ApJ*, **217**, L77

Bahcall, N. A. 1977b, *ApJ*, **218**, L93

Bahcall, N. A. 1981, *ApJ*, **247**, 787

Bahcall, N. A. 1988, *ARA&A*, **26**, 631

Bahcall, N. A. 1991, *ApJ*, **376**, 43

Bahcall, N. A. 1995, in *Dark Matter*, AIP Conf. Proceedings 336, ed. S. Holt & C. Bennet (New York: AIP), 201

Bahcall, N. A. 1996, in *Astrophysical Quantities*, ed. A. Cox (New York: AIP)

Bahcall, N. A. & Burgett, W. 1986, *ApJ*, **300**, L35

Bahcall, N. A. & Cen, R. 1992, *ApJ*, **398**, L81

Bahcall, N. A. & Cen, R. 1993, *ApJ*, **407**, L49

Bahcall, N. A. & Cen, R. Y. 1994, *ApJ*, **426**, L15

Bahcall, N. A. & Chokshi, A. 1991, *ApJ*, **380**, L9

Bahcall, N. A. & Chokshi, A. 1992, *ApJL*, **385**, L33

Bahcall, N. A. & Lubin, L. M. 1994, *ApJ*, **426**, 513

Bahcall, N. A. & Oh, P. 1996, *ApJ*, **462**, L43

Bahcall, N. A. & Soneira, R. M. 1983, *ApJ*, **270**, 20

Bahcall, N. A. & Soneira, R. M. 1984, *ApJ*, **277**, 27

Bahcall, N. A. & West, M. 1992, *ApJL*, **392**, 419

Bahcall, N. A., Gramann, M. & Cen, R. 1994, *ApJ*, **436**, 23

Bahcall, N. A., Lubin, L. M. & Dorman, V. 1995, *ApJ*, **447**, L81

Bahcall, N. A., Miller, N. & Udomprasert, P. 1998, in preparation

Bautz, L. P. & Morgan, W. W. 1970, *ApJ*, **162**, L149

Birkinshaw, M., Gull, S. F. & Hardebeck, H. E. 1984, *Nature*, **309**, 34

Broadhurst, T. J., Ellis, R., Koo, D. & Szalay, A. 1990, *Nature*, **343**, 726

Burg, R., Giacconi, R., Forman, W. & Jones, C. 1994, *ApJ*, **422**, 37

Cen, R., Bahcall, N. A. & Gramann, M. 1994, *ApJ*, **437**, L51

Chincarini, G., Rood, H. J. & Thompson, L. A. 1981, *ApJ*, **249**, L47

Cole, S., Fisher, K. B. & Weinberg, D. H. 1994, *MNRAS*, **267**, 785

da Costa, L. N., et al. 1988, *ApJ*, **327**, 544

Dalton, G. B., Efstathiou, G., Maddox, S. J. & Sutherland, W. 1992, *ApJ*, **390**, L1

David, L. P., Slyz, A., Jones, C., Forman, W., Vrtilek, S. & Arnaud, K. 1993, *ApJ*, **412**, 479

de Lapparent, V., Geller, M. & Huchra, J. 1986, *ApJ*, **302**, L1

Dekel, A. 1994, *ARA&A*, **32**, 371

Dressler, A. 1978, *ApJ*, **226**, 55

Dressler, A. 1980, *ApJ*, **236**, 351

Dressler, A. 1984, *ARA&A*, **22**, 185

Edge, A. & Stewart, G. C., 1991, *MNRAS*, **252**, 428

Edge, A., Stewart, G. C., Fabian, A. C. & Arnaud, K. A. 1990, *MNRAS*, **245**, 559

Ellingson, E., Yee, H. K. C. & Green, R. F. 1991, *ApJ*, **371**, 45

Evrard, A. E. 1990, *ApJ*, **363**, 349

Fabian, A. C. 1992, in *Clusters and Superclusters of Galaxies*, NATO ASI Series No. 366, ed. A. C. Fabian (Dordrecht: Kluwer Academic), p. 151

Fischer, P. & Tyson, J. A. 1997, *AJ*, **114**, 14

Fischer, P., Bernstein, G., Rhee, G. & Tyson, J. A. 1997, *AJ*, **113**, 521

Geller, M. J. 1990, in *Clusters of Galaxies*, STScI Symposium No. 4, ed. W. R. Oegerle et al. (Cambridge: Cambridge Univerity Press), p. 25

Giovanelli, R., et al. 1997, *AJ*, **113**, 22

Giovanelli, R., Haynes, M. & Chincarini, G. 1986, *ApJ*, **300**, 77

Gregory, S. A. & Thompson, L. A. 1978, *ApJ*, **222**, 784

Gregory, S. A., Thompson, L. A. & Tifft, W. 1981, *ApJ*, **243**, 411

Henry, J. P. & Arnaud, K. A. 1991, *ApJ*, **372**, 410

Henry, J. P., Gioia, I. M., Maccacaro, T., Morris, S. L., Stocke, J. T. & Walter, A. 1992, *ApJ*, **386**, 408

Herbig, T., Lawrence, C. R., Readhead, A. C. S. & Gulkis, S. 1995, *ApJL*, **449**, L1

Hoessel, J. G., Gunn, J. E. & Thuan, T. X. 1980, *ApJ*, **241**, 486

Hughes, J. P. 1989, ApJ, **337**, 21

Jones, C. & Forman, W. 1984, *ApJ*, **276**, 38

Jones, C. & Forman, W. 1992, in *Clusters and Superclusters of Galaxies*, NATO ASI Series, No. 366, ed. A. C. Fabian (Dordrecht: Kluwer Academic), p. 49

Jones, M., Saunders, R., Alexander, P., Birkinshaw, M. & Dillon, N. 1993, *Nature*, **365**, 320

Kaiser, N. 1984, *ApJ*, **284**, L9

Kaiser, N. & Squires, G. 1993, *ApJ*, **404**, 441

King, I. 1972, *ApJ*, **174**, L123

Landy, S., Shectman, S., Lin, H., Kirshner, R., Oemler, A. & Tucker, A. 1996, *ApJ*, **456**, L1

Lauer, T. & Postman, M. 1994, *ApJ*, **425**, 418

Lilje, P. B. & Efstathiou, G. 1988, *MNRAS*, **231**, 635

Lubin, L. & Bahcall, N. A. 1993, *ApJ*, **415**, L17

Lubin, L., Cen, R., Bahcall, N. A. & Ostriker, J. P. 1996, *ApJ*, **460**, 10

Lumsden, S. L., Nichol, R. C., Collins, C. A. & Guzzo, L. 1992, *MNRAS*, **258**, 1

Mushotzky, R. & Loewenstein, M. 1997, *ApJ*, **481**, L63

Nichol, R., Collins, C. A., Guzzo, L. & Lumsden, S. L. 1992, *MNRAS*, **255**, 21

Oemler, A. 1974, *ApJ*, **194**, 1

Oort, J. 1983, *ARA&A*, **21**, 373

Ostriker, J. P., Peebles, P. J. E. & Yahil, A. 1974, *ApJ*, **193**, L1

Peacock, J. & West, M. 1992, *MNRAS*, **259**, 494

Peebles, P. J. E. 1980, *The Large Scale Structure of the Universe* (Princeton: Princeton University Press)

Peebles, P. J. E. 1993, *Principles of Physical Cosmology* (Princeton: Princeton University Press)

Postman, M. & Geller, M. 1984, *ApJ*, **281**, 95

Romer, A. K., et al. 1994, *Nature*, **372**, 75

Rood, H. J. 1988, *ARA&A*, **26**, 245

Rood, H. J. & Sastry, G. N. 1971, *PASP*, **83**, 313

Sarazin, C. L. 1986, *RMP*, **58**, 1

Schechter, P. L. 1976, *ApJ*, **203**, 297

Shaver, P. 1988, in *Large-Scale Structure of the Universe*, IAU Symposium No. 130, ed. J. Audouze et al. (Dordrecht: Reidel), p. 359

Shectman, S., Landy, S., Oemler, A., Tucker, A., Lin, H. & Kirshner, R. 1996, *ApJ*, **470**, 172

Smail, J., Ellis, R., Fitchett, M. & Edge, A.

1995, *MNRAS*, **273**, 277

Strauss, M. & Willick, J. 1995, *Phys. Rep.*, **261**, 271

Struble, M. & Rood, H. 1991, *ApJS*, **77**, 363

Sunyaev, R. A. & Zeldovich, Ya. B., 1972, *Comments Astrophys. Space Phys.*, **4**, 173

Szalay, A. & Schramm, D. N. 1985, *Nature*, **314**, 718

Tyson, J. A., Valdes, F. & Wenk, R. A. 1990, *ApJ*, **349**, L1

Walker, T. P., et al. 1991, *ApJ*, **376**, 51

White, D. & Fabian, A. 1995, *MNRAS*, **273**, 72

White, S. D. M., Navaro, J. F., Evrard, A. E. & Frenk, C. S. 1993 *Nature*, **366**, 429

Wilbanks, T. M., Ade, P. A. R., Fischer, M. L., Holzapfel, W. L. & Lange, A. 1994, *ApJ*, **427**, L75

Yee, H. K. C. 1992, in *Clusters and Superclusters of Galaxies*, NATO ASI Series No. 366, ed. A. C. Fabian (Dordrecht: Kluwer Academic), p. 293

Yee, H. K. C. & Green, R. F. 1987, *ApJ*, **319**, 28

Zwicky, F. 1957, *Morphological Astronomy* (Berlin: Springer)

5 Redshift surveys of the local universe

Michael A. Strauss

Abstract

Progress in studying the large-scale structure of the universe through redshift surveys of galaxies is reviewed. Of the many statistical methods used to describe the galaxy distribution, the focus here is on the power spectrum, and the factors which complicate (and make interesting!) its interpretation, such as redshift-space distortions, nonlinear effects, and the relative biasing of galaxies and dark matter. Also discussed are two large redshift surveys which are just starting, the Sloan Digital Sky Survey (SDSS), and the Two-Degree Field Redshift Survey (2dF), which promise to increase the number of measured redshifts of galaxies in uniform surveys by more than an order of magnitude.

5.1 Introduction

The past two decades have seen an explosion of our knowledge in many areas of observational cosmology. One of the most significant has been increases in our understanding of the distribution of galaxies in the nearby universe (defined in the context of the present review as $z \lesssim 0.1$, where cosmological and evolutionary corrections can be neglected, for the most part). In addition to allowing us to do *cosmography*, whereby the structures and forms the galaxies find themselves in are mapped and cataloged, the several orders of magnitude increase in the number of galaxies with measured redshifts over this time period allows us to do quantitative *cosmology*, whereby we put specific constraints on models for structure formation and the various parameters which are input to the Friedman–Robertson–Walker metric. In addition to the much larger number of measured redshifts in complete samples of galaxies available now, there have been great advances in our theoretical understanding of the various statistics that have been measured from redshift surveys.

This review does not attempt to give a complete and thorough summary of the entire subject of what can be learned from redshift surveys. The field was reviewed in a comprehensive article which covers material through the end of 1994 (Strauss & Willick 1995, hereafter SW), and much of the discussion found therein is not duplicated here. Other recent reviews that discuss redshift surveys include those of Giovanelli & Haynes (1991), Dekel (1994), Borgani (1995), Efstathiou (1996), and Guzzo (1996). The emphasis in this review will be on developments which have occurred since the writing of SW, with special emphasis on future redshift surveys and what they will be able to measure.

5.2 Varieties of redshift surveys

In order to be useful for any sort of statistical work, a redshift survey must have a well-defined selection function in the most general sense of the term. That is, the selection criteria of the galaxies whose redshifts are included must be objective and quantifiable. This is not the same as saying that the redshift survey must be complete in some sense, just that its incompleteness follows some known rule(s), such that the fraction of galaxies with redshifts is known, at least statistically, as a function of the observational properties of a galaxy. In practice, this means that the following must be known for any redshift survey, and the galaxies contained therein, if it is to be useful:

- The region of sky covered by the survey.
- The observational criterion or criteria by which galaxies have been selected for redshifts, such as apparent magnitude in a given band, diameter, surface brightness cuts, emission-line strength,[1] etc.
- The value of the relevant quantity or quantities for all objects in the sample.
- Limits in the above quantity or quantities. This may very well be a function of position on the sky, such as in the case of an apparent magnitude-limited survey in optical bands, in the presence of Galactic extinction. These limits determine the depth of the survey, which can be quantified in terms of the expected number distribution of galaxies as a function of redshift.
- The fraction of galaxies for which redshifts are actually obtained. Again, this could be a function of position on the sky, or could be a fixed fraction over the survey area. If the fraction is close to 100%, we call this a complete redshift survey.
- The positions and redshifts of the individual galaxies.

Speaking loosely, then, a survey is characterized by its solid angle coverage, its depth, its sampling rate, and the method of selection of the galaxies. Comprehensive lists of redshift surveys are given in the various reviews listed in the Introduction, including SW; we will not repeat this here. However, we briefly summarize some of the redshift

---

[1] Relevant, e.g., in objective prism surveys for emission-line galaxies or quasars.

surveys we will find ourselves referring to through this review, with apologies to those whose surveys we do not have space to discuss here.

There have been a few surveys which have covered close to the entire celestial sphere; as we will see, these have been very important both for cosmography, and for making dynamical predictions of the peculiar velocity field in the nearby universe. Early work in this direction was done by Yahil, Sandage & Tammann (1980) with the Revised Shapley–Ames catalog of galaxies (cf. Sandage & Tammann 1981), but this sample only extended to redshifts of 4000 $\mathrm{km\,s^{-1}}$, and was affected strongly by the Galactic zone of avoidance. The *Infrared Astronomical Satellite*, or IRAS, flew in 1983, and scanned the entire sky at $\sim 1'$ resolution in four broad bands centered at 12, 25, 60, and 100 µm (cf. IRAS Point Source Catalog Explanatory Supplement 1988 for details). An ordinary spiral galaxy with a moderate amount of star formation emits strongly at 60µm as thermal emission from dust heated by the interstellar radiation field. Because of the all-sky nature of the IRAS survey, and the transparency of the dust of the Milky Way to 60 µm radiation, a galaxy sample selected at 60 µm has uniform and full-sky coverage. Two groups have carried out extensive full-sky redshift surveys based on these data: one centered in Berkeley, doing redshift surveys complete at 60 µm to 1.936 Jy and subsequently to 1.2 Jy (Strauss et al. 1990, 1992c; Fisher et al. 1995a), and the other a mostly British collaboration which obtained redshifts of 1 in 6 galaxies to a flux limit of 0.6 Jy (Rowan-Robinson et al. 1990; Lawrence et al. 1998). This latter collaboration, referred to as QDOT for the initials of the institutions of the investigators, is currently extending their effort to measure redshifts for a *complete* sample of galaxies to 0.6 Jy (15,500 galaxies) at 60 µm. (See http://www-astro.physics.ox.ac.uk/~wjs/pscz.html for the latest details.) In practice, the IRAS surveys are limited by the effects of Galactic extinction at very low latitudes (if you can't see the galaxy optically, you certainly cannot measure a redshift!) and systematic effects in the IRAS Point Source Catalog in regions of very high source density (mainly confusion and hysteresis); these surveys thus cover between 80 and 90% of the sky.

Elliptical galaxies are very faint for the most part in the infrared bands, and are therefore essentially absent from the IRAS surveys. Given that the cores of clusters of galaxies are dominated by elliptical galaxies (e.g., Dressler 1980, 1984; Postman & Geller 1984; Whitmore, Gilmore & Jones 1993), the number density of galaxies in the cores of clusters is systematically underestimated in IRAS, as is in fact borne out in direct comparisons of IRAS and optically selected samples (cf. Strauss et al. 1992a; Santiago & Strauss 1992). This, together with the rather sparse sampling of the galaxy distribution by the IRAS satellite, motivated the compilation of the *Optical Redshift Survey* (ORS; Santiago et al. 1995, 1996; Hermit et al. 1996), which selects galaxies from the Uppsala Galaxy Catalogue (Nilson 1973), the ESO Galaxy Catalogue (Lauberts 1982) and its photometric counterpart (Lauberts & Valentijn 1989) and the Extension to the Southern Galaxy Catalogue (Corwin & Skiff 1996). Galactic extinction restricts the survey to Galactic latitudes $|b| > 20°$. The depth is comparable to that of the original CfA survey (Davis et al. 1980; Huchra et al. 1983) but with 4.5 times the

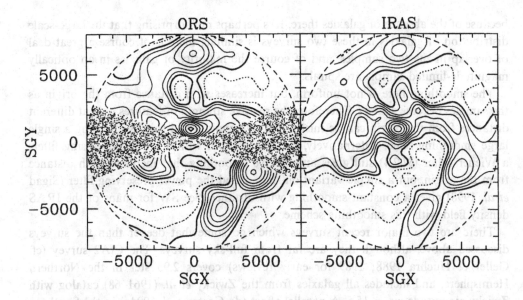

Figure 5.1. Galaxy isodensity contours in the Supergalactic plane for the ORS (*left*) and IRAS 1.2 Jy redshift surveys (*right*). The Local Group is at the center of each map. The smoothing in the two cases is the same, and increases with distance from the center; therefore, the relative strength of features at different distances from the center can be misleading. The heavy contour is at the mean density; dotted contours are underdense relative to the mean. The zone of avoidance is indicated in the case of the ORS sample.

sky coverage. Santiago et al. (1996) detail the effort required to tie the three parts of the survey together, and the corrections made for Galactic extinction.

The Local Group is a member of the Local Supercluster, a highly flattened structure which extends at least to 3000 km s^{-1} from us. Indeed, many of the other dramatic superclusters in the nearby universe are found in, or at least are intersected by, the plane defined by the Local Supercluster, including the Perseus–Pisces Supercluster, the Hydra–Centaurus and Pavo–Indus–Telescopium Superclusters (almost certainly a single structure bisected by the zone of avoidance) and the Coma–A1367 Supercluster. Figure 5.1 shows isodensity contours of the galaxy distribution in the Supergalactic plane, in the IRAS 1.2 Jy (*left*) and ORS (*right*). The smoothing is Gaussian with $\sigma = 400$ km s^{-1} at low redshift, increasing like the IRAS mean interparticle spacing at greater distances (thus the smoothing is the same in the two panels). Mean density is indicated with a heavy line; contours above mean density are spaced logarithmically, with every third contour representing a factor of two in density. Dotted contours are at 0.66 and 0.33 of the mean density. Fingers of God associated with rich clusters have been collapsed to a common redshift; otherwise the galaxies are placed at the distances indicated by their redshifts in the Local Group frame (i.e., no correction for peculiar velocities have been made). The low-latitude regions of the ORS map are masked out, because of the absence of galaxies there. It is perhaps not surprising that the large-scale distribution of galaxies in these two surveys is similar; there is of course a great deal of overlap in the two samples, and of course the majority of galaxies in an optically magnitude-limited sample are spirals.

The smoothing here is not uniform, but increases with distance from the origin as the samples become sparser. Therefore, the relative strength of features seen at different distances from the origin can be misleading. We could avoid this by choosing a single large smoothing scale. Alternatively, one can smooth with a noise-suppressing filter; a Wiener Filter (§7.2.4) still gives an effective smoothing that increases with distance from the origin, but a simple variant on that, called the power-preserving filter (Sigad et al. 1998) gives a constant smoothing with distance. See SW for maps of the IRAS density field with this smoothing scheme.

There are two major recent surveys which go somewhat deeper than the surveys discussed above, although they are far from full-sky surveys. The CfA2 survey (cf. Geller & Huchra 1988; 1989 for early reviews) covers 2.95 ster in the Northern Hemisphere, and includes all galaxies from the Zwicky et al. (1961-68) catalog with Zwicky magnitude $m_Z \leq 15.5$. A parallel effort (da Costa et al. 1994a) in the Southern Hemisphere is covering 1.13 ster to the same depth of the CfA2 survey, based on scans of sky survey plates. The two surveys have been analyzed together to determine the power spectrum of galaxies (da Costa et al. 1994b) and the small-scale velocity dispersion of galaxies (Marzke et al. 1995).

On smaller angular scales, but going deeper, are several surveys. Shectman et al. (1996) have used a multi-object spectrograph on the Las Campanas 2.5-m Du Pont telescope to obtain redshifts for 26,418 galaxies to $r \approx 17.5$ (the Las Campanas Redshift Survey, or LCRS). The photometry is based on CCD drift-scan data obtained by the same workers on the Swope 40-in telescope at Las Campanas. This sample is not complete: fields in which to do spectroscopy were laid down on a regular grid in a series of six $\sim 90° \times 1.5°$ wide strips across the sky, and the sampling rate was simply the ratio of the number of galaxies available to their magnitude limit, to the number of fibers of the spectrograph, averaging roughly 70% over their fields. This survey, covering a total of ~ 700 square degrees, has roughly a factor of two more redshifts than any other single redshift survey of galaxies.

These surveys are best viewed not as contour plots as in Fig. 5.1, but rather in the form of pie diagrams. Figure 5.2 shows the redshift distribution in the LCRS survey; in each segment of the pie, three of the 1.5° slices are plotted on top of one another. The angular coordinate is right ascension, and the radial component is redshift. Perhaps the most striking feature of this map is the fact that one does *not* see coherent structures stretching across the survey volume (compare, for example, with the famous CfA2 slice of de Lapparent, Geller & Huchra 1986). Bob Kirshner of the LCRS team has called this "the end of greatness", in the sense that surveys are now probing a volume appreciably larger than the largest structures in the universe. We discuss below quantitative measures of structure on the largest scales with this survey.

Redshift surveys of the local universe

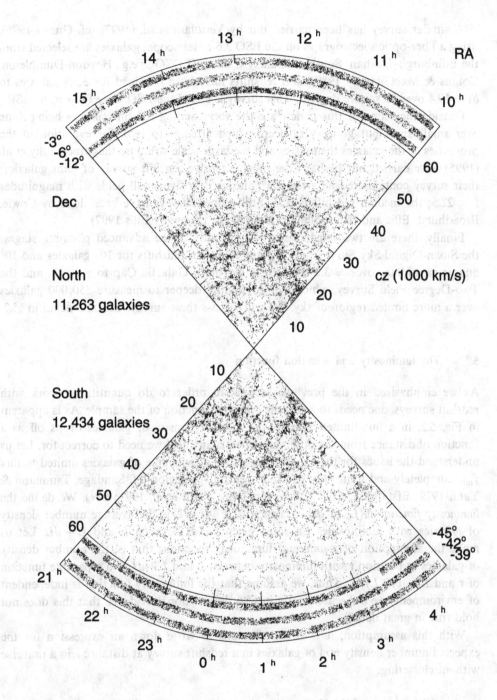

Figure 5.2. A redshift pie diagram of galaxies in the LCRS, kindly supplied by Huan Lin. The sky distribution of this survey is shown below in Fig. 5.7. The three slices in the Northern and Southern Galactic Caps are each shown projected on top of one another.

A similar survey has been carried out by Vettolani et al. (1997) (cf. Guzzo 1996), using a fiber-optic spectrograph on the ESO 3.6-m telescope; galaxies are selected from the Edinburgh–Durham Southern Galaxy Catalog (EDSGC; e.g., Heydon-Dumbleton, Collins & MacGillivray 1989), and redshifts have been obtained for 3348 galaxies to $b_J = 19.4$ over 36 square degrees. This survey is called the ESO Slice Project, or ESP.

Outside the scope of this review are the very exciting redshift surveys being done over small areas but going very deep, which attempt to look for evolution in the properties of the galaxies themselves and the large-scale structure they trace. Lilly et al. (1995) have carried out perhaps the most extensive redshift survey of faint galaxies; their survey contains redshifts for 591 galaxies over five small fields with magnitudes $I < 22.5$; the median redshift is 0.56. Still deeper surveys have been done by Cowie, Broadhurst, Ellis and their collaborators (cf. the review by Ellis 1997).

Finally, there are two redshift surveys currently in the advanced planning stages: the Sloan Digital Sky Survey, which plans to obtain redshifts for 10^6 galaxies and 10^5 quasar candidates over ≈ 3.2 ster in the Northern Galactic Cap to $r' \approx 18$, and the Two-Degree Field Survey, which will go slightly deeper to measure 250,000 galaxies over a more limited region of sky. We will discuss these surveys in some detail in §5.7.

5.3 The luminosity and selection function

As we emphasized in the previous section, in order to do quantitative work with redshift surveys, one needs to know the selection function of the sample. As is apparent in Fig. 5.2, in a flux-limited sample, the number density of galaxies drops off as a function of distance from the observer, which is an effect we need to correct for. Let us understand the issues for the simplest case, that of a sample of galaxies limited to flux $f_{\min}$ completely and uniformly selected over a given region (cf. Sandage, Tammann & Yahil 1979; Efstathiou, Ellis & Peterson 1988; Yahil et al. 1991; SW). We define the *luminosity function* $\Phi(L)$ of galaxies such that $\Phi(L) dL$ is the average number density of galaxies with luminosities (in some given band) between L and $L + dL$. Let us make the assumption of a *universal* luminosity function; that is, the number density of galaxies at position **r** with luminosity between L and $L + dL$ is a separable function of **r** and L: $\rho(\mathbf{r})\Phi(L) dL$. Thus, we assume that the luminosity function is independent of environment (we'll discuss briefly in §5.6 observational evidence that this does not hold true in great detail).

With this assumption, it is straightforward to write down an expression for the expected number density $n(r)$ of galaxies in a redshift survey at distance r in a universe without clustering:

$$n(r) = \int_{4\pi r^2 f_{\min}}^{\infty} \Phi(L) dL . \tag{5.1}$$

This suggests a simple estimator for the observed fractional overdensity of galaxies $\delta(\mathbf{r}) \equiv (\rho(\mathbf{r}) - \langle\rho\rangle)/\langle\rho\rangle$ smoothed with a window $W(x)$:

$$\delta(\mathbf{r}) = \frac{1}{\int d^3\mathbf{r}\, W(r)} \sum_{\text{galaxies } i} \frac{W(|\mathbf{r}-\mathbf{r}_i|)}{n(r_i)} - 1, \qquad (5.2)$$

where $n(r)$ is given by equation (5.1) and $r_i \equiv |\mathbf{r}_i|$, of course. What this means, in effect, is that for the purposes of deriving the density field δ, each galaxy in the sample is assigned a weight given by the inverse of $n(r)$ at the distance of that galaxy.

What is the best way to calculate $n(r)$? It is straightforward, given a fit to the luminosity function. There is a lengthy literature of determinations of the luminosity function, which is reviewed in the various articles listed at the beginning of this section. The classic method, reviewed in, e.g., Felten (1977), simply involves binning the galaxies in a redshift survey by absolute luminosity and dividing each by the effective volume probed at that luminosity. However, this method gives unbiased results only in the limit that the number of galaxies per unit redshift is unaffected by clustering, which is exactly the quantity we ultimately wish to measure. Here we present a method of determining the luminosity function which does not require any assumption about the large-scale homogeneity of galaxies; its history can be traced through Sandage, Tammann & Yahil (1979), Nicoll & Segal (1983), Efstathiou, Ellis & Peterson (1988), Saunders et al. (1990), and Yahil et al. (1991). We will take a maximum likelihood approach and ask, *given* that we know that a given galaxy i is at a distance r_i, what is the likelihood that its observed flux will lie between f_i and $f_i + df$? The answer clearly depends on the luminosity function, thus this approach will give us a handle on the luminosity function itself. Indeed, this likelihood $\mathscr{L}_i$ is given by the luminosity function at $L_i = 4\pi r_i^2 f_i$, normalized by the integral of the luminosity function over all luminosities it could have, given the flux limit. Formally:

$$\mathscr{L}_i = \frac{\Phi(L_i)\, dL}{\int_{4\pi r_i^2 f_{\min}}^{\infty} \Phi(L_i)\, dL} \propto \left. \frac{dn(r)/dr}{n(r_i)} \right|_{r=r_i(f_i/f_{\min})^{1/2}}, \qquad (5.3)$$

where the proportionality follows directly from equation (5.1). The constant of proportionality is just $dL/(8\pi r f_i)$, which is independent of the parameters of the selection function, and therefore does not concern us as we maximize the likelihood below. Given the likelihood for observing any one galaxy, the likelihood for an entire sample is given by the product of this expression for all galaxies in the sample. The maximum likelihood method then consists of the following: (i) choose a parametric form for $n(r)$; (ii) calculate the likelihood function over all the galaxies in the sample as a function of the parameters, and find its maximum (as is often done in this sort of exercise, the quantity actually maximized in practice is the logarithm of the likelihood function). This determines $n(r)$, from which the luminosity function follows as a simple derivative. A variant of this approach involves specifying the luminosity function not in terms of a smooth functional form, but as a series of constants in bins. Nicoll & Segal (1983), Efstathiou, Ellis & Peterson (1988), and Koranyi & Strauss (1997) show how the likelihood can be maximized with respect to the values of the constants through a straightforward iterative procedure.

It is clear how the likelihood procedure outlined above is independent of density inhomogeneities: the *positions* of galaxies are taken as a prior, and one asks for the likelihood of observing their fluxes. Thus the results are not biased by the distribution of the positions of the galaxies. But a consequence of this is that the *normalization* of $\Phi(L)$ (or equivalently, of $n(r)$) is not determined. With this in mind, we explicitly drop the normalization of $n(r)$ to define the *selection function*:

$$\phi(r) \equiv \begin{cases} \frac{n(r)}{n(r_s)}, & r > r_s \\ 1 & r \le r_s, \end{cases} \tag{5.4}$$

where r_s is some small fiducial distance, typically 500 km s^{-1}, below which the selection function is set to unity. That is, the selection function quantifies the fraction of the luminosity function seen at a distance r, relative to the numbers at r_s. This of course begs the question of how to normalize the selection function itself in order to calculate $\delta(\mathbf{r})$. The usual assumption is that the entire volume covered by a given redshift survey is a fair sample of the universe; that is, the mean density of galaxies in this volume differs negligibly from the true mean density. Given this, there are several ways one can calculate the mean density as a weighted sum over the galaxies in the sample (Davis & Huchra 1982). The simplest is to define

$$n(r_s) = \frac{1}{V} \sum_{\text{galaxies } i} \frac{1}{\phi(r_i)}, \tag{5.5}$$

where the sum is over all galaxies in the sample between r_s and some outer radius beyond which the sample gets too sparse to be useful, and V is the volume enclosed. Davis & Huchra (1982) derive a minimum variance version of equation (5.5) that takes the known clustering of galaxies into account, although in practice, it requires knowing the strength of the correlation function on the largest scales, where it is most poorly understood. Another approach is that adopted by Saunders et al. (1990), who note that in the approximation that the sample is uniform, the total number $\mathcal{N}$ of galaxies in the sample should be given by $\int d^3\mathbf{r}\, n(r)$ (where the integral is over the solid angle and depth of the survey), which allows one to normalize $n(r)$.

It is actually quite straightforward to modify the maximum likelihood method sketched out here to the more general case of selection depending on more complicated criteria, such as both a magnitude and diameter cut, or extinction as a function of direction; some of these complications are described in Santiago et al. (1996). Another interesting issue discussed in that paper, and also in Mancinelli (1996), is the effect of flux errors on the derived luminosity function, selection function, and density field. The derived luminosity function is given by the true luminosity function convolved with the flux errors, not surprisingly. This gives a selection function which falls less rapidly with distance than in the case without flux errors. One might think that this could cause the derived density field to be systematically biased as a function of distance, but there is a competing effect which goes in the opposite direction, namely Malmquist bias. Flux errors scatter galaxies over the flux limit of the sample. Because the number of galaxies is a monotonically decreasing function of flux, more galaxies scatter into the sample than out of it. If the flux errors are proportional to the flux itself (such as one gets with magnitude errors which are constant with magnitude), these two effects cancel exactly; the derived density field is unbiased, even though the luminosity function is biased.[2]

5.4 Clustering statistics

As Figs. 5.1 and 5.2 make clear, galaxies are not uniformly distributed in space; indeed, one of the main motivations for doing redshift surveys is to quantify the observed clustering and, from it, draw inferences about larger cosmological questions such as are addressed throughout this book: What is the nature of dark matter? What is the value of the Cosmological Density Parameter? How did galaxies form? We will be able to address only some of these issues here. But let us ask how best to quantify in a statistical way the clustering that is seen here.

The current dominant paradigm for the formation of large-scale structure postulates that it grows by the process of gravitational instability from an early epoch when the fluctuations were of very low amplitude (Ch. 2; textbooks treating this include Kolb & Turner 1990; Peebles 1993; Padmanabhan 1993; Coles & Lucchin 1995). This basic picture was given great support by the fact that the fluctuations in the Cosmic Microwave Background (CMB) as detected by the Cosmic Background Explorer (COBE; Smoot et al. 1991) are within a factor of two of those expected from extrapolation of the present-day observed clustering of galaxies (e.g., Wright et al. 1992; cf. Ch. 3). Moreover, if the fluctuations arise from inflationary processes in the very early universe (e.g., Kolb & Turner 1990), they turn out to be *random phase*. This means the following. Given the density field $\delta(\mathbf{r})$ at some early time, one can take its Fourier Transform to obtain the complex quantity $\tilde{\delta}(\mathbf{k}) = |\tilde{\delta}(\mathbf{k})| e^{i\theta}$, where θ is the phase of each mode. Inflationary models predict the quantity θ to be uniformly distributed between 0 and 2π.[3]

In any case, if the phases are random, then the Central Limit Theorem implies that the distribution function of the density field is Gaussian.[4] What this means is that all its reduced moments of third order and higher vanish (after all, they are defined relative to a Gaussian). That is, one can give a full statistical description of the density field by specifying its second moment with respect to the only relevant independent variable, the scale (the Cosmological Principle says that space is isotropic, which is confirmed to fantastic accuracy with the COBE data, so there is no directional dependence to the clustering in a sufficiently large sample). There are a number of ways we might quantify

---

[2] The caveat that the flux errors must be proportional to flux is not made clear in Santiago et al. (1996). However, there is no condition on the Gaussianity of the error distribution.

[3] Examples of models in which the phases are not random include those with gravitating seeds (such as cosmic string loops or textures) or explosion models. A general discussion of the large-scale structure implications of these models can be found in Weinberg & Cole (1992).

[4] However, the converse is not true: a Gaussian distribution function for the density field does not imply random phases.

this. One way is simply to calculate the statistic implied by this discussion: the second moment of the density distribution as a function of scale. That is, one asks for the variance in δ, averaged over spheres of a given radius r; this quantity will be referred to as $\sigma^2(r)$. Alternatively, one can calculate the *correlation function* of the density field: $\xi(r) \equiv \langle \delta(\mathbf{x})\delta(\mathbf{x}+\mathbf{r}) \rangle$, where the averaging is over position $\mathbf{x}$ and over direction of $\mathbf{r}$. Finally, one can define the *power spectrum* of the density field, which is the modulus squared of $\tilde{\delta}$:

$$\langle \tilde{\delta}(\mathbf{k})\tilde{\delta}^*(\mathbf{k}') \rangle = (2\pi)^3 P(k) \delta_D(\mathbf{k}-\mathbf{k}') , \qquad (5.6)$$

where the averaging is over directions of $\mathbf{k}$, and δ_D is the Dirac δ function.[5]

These three statistics are related to each other in straightforward ways:

$$\xi(r) = \frac{1}{2\pi^2} \int dk\, k^2 P(k) \frac{\sin kr}{kr} \qquad (5.7)$$

and

$$\sigma^2(r) = \frac{1}{2\pi^2} \int dk\, k^2 P(k) \widetilde{W}^2(kr) , \qquad (5.8)$$

where $\widetilde{W}(x)$ is given by a spherical Bessel function for a spherical tophat window:

$$\widetilde{W}(x) = \frac{3(\sin x - x\cos x)}{x^3} . \qquad (5.9)$$

How does this relate to the galaxy distribution? As we said before, we believe (although do not have definitive proof) that clustering grew through the process of gravitational instability. In linear perturbation theory it is straightforwardly shown (cf. the textbooks referenced above) that the amplitude of perturbations grows as a function of time, independent of the wavelength of the perturbations. What this means is that the power spectrum and related statistics change in *amplitude*, but not in *shape*, as perturbations grow. Therefore, measurements of the shape of the power spectrum as a function of k today are a direct measure of its shape in the past. That prospect is very exciting, because the shape of the power spectrum is commonly modelled to be due to two components. One is the *primordial* power spectrum, i.e., that laid down, perhaps during inflation, when the universe was very young. This is modified by the differing growth of perturbations of super-horizon and sub-horizon scales before and after the epoch of matter-radiation equality (cf. Kolb & Turner 1990; Efstathiou 1991 for reviews). The details of this are determined by the nature of the dark matter (i.e., hot or cold) and how much of it there is (as this determines the matter–radiation equality epoch).

Thus, in principle, measurements of the galaxy power spectrum can tell us a great deal about the early universe and the dark matter. However, there are some real complications that come in. The first is that we are observing galaxies, while it is the distribution of dark matter that the theories predict. Thus we need a model for the

---

[5] The normalization, and even the definition of $P(k)$ depends on one's Fourier Transform convention in defining $\tilde{\delta}$; compare equation (5.6) with that in Peebles (1980), for example.

relative distribution of galaxies and dark matter. The simplest assumption is that the distribution of galaxies and dark matter are the same [i.e., $\delta_{\text{galaxies}}(\mathbf{r}) = \delta_{\text{dark matter}}(\mathbf{r})$], but it was realized in the mid-1980s (Kaiser 1984; Davis et al. 1985; Bardeen et al. 1986; Dekel & Rees 1987; see §7.6.7) that there is no a priori reason that this might be true. Indeed, one could explain a number of observations (such as the relative strength of the cluster and galaxy correlation functions, and the pairwise velocity dispersion of galaxies on small scales) if it were false. The simplest model of biasing has the galaxy and dark-matter density fields differing by a constant factor b. That is,

$$\delta_{\text{galaxies}}(\mathbf{r}) = b\, \delta_{\text{dark matter}}(\mathbf{r}), \tag{5.10}$$

independent of smoothing scale, although reality is almost certainly more complicated than this. In any case, to the extent that this *linear biasing* holds, the shape of the derived galaxy power spectrum of galaxies will still be the same as that of the dark matter.

The next complication is due to nonlinear evolution of the power spectrum. The statement that density perturbations grow at a rate which is independent of scale holds only when the perturbations are in the linear regime, i.e., $|\delta| \ll 1$. When this condition no longer holds, all modes no longer grow independently, and the growth rate does indeed become a function of scale, meaning that the nonlinear power spectrum no longer keeps the same shape as its linear progenitor. In practice, because density perturbations are generically an increasing function of k (at least for power spectra $P(k) \propto k^n, n \geq -3$, cf. equation [5.8]), this means that the power spectrum is modified by nonlinear effects on small scales. To a certain degree the growth of the power spectrum on small scales can be calculated analytically (e.g., Jain & Bertschinger 1994) or phenomenologically (Hamilton et al. 1991; Jain, Mo & White 1995) and, indeed, there is a quite extensive literature on extensions of linear theory for the growth of perturbations into the nonlinear regime (cf. the review by Sahni & Coles 1995).

As we will see below, the power spectrum measured within the necessarily finite volume of any given redshift survey is not identical to the theoretical ideal of that measured in an infinite volume, essentially because of the difficulties of defining the continuous Fourier Transform in a finite volume. Indeed, the measured power spectrum is a convolution of the "true" power spectrum with the Fourier Transform of the observing volume, which tends to depress the power spectrum on large scales.

Finally, what we observe for each galaxy in a redshift survey is a redshift, not a distance. Only in the approximation that peculiar velocities are negligible are the two the same (cf. equation [5.27]). Peculiar velocities have two effects on the power spectrum as measured in redshift space. On small scales, the pairwise velocity dispersion of galaxies spreads galaxies out in redshift space relative to their distribution in real space (think of a cluster of galaxies stretched out into a "Finger of God" in redshift space), thereby decreasing the apparent amplitude of clustering on small scales. On large scales, the dominant effect is due to coherent streaming of galaxies towards overdensities, giving a compression in redshift space, and therefore amplifying the apparent clustering (Kaiser 1987).

184 M. A. Strauss

With all these effects acting, the interpretation of the observed power spectrum is thus quite non-trivial, and the next section of this review concentrates on the details of how people have tried to take these various effects into account and, indeed, to take advantage of them to get additional information out of the available data.

5.5   Measurements of the power spectrum

5.5.1   Techniques

For two decades, the standard way to quantify the clustering seen in galaxy surveys was through the use of the N-point correlation functions, especially the two-point correlation function $\xi(r)$ (Peebles 1980; SW). However, much of the recent developments in the field have been on analyses of its Fourier Transform $P(k)$ (cf. equation [5.7]), and we will emphasize this in this review.

Recent work on techniques for measuring the galaxy power spectrum follow one of two approaches: working with the galaxy data themselves, and working with binned or pixelized versions of the data (cf., Tegmark 1997b). Although binning destroys clustering information on scales smaller than the bin size, this approach becomes computationally much more feasible for very large datasets, especially when doing rigorous model-testing of structure formation models with respect to data.

We start with the standard unbinned approach. We repeat the analysis of SW, correcting their equations for erroneous factors of V.

The Fourier Transform of the unsmoothed galaxy density field is

$$\tilde{\delta}(\mathbf{k}) = \frac{1}{n} \sum_i \frac{1}{\phi(r_i)} e^{i\mathbf{k}\cdot\mathbf{r}_i} - W(\mathbf{k}) , \tag{5.11}$$

where

$$W(\mathbf{k}) \equiv \int_V d^3r\, e^{i\mathbf{k}\cdot\mathbf{r}} \tag{5.12}$$

is the Fourier Transform of the survey volume. Our estimator of the power spectrum is then

$$\Pi(\mathbf{k}) \equiv \frac{1}{V} \tilde{\delta}(\mathbf{k})\, \tilde{\delta}(\mathbf{k})^* , \tag{5.13}$$

where the factor of $1/V$ on the right-hand side gets the units right. Several lines of algebra (e.g., Fisher et al. 1993; Vogeley 1995) show that the expectation value of this estimator is given by

$$\langle \Pi(\mathbf{k}) \rangle = \int d^3k'\, P(k') G(\mathbf{k}-\mathbf{k}') + \frac{1}{nV} \int d^3r \frac{1}{\phi(r)} , \tag{5.14}$$

where

$$G(\mathbf{k}-\mathbf{k}') \equiv \frac{1}{(2\pi)^3 V} |W(\mathbf{k}-\mathbf{k}')|^2 . \tag{5.15}$$

In the limit of an infinitely large volume, G approaches a Dirac delta function, as it must. Thus the power spectrum estimator is given by the true power spectrum convolved with an expression involving the Fourier Transform of the volume, plus a shot noise term. Because one normalizes the density field assuming that the mean density inside the surveyed volume is equal to the true global mean, there is an additional correction factor to equation (5.14) to compensate for the resulting loss of power (cf., Tegmark et al. 1998); this term is important for measurements of the power spectrum on scales approaching that of the survey itself.

The quantity in equation (5.14) is still a function of $\mathbf{k}$, and thus must be averaged over solid angle in $\mathbf{k}$-space in order to calculate a quantity dependent only on k. This is a straightforward procedure for values of k probing scales appreciably smaller than the survey dimensions. However, when $1/k$ becomes comparable to the smallest dimension of the survey, this averaging can mix together modes with very different convolutions with the survey window (simply because G in equation (5.15) becomes quite anisotropic for surveys with restricted geometries). A related problem is that an appreciable covariance can develop between determinations of the power spectrum for different values of k. Different workers have found different ways of dealing with these problems. Fisher et al. (1993), working with the IRAS 1.2 Jy survey, have close to full-sky coverage, and therefore are not affected much by the anisotropy of the window function. They measure $\Pi(\mathbf{k})$ within cylinders embedded within the survey volume, whose long axis of length $2R$ is parallel to the vector $\mathbf{k}$. Choosing kR to be an integral multiple of π means that $\delta(\mathbf{k})$ now scales exactly with the mean density, and thus errors in the mean density affect only the *amplitude*, and not the shape, of $P(k)$. More generally, one should choose values of k to measure the power spectrum that are separated roughly by $2\pi/L$, where L is the characteristic dimension of the survey, in order to give estimates that are statistically independent. Very recently, Hamilton (1997a,b) has made these ideas rigorous, finding window functions which yield statistically independent measurements of $P(k)$, at least for a full-sky survey.

Feldman, Kaiser & Peacock (1994) approach the problem from a different viewpoint, by asking for a weighting to equation (5.11) that allows them to measure $P(k)$ with the minimum variance. For the case of a full-sky sample, and assuming that the different Fourier modes have random phases, they derive the weights:

$$w_i = \frac{1}{1 + n\phi(r_i)P(k)}. \tag{5.16}$$

With this weight function, the variance in the estimate of the power spectrum is given by

$$\sigma^2[P(k)] = \frac{(2\pi)^3}{V_k \int d^3\mathbf{r}\, [n w \phi(r)]^2}, \tag{5.17}$$

where V_k is the volume in k-space occupied by the bin in question. This expression assumes that the bins in k are spaced far enough apart that the covariance is negligible.

Tegmark (1995) has carried this type of analysis further, by asking for the optimal weighting of $\Pi(\mathbf{k})$ as a function of amplitude and direction of $\mathbf{k}$, given the survey geometry. One wants to measure the power spectrum with as much resolution in k as possible, without introducing large amounts of covariance between adjacent values. Of course, the larger the binning in k, the smaller the statistical error bars, and therefore one needs to balance the desire for resolution against the desire for minimum covariance. Tegmark introduces a weighting function $w(\mathbf{r})$ that accomplishes this, which is the ground-state solution to the Schrödinger equation with potential given by the inverse of the selection function:

$$\left[-\frac{1}{2}\nabla^2 + \frac{\gamma}{\phi(\mathbf{r})}\right] w(\mathbf{r}) = E\, w(\mathbf{r}), \tag{5.18}$$

where γ is a parameter which determines the resolution in k of the determination of the power spectrum, at the expense of signal-to-noise ratio.

As the redshift survey data improve, there is clearly a need for more than the mostly qualitative comparisons that have been carried out to date, of observed power spectra with cosmological models. In particular, we need rigorous expressions for the covariance matrix of the power spectrum, the distribution function of the errors, and so on, which allow for arbitrarily complicated survey geometry. In the last few years, it has been realized that this is possible if one pixelizes the redshift data either in real space or Fourier space; one of course throws out information on scales smaller than the pixels, but this allows one to write exact likelihood expressions for the data as a function of the power spectrum on large scales.

Vogeley & Szalay (1996) and Tegmark, Taylor & Heavens (1997) have taken this approach. While the Fourier modes are the ideal basis in which to expand the density field in the ideal case of an infinite survey, this is not the case in the realistic case of a survey covering a finite area of sky, with a radial selection function. In particular, the Fourier modes are not orthonormal over the survey volume. Thus the above authors expand the observed density field in orthonormal eigenmodes which maximize the signal-to-noise ratio, given the survey geometry and selection function, using the Karhunen–Loève (K–L) Transform. These modes are linear combinations of the standard Fourier modes which enter the power spectrum, and can also be expressed as linear combinations of the counts in cells in the survey volume.

Following Vogeley & Szalay, divide a survey volume into a series of M volume elements centered at positions $\mathbf{x}_i$ with volumes V_i. Let $f(\mathbf{x}_i)$ be the counts of galaxies observed within cell i. We expand the $f(\mathbf{x}_i)$ in a series of orthonormal basis vectors $\mathbf{\Psi}_j$, i.e.,

$$f(\mathbf{x}_i) = \sum_j \mathbf{\Psi}_j(\mathbf{x}_i) B_j. \tag{5.19}$$

The K–L Transform uses the basis vectors which satisfy the eigenvalue problem:

$$\mathbf{R}\mathbf{\Psi}_j = \lambda_j \mathbf{\Psi}_j, \tag{5.20}$$

where $\mathbf{R}$ is the correlation matrix of the f's: $R_{ij} = \langle f(\mathbf{x}_i) f(\mathbf{x}_j) \rangle$, and the eigenvalues are

$\lambda_j = \langle B_j^2 \rangle$. The relation to the power spectrum is clear: the matrix R_{ij} has elements given by the sum of the correlation function between volume V_i and V_j, and contributions from shot noise. Given the K–L expansion of an observed density field, the best-fit power spectrum can be found by the methods of maximum likelihood.

Vogeley & Szalay point out that the K–L Transform is a maximally efficient representation of the data, in the sense that if the eigenmodes are ordered in decreasing order of their eigenvalues, the truncation of the eigenvectors to the first N gives a representation of the data that differs from the truth by as small an amount as possible. Another way to say this is that this gives an expansion of the data in modes of decreasing signal-to-noise ratio. A dramatic demonstration of the power of the K–L Transform to compress data, in quite a different astronomical context, is that of Connolly et al. (1995), who show that the optical spectral energy distributions of galaxies can be well represented by their first three eigenvectors.

This approach to analysis of redshift surveys promises the greatest advantage over the standard power spectrum analysis in the case of surveys with sharp boundaries, especially those covering a fraction of the sky with anisotropic sky coverage, such as slice surveys or pencil-beam surveys. The LCRS is an example of a slice survey. It has not yet been analyzed using the methods of the K–L Transform (although I understand that Vogeley & Szalay, private communication, intend to do so), but Landy et al. (1996) do an analysis of the power spectrum of this survey in the same spirit. Their survey consists of six narrow slices, and thus recognizing that there is little information on the large-scale power spectrum in a direction perpendicular to the slice, they calculate in effect the two-dimensional analog of the power spectrum on the density field of the slices, collapsed along this narrowest direction.

The K–L-based maximum likelihood technique briefly outlined above fits directly to the counts of galaxies in bins. The power spectrum is a quantity quadratic in the galaxy counts, and Tegmark (1997a), Tegmark & Hamilton (1997), Hamilton (1997a,b), and Tegmark et al. (1998) have developed techniques to fit the power spectrum based on the second moments of these counts. The K–L modes, linear in the density field, keep the phase information, and therefore are useful for extracting information, e.g., on redshift-space distortions (§5.5.3), which one cannot do with these quadratic methods. However, for a variety of technical reasons having to do with the manipulation of very large covariance matrices (cf., the discussion in Tegmark et al. 1998), the quadratic methods lend themselves better to analyses of large redshift surveys over a large range of scales. At this writing, the K–L and quadratic pixelized methods have been applied to analyses of the CMB (Tegmark, Taylor & Heavens 1997; Tegmark & Hamilton 1997), but not yet to any existing redshift survey.

5.5.2 Results

The power spectrum of various redshift surveys of galaxies has been measured by a large number of groups (Baumgart & Fry 1991; Peacock & Nicholson 1991; Park,

Gott & da Costa 1992, Vogeley et al. 1992; Fisher et al. 1993, Feldman, Kaiser & Peacock 1994; Park et al. 1994; da Costa et al. 1994b; Tadros & Efstathiou 1995, 1996; Vogeley 1995; Landy et al. 1996; Lin et al. 1996). A summary of results through 1994 is given by Peacock & Dodds (1994), and is reviewed in SW. For the most part, this large range of determinations of $P(k)$ is remarkable for the uniformity of the results they give. The amplitude of the power spectrum determined from different surveys often differs significantly, even for galaxies selected in the same way (compare Fisher et al. 1993, Feldman, Kaiser & Peacock 1994, and Tadros & Efstathiou 1995 for IRAS-selected galaxies), but there seems to be broad agreement between groups on the shape of the power spectrum, at least within the rather large error bars. This is especially remarkable, given the very different geometries, and therefore window functions, of the different surveys (cf. equation [5.14]). The observed power spectrum is a power law, $P(k) \propto k^n$, $n \approx -2$ on small scales, changing to $n \approx -1$ on scales above $\lambda \equiv 2\pi/k \sim 30\,h^{-1}$ Mpc. There is tentative evidence for a flattening of $P(k)$ on the largest scales on which it has been measured, $\lambda \sim (100-200)\,h^{-1}$ Mpc. The best evidence for this comes not from a redshift survey, but rather the APM photometric survey of 2×10^6 galaxies over the Southern Galactic Cap (Maddox et al. 1990a,b,c, 1996). Baugh & Efstathiou (1993, 1994; cf. Gaztañaga 1995) discuss the determination of the spatial power spectrum from the angular correlation function; the evidence for a flattening of the power spectrum is fairly unambiguous.

The power spectra derived from the surveys described above are shown in Fig. 5.3, which corrects the published power spectra in redshift space for distortions as predicted in linear theory using equation (5.22) below, following Kolatt & Dekel (1997). With the exception of the CfA2+SSRS2 curve, all curves shown are fits of the power spectra to simple functional forms. Thus this figure does not give a sense of the size of the error bars (compare with Fig. 5.6 below). The amplitudes of the different power spectra are indeed quite different from one another,[6] varying by almost a factor of 3 at $k = 0.1\,h\,\mathrm{Mpc}^{-1}$, but, as remarked above, the shapes are remarkably similar.

Before 1992, theoretical predictions for the form of the power spectrum had an important freedom. Current models for the origin of density fluctuations invoking quantum processes in the early universe do not constrain the amplitude of these fluctuations (essentially because we do not yet have a detailed enough model of the relevant particle physics). Thus the normalizations of the power spectra were essentially unconstrained. Now, however, with observations of the fluctuations of the CMB with COBE (cf. Bennett et al. 1996; Górski et al. 1996 for the latest results), this normalization is tied down to 10% accuracy for any given model[7] (cf. Bunn & White 1997 and references therein).

---

[6] Tadros & Efstathiou (1995) go a long way towards reconciling the amplitudes of the two IRAS surveys shown in Fig. 5.3.

[7] This is true only to the extent that the fluctuations are interpreted as being due solely to the Sachs–Wolfe effect (Sachs & Wolfe 1967) on large scales. However, there exist models in which a substantial contribution to the fluctuations comes in the form of gravitational waves (e.g., Davis et al. 1992), making the relationship between the COBE fluctuations and the normalization of the power spectrum ambiguous.

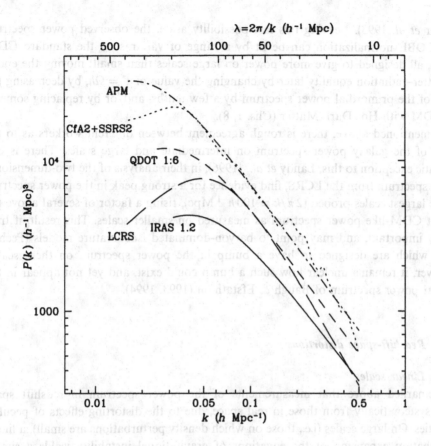

Figure 5.3. Real-space correlation functions of some of the surveys discussed in the text, after a similar figure in Kolatt & Dekel (1997). The CfA2+SSRS2 line is drawn from interpolation between datapoints given in da Costa et al. (1994b); the others are parameterized fits to the observed power spectra. The APM curve is from Baugh & Efstathiou (1993), QDOT from Feldman et al. (1994), LCRS from Lin et al. (1996), and IRAS 1.2 from Fisher et al. (1993). Each curve extends to the largest scales on which the power spectrum was determined.

In any case, we can now make definite predictions, including normalization, for the power spectra of different models. If the biasing of galaxies relative to mass is independent of scale on large scales, then the measured galaxy power spectrum definitely rules out the standard Cold Dark Matter model (SCDM), as defined, e.g., in Davis et al. (1985): adiabatic fluctuations with primordial spectral index $n = 1$, $\Omega = 1$, and $H_0 = 50 \,\mathrm{km\,s^{-1}\,Mpc^{-1}}$. If biasing is *local*, i.e., if the probability that a galaxy form at a given place is a function of the physical properties of that region of space within a few Mpc, then the biasing is indeed probably independent of scale (Weinberg 1995; Kauffmann, Nusser & Steinmetz 1997); if this assumption does not hold, one can find models in which SCDM fits the observed power spectrum on large scales

(Bower et al. 1993). Leaving this last possibility aside, the observed power spectrum with COBE normalization can be fit by a range of variants on the standard CDM model, all designed to give more power on large scales than small: moving the epoch of matter–radiation equality later by changing the value of $\Gamma = \Omega h$, by decreasing the index of the primordial power spectrum by a few tenths, and/or by replacing some of the CDM with Hot Dark Matter (Chs. 1, 8).

As mentioned above, there is rough agreement between different workers as to the shape of the galaxy power spectrum on intermediate and large scales. There is one dramatic exception to this. Landy et al. (1996), in their analysis of the two-dimensional power spectrum from the LCRS, find evidence for a strong peak in the power spectrum on the largest scales probed ($2\pi/k \approx 100\, h^{-1}$ Mpc), rising a factor of several above the best-fit CDM-like power spectrum as measured on smaller scales. This result, if true, is very important, and may point to baryon-dominated isocurvature models (Peebles 1987) which are designed to have a bump in the power spectrum on these scales. However, it remains unclear how such a bump could exist, and yet not appear in the angular power spectrum of Baugh & Efstathiou (1993, 1994).

5.5.3 Redshift-space distortions

5.5.3.1 Linear scales

We remarked above that measurements of the power spectrum in redshift space differ systematically from those in real space, due to the distorting effects of peculiar velocities. On large scales (i.e., those on which density perturbations are small), a linear perturbation expansion of the equations of gravitational instability yields a simple relation between the peculiar velocity and density fields (Peebles 1980):

$$\nabla \cdot \mathbf{v}(\mathbf{r}) = -\Omega^{0.6} \delta(\mathbf{r}) \tag{5.21}$$

(cf. Chs. 2, 7). Given this, one can calculate that the effect of coherent infall into overdensities is to multiply the Fourier modes by a fixed operator independent of scale (Kaiser 1987):

$$\tilde{\delta}(\mathbf{k}) \Rightarrow (1 + \beta \mu^2) \tilde{\delta}(\mathbf{k}) \; ; \tag{5.22}$$

in real space, this becomes the operator equation:

$$\delta(\mathbf{r}) \Rightarrow \left[1 + \beta \left(\frac{\partial}{\partial z}\right) \nabla^{-2}\right] \delta(\mathbf{r}) , \tag{5.23}$$

where $\beta \equiv \Omega^{0.6}/b$ is the proportionality constant between the *galaxy* density field and the divergence of the peculiar velocity field (cf. equation [5.10]), and μ is the cosine of the angle between the wavevector $\mathbf{k}$ and the line of sight. It is a straightforward calculation to propagate the effects of equation (5.22) through to the power spectrum; the power spectrum gets multiplied by a factor:

$$K(\beta) = \left(1 + \frac{2}{3}\beta + \frac{1}{5}\beta^2\right), \qquad (5.24)$$

which is an appreciable correction; $K(\beta = 1) = 1.87$.

On small scales, linear theory breaks down, and peculiar velocities are dominated by the pairwise velocities of galaxies in groups and clusters. The effect of this, as we saw above, is to reduce the amplitude of clustering in redshift space relative to real space. As we will see in a moment, the amplitude of the small-scale velocity dispersion of galaxy pairs remains quite unclear, making it difficult to predict a priori on what scales the redshift distortions make the transition from nonlinear to linear behavior (cf. the discussion in Brainerd & Villumsen 1993; Gramann, Cen & Bahcall 1993; Fisher et al. 1994b). These effects can be measured by the fact that they make the clustering in redshift space anisotropic; that is, the radial direction is the only one affected by distortions. This can be seen directly by measuring the correlation function not as a function simply of redshift-space separation, as is usually done, but rather as a function of the separation both perpendicular and along the line of sight; redshift-space distortions will make the contours of ξ in this space anisotropic. This has been carried out by a number of workers with various datasets (Hamilton 1993, Fisher et al. 1994b; Loveday et al. 1996). One can carry out similar analyses on the power spectrum (Cole, Fisher & Weinberg 1994, 1995; Fisher & Nusser 1995; Taylor & Hamilton 1996; Hamilton 1995), or on a spherical expansion of the density field (Fisher, Scharf & Lahav 1994; Heavens & Taylor 1995); these various approaches are reviewed in SW (see also §7.6.6). All these results have been limited by the fact that existing datasets still do not cover enough volume that the underlying real-space clustering on large scales can be adequately modelled as isotropic; the surveys cover only a relatively small number of superclusters, whose orientations relative to the line of sight do not necessarily average out. This is reflected in the range of values of β and the errors that people quote in their detection of this effect; most workers detect the distortions due to peculiar velocities on large scales at only the 3-σ level or so. Results from the correlation function and power spectrum redshift anisotropies give $\beta = 0.4 - 0.7$, with error bars 1/3 as large as the signal; interestingly, the spherical expansion method quoted above gives appreciably larger values, $\beta = 1$ for the same data. In any case, the surveys discussed in §5.7 promise to survey a large enough volume to give much smaller error bars, and should be among the most exciting science done with these samples.

There is another technical problem in these analyses. On the face of it, equation (5.22) seems nonsensical; what does μ, the cosine of the angle between the line of sight (defined in real space, of course) and the vector $\mathbf{k}$ mean? One is using the "distant observer" approximation, wherein the sample is approximated to be far from the observer, allowing the quantity μ to be defined (cf. the discussion in Cole, Fisher & Weinberg 1994). This means that in practice, one must either work with a fraction of the galaxy pairs available (especially in a full-sky redshift survey, where the distant observer approximation is most grossly violated), or work with a quantity which is related to the theorist's ideal in a very complicated way. Zaroubi & Hoffman (1996) show that even in the limit of an infinite universe, the Fourier modes of the density field measured in redshift space are coupled. Recently, Hamilton & Culhane (1996) have suggested a generalization of equation (5.23) that allows not having to invoke the distant observer approximation:

$$\delta(\mathbf{r}) \Rightarrow \left[1 + \beta \left(\frac{\partial^2}{\partial r^2} + \frac{\partial \ln r^2 \phi(r)}{\partial \ln r} \frac{\partial}{r \partial r}\right) \nabla^{-2}\right] \delta(\mathbf{r}), \qquad (5.25)$$

which is an operator equation both in real space (as shown here) and k-space (Zaroubi & Hoffman 1996). This approach has yet to be applied to real data.

5.5.3.2 Nonlinear scales

On small scales, it is the pairwise velocity dispersion of galaxies, especially in virialized systems, that dominates the redshift-space distortions. The small-scale velocity dispersion of galaxies is interesting for two reasons. First, given various assumptions about the stability of galaxy pairs on small scales, it can be related to the two-point and three-point correlation functions and to the value of Ω (the Cosmic Virial Theorem) (Peebles 1976a,b, 1980; cf. the discussion in Fisher et al. 1994b; Bartlett & Blanchard 1995; Kepner, Summers & Strauss 1997). Second, it is a quantity that can be predicted for various cosmological models from N-body simulations, and has been used in the literature in the past as a strong, although not completely unambiguous, discriminator between models (for the SCDM model alone, which seems to overpredict the small-scale velocity dispersion by a large factor, one can follow the controversy through Davis et al. 1985; Couchman & Carlberg 1992; Cen & Ostriker 1993; Brainerd & Villumsen 1994; Zurek et al. 1994; and Brainerd et al. 1996).

The measurement of this small-scale pairwise velocity dispersion is quite straightforward because the correlation function is strong on small scales, the approximation of isotropy is a good one on these scales, and the effects of peculiar velocities are large. Indeed, measurements of this quantity have been done from the anisotropy of the correlation function for many surveys (cf. Davis & Peebles 1983b; Mo, Jing & Börner 1993; Fisher et al. 1994b; Marzke et al. 1995; Guzzo et al. 1996, 1997; Somerville, Davis & Primack 1997). However, the results have varied over a large range. The problem is that this statistic, as its name implies, is pair-weighted, which means that the densest regions (the rarest, richest clusters) where the velocity dispersion is the highest, tend to dominate the measurement. Therefore, a given measurement of the velocity dispersion is very sensitive to the presence or absence of the richest clusters in the survey volume. There have been several attempts to invent variants on the statistic that are less sensitive to this problem, and thus measure the velocity dispersion in the field. The point is that there is a fair amount of evidence from observations of the peculiar velocity field of galaxies (cf. Ch. 6) that indicates that the velocity field outside of clusters is very quiet (e.g., Sandage 1986; Brown & Peebles 1987; Groth, Juszkiewicz & Ostriker 1989; Burstein 1990; Ostriker & Suto 1990; Strauss, Cen & Ostriker 1993), and that a measure of this should be a sharper discriminator of models than the classic velocity dispersion measure.

5.5.3.3 Correcting the density field for peculiar velocities

We have discussed the relationship between the clustering of galaxies in real space and redshift space, and found that statistical measures of clustering, such as the power spectrum, differ systematically in the two cases. However, there are situations in which we want to do more than simply ask for corrections to statistical quantities measured in redshift space. In particular, it is quite straightforward to measure the density field $\delta(\mathbf{s})$ of galaxies in redshift space,[8] but how might we correct the resulting map for peculiar velocities? In linear theory, we have an answer to this question, due to the direct relation between the density and peculiar velocity field, in the form of equation (5.21) or its integral equivalent:

$$\mathbf{v}(\mathbf{r}) = \frac{\beta}{4\pi} \int \frac{\delta(\mathbf{r}')(\mathbf{r}' - \mathbf{r})d^3\mathbf{r}'}{|\mathbf{r}' - \mathbf{r}|^3}, \qquad (5.26)$$

where we have assumed linear biasing as before. This equation offers an approach to correcting the distribution of galaxies explicitly for peculiar velocities if one has a moderately deep, full-sky redshift survey, at least on scales large enough that linear theory is likely to hold (Yahil et al. 1991). One measures the quantity δ of the galaxies, and solves for the resulting velocity field using equation (5.26). One then corrects the redshifts of each galaxy accordingly:[9]

$$r = cz - \hat{\mathbf{r}} \cdot \mathbf{v}(\mathbf{r}), \qquad (5.27)$$

where $\hat{\mathbf{r}}$ is the unit vector in the direction of the galaxy in question, and $\mathbf{v}$ is the peculiar velocity predicted by equation (5.26) at that position. Of course, this changes the position of each galaxy, and therefore the density field, and thus this process must be done iteratively until convergence. There are a number of points to be made here:

- The corrections clearly depend on an assumed value of β, and this analysis in and of itself yields no estimate of β. Therefore, in practice, one produces a separate density field solution for each value of β one might be interested in.
- Equation (5.26) is only valid on linear scales, and therefore one must, in practice, smooth the density field on small scales in applying this equation. A typical smoothing that is used is a Gaussian with $\sigma = 500 \text{ km s}^{-1}$. A related problem is that clusters of galaxies typically have quite large velocity dispersions, which are very far from being describable by linear theory. In practice, one collapses the galaxies associated with the prominent clusters to a single redshift, to suppress this behavior.

[8] s is the standard notation to refer to redshift space, in contradistinction to the real space r.
[9] Keep in mind that $H_0 \equiv 1$ with our units.

- In a flux-limited redshift survey, the shot noise in the density field necessarily increases as a function of distance from the Local Group. Therefore one needs to carry out some sort of adaptive smoothing as a function of distance to suppress the shot noise. One possibility is to simply have a smoothing length that increases as the mean intergalaxy separation. Another is to expand the density field not in Cartesian coordinates, but rather in spherical harmonics (cf. Nusser & Davis 1994; Fisher et al. 1995a,b). A third possibility is to filter the density field with a Wiener or related filter. These filters are optimal in the sense of suppressing the shot noise while giving the minimum variance difference between the derived and true density fields.
- In principle, the integral in equation (5.26) extends to infinity, while redshift surveys are clearly finite. This is not as serious a problem as it may seem. Redshifts are measured in the rest frame of the earth, and are easily corrected to the heliocentric frame, and to the rest frame of the barycenter of the Local Group (cf. Yahil, Tammann & Sandage 1977). We of course know that the Local Group is moving with respect to the rest frame of the CMB at ≈ 620 km s^{-1} (e.g., Kogut et al. 1993), but we can use a full-sky redshift survey and equation (5.26) to *predict* this motion (e.g., Strauss et al. 1992b; SW). With this predicted peculiar velocity at $\mathbf{r} = 0$, equation (5.27) is modified to:

$$r = cz_{\text{Local Group}} - \hat{\mathbf{r}} \cdot (\mathbf{v}(\mathbf{r}) - \mathbf{v}(0)) \ . \tag{5.28}$$

Here it is apparent that any bulk flow induced by density fluctuations outside the volume surveyed cancel out; all that is relevant are higher-order (and therefore intrinsically weaker) multipoles of the large-scale velocity field.
- The redshift–distance relation along any line of sight (as given by equation [5.28]) is not necessarily monotonic. In particular, in the vicinity of clusters of galaxies, one can get *triple-valued zones*, in which a single redshift can correspond to three distinct distances. This is illustrated in Fig. 5.4, which shows the relation between the redshift and distance (equation [5.28]) along a line of sight that passes close to a large mass concentration. Infall into the cluster causes the redshift to be greater than the distance on the near side of the cluster, and further than the distance on the far side. Thus a galaxy at $cz = 1200$ km s^{-1} can be at three distinct distances, as indicated by stars. Methods for dealing with this are described in Yahil et al. (1991) and SW.
- There are methods other than the iterative method outlined here for correcting the density field for peculiar velocities. In particular, Nusser & Davis (1994) and Fisher et al. (1995b) describe non-iterative methods which involve expanding the density field in spherical harmonics and correcting each mode individually for peculiar velocities. The latter authors compare various different methods by means of of N-body simulations, and conclude that all do roughly equally well.

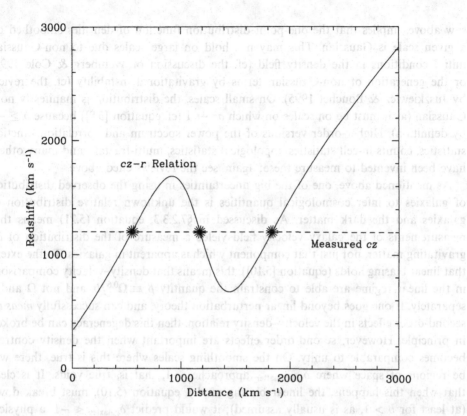

Figure 5.4. The redshift–distance diagram in the vicinity of a cluster. Infall into the cluster causes a region in which the redshift does not climb monotonically with distance. A galaxy with redshift 1200 km s^{-1} in this direction could therefore be at three distinct distances, as indicated by the stars.

Although these techniques do indeed correct the galaxy density field for redshift distortions, they simultaneously make a prediction for the velocity field at every point in space. This is very interesting, for it allows a direct comparison with the observed peculiar velocity field, which allows gravitational instability theory (in the form of equation [5.21] or equation [5.26]) to be tested, and to constrain the value of β (reviewed in detail in §7.6; Dekel 1994; SW).

5.6 The relative distribution of galaxies and dark matter

This review has concentrated on the power spectrum as a measure of the clustering of galaxies. There is quite a variety of other useful statistics with which to measure the clustering properties of galaxies, as reviewed in SW and Borgani (1995). The power spectrum is a complete statistical description of the distribution of galaxies to the extent that the phases of the Fourier modes are randomly distributed, which, as we saw above, implies that the one-point distribution function of densities smoothed on a given scale is Gaussian. This may not hold on large scales due to non-Gaussian initial conditions in the density field (cf. the discussion of Weinberg & Cole 1992) or the generation of non-Gaussian terms by gravitational instability (cf. the review by Juszkiewicz & Bouchet 1995). On small scales, the distribution is manifestly non-Gaussian (as it must be on scales on which $\sigma^2 \to 1$ (cf. equation [5.8]) because $\delta \geq -1$ by definition). Higher-order versions of the power spectrum and correlation function statistics, counts-in-cell statistics, topological statistics, multi-fractals, and many others have been invented to measure these; again, see the reviews cited above.

As mentioned above, one of the big uncertainties in using the observed distribution of galaxies to infer cosmological quantities is the unknown relative distribution of galaxies and the dark matter. As discussed in §7.2.3.3, equation (5.21) means that measurements of the galaxy velocity field yields a measure of the distribution of *all* gravitating matter, not just that component which is apparent in galaxies. To the extent that linear biasing holds (equation [5.10]), this means that density–velocity comparisons in the linear regime are able to constrain the quantity $\beta \equiv \Omega^{0.6}/b$, and not Ω and b separately. If one goes beyond linear perturbation theory, and can successfully *measure* second-order effects in the velocity–density relation, then this degeneracy can be broken in principle. However, second-order effects are important when the density contrast becomes comparable to unity. On the smoothing scales where this is true, there will be regions of space where $\delta_{\text{dark matter}}$ approaches -1, that is, true voids. It is clear that when this happens, the linear biasing model, equation (5.10), must break down (at least for $b > 1$, as is usually assumed); it would predict $\delta_{\text{galaxies}} < -1$, a physical impossibility. Therefore, if we are to allow ourselves the generalization to a nonlinear relation between velocity and density, we must also allow ourselves a nonlinear relation between dark matter and galaxy density contrasts, with the resultant increase in the number of parameters, and the degeneracy between Ω and the biasing parameter(s) remains. The results quoted above in §5.5.2 should be kept in mind, however, when considering these complications: to the extent that the biasing process is *local*, that is, that the process of galaxy formation is a function of the physical properties of matter within a few Mpc of the galaxy in question, the biasing parameter, as measured by the ratio of the power spectra of the galaxies and the dark matter, appears to be independent of scale (Weinberg 1995; Kauffmann, Nusser & Steinmetz 1996).

Another interesting handle on biasing comes through measurements of higher-order correlation functions; their scaling with scale, it turns out, depends non-trivially on the biasing factor(s) (Fry & Gaztañaga 1993; Juszkiewicz et al. 1995; Fry 1994; Gaztañaga 1995; Mo, Jing & White 1996). The observed amplitudes of the high-order correlations as a function of scale are sufficiently close to those predicted from gravitational instability theory without the complications of biasing as to make several of the above papers argue for the biasing being quite small (i.e., $|b - 1| \approx 0$).

There is a more indirect approach to the biasing question, however. Although we scarcely understand the process of galaxy formation in any detail, we imagine that there existed astrophysical processes in the early universe which caused the relative distribution of galaxies and dark matter to differ from one another. If this is the case, we would expect generically that because the astrophysics of formation of galaxies of different types should be different, these various galaxies should be distributed differently with respect to each other. That is, there should exist a *relative* biasing of galaxies of different types. This is of course straightforward to measure from redshift surveys.

There is one such biasing which has been known about since the time of Hubble: although elliptical galaxies make up only 10–20% of the galaxies in the field, they dominate completely in the cores of clusters of galaxies. That is, the density fields of elliptical and spiral galaxies are quite different in the densest regions (cf. Dressler 1980, 1984; Postman & Geller 1984; Whitmore, Gilmore & Jones 1993), something that is quite apparent in redshift-space maps of spirals and ellipticals separately showing the relative distribution of each (e.g., Giovanelli, Haynes & Chincarini 1986; Huchra et al. 1990). It remains quite unclear how this relative biasing continues into the field.

There are two ways one might imagine measuring the relative biasing of two samples of objects. The first is to compare them statistically: compute the power spectrum, the correlation function or higher-order statistics for each, and determine the biasing accordingly. Thus, for linear biasing, the ratio of the two power spectra is proportional to the relative biasing parameter squared, while the S_3 parameter (defined as $\langle \delta^3 \rangle / \langle \delta^2 \rangle^2$ smoothed on some scale) should scale as $1/b$. SW give a brief review of the extensive literature on determinations of relative biasing between samples using this approach. However, the statement $P(k)_{\text{galaxies}} = b^2 P(k)_{\text{dark matter}}$ is clearly a much weaker statement than is equation (5.10) (for the latter to hold, one not only needs a specific relation between the amplitudes of the Fourier modes of δ, but also that the phases of the modes agree), and if two samples are contained in the same volume, it is a much more powerful tool simply to compare the density fields of the two samples directly. There is much work that has gone in this direction in the guise of looking for a population of objects that "fills the voids" defined by the distribution of ordinary galaxies. One elegant statistical approach to quantifying relative biasing which takes the phase information into account is the cross-correlation statistic: one asks for the mean number of galaxies of sample 2 in excess of random a given distance from galaxies of sample 1. Another approach is to take equation (5.10) literally, and compare the density fields of the two samples point by point (cf. Strauss et al. 1992a; Santiago & Strauss 1992).

Despite a great deal of work on this question using a large variety of samples, the results can be summarized in a few sentences. On scales larger than $5h^{-1}$ Mpc or so, there is no direct evidence for relative biasing that is nonlinear or a function of scale, although the limits on such effects are rather weak.[10] Late-type galaxies show a weaker clustering signal than do early type galaxies, by a factor of 1.5 to 2.0 in the two-point correlation statistics. This manifests itself in a number of ways: one can find papers in

---

[10] On small scales, at least, the slopes of the two-point correlation functions of ellipticals and spirals are different (Davis & Geller 1976; Giovanelli, Haynes & Chincarini 1986; Guzzo et al. 1997).

the literature that show that clustering strength increases with the luminosity, central surface brightness, redness, radio power, and mass of the population of galaxies, and decreases with their infrared emission and emission-line strength. Of course, all these quantities are correlated with morphological type, and so it remains unclear whether these trends are separate from the correlation of clustering strength with Hubble type. Moreover, there has yet to be a sample which has been shown clearly to fill the voids which are so prominent in the galaxy distribution.

If clustering is indeed a function of galaxy luminosity, the assumption of the universal luminosity function which went into the definition and use of the selection function (§5.3) is clearly not valid. For this reason, as the data get better and the precision with which we measure statistics such as the power spectrum improve, we will either have to include elaborate models for the dependence of clustering strength on luminosity in our analyses, or we will have to do analyses on subsamples of our data over narrow slices in luminosity.

5.7 Surveys for the future

The progress made over the last decade in this field is striking. We are now measuring, or at least constraining, quantities such as the large-scale biasing, the power spectrum on the largest scales, and the value of Ω, which we had very little hope of getting a handle on at the time people started carrying out large-scale redshift surveys at the end of the 1970s. Part of this progress has been a realization, as is so common in astrophysics, that as we learn more about the systems we study, we realize how naïve and over-simplified our analyses and assumptions have been, and how much more complicated (and interesting!) reality is. In the early 1980s, for example, people were modelling infall into the Virgo cluster (cf. Davis & Peebles 1983a) as if it were an isolated overdensity in a uniformly distributed sea of galaxies. A glance at Figs. 5.1 and 5.2 tells us that this is far from a valid model![11]

That having been said, we are really only starting to be able to measure cosmological quantities of interest. For example, the values of β determined from redshift-space distortions, analyses of the peculiar velocity field (§7.6.6; Shaya, Peebles & Tully 1995), and the mass-to-light ratio of virialized systems (e.g., Bahcall, Lubin & Dorman 1995) vary from 0.3 to over unity, with little convergence between methods in sight at the moment (cf. Fig. 20 of SW). The amplitude of the galaxy power spectrum on large scales is uncertain by at least a factor of two, and probably more, and the relative biasing of galaxies of different types (§5.6) is only partially to blame for the ambiguity. In other words, we are only just starting to do real quantitative cosmology with large-scale surveys, and there is much exciting, and unanticipated, science waiting to be discovered when we start measuring quantities to 10% accuracy (cf. the flurry of theoretical and observational activity prompted by the recent convergence of measurements of H_0 to

---

[11] Although Virgocentric infall remains a useful cosmological probe; see §11.2.3.

within 25%, as summarized in the contributions of Tammann and Freedman to Turok 1997). To beat down the errors further requires much more massive redshift surveys, with:

- much larger volume surveyed;
- much larger number of galaxies with redshifts;
- much tighter control of statistical and especially systematic errors of the photometric quantities by which galaxies are selected.

Two such surveys in the advanced planning stages are discussed here: the Sloan Digital Sky Survey and the Two-Degree Field Survey. As I am involved with the former, and thus much more familiar with it, I will put the greater emphasis on it here.

5.7.1 The Sloan Digital Sky Survey

The SDSS is a collaboration between Princeton University, the Institute for Advanced Study, the University of Chicago, the Fermi National Accelerator Laboratory, the University of Washington, John's Hopkins University, the United States Naval Observatory, and the Japan Promotion Group. It is based on a dedicated 2.5-m large-field optical telescope at Apache Point in south-eastern New Mexico, which saw first light in May 1998. The telescope has two main instruments: a photometric camera with thirty 2048×2048, and twenty-four 2048×400 SIte CCD chips on its focal plane, and a pair of double multi-object spectrographs, together taking 640 fibers of $3''$ aperture. The purpose of this survey is several-fold. With the photometric camera, the one-quarter of the sky (roughly 10 000 square degrees) centered on the Northern Galactic Cap will be surveyed in drift-scan mode through five broad-band filters (u', g', r', i', z', a new photometric system (cf. Fukugita et al. 1996) with effective wavelengths of 3540Å, 4760Å, 6280Å, 7690Å, and 9250Å, respectively) almost simultaneously, with an effective exposure time of 55 s for each. Stellar objects will be detected at 5σ to $r' \approx 23$. From the resulting list of objects detected in the photometric survey, galaxies will be selected to a photometric limit of $r' \approx 18$, and quasar candidates will be selected from the stellar objects by their distinctive colors to a magnitude and a half fainter. These objects, with a density of roughly 120 per square degree, will be observed with the multi-object spectrograph to obtain spectra from 3900–9100Å with a resolution of 2000. Over the course of the survey, which is expected to take five years, we will thus obtain spectra of roughly 10^6 galaxies and 1.5×10^5 quasar candidates, of which we hope 60–70% will be bona fide quasars. Finally, during the fall months, when the Northern Galactic Cap is unreachable, we plan to repeatedly scan an equatorial strip 2.5° wide and 90° long in the South Galactic Cap, centered on $\delta = 0°$. This will allow us to do photometry over 225 square degrees to a photometric limit roughly two magnitudes fainter than in the North. Further photometry (only a single pass) will be done on two "outrigger" great circle strips centered roughly at $\delta = +15°$ and $\delta = -10°$ (cf. Fig. 5.7) to maximize the number of baselines for measurements of the power spectrum on the very largest scales.

This is a very quick overview of the plans for the SDSS. Reviews can be found by Gunn & Knapp (1993) and Gunn & Weinberg (1995), and a great deal of technical detail can be found in the text of the proposal to NASA (at http://www.astro.princeton.edu/BBOOK/). Here, let us summarize the goals and prospects for large-scale structure studies with the data from the SDSS. We have room here to touch upon only a few scientific issues which the SDSS will impact; for many more (both in large-scale structure, and in many other fields of extragalactic and Galactic astronomy), see the Web site above.

The SDSS redshift survey will survey to a depth that has been probed previously; the depth is comparable to that of the LCRS and ESP surveys discussed above in §5.2. Moreover, like the LCRS, the survey galaxies will be selected from photometric CCD data. However, the the volume covered by the SDSS redshift survey will be enormously greater; the solid angle coverage of the SDSS will be 14 times that of the LCRS.

Large-scale structure studies are a major goal of the SDSS. Thus a great deal of emphasis has been put on controlling systematic errors in the survey. The focal plane of the photometric camera is shown in Fig. 5.5. The camera will drift-scan at sidereal rate along great circles in the sky and thus observe six parallel strips of the sky simultaneously; two passes separated by roughly 13' will cover a strip 2.5° wide, with roughly 2' overlap between strips. Astrometric calibration is done with the astrometric chips, as described in the caption to Fig. 5.5. Photometric calibration will be done with a separate dedicated 20-in telescope at the Apache Point site, which will spend the night obtaining photometry of a series of roughly 30 primary standards through the SDSS filters to measure the extinction and photometricity of the sky every hour, and then tie this solution to the photometry of the 2.5-m telescope by observing secondary patches in regions of the sky covered by the strips. The photometric calibration will be checked a posteriori with the overlaps between survey strips, and probably also with a series of strips taken perpendicular to the main great circle scans of the sky.

We of course want a sample of galaxies with accurate photometry *as they would be seen from outside our Galaxy.* Therefore we plan to measure Galactic extinction using our multi-color data. In particular, we will use the colors of distant hot halo subdwarfs (selected by their colors, and confirmed spectroscopically), as well as number counts and color distributions of the faint galaxy populations we find, possibly supplemented by HI maps from Stark et al. (1992) and Burton & Hartmann (1994), and the long-wavelength DIRBE maps from Schlegel, Finkbeiner & Davis (cf. Schlegel 1995).

Galaxies will be selected for spectroscopy from the photometrically calibrated images after an a priori correction for reddening either from Burstein & Heiles (1982) or Schlegel (1995).[12] We wish to select galaxies in as uniform a way as possible, minimizing the effects of redshift. Ideal would be to measure "total" fluxes for galaxies, but

---

[12] Given the subtleties of the determination of reddening using the methods mentioned above, we will be unable to create an improved reddening map "on the fly", thus our final large-scale structure analyses will require correction for the difference between our a priori and final reddening maps.

Redshift surveys of the local universe

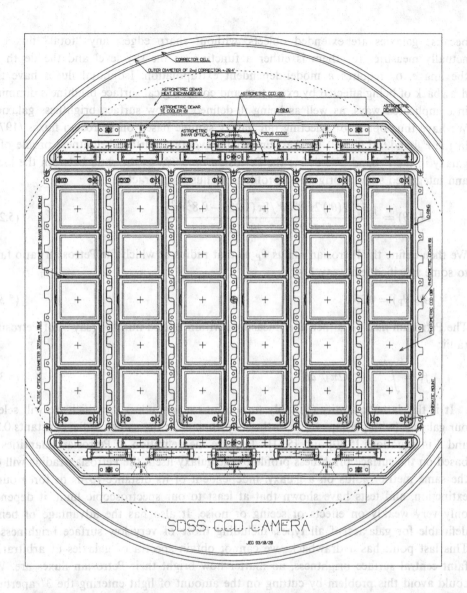

Figure 5.5. A schematic of the focal plane of the photometric camera of the SDSS. The camera works in drift-scan mode, scanning in the vertical direction, and thus traces out six parallel strips on the sky. The radius of the focal plane (as indicated by the dashed circle) is roughly 2.5 degrees in diameter. Each of the large CCDs is 2048 × 2048, with 0.4″ pixel size. Each of the five rows of CCDs has a different photometric filter in front of it, so that a given area of sky is scanned in r', i', u', z', and g', in order, over a span of roughly 8 minutes. There are twenty-two 2048 × 400 CCDs with r' filters and 4.2 mag neutral density filters, at the leading and trailing edges of the arrays. These chips will saturate at $r' = 6.6$ (rather than $r' \approx 14$ for the large chips), allowing stars in astrometric catalogs (especially HIPPARCOS, cf. Kovalevsky *et al.* 1995) to be tied to the SDSS images. Finally, the two 2048 × 400 CCDs at the top and bottom of the array are used to measure and adjust the focus.

because galaxies are extended objects, without sharp edges, any "total" flux one actually measures for them is either a function of the sky level and the depth of the image, or requires a model-dependent extrapolation. Isophotal fluxes have the drawback of being affected by extinction and cosmological surface-brightness dimming in complicated ways, as well as being ill-defined for low surface brightness galaxies. We have thus opted for selecting our galaxy sample based on Petrosian fluxes (1976) in the r' band.[13] Let $I(\theta)$ be the azimuthally averaged surface brightness profile of a galaxy in the r' band. We define the Petrosian *ratio* $\mathcal{R}_P$ as the ratio between the local and integrated surface brightness profile at radius θ; in practice:

$$\mathcal{R}_P(\theta) = \frac{\int_{0.8\theta}^{1.25\theta} I(\theta') 2\pi \theta' d\theta' / [\pi (1.25^2 - 0.8^2)\theta^2]}{\int_0^\theta I(\theta') 2\pi \theta' d\theta' / [\pi \theta^2]}. \qquad (5.29)$$

We then define the Petrosian *radius* θ_P as that radius at which the Petrosian ratio falls to some specified level, say 1/4:

$$\mathcal{R}_P(\theta_P) = 0.25. \qquad (5.30)$$

The Petrosian *flux* f_P is then that measured within a fixed number (say 3) of Petrosian radii:

$$f_P = \int_0^{3\theta_P} I(\theta') 2\pi \theta' d\theta'. \qquad (5.31)$$

It is this latter quantity (suitably turned into magnitudes) on which we will select our galaxies (with perhaps some further adjustments in the values of the constants 0.25 and 3 in equations [5.30] and [5.31]). Because the definition of Petrosian quantities is based on the surface brightness profile of the galaxy itself, the Petrosian radius will be the same metric radius on a galaxy, independent of its distance to us or foreground extinction, and tests have shown that at least to our spectroscopic limit, it depends only very weakly on effects of seeing or noise. It also has the advantage of being definable for galaxies of all types, including those of very low surface brightness.[14] This last point has a drawback; we cannot obtain spectra of galaxies of arbitrarily faint central surface brightness, no matter how bright their Petrosian fluxes are. We could avoid this problem by cutting on the amount of light entering the 3″ aperture of the fibers. However, such a cut would be very difficult to model as a function of redshift. Therefore, we instead include a secondary cut on Petrosian surface brightness

---

[13] We have looked into the possibility of a joint selection in r' and one of the bluer bands. However, the larger extinction in the atmosphere and the lowered throughput at u' makes this band impractical, and simulations have shown that the galaxy populations selected in g' would be almost indistinguishable from those at r', both in redshift distribution and morphological mix.

[14] One important exception to this last statement is galaxies with power-law surface brightness profiles, as many cD galaxies have (e.g., Postman & Lauer 1995, and references therein); for such galaxies, the Petrosian ratio asymptotes to a constant that can be above the level at which the Petrosian radius is defined. In such cases, we will switch over to a model magnitude. Another potential problem is galaxies with hierarchical structure (e.g., a Seyfert nucleus in a disk galaxy) for which the Petrosian ratio may not be monotonic. Given our chosen value of the value of Petrosian ratio at which the Petrosian radius is defined, this is a problem only for a very small fraction of galaxies.

at $\mu_P \leq 22$ in r'. This quantity is defined as follows: we define a Petrosian half-light radius θ_{50}, such that:

$$\int_0^{\theta_{50}} I(\theta') 2\pi \theta' d\theta' = 0.5 f_P ; \qquad (5.32)$$

the Petrosian surface brightness is then:

$$\mu_P = \frac{0.5 f_P}{\pi \theta_P^2} . \qquad (5.33)$$

Simulations show that the resulting galaxy sample has a distribution of 3" aperture magnitudes with a sharp cut-off at $r' = 19.5$, at which point the signal-to-noise ratio is still adequate to obtain redshifts for our exposure time (~ 45 min). It is unfortunate that the limitations of exposure time and the size of our aperture do not allow us to obtain redshifts for the very low surface brightness population of our galaxies, which show an intriguingly different large-scale distribution from those of "ordinary" galaxies (cf. Mo, McGaugh & Bothun 1994).

There is another class of galaxies to be targeted spectroscopically in the SDSS. Luminous red elliptical galaxies tend to be metal-rich, and thus have strong and prominent metal absorption lines. This means that redshifts can be measured for them at a lower signal-to-noise ratio than for typical absorption-line galaxies. These objects are interesting in their own right, because they are often associated with clusters; indeed, brightest cluster galaxies make up the most luminous and red population of galaxies, and thus targeting the luminous red ellipticals allows us to obtain redshifts for a deep sample of clusters. We will estimate redshifts photometrically from the five-color data for all galaxies using the methods of Connolly et al. (1995) (which should be especially accurate for quiescent red ellipticals), and thereby determine a luminosity. Cuts will be made in luminosity and K-corrected color (the most luminous elliptical galaxies are very uniform in color, which is why the photometric redshift determination is so accurate for this class of objects), yielding a spectroscopic sample of $\sim 10^5$ objects volume-limited roughly to a redshift of 0.45.

The galaxy selection criteria have been described in quite a bit of detail in order to emphasize the care that is being taken to obtain a uniform sample, as free as possible of biases and systematic errors. The scientific goals of the project range from large-scale structure studies, which are the subject of this review, through quasars and the global properties of galaxies, to Galactic structure and interstellar extinction. Again, see the NASA proposal at the URL quoted above for more detail.

The large-scale structure goals of the survey are manifold. The most obvious is, of course, the measurement of the power spectrum on very large scales. Figure 5.6 shows the measured power spectrum of IRAS galaxies from Fisher et al. (1993), together with a prediction of what the SDSS will see ($\Gamma = 0.3$ CDM), with error bars following the formalism of Feldman, Kaiser & Peacock (1994). Thus the SDSS will measure the galaxy power spectrum on scales on which the Sachs–Wolfe effect is directly measured by COBE. Indeed, it will measure $P(k)$ with great precision on smaller scales in the

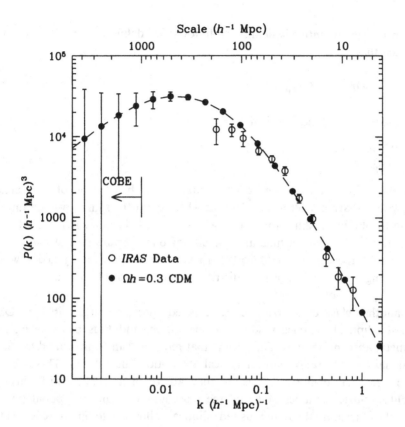

Figure 5.6. The observed power spectrum of IRAS galaxies from Fisher et al. (1993) (open circles) is shown on top of the linear theory $\Gamma \equiv \Omega h = 0.3$ CDM power spectrum, with error bars at selected points expected for the SDSS volume, using the formulae of Feldman, Kaiser & Peacock (1994). The physical scale corresponding to the 10° resolution of the COBE satellite is indicated.

CMB which are now just starting to be probed by balloon-borne and ground-based missions, and which will be studied in detail by the next generation of satellites (cf. Bennett et al. 1995; Mandolesi 1995). If we have sufficient control on systematic errors, the red-luminous elliptical sample described above should be able to probe $P(k)$ on the larger scales with appreciably smaller error bars than are shown in Fig. 5.6.

The redshift-space distortions discussed above in §5.5.3 have allowed determinations of β to an accuracy of 50% or so, and even that has required a great deal of modelling (cf. the discussion in Fisher et al. 1994b). As we have mentioned above, the principal limitation of these analyses is the smallness of the volume probed; for this statistic, we are far from a fair sample.[15] The SDSS will probe a very large number

---

[15] As this example makes clear, the definition of a "fair sample" of the universe depends very much on the sort of statistic one is trying to measure. Thus current samples are more than large enough to define the mean number density of luminous galaxies, but are certainly not adequate to measure the power spectrum on the largest scales.

of superclusters at (presumably) random inclinations, greatly reducing the statistical errors in this analysis. Moreover, one can probe for more subtle effects which could only be set a priori in the analysis of Fisher et al. (1994b), such as the amplitude of the mean streaming as a function of scale, and the detailed shape of the pairwise velocity distribution function on small scales. There is even the possibility of measuring the second moment of the velocity distribution function on *linear* scales (cf. Fisher 1995), which has the possibility of breaking the degeneracy between Ω and b.

But perhaps the most exciting large-scale structure studies to be done with the SDSS will involve measurements of the distribution of galaxies as a function of the physical properties of the galaxies (such as morphological type, color, luminosity, strength of emission or absorption lines, surface brightness, and so on). As we have discussed in §5.6, it is now clear that there exists a relative biasing between galaxies of different types, but we as yet cannot characterize its nature as a function of scale or its universality. It is now becoming clear that the simple linear biasing model of equation (5.10) cannot hold true in detail, and that therefore different analyses of large-scale structure in a sense are sensitive to different moments of the relationship between the galaxy and dark-matter distribution. The analyses mentioned above, as well as many others, will be done not only on the full sample of galaxies in the SDSS, but also on subsamples divided by the physical properties of galaxies (the sample will be large enough to allow us to do this!); comparison of the results will give us strong constraints on the relative biasing of different populations of galaxies.

5.7.2 The Two-Degree Field

The Two-Degree Field (2dF) refers to an instrument mounted on the prime focus of the Anglo–Australian Telescope. It is a fiber-fed multi-object spectrograph with 400 fibers and a two-degree diameter field, as its name implies. The fiber coupler is doubled, so that one field can be prepared with a robot arm while a second one is observed, minimizing overhead. This is a general purpose observatory instrument which is to be used for a variety of surveys, but a collaboration of British and Australian astronomers is planning to use it for a redshift survey of galaxies selected from the APM galaxy catalog (cf. Maddox et al. 1990a,b,c); thus, unlike the SDSS, the 2dF survey does not attempt to create a galaxy catalog from scratch. The survey will cover 1700 square degrees, and will obtain redshifts of galaxies to $B = 19.7$[16] as measured by the APM. It will consist of three parts: contiguous strips of $75° \times 12.5°$ and $65° \times 7.5°$ in the Southern and Northern Galactic Caps, respectively; and 100 random circular fields of radius $2°$ over the Southern Galactic Cap. The survey will contain roughly 250,000 galaxies, and will take roughly 90 nights of AAT dark time. The survey geometry is chosen to maximize sensitivity to large-scale structure; indeed, the 100 random fields give a variety of baselines to probe the power spectrum on the largest scales. The

---

[16] This is slightly deeper than the SDSS, given typical galaxy colors of $r' - B \approx -1$ (Frei & Gunn 1994).

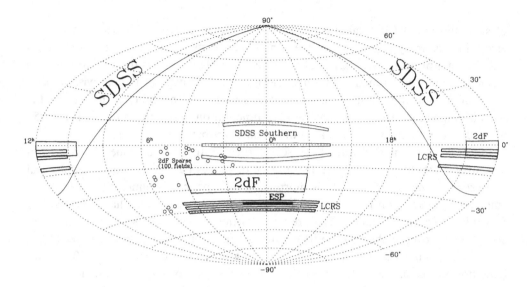

Figure 5.7. An Aitoff projection of the sky in right ascension and declination, showing the sky coverage of the various surveys discussed in the text.

science goals of the 2dF survey are similar to those of the large-scale structure goals of the SDSS, although the two surveys differ on the approach to galaxy catalogs and target selection, the sky coverage, the depth of the survey, and other details.

There will be an extension to the 2dF survey to measure redshifts for ~ 6000 galaxies to $R = 21$, to extend large-scale structure studies to redshifts of 0.3. The SDSS will be able to do such studies using the luminous red elliptical sample mentioned above, as well as from the photometric sample (which of course extends quite a bit fainter than the spectroscopic sample). The 2dF instrument is described in Taylor (1995), and the the plan for the redshift survey can be found at http://qso.lanl.gov/~heron/Colless/colless_heron/colless_heron.html.

5.8 Conclusions

This chapter reviewed recent progress in measurements of large-scale structure with redshift surveys of galaxies in the nearby universe, with emphasis on the power spectrum of the galaxy density field. There are a number of complicating factors which separate the ideal linear power spectrum of the dark matter from the quantity measured, such as redshift-space distortions, nonlinear effects, and the relative biasing of galaxies and dark matter, but in each case, these complications include in themselves valuable cosmological information. In conclusion, Fig. 5.7, modified from a similar figure in Guzzo (1996), shows the coverage on the sky of the various surveys we have discussed. This is certainly not complete; the IRAS surveys are not shown here, nor are the CfA2 and SSRS2 surveys (cf. da Costa et al. 1994a) or the Perseus–Pisces redshift survey

(Haynes & Giovanelli 1989). However, all the surveys shown here are probing, or will probe, the large-scale distribution of galaxies to redshifts of 0.1 over substantial solid angles, with more careful and uniform sample selection than has been possible before now. We are just starting to measure cosmologically important parameters with redshift surveys with errors of 50% or less; this next generation of surveys holds out great hope to reduce these errors substantially, with the potential of new and unexpected insights into our picture about the formation of large-scale structure and galaxies.

Acknowledgments

MAS wishes to thank his colleagues on the various redshift surveys discussed here, as well as his SDSS colleagues whose work is summarized above. Avishai Dekel, Michael Vogeley and David Weinberg all made useful comments on a previous draft of this paper. MAS also thanks Huan Lin of the LCRS collaboration for permission to show Fig. 5.2, Tsafrir Kolatt for the input files to Fig. 5.3, Luigi Guzzo for his SM macros which led to Fig. 5.7, and Max Tegmark for many useful discussions. Finally, MAS is happy to acknowledge the financial support of an Alfred P. Sloan Foundation fellowship.

References

Bahcall, N. A., Lubin, L. M. & Dorman, V. 1995, *ApJ*, **447**, L81

Bardeen, J., Bond, J. R., Kaiser, N. & Szalay, A. 1986, *ApJ*, **304**, 15

Bartlett, J. G. & Blanchard, A. 1996, *A&A*, **307**, 1

Baugh, C. M. & Efstathiou, G. 1993, *MNRAS*, **265**, 145

Baugh, C. M. & Efstathiou, G. 1994, *MNRAS*, **267**, 323

Baumgart, D. J. & Fry, J. N. 1991, *ApJ*, **375**, 25

Bennett, C. L., et al. 1995, *BAAS*, **187**, #71.09

Bennett, C. L., et al. 1996, *ApJ*, **464**, L1

Borgani, S. 1995, *Phys. Rep.*, **251**, 1

Bower, R. G., Coles, P., Frenk, C. S. & White, S. D. M. 1993, *ApJ*, **405**, 403

Brainerd, T. G., Bromley, B. C., Warren, M. S. & Zurek, W. 1996, *ApJ*, **464**, L103

Brainerd, T. G. & Villumsen, J. V. 1993, *ApJ*, **415**, L67

Brainerd, T. G. & Villumsen, J. V. 1994, *ApJ*, **436**, 528

Brown, M. E. & Peebles, P. J. E. 1987, *ApJ*, **317**, 588

Bunn, E. F. & White, M. 1997, *ApJ*, **480**, 6

Burstein, D. 1990, *Rep. Prog. Phys.*, **53**, 421

Burstein, D. & Heiles, C. 1982, *AJ*, **87**, 1165

Burton, W. B. & Hartmann, D. 1994, *Ap&SS*, **217**, 189

Cen, R. Y. & Ostriker, J. P. 1993, *ApJ*, **417**, 415

Cole, S., Fisher, K. B. & Weinberg, D. 1994, *MNRAS*, **267**, 785

Cole, S., Fisher, K. B. & Weinberg, D. 1995, *MNRAS*, **275**, 515

Coles, P. & Lucchin, F. 1995, *The Origin and Evolution of Cosmic Structure* (New York: John Wiley and Sons)

Connolly, A. J., Csabai, I., Szalay, A. S., Koo, D. C., Kron, R. G. & Munn, J. A. 1995, *AJ*, **110**, 1071

Corwin, H. G. & Skiff, B. A. 1996, *Extension to the Southern Galaxies Catalogue*, in preparation

Couchman, H. M. P. & Carlberg, R. G. 1992, *ApJ*, **389**, 453

da Costa, L. N. et al. 1994a, *ApJ*, **424**, L1

da Costa, L. N., Vogeley, M.S., Geller, M.J., Huchra, J. P. & Park, C. 1994b, *ApJ*, **437**, L1

Davis, M., Efstathiou, G., Frenk, C. S. & White, S. D. M. 1985, *ApJ*, **292**, 371

Davis, M. & Geller, M. J. 1976, *ApJ*, **208**, 13

Davis, M. & Huchra, J. P. 1982, *ApJ*, **254**, 437

Davis, M. & Peebles, P. J. E. 1983a, *ARA&A*, **21**, 109

Davis, M. & Peebles, P. J. E. 1983b, *ApJ*, **267**, 465

Davis, M., Tonry, J., Huchra, J. & Latham, D. W. 1980, *ApJ*, **238**, L113

Davis, R. L., Hodges, H. M., Smoot, G. F., Steinhardt, P. J. & Turner, M. S. 1992, *PRL*, **69**, 1856

Dekel, A. 1994, *ARA&A*, **32**, 371

Dekel, A. & Rees, M. J. 1987, *Nature*, **326**, 455

de Lapparent, V., Geller, M. J. & Huchra, J. P.

1986, *ApJ*, **302**, L1

Dressler, A. 1980, *ApJ*, **236**, 351

Dressler, A. 1984, *ARA&A*, **22**, 185

Efstathiou, G. 1991, in *Physics of the Early Universe*, eds. J. A. Peacock, A. F. Heavens & A. T. Davies (Edinburgh: SUSSP), p. 361

Efstathiou, G. 1996, *Les Houches Lectures*, ed. R. Schaeffer (Netherlands: Elsevier Science Publishers), p. 133

Efstathiou, G., Ellis, R. S. & Peterson, B. S. 1988, *MNRAS*, **232**, 431

Ellis, R. S. 1997, in *Critical Dialogues in Cosmology*, ed. N. Turok (Singapore: World Scientific), p. 461

Feldman, H., Kaiser, N. & Peacock, J. 1994, *ApJ*, **426**, 23

Felten, J. E. 1977, *AJ*, **82**, 861

Fisher, K. B. 1995, *ApJ*, **448**, 494

Fisher, K. B., Davis, M., Strauss, M. A., Yahil, A. & Huchra, J. P. 1993, *ApJ*, **402**, 42

Fisher, K. B., Davis, M., Strauss, M. A., Yahil, A. & Huchra, J. P. 1994a, *MNRAS*, **266**, 50

Fisher, K. B., Davis, M., Strauss, M. A., Yahil, A. & Huchra, J. P. 1994b, *MNRAS*, **267**, 927

Fisher, K. B., Huchra, J. P., Davis, M., Strauss, M. A., Yahil, A. & Schlegel, D. 1995a, *ApJS*, **100**, 69

Fisher, K. B., Lahav, O., Hoffman, Y., Lynden-Bell, D. & Zaroubi, S. 1995b, *MNRAS*, **272**, 885

Fisher, K. B. & Nusser, A. 1995, *MNRAS*, **279**, L1

Fisher, K. B., Scharf, C. A. & Lahav, O. 1994, *MNRAS*, **266**, 219

Frei, Z. & Gunn, J. E. 1994, *AJ*, **108**, 1476

Fry, J. N. 1994, *PRL*, **73**, 215

Fry, J. N. & Gaztañaga, E. 1993, *ApJ*, **413**, 447

Fukugita, M., Ichikawa, T., Gunn, J. E., Doi, M., Shimasaku, K. & Schneider, D. P. 1996, *AJ*, **111**, 1748

Gaztañaga, E. 1995, *ApJ*, **454**, 561

Geller, M. J. & Huchra, J. P. 1988, in *Large-Scale Motions in the Universe*, eds. V. C. Rubin & G. V. Coyne (Princeton: Princeton University Press), p. 3

Geller, M. J. & Huchra, J. P. 1989, *Science*, **246**, 897

Giovanelli, R., Haynes, M. P. & Chincarini, G. L. 1986, *ApJ*, **300**, 77

Giovanelli, R. & Haynes, M. P. 1991, *ARA&A*, **29**, 499

Górski, K. M. *et al.* 1996, *ApJ*, **464**, L5

Gramann, M., Cen, R. & Bahcall, N. A. 1993, *ApJ*, **419**, 440

Groth, E. J., Juszkiewicz, R. & Ostriker, J. P. 1989, *ApJ*, **346**, 558

Gunn, J. E. & Knapp, G. 1993, in *Sky Surveys*, ed. B. T. Soifer, Astronomical Society of the Pacific Conference Series # 43, p. 267

Gunn, J. E. & Weinberg, D. H. 1995, in *Wide-Field Spectroscopy and the Distant Universe*, eds. S. J. Maddox & A. Aragón-Salamanca (Singapore: World Scientific), p. 3

Guzzo, L. 1996, in *Mapping, Measuring, and Modelling the Universe*, eds. P. Coles & V. Martinez, (ASP Conference Series # 94), p. 157

Guzzo, L., Fisher, K. B,, Strauss, M. A., Giovanelli, R. & Haynes, M. P. 1996, *Astrophysics Letters and Communications*, **33**, 231

Guzzo, L., Strauss, M. A., Fisher, K. B., Giovanelli, R. & Haynes, M. P. 1997, *ApJ*, **489**, 37

Hamilton, A. J. S. 1993, *ApJ*, **406**, L47

Hamilton, A. J. S. 1995, in *Clustering in the Universe*, Proc. 30th Rencontres de Moriond, eds. S. Maurogordato, C. Balkowski, C. Tao & J. Trân Thanh Vân (Gif-sur-Yvette: Editions Frontières), p. 143

Hamilton, A. J. S. 1997a, *MNRAS*, **289**, 285

Hamilton, A. J. S. 1997b, *MNRAS*, **289**, 295

Hamilton, A. J. S. & Culhane, M. 1996, *MNRAS*, **278**, 73

Hamilton, A. J. S., Kumar, P., Lu, E. & Matthews, A. 1991, *ApJ*, **374**, L1

Haynes, M. P. & Giovanelli, R. 1989, in *Large Scale Motions in the Universe*, eds. V. C.

Rubin & G. V. Coyne (Princeton: Princeton University Press), p. 31

Heavens, A. F. & Taylor, A. N. 1995, *MNRAS*, **275**, 483

Hermit, S., Lahav, O., Santiago, B. X., Strauss, M. A., Davis, M., Dressler, A. & Huchra, J. P. 1996, *MNRAS*, **283**, 709

Heydon-Dumbleton, N. H., Collins, C. A. & MacGillivray, H. T. 1989, *MNRAS*, **238**, 379

Huchra, J. P., Davis, M., Latham, D. & Tonry, J. 1983, *ApJS*, **52**, 89

Huchra J. P., Geller, M. J., de Lapparent, V. & Corwin, H. G. 1990, *ApJS*, **72**, 433

IRAS Catalogs & Atlases, Explanatory Supplement 1988, eds. C. A. Beichman, G. Neugebauer, H. J. Habing, P. E. Clegg & T. J. Chester (Washington D.C.: U.S. Government Printing Office)

Jain, B. & Bertschinger, E. 1994, *ApJ*, **431**, 495

Jain, B., Mo, H. J. & White, S. D. M. 1995, *MNRAS*, **276**, L25

Juszkiewicz, R. & Bouchet, F. R. 1995, in *Clustering in the Universe*, Proc. 30th Rencontres de Moriond, eds. S. Maurogordato, C. Balkowski, C. Tao & J. Trân Thanh Vân (Gif-sur-Yvette: Editions Frontières), p. 167

Juszkiewicz, R., Weinberg, D., Amsterdamski, P., Chodorowski, M. & Bouchet, F. 1995, *ApJ*, **442**, 39

Kaiser, N. 1984, *ApJ*, **284**, L9

Kaiser, N. 1987, *MNRAS*, **227**, 1

Kauffmann, G., Nusser, A. & Steinmetz, M. 1997, *MNRAS*, **286**, 795

Kepner, J. P., Summers, F. J. & Strauss, M. A. 1997, *New Astronomy*, **2**, 165

Kogut, A., et al. 1993, *ApJ*, **419**, 1

Kolb, E. W. & Turner, M. S. 1990, *The Early Universe* (Redwood City: Addison-Wesley)

Kolatt, T. & Dekel, A. 1997, *ApJ*, **479**, 592

Koranyi, D. M. & Strauss, M. A. 1997, *ApJ*, **477**, 36

Kovalevsky, J. et al. 1995, *A&A*, **304**, 34

Landy, D. S., Shectman, S. A., Lin, H., Kirshner, R. P., Oemler, A. A. & Tucker, D. 1996, *ApJ*, **456**, L1

Lauberts, A. 1982, *The ESO/Uppsala Survey of the ESO(B) Atlas* (München: European Southern Observatory)

Lauberts, A. & Valentijn, E. A. 1989, *The Surface Photometry Catalogue of the ESO-Uppsala Galaxies* (München: European Southern Observatory)

Lawrence, A. et al. 1998, in preparation

Lilly, S. J., Le Fevre, O., Crampton, D., Hammer, F. & Tresse, L. 1995, *ApJ*, **455**, 50

Lin, H., et al. 1996, *ApJ*, **471**, 617

Loveday, J., Efstathiou, G., Maddox, S. J. & Peterson, B. A. 1996, *ApJ*, **468**, 1

Maddox, S. J., Efstathiou, G., Sutherland, W. J. & Loveday, J. 1990a, *MNRAS*, **242**, 43P

Maddox, S. J., Efstathiou, G., Sutherland, W. J. & Loveday, J. 1990b, *MNRAS*, **243**, 692

Maddox, S. J., Efstathiou, G. & Sutherland, W. J. 1990c, *MNRAS*, **246**, 433

Maddox, S. J., Efstathiou, G. & Sutherland, W. J. 1996, *MNRAS*, **283**, 1227

Mancinelli, P. 1996, PhD Thesis (SUNY Stony Brook)

Mandolesi, N. 1995, *Planetary & Space Science*, **43**, 1459

Marzke, R. O., Geller, M. J., da Costa, L. N. & Huchra, J. P. 1995, *AJ*, **110**, 477

Mo, H. J., Jing, Y. P. & Börner, G. 1993, *MNRAS*, **264**, 825

Mo, H. J., Jing, Y. P. & White, S. D. M. 1996, *MNRAS*, **284**, 189

Mo, H. J., McGaugh, S. S. & Bothun, G. D. 1994, *MNRAS*, **267**, 129

Nicoll, J. F. & Segal, I. E. 1983, *A&A*, **118**, 180

Nilson, P. 1973, *The Uppsala General Catalogue of Galaxies*, Ann. Uppsala Astron. Obs. Band 6, Ser. V:A. Vol. 1

Nusser, A. & Davis, M. 1994, *ApJ*, **421**, L1

Ostriker, J. P. & Suto, Y. 1990, *ApJ*, **348**, 378

Padmanabhan, T. 1993, *Structure Formation in the Universe* (Cambridge: Cambridge University Press)

Park, C., Gott, J. R. & da Costa, L. N. 1992,

ApJ, **392**, L51

Park, C., Vogeley, M. S., Geller, M. J. & Huchra, J. P. 1994, ApJ, **431**, 569

Peacock, J. A. & Dodds, S. J. 1994, MNRAS, **267**, 1020

Peacock, J. A. & Nicholson, D. 1991, MNRAS, **253**, 307

Peebles, P. J. E. 1976a, ApJ, **205**, L109

Peebles, P. J. E. 1976b, A&SS, **45**, 3

Peebles, P. J. E. 1980, *The Large Scale Structure of the Universe* (Princeton: Princeton University Press)

Peebles, P. J. E. 1987, *Nature*, **327**, 210

Peebles, P. J. E. 1993, *Principles of Physical Cosmology* (Princeton: Princeton University Press)

Petrosian, V. 1976, ApJ, **209**, L1

Postman, M. & Geller, M. J. 1984, ApJ, **281**, 95

Postman, M. & Lauer, T. R. 1995, ApJ, **440**, 28

Rowan-Robinson, M., et al. 1990, MNRAS, **247**, 1

Sachs, R. K. & Wolfe, A. M. 1967, ApJ, **147**, 73

Sahni, V. & Coles, P. 1995, *Physics Reports*, **262**, 1

Sandage, A. 1986, ApJ, **307**, 1

Sandage, A. & Tammann, G. A. 1981, *A Revised Shapley-Ames Catalog of Bright Galaxies* (Washington DC: Carnegie Institute of Washington)

Sandage, A., Tammann, G. A. & Yahil, A. 1979, ApJ, **232**, 352

Santiago, B. X. & Strauss, M. A. 1992, ApJ, **387**, 9

Santiago, B. X., Strauss, M. A., Lahav, O., Davis, M., Dressler, A. & Huchra, J. P. 1995, ApJ, **446**, 457

Santiago, B. X., Strauss, M. A., Lahav, O., Davis, M., Dressler, A. & Huchra, J. P. 1996, ApJ, **461**, 38

Saunders, W., et al. 1990, MNRAS, **242**, 318

Schlegel, D. 1995, PhD Thesis, University of California, Berkeley

Shaya, E. J., Peebles, P. J. E. & Tully, R. B. 1995, ApJ, **454**, 15

Shectman, et al. 1996, ApJ, **470**, 172

Sigad, Y., Eldar, A., Dekel, A., Strauss, M. A. & Yahil, A. 1998, ApJ, **495**, 516

Smoot, G. F., et al. 1991, ApJ, **371**, L1

Somerville, R. S., Davis, M. & Primack, J. R. 1997, ApJ, **479**, 616

Stark, A. A., et al. 1992, ApJS, **79**, 77

Strauss, M. A., Cen, R. Y. & Ostriker, J. P. 1993, ApJ, **408**, 389

Strauss, M. A., Davis, M., Yahil, A. & Huchra, J. P. 1990, ApJ, **361**, 49

Strauss, M. A., Davis, M., Yahil, A. & Huchra, J. P. 1992a, ApJ, **385**, 421

Strauss, M. A., Huchra, J. P., Davis, M., Yahil, A., Fisher, K. B. & Tonry, J. 1992b, ApJS, **83**, 29

Strauss, M. A. & Willick, J. A. 1995, *Phys. Rep.*, **261**, 271 (SW)

Strauss, M. A., Yahil, A., Davis, M., Huchra, J. P. & Fisher, K. B. 1992c, ApJ, **397**, 395

Tadros, H. & Efstathiou, G. 1995, MNRAS, **276**, L45

Tadros, H. & Efstathiou, G. 1996, MNRAS, **282**, 1381

Taylor, A. N. & Hamilton, A. J. S. 1996, MNRAS, **282**, 767

Taylor, K. 1995, in *Wide-Field Spectroscopy and the Distant Universe*, eds. S. J. Maddox & A. Aragón-Salamanca (Singapore: World Scientific), p. 15

Tegmark, M. 1995, ApJ, **455**, 429

Tegmark, M. 1997a, *Phys. Rev. D.*, **55**, 5895

Tegmark, M. 1997b, in *Ringberg Workshop on Large-Scale Structure*, ed. D. Hamilton (in press)

Tegmark, M. & Hamilton, A. J. S. 1997, preprint (astro-ph/9702019)

Tegmark, M., et al. 1998, ApJ, **499**, 555

Tegmark, M., Taylor, A. & Heavens, A. 1997, ApJ, **480**, 22

Turok, N., editor, 1997, *Critical Dialogues in Cosmology* (Singapore: World Scientific)

Vettolani, G. et al. 1997, *A&A*, **325**, 954

Vogeley, M. S. 1995, in *Clustering in the Universe*, Proc. 30th Rencontres de Moriond, eds. S. Maurogordato, C. Balkowski, C. Tao & J. Trân Thanh Vân (Gif-sur-Yvette: Editions Frontières), p. 13

Vogeley, M. S., Park, C., Geller, M. J. & Huchra, J. P. 1992, *ApJ*, **391**, L5

Vogeley, M. S. & Szalay, A. S. 1996, *ApJ*, **465**, 34

Weinberg, D. H. 1995, in *Wide-Field Spectroscopy and the Distant Universe*, eds. S. J. Maddox & A. Aragón-Salamanca (Singapore: World Scientific), p. 129

Weinberg, D. H. & Cole, S. 1992, *MNRAS*, **259**, 652

Whitmore, B. C., Gilmore, D. M. & Jones, C. 1993, *ApJ*, **407**, 489

Wright, E. L., et al. 1992, *ApJ*, **396**, L13

Yahil, A., Sandage, A. & Tammann, G. A. 1980, *ApJ*, **242**, 448

Yahil, A., Strauss, M. A., Davis, M. & Huchra, J. P. 1991, *ApJ*, **372**, 380

Yahil, A., Tammann, G. & Sandage, A. 1977, *ApJ*, **217**, 903

Zaroubi, S. & Hoffman, Y. 1996, *ApJ*, **462**, 25

Zurek, W. H., Quinn, P. J., Salmon, J. K. & Warren, M. S. 1994, *ApJ*, **431**, 559

Zwicky, F., Herzog, E., Wild, P., Karpowicz, M. & Kowal, C. 1961–68, *Catalog of Galaxies and of Clusters of Galaxies* (Pasadena: California Institute of Technology)

6 Measurement of galaxy distances

Jeffrey A. Willick

Abstract

Six of the principal galaxy distance indicators are discussed: Cepheid variables, the Tully–Fisher relation, the D_n–σ relation, surface-brightness fluctuations, brightest-cluster galaxies, and Type Ia supernovae. The role they play in peculiar-velocity surveys and Hubble constant determination is emphasized. Past, present, and future efforts at constructing catalogs of redshift-independent distances are described. The chapter concludes with a qualitative overview of Malmquist and related biases.

6.1 Introduction

The measurement of galaxy distances is one of the most fundamental problems in astronomy. To begin with, we would simply like to know the scale of the cosmos; we do so by determining the distances to galaxies. Beyond this, galaxy distances are the key to measuring the Hubble constant H_0, perhaps the most important piece of information for testing the validity of the Big Bang model. Finally, galaxy distances are necessary if we are to study the large-scale peculiar-velocity field. Peculiar-velocity analysis is among the most promising techniques for confirming the gravitational instability paradigm for the origin of large-scale structure, deducing the relative distributions of luminous and dark matter, and constraining the value of the cosmological density parameter Ω_0 (see Ch. 7). This chapter describes a number of the methods used for measuring galaxy distances, and discusses their application to the H_0 and peculiar-velocity problems. Their relevance to the determination of other cosmological parameters is discussed when appropriate. The goal of this chapter is not to present an exhaustive review of galaxy distance measurements, but rather to provide a summary of where matters stand, and an indication of what the next few years may bring.

6.1.1 Peculiar velocities versus H_0

What it means to "measure a galaxy's distance" depends on whether one is interested in studying peculiar velocities or determining the value of the Hubble constant. A galaxy's peculiar velocity may be estimated given its "distance" in $\mathrm{km\,s^{-1}}$ – the part of its radial velocity due solely to the Hubble expansion. The same object provides an estimate of H_0 only if one can measure its distance in metric units such as megaparsecs. What this means in practice is that accurate peculiar-velocity studies may be carried out *today*, despite the fact that H_0 remains undetermined at the $\sim 20\%$ level.

Another basic distinction between velocity analysis and the search for H_0 concerns the distance regimes in which they are optimally conducted. Peculiar-velocity surveys are best carried out in the "nearby" universe, where peculiar-velocity errors are comparable to, or less than, the peculiar velocities themselves. The characteristic amplitude of the radial peculiar velocity, v_p, is a few hundred $\mathrm{km\,s^{-1}}$ at all distances, whereas the errors we make in estimating v_p grow linearly with distance (§6.3). It turns out that the "break-even" point occurs at distances of ~ 5000 $\mathrm{km\,s^{-1}}$. Although we may hope to glean some important information (such as bulk flow amplitudes) on larger scales, our ability to construct an accurate picture of the velocity field is restricted to the region within about $50\,h^{-1}$ Mpc. In the Hubble constant problem, by contrast, peculiar velocities are basically a nuisance. We would like them to be a small fraction of the expansion velocity, so that we incur as small as possible an error by neglecting them. This is best achieved by using comparatively *distant* objects, $d \gtrsim 7000$ $\mathrm{km\,s^{-1}}$, as tracers of the expansion.

On the other hand, to obtain the absolute distances needed to measure H_0, we must first calibrate our distance indicators *locally* ($\lesssim 2000$ $\mathrm{km\,s^{-1}}$). This is because the distance indicators capable of reaching the "far field" ($\gtrsim 7000$ $\mathrm{km\,s^{-1}}$) of the Hubble flow generally have no a priori absolute calibration (cf. §6.1.2). The only reliable distance indicator that can bridge the gap between the Milky Way and the handful of Local Group galaxies whose absolute distances are well known, and galaxies beyond a few Mpc, is the Cepheid variable method (§6.2), which is limited to distances $\lesssim 2000$ $\mathrm{km\,s^{-1}}$. As a result, *Hubble constant measurement is inherently a two-step process:* local calibration in galaxies with Cepheid distances, followed by distance measurements in the far field where the effect of peculiar velocities is small. The local calibration step is unnecessary in peculiar-velocity studies.

Although peculiar-velocity surveys and H_0 measurement differ in the ways just discussed, the two problems are, ultimately, closely related. Many distance indicator methods have been and are being used for both purposes. Indeed, a distance indicator calibrated in $\mathrm{km\,s^{-1}}$ may be turned into a tool for measuring H_0 simply by knowing the distances in Mpc to a few well-studied objects to which it has been applied. This chapter will thus be organized not around the peculiar velocity–H_0 distinction, but rather around methods of distance estimation.

6.1.2 Distance indicators

Measuring the distance to a galaxy almost always involves one of the following properties of the propagation of light: (1) The apparent brightness of a source falls off inversely with the square of its distance; (2) The angular size of a source falls off inversely with its distance. As a result, we can determine the distance to an object by knowing its intrinsic luminosity or linear size, and then comparing with its apparent brightness or angular size, respectively. If all objects of a given class had approximately the same absolute magnitude, we could immediately determine their distances simply by comparing with their apparent magnitudes. Such objects are called *standard candles*. Similarly, classes of objects whose intrinsic linear sizes are all about the same are known as "standard rulers." True standard candles or rulers are, however, extremely rare in astronomy. It is much more often the case that the objects in question possess another, *distance-independent* property from which we infer their absolute magnitudes or diameters. For example, the rotation velocities of spiral galaxies are good predictors of their luminosities (§6.3), while the central velocity dispersions and surface brightnesses of ellipticals together are good predictors of their diameters (§6.4). Whether standard candles or rulers, or members of the more common second category, objects whose absolute magnitudes or diameters we can somehow ascertain are known as *Distance Indicators*, or DIs.

Absolute calibration of most DIs is not straightforward. One discovers that a particular distance-independent property is a good predictor of absolute magnitude because it is well correlated with the *apparent* magnitudes of objects lying at a common distance – in a rich cluster of galaxies, for example. Such data may be used to determine the mathematical form of the correlation (e.g., linear with a given slope). However, the cluster distance in most cases is not accurately known. Thus, the predicted absolute magnitude corresponding to a given value of the distance-independent property – the "zero point" of the DI – remains undetermined up to a constant, assuming one has no rigorous, a priori physical theory of the correlation, as is usually the case (but see below). Any distances obtained from the DI at this point will be in error by a fixed scale factor. This situation is obviously unacceptable for the Hubble constant problem, in which absolute distances are required. The remedy is to determine the zero point of the DI by applying it to galaxies whose true distances have been determined by an independent technique (e.g., Cepheid variables), as discussed above. Such a DI is said to be "empirically" calibrated.

For peculiar-velocity surveys, the situation is simpler because absolute calibration is not required. However, the DI must still be calibrated such that it yields distances in $km\,s^{-1}$, the radial velocity due to Hubble flow. For this, one must apply the DI to many galaxies, widely enough distributed around the sky and at large enough distances that peculiar velocities tend to cancel out. Only then can redshift be taken as a good indicator on average of distance in $km\,s^{-1}$, and a calibration in velocity units thereby obtained (Willick *et al.* 1995, 1996). Empirical DI calibration, in this sense, is needed even for peculiar-velocity work.

DIs of this sort tend to make some people nervous. They argue that a good distance estimation method should be based on solid, calculable physics. There are, in fact, a few such techniques. One involves exploitation of the Sunyaev–Zeldovich effect in clusters, in which comparison of cosmic microwave background distortions and the X-ray emission produced by hot, intracluster gas yields the physical size of the cluster (cf. Rephaeli 1995 for a comprehensive review). Another method involves modelling time delays between multiple images of gravitationally lensed background objects (see Ch. 10). Other DIs for which theoretical absolute calibration may be possible are Type II supernovae, whose expansion velocities may be related to luminosities (Montes & Wagoner 1995; Eastman, Schmidt & Kirshner 1996), and Type Ia supernovae, whose luminosities may be calculated from theoretical modelling of the explosion mechanism (Fisher, Branch, & Nugent 1993). Such approaches are indeed promising, and will undoubtedly contribute to the measurement of H_0 over the next decade. However, at present these methods should be considered preliminary. Some of the underlying physics remains to be worked out, and many of the underlying assumptions will need to be tested. Furthermore, the data needed to implement such techniques are currently rather scarce. With the exception of Type Ia supernovae (discussed in §6.6 in their traditional, empirical context), these methods will not be discussed further in this chapter.

The focus, instead, is on methods that require empirical calibration. These DIs arise from astrophysical correlations Nature was kind enough to provide us with, but mischievous enough to deny us a full understanding of. The canonical wisdom, which states that we need hard physical theory that explains a DI in order to trust it, is a bit too exacting given our present theoretical and observational capabilities. We should conditionally trust our empirical DIs while recognizing the uncertainties involved. In particular, we must remember that since they possess no a priori absolute calibration, they must (for measuring H_0) be carefully calibrated locally. We must also remain open to the possibility that they may not behave identically in different environments and at different redshifts. Our belief in their utility should be tempered by a healthy skepticism about their universality, and the distance estimates we make with them subjected to continuing consistency checks.

6.2   Cepheid variables

Cepheid variable stars have been fundamental to unlocking the cosmological distance scale since Henrietta Leavitt used them in 1912 to estimate the distances to the Magellanic Clouds. Of the various DIs discussed in this chapter, the Cepheid method is the only one involved in the Hubble constant but not the peculiar-velocity problem. Indeed, it is probably safe to say that the *raison d'être* for Cepheid observations is the ultimate determination of H_0. They will do so, however, in conjunction with, not independently of, the secondary distance indicators discussed in later sections.

Cepheids are post-main sequence stars that occupy the instability strip in the H–R diagram. They pulsate according to a characteristic "sawtooth" pattern, with periods that can range from a few days to a good fraction of a year. Cepheids exhibit an excellent correlation between mean luminosity (averaged over a pulsation cycle) and pulsation period. This correlation is shown in Fig. 6.1 for Cepheids recently measured by the Hubble Space Telescope (HST) in the nearby galaxy M101 (solid points), and also for Cepheids in the Large Magellanic Cloud (LMC) as they would appear if the LMC lay at the distance of M101. It is apparent that the correlation is extremely similar for the two galaxies. Modern calibrations of the Cepheid *Period–Luminosity* (P–L) relation in the V and I bandpasses are

$$M_V = -2.76\,[\log(P) - 1.0] - 4.16 \tag{6.1}$$

and

$$M_I = -3.06\,[\log(P) - 1.0] - 4.87 \tag{6.2}$$

(Ferrarese *et al.* 1996). The absolute zero points of these P–L relations have been obtained by observing Cepheids in the the Large and Small Magellanic Clouds, whose distances are known from main sequence fitting (Kennicut, Freedman & Mould 1995). Equations (6.1) and (6.2) show that Cepheid variables *are intrinsically bright stars*. Even short-period ($P \simeq 10$ days) Cepheids have absolute magnitudes $M_V < -4$, and long-period ($P \simeq (50-100)$ days) Cepheids are 2–3 magnitudes brighter still. It follows that individual Cepheid stars can be observed at relatively large distances. Indeed, with the HST Cepheids can be observed out to the distance of the Virgo cluster and possibly beyond. To be useful as distance indicators, however, Cepheids cannot be merely *detected*. Because they are found in crowded fields, they must be well above the limit of detectability at all phases in order to be accurately photometered. These stringent requirements place a limit of $m_V \simeq 26$ mag, much brighter than the HST detection limit of ~ 30 mag, for distance scale work using Cepheids.

Cepheids yield distances to their host galaxies by comparison of their absolute magnitudes, inferred from the P–L relation, with their observed apparent magnitudes. Specifically, the distance to the host galaxy is obtained by fitting equations (6.1) and (6.2), plus a distance modulus offset $\mu = 5\log(d/10)$ (where d is distance in parsecs), to the observed m_V and m_I versus $\log(P)$ diagram. (The same exercise may of course be carried out in other bandpasses as well.) An important advance has been made in recent years by Freedman, Madore, and coworkers, who have developed a method for correcting for extinction in the host galaxies (Freedman & Madore 1990; Freedman, Wilson & Madore 1991). In brief, the photometry is done in several bandpasses, and the magnitudes corrected for an assumed value of the extinction within the host galaxy. The distance modulus is determined for each bandpass, as described above. The value of extinction which brings the distance moduli in the various bands into agreement is assumed to be the correct one. This technique works best when data for a wide range of wavelengths, including if possible the near infrared, are available.

The great utility of Cepheids has been recognized in the designation of an HST Key Project to measure Cepheid distances for 20 nearby galaxies. This program, led by Wendy Freedman, Robert Kennicut, and Jeremy Mould, produced its first results

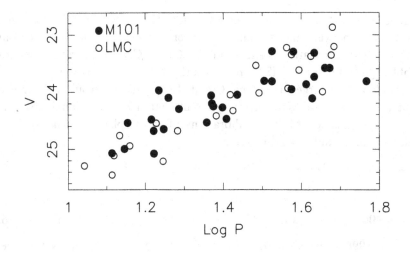

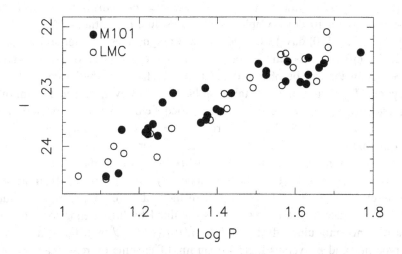

Figure 6.1. Cepheid variable Period–Luminosity (P–L) relations for the V (*upper*) and I (*lower*) bandpasses. Data for M101 and the Large Magellanic Cloud are shown. Adapted from Ferrarese *et al.* (1996).

in late 1994. As of this writing (July 1996), Cepheid distances from the Key Project are available for only a handful of galaxies. Distances for the remaining galaxies are expected to become available over the next few years. The results that have received the greatest attention to date involve the Virgo cluster galaxy M100, in which over 50 Cepheid variables have now been accurately measured (Freedman *et al.* 1994; Mould *et al.* 1995; Ferrarese *et al.* 1996). Fitting the universal P–L relations above to the M100 data yields a distance of 16.1 ± 1.3 Mpc. When combined with a suite of assumptions concerning the morphology and peculiar velocity of the Virgo cluster, this distance suggests a Hubble constant of about 85 km s^{-1} Mpc^{-1} (Freedman et al. 1994).

Unfortunately, the Hubble constant estimate obtained from M100 has received undue attention. This is understandable, given that determination of H_0 is the long-term aim of the Key Project. And, of course, values of H_0 in excess of ~ 75 km s^{-1} Mpc^{-1} are difficult to square with most estimates of the age of the universe based on its oldest constituents. But as the Key Project group has emphasized (Kennicutt, Freedman & Mould 1995), a single galaxy in the Virgo cluster with a good Cepheid distance does not allow one to estimate the Hubble constant with any accuracy. In fact, the Virgo cluster is a poor laboratory in which to estimate H_0 no matter how many galaxies one has Cepheid distances for. The reasons are simple: Virgo's depth is a good fraction ($\sim 30\%$) of its distance, and its peculiar velocity is likely to be a good fraction (~ 20–30%) of its Hubble velocity. The velocity/distance ratio of any single Virgo object, or even group of objects, may therefore be a poor approximation of H_0, and it is difficult to gauge the systematic errors that affect it.

Thus, Cepheid variables will not themselves be used to measure H_0. Instead, they will be used to obtain accurate distances for several tens of galaxies within about $20\,h^{-1}$ Mpc. These galaxies will in turn serve as calibrators for the *secondary distance indicators,* such as Type Ia supernovae and the Tully–Fisher relation, that are applicable in the far field of the Hubble flow (and occupy the remainder of this chapter). Initial steps in this direction have already been taken by Sandage, Tammann, and coworkers (Sandage et al. 1996), who used HST Cepheid distances (their own, not those of the Key Project) to calibrate historical and contemporary Type Ia supernovae. When they apply this calibration to distant Type Ia SNe (Tammann & Sandage 1995), they derive $H_0 = 56$–58 km s^{-1} Mpc^{-1} (the lower value applies to B-band, and the higher value to V-band, measurements; Sandage et al. 1996). There is considerable controversy, however, surrounding the calibration of the historical photometry used in the SNe Ia calibration. Furthermore, the Sandage group has neglected the correlation between the peak luminosity of SNe Ia and the width of their light curves, an effect which now appears important (§6.6). Until these issues are resolved, and agreement between the Sandage and HST Key Project groups on local Cepheid distances achieved, estimates of H_0 based on this approach should be considered preliminary.

6.3 The Tully–Fisher relation for spiral galaxies

It has been said that the Tully–Fisher (TF) relation is the "workhorse" of peculiar-velocity surveys. One can anticipate a time in the not-so-distant future when more accurate techniques may supplant it, but for the next few years at least, the TF relation is likely to remain the most widely used distance indicator in cosmic velocity studies. Its role in such studies to date has been, in fact, too large to be reviewed here, and interested readers are referred to Ch. 7 of Strauss & Willick (1995). Several recent developments are discussed in §6.8 below.

The TF relation is one of the most fundamental properties of spiral galaxies. It is the empirical statement of an approximately power-law relation between luminosity and rotation velocity,

$$L \propto v_{\rm rot}^{\alpha}, \tag{6.3}$$

or, using the logarithmic formulation preferred by working astronomers,

$$M = A - b\eta. \tag{6.4}$$

In equation (6.4), $M = -2.5\log(L) + {\rm const.}$ is the absolute magnitude, and the *velocity width parameter* $\eta \equiv \log(2v_{\rm rot}) - 2.5$, where $v_{\rm rot}$ is expressed in $\rm km\,s^{-1}$, is a useful dimensionless measure of rotation velocity.

An important fact, not always sufficiently appreciated, is that the power-law exponent α does not have a unique value. The details of both the photometric and spectroscopic measurements affect it. A typical result found in contemporary studies is $\alpha \simeq 3$. The corresponding value of the "TF slope," b, is ~ 7.5. However, slight changes in the details of measurement can result in significant changes in b. This is illustrated in Fig. 6.2, in which TF relations in four bandpasses are plotted. The optical bandpasses (B, R, and I) all represent data from the sample of Mathewson, Ford & Buchhorn (1992). In each case the slope is < 7. This is because they defined their velocity widths rather differently than most observers, with the result that small velocity widths are made smaller still relative to other width measurement systems, while large velocity widths are unchanged in comparison with other systems. If one were to transform the Mathewson et al. widths to those of standard systems, one would obtain an I-band TF slope of ~ 7.7, comparable to other I-band samples (Willick et al. 1997). Note, however, that the slope increases steadily from B to I. This reflects a general trend of increasing TF slope toward larger wavelengths, which was noted over a decade ago (Bottinelli et al. 1983). The H-band data come from the compilation of Aaronson et al. (1982), as reanalyzed by Tormen & Burstein (1995) and Willick et al. (1996). The H-band TF slope is considerably greater than its optical counterparts. In part this reflects the trend just noted. However, to a greater extent it is due to the relatively small apertures within which the H-band photometry is done: the TF slope increases as the photometric aperture size is *decreased* (Willick 1991). Infrared TF samples in which *total* magnitudes were measured exhibit TF slopes not much in excess of the optical values (Bernstein et al. 1994).

The numerical value of the TF zero point A has no absolute significance in itself, reflecting mainly the photometric system in which the TF measurements are done. Absolute magnitudes measured in different bandpasses can differ numerically by a few magnitudes for a given object. Clearly, this must have no meaning for distance measurements, and it is the zero point that absorbs such differences. For any given measurement system, however, the value of A is highly significant. To see this, consider how the TF relation is used to infer distances and peculiar velocities. Given a measured apparent magnitude m and width parameter η, one infers the distance modulus to the galaxy as $\mu = m - (A - b\eta)$. The corresponding distance $d \propto 10^{0.2\mu}$. It follows that an

Measurement of galaxy distances

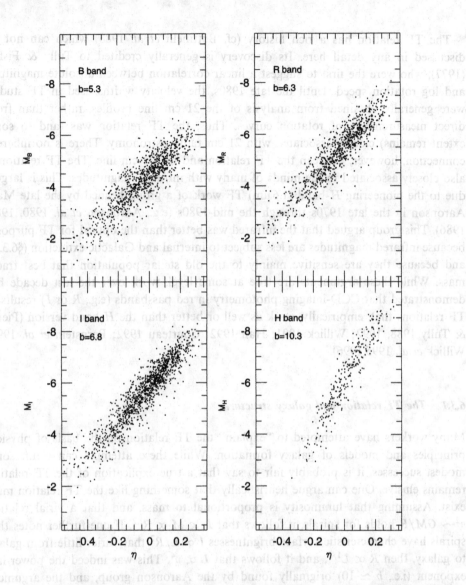

Figure 6.2. Tully–Fisher relations in four bandpasses. The absolute magnitudes are given in units such that $d = 10^{0.2(m-M)}$ is the galaxy distance in km s^{-1}. Adapted from Strauss & Willick (1995).

error δA in the TF zero point corresponds to a fractional distance error $f = 10^{-0.2\delta A}$. A Hubble constant inferred from such distances will then be off by a factor f^{-1}. For peculiar velocities, calibration of the TF relation consists in choosing A such that $d \equiv 10^{0.2[m-(A-b\eta)]}$ gives a galaxy's distance in km s^{-1}. There is no requirement that it yield the distance in Mpc. Nonetheless, as mentioned above, zero-point calibration errors are still possible. Errors in A produce distances in km s^{-1} that differ by a fraction f from the true Hubble velocity, with resultant peculiar-velocity errors $\delta v_p = -fd$.

The TF relation has a rich history (cf. Bottinelli et al. 1983) which can not be discussed in any detail here. Its discovery is generally credited to Tully & Fisher (1977), who were the first to suggest a linear correlation between absolute magnitude and log rotation speed. Until the late 1980s, the velocity widths used in TF studies were generally obtained from analysis of the 21 cm line profiles, rather than from direct measurements of rotation curves. Thus, the TF relation was (and to some extent remains) closely associated with 21 cm radio astronomy. There is no inherent connection, however, between the TF relation and the 21 cm line. The TF relation is also closely associated in the minds of many with infrared magnitudes. This is largely due to the pioneering H-band (1.6 μm) TF work of a group headed by the late Marc Aaronson in the late 1970s through the mid-1980s (e.g., Aaronson et al. 1980, 1982, 1986). This group argued that the infrared was better than the optical for TF purposes because infrared magnitudes are less subject to internal and Galactic extinction (§6.3.2), and because they are sensitive mainly to the old stellar population that best traces mass. While these arguments are true at some level, work over the last decade has demonstrated that CCD-imaging photometry in red passbands (e.g., R or I) results in TF relations that, empirically, work as well or better than the H-band version (Pierce & Tully 1988, 1992; Willick 1991; Han 1992; Courteau 1992; Bernstein et al. 1994; Willick et al. 1995, 1996).

6.3.1 The TF relation and galaxy structure

Many workers have attempted to "explain" the TF relation on the basis of physical principles and models of galaxy formation. While these attempts can claim some modest successes, it is probably fair to say that a true explication of the TF relation remains elusive. One can argue heuristically that something like the TF relation must exist. Assuming that luminosity is proportional to mass, and that a virial relation $v^2 \sim GM/R$ holds for spirals, it follows that $L \propto M \propto Rv^2$. If one further notes that spirals have characteristic surface brightnesses $I \propto L/R^2$ that varies little from galaxy to galaxy, then $R \propto L^{1/2}$, and it follows that $L \propto v^4$. This was indeed the power-law exponent (i.e., $b \simeq 10$) originally found by the Aaronson group, and the argument seemed reasonable to them (Aaronson, Huchra & Mould 1979).

However, quite a few loose ends remain. First, as noted above, contemporary measures of the TF slope suggest that the exponent is closer to 3 than to 4. The aperture- and wavelength-dependences noted above tell us that the TF slope is not determined strictly by idealized dynamics, but depends also on the details of the distribution – in both space and wavelength – of the starlight emitted by the galaxy. Furthermore, while a number of theoretical approaches can approximately predict the TF slope, no realistic model has successfully accounted for its rather small (~ 0.3 mag; see below) intrinsic scatter (Eisenstein & Loeb 1996).

Another, more fundamental, problem is that the TF relation is evidently connected with the phenomenon of flat rotation curves (RCs) exhibited by most spiral galaxies.

Were the RCs not flat, there would be no well-defined rotation velocity, and one would expect the TF relation to require a very specific type of velocity width measurement. In fact, a well-defined TF relation is found regardless of the specific algorithm for measuring rotation velocity (although slight variations of slope and zero point arise as a result of algorithmic differences). Whether one measures HI profile widths, asymptotic rotation velocities, "isophotal" rotation velocities (Schlegel 1996), or maximum rotation velocities, basically similar TF relations result. Because the origin of flat rotation curves is connected with the nature of dark matter, it follows that we cannot fully understand the TF relation until we understand how galaxies form in their dark matter halos.

6.3.2 Applying the TF relation: a few details

Widely appreciated by practitioners of the TF relation, but often hidden to the wider astronomical public, are the careful correction procedures applied to the magnitudes and velocity widths that go into the TF relation. Probably the most important step is correction for projection of the disk on the plane of the sky. The observed velocity width is smaller by a factor $\sin(i)$, where i is the galaxy inclination, than the intrinsic value. Observers correct for this by estimating i from the apparent ellipticity of the galaxy disk. Modern CCD observations allow one to fit elliptical isophotes to the galaxy image; these isophotes typically converge to a constant ellipticity ε in the outer regions. When CCD surface photometry is not available (as is the case for many of the older infrared data), one simply takes $\varepsilon = 1 - b/a$, where a and b are the major and minor axis diameters of the galaxy obtained from photographic data. Whichever method is used, the inclination i is taken to be a function of ε. A typical formula employed is

$$\cos^2 i = \begin{cases} \frac{(1-\varepsilon)^2 - (1-\varepsilon_{max})^2}{1-(1-\varepsilon_{max})^2}, & \varepsilon < \varepsilon_{max}; \\ 0, & \varepsilon \geq \varepsilon_{max}, \end{cases} \quad (6.5)$$

where $\varepsilon_{max} \simeq 0.8$ is the ellipticity exhibited by an edge-on spiral. It is apparent that formulae such as equation (6.5) are at best approximations, hopefully valid in a statistical sense. However, they are usually the best we can do, and are certainly far better than doing nothing. Still, the inclination correction to the widths can go seriously awry at small inclinations, and most TF samples exclude galaxies with $i \lesssim 40°$.

Another tricky detail of the TF relation is correcting for *internal extinction*. As a spiral galaxy tilts toward edge-on orientation, it becomes fainter. Since spirals are viewed at a range of orientations, it is important to correct for this effect. The most widely used correction is to brighten the raw magnitudes by an amount $C_{int} \times \log(a/b)$, where C_{int} is the *internal extinction coefficient*. Studies have shown that C_{int} is bandpass-dependent, as one might expect. However, in the optical red (R and I bandpasses), the wavelength-dependence is very weak, and $C_{int} \simeq 1$ is a good approximation (Burstein, Willick & Courteau 1995; Willick et al. 1996, 1997). A controversial question is whether internal extinction depends on any galaxian property other than axial ratio. Giovanelli et al. (1995) argued that it is luminosity-dependent, but Willick et al. (1996) reached the opposite conclusion through a TF-residual analysis. This issue merits further consideration in the future.

6.3.3 The TF scatter

Of great importance to applications of the TF relation is its scatter σ_{TF}, the rms magnitude dispersion about the mean relation $M(\eta)$. This scatter is composed of three basic contributions: magnitude and velocity width measurement errors, and intrinsic or "cosmic" scatter. Of the three, recent analyses have suggested that the second and third are about equally important, contributing ~ 0.25–0.30 mag each (Willick et al. 1996). Photometric measurement errors are quite small in comparison. Thus, the overall TF scatter is about 0.4 mag. It is significant that σ_{TF} determines not only random distance errors ($\delta d/d \simeq 0.46\,\sigma_{TF}$), but also systematic errors associated with statistical bias effects (§6.9). Knowing σ_{TF} is therefore crucial for assessing the reliability of TF studies. (An analogous statement applies to the scatter of the other DIs discussed in this chapter as well.)

Estimates of σ_{TF} have varied widely in the last decade. Bothun & Mould (1987) suggested that σ_{TF} could be made as small as $\lesssim 0.25$ mag with a velocity width-dependent choice of photometric aperture. Pierce & Tully (1988) also found $\sigma_{TF} \simeq 0.25$ using CCD data in the Virgo and Ursa Major clusters. Willick (1991) and Courteau (1992) found somewhat higher but still small values of the TF scatter ($\sigma_{TF} = 0.30$–0.35 mag). Bernstein et al. (1994) found the astonishingly small value of 0.1 mag for the Coma Cluster TF relation using I-band CCD magnitudes and carefully measured HI velocity widths.

Unfortunately, these relatively low values have not been borne out by later studies using more complete samples. Willick et al. (1995, 1996, 1997) calibrated TF relations for six separate samples comprising nearly 3000 spiral galaxies, and found typical values of $\sigma_{TF} \simeq 0.4$ mag for the CCD samples. Willick et al. (1996) argued that the large sample size and a relatively conservative approach to excluding outliers drove up earlier, optimistically low estimates of the TF scatter. Other workers, notably Sandage and collaborators (e.g., Sandage 1994; Federspiel, Sandage & Tammann 1994) have taken an even more pessimistic view of the accuracy of the TF relation, suggesting that typical spirals scatter about the TF expectation by 0.6–0.7 mag.

How can one reconcile this wide range of values? At least part of the answer lies in different workers' preconceptions and preferences. Those excited at the possiblity of finding a more accurate way of estimating distances tend to find low ($\sigma_{TF} \lesssim 0.3$ mag) values. Those who doubt the credibility of TF distances tend to find high ($\sigma_{TF} \gtrsim 0.5$ mag) ones. It is possible to arrive at such discrepant results in part because the samples differ so dramatically. Perhaps it is only justified to speak of a particular value of the TF scatter for a given set of sample selection criteria; hopefully, this issue will be clarified in the years to come.

There is one galaxian property with which the TF scatter demonstrably appears to vary, however, and that is luminosity (velocity width). Brighter galaxies exhibit a smaller TF scatter than fainter ones (Federspiel, Sandage & Tammann 1994; Freudling et al. 1995; Willick et al. 1997). Part of this effect is undoubtedly due to the fact that the errors in $\eta = \log\Delta v - 2.5$ go as $(\Delta v)^{-1}$, if errors in Δv itself are roughly constant as is most likely the case. Such velocity-width errors translate directly into a TF scatter that increases with decreasing luminosity. A careful study of whether the *intrinsic* TF scatter varies with luminosity has not yet been carried out.

6.3.4 Future directions

An intriguing recent development has been application of the TF relation to relatively high-redshift galaxies. This has been made possible by the advent of large-aperture telescopes capable of measuring rotation curves out to redshifts of $z \simeq 1$. Vogt et al. (1996) measured rotation curves and magnitudes for nine field galaxies in the redshift range $0.1 \lesssim z \lesssim 1$ using the Keck 10-m telescope. They found such objects obey a TF relation similar to that of local objects, with only a modest shift ($\Delta M_B \lesssim 0.6$ mag) toward brighter magnitudes. This is illustrated in Fig. 6.3, in which the Vogt et al. data are plotted along with the TF relation derived by Pierce & Tully (1992). However, a very different conclusion has been reached by Rix et al. (1996), who combined photometry with fiber-optic spectroscopy of spirals at moderate ($z \simeq 0.25$) redshift. Rix et al. conclude that even at such modest look-back times, spiral galaxies are significantly (~ 1.5 mag) brighter than their local counterparts. If the TF relation is to be applied to problems such as peculiar velocities at high redshift or estimation of the cosmological parameters (e.g., Ω), its evolution with redshift will have to be understood. This is an observational problem which deserves, and will undoubtedly receive, considerably more attention in the near future.

6.4 Fundamental Plane relations for elliptical galaxies

If the TF relation has been the workhorse of modern velocity field studies, the D_n–σ relation has been a short step behind. The closest analogue to the TF relation for elliptical galaxies is actually the predecessor of D_n–σ, the Faber–Jackson (FJ) relation. FJ expresses the power-law correlation between an elliptical galaxy's luminosity and its internal velocity dispersion,

$$L \propto \sigma_e^\alpha, \tag{6.6}$$

where the exponent α was found empirically to be $\sim 4 \pm 1$ (Faber & Jackson 1976; Schechter 1980; Tonry & Davis 1981). Although discovered around the same time, and viewed as closely related in physical origin, TF and FJ were not equally good distance indicators. It was clear from the outset that the scatter in the FJ relation

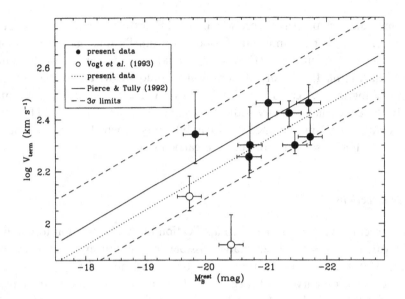

Figure 6.3. Rotation velocity versus absolute magnitude for spiral galaxies at a median redshift of ∼0.5. This figure has been adapted from Vogt et al. (1996).

was about twice that of the TF relation, on the order of 0.8 mag. Thus, while the TF relation flourished in the early 1980s as a tool of distance measurement (§6.3), elliptical galaxy surveys focused more on the stuctural and dynamical implications of the FJ relation.

These surveys bore unexpected fruit, however, in the latter part of the 1980s. Two groups conducting surveys of ellipticals arrived independently at a new result: the FJ correlation could be considerably tightened by the addition of a third parameter, namely, surface brightness (Djorgovski & Davis 1987; Dressler et al. 1987). The new correlation has become known as the D_n–σ relation: a power-law correlation between the *luminous diameter* D_n and the internal velocity dispersion,

$$D_n \propto \sigma_e^\gamma, \qquad (6.7)$$

where $\gamma = 1.20 \pm 0.10$ (Lynden-Bell et al. 1988). (D_n is defined as the diameter within which the galaxy has a given mean surface brightness. As such, it implicitly incorporates the third parameter into the correlation.)

More broadly, D_n–σ and its variants may be viewed as manifestations of the *Fundamental Plane (FP) of Elliptical Galaxies,* a planar region in the three-dimensional space of structural parameters in which normal ellipticals are found. One expression of the FP relates effective diameter to internal velocity dispersion and central surface brightness,

$$R_e \propto \sigma_e^\alpha I_e^{-\beta}. \qquad (6.8)$$

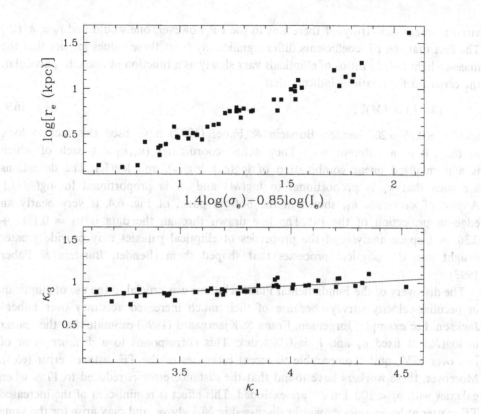

Figure 6.4. Two versions of the Fundamental Plane for the Virgo and Coma ellipticals studied by Bender, Burstein & Faber (1992). Further details are given in the main text. (The data used for these figures were kindly provided by D. Burstein.)

An early determination of the parameters α and β using B-band photometry gave $\alpha \simeq 1.4$, $\beta \simeq 0.9$ (Faber et al. 1987). More recently, Bender, Burstein & Faber (1992) found $\alpha = 1.4$, $\beta = 0.85$ using B-band data for a sample of Virgo and Coma cluster ellipticals; the upper panel of Fig. 6.4 shows the FP for this sample. A recent R-band FP analysis by the EFAR group (Wegner et al. 1996) is $\alpha = 1.23$, $\beta = 0.72$. A measurement based on Gunn r-band photometry (Jorgensen, Franx & Kjaergaard 1996) yields a similar value of α (1.24 ± 0.07) but a somewhat different value of β (0.82 ± 0.02), perhaps due to the slightly different bandpass used. Pahre, Djorgovski & de Carvalho (1995) have recently carried out the first analysis of the FP using K-band photometry, finding $\alpha = 1.44 \pm 0.04$, $\beta = 0.79 \pm 0.04$.

The two-dimensionality of the loci in parameter space occupied by ellipticals actually makes the FP relations, including D_n–σ, somewhat less mysterious than the one-dimensional TF sequence. As noted by Faber et al. (1987), such two-dimensionality is expected on virial equlibrium grounds alone. Unlike the TF relation, therefore, the FP is not obviously related to the relative distribution of luminous and dark matter. If the virial theorem were truly *all* there was to the FP, however, one would find $r_e \propto \sigma_e^2 I_e^{-1}$. The fact that the FP coefficients differ significantly from these values implies that the mass-to-light (M/L) ratios of ellipticals vary slowly as a function of mass. In particular, the observed FP relations indicate that

$$(M/L) \propto (M)^\epsilon, \tag{6.9}$$

with $\epsilon \simeq 0.15$–0.20. Bender, Burstein & Faber (1992) have used this fact to look at the FP in a different way. They define coordinates $(\kappa_1, \kappa_2, \kappa_3)$, each of which is a normalized linear combination of $\log(\sigma_e)$, $\log(r_e)$, and $\log(I_e)$. The definitions are such that κ_1 is proportional to $\log(M)$ and κ_3 is proportional to $\log(M/L)$. A plot of κ_3 versus κ_1, shown in the bottom panel of Fig. 6.4, is very nearly an edge-on projection of the FP. The line drawn through the data is $\kappa_3 = 0.15\kappa_1 + 0.36$. A κ-space analysis of the properties of elliptical galaxies may provide greater insight into the physical processes that shaped them (Bender, Burstein & Faber 1993).

The discovery of the Fundamental Plane relations was crucial to the use of ellipticals in peculiar-velocity surveys because of their much increased accuracy over Faber–Jackson. For example, Jorgensen, Franx & Kjaergaard (1996) estimate that the scatter in $\log(R_e)$ at fixed σ_e and I_e is 0.084 dex. This corresponds to a distance error of just over 19%, quite comparable to recent estimates of the TF distance error (§6.3). Moreover, these workers have found that the distance error is reduced to 17% when galaxies with $\sigma_e < 100$ km s^{-1} are excluded. This effect is reminiscent of the increased TF scatter at lower velocity widths discussed in §6.3 above, and may arise for the same reason. Pahre, Djorgovski & de Carvalho (1995) find a distance error of 16.5% from the K-band FP.

The D_n–σ relation occupies a special place in the history of peculiar-velocity surveys because it was used in the first detection of very large-scale streaming by the "7-Samurai" group (Dressler *et al.* 1987; Lynden-Bell *et al.* 1988). In the 7-Samurai survey, a full-sky sample of elliptical galaxies revealed a streaming motion of amplitude ~ 500 km s^{-1} that was coherent across the entire sky to a depth of ~ 40 Mpc. Subsequent studies of spiral galaxies (Willick 1990; Han & Mould 1992; Mathewson, Ford & Buchhorn 1992; Courteau *et al.* 1993) have lent confirmation to this result, although the coherence length of the flow remains controversial. Since the late 1980s no new results concerning the peculiar-velocity field have been obtained using elliptical galaxy data. However, this situation will change in the coming years as several large surveys of elliptical galaxies (e.g., Wegner *et al.* 1996) come to fruition.

Like the TF relation, the FP relations are now being studied at appreciable redshift. Bender, Ziegler & Bruzual (1996) studied a sample of cluster ellipticals at $z = 0.37$. They have found evidence for mild (~ 0.5 mag) evolution toward brighter magnitudes at such redshifts, comparable to the result found by Vogt *et al.* (1996) for the TF relation.

Because it is difficult to find Cepheids in nearby elliptical galaxies, there has been little attempt to provide absolute calibrations of the D_n–σ and FP relations. As a result, elliptical galaxy distances have not figured prominently in the Hubble constant problem. However, this situation may change in the near future, if surface brightness fluctuation distances (discussed in the next section) can provide an absolute calibration for the D_n–σ and FP relations.

6.5 Surface brightness fluctuations

The early 1990s saw the revival of an old idea with modern technology: determining distances from the "graininess" of a galaxy image. The basic idea is simple. Galaxies are made up of stars. The discrete origin of galaxian luminosity is detectable in the pixel-to-pixel intensity fluctuations of the galaxy image. Such fluctuations derive from Poisson statistics of two sorts: (1) photon number fluctuations $(\delta N/N)_\gamma$, and (2) star number fluctuations $(\delta N/N)_s$. The first is distance-independent, but $(\delta N/N)_s$ decreases with distance, as the solid angle subtended by a pixel encompasses more and more individual stars. Consequently, the pixel-to-pixel intensity fluctuations in a nearby galaxy are greater than in a more distant galaxy. If this effect can be calibrated, it can be used as a distance indicator.

Though originally proposed by Baum (1955), it was not until the late 1980s that this idea was put into practice, made possible by the advent of CCD detectors and telescopes with improved seeing. Tonry and coworkers (Tonry & Schneider 1988; Tonry, Ahjar & Luppino 1989, 1990; Tonry & Schecter 1990; Tonry 1991; Tonry et al. 1997) pioneered this technique, which has come to be known as the Surface Brightness Fluctuation (SBF) method. The method can, in principle, be applied to any type of galaxy. In practice, late-type ($\gtrsim$ Sb) galaxies have too many sources of fluctuations over and above Poisson statistics, such as spiral structure and dust lanes, to apply the method to them. The method is thus preferentially applied to ellipticals and the bulges of early-type spirals.

The distance at which SBF may be applied goes inversely with the seeing. It is possible to measure distances out to ~ 4000 km s^{-1} with a 2.4-m telescope, ~ 2-hour exposures, and ~ 0.5 arcsec seeing. This is an effective limit for current ground-based observations. As half-arcsec seeing is infrequently achieved at even the best sites (such as Mauna Kea and Las Campanas), 3000 km s^{-1} is a practical limit for complete SBF surveys. In principle, the HST is capable of yielding SBF distances for objects as distant as 10,000 km s^{-1}. However, the required exposure times are such that few galaxies at such distances are likely to be observed for this purpose.

During the 1990s, Tonry, Dressler, and coworkers have been conducting an SBF survey of ~ 400 early-type galaxies within ~ 3000 km s^{-1}(Dressler 1994; Tonry et al. 1997). The data suggest that median SBF distance errors are $\sim 8\%$ within this distance range; the most well-observed objects have distance errors of $\sim 5\%$. Such accuracy is considerably better than most of the secondary distance indicators discussed here, with the possible exception of Type Ia supernovae (§6.6).

6.5.1 Calibration of SBF

It may appear from the brief description above that SBF is a purely "geometrical method," like parallax. If this were true, it would free the method from the nagging questions that plague other DIs: are they really universal, or do they depend on galaxy type, age, environment, and so forth? In reality, however, SBF is dependent on the stellar populations in the galaxies to which it is applied. Not only does this mean that we need to be cautious with regard to its universality, but, also, that it is difficult to derive an absolute calibration of SBF from first principles. Like the other DIs considered in this chapter, absolute distances obtained from the SBF technique are tied to the Cepheid distance scale. If the latter were to change, so would the SBF distances. In particular, estimates of H_0 derived from SBF studies (see below) may well require revision as the HST Key Project (§6.2) yields new results.

The stellar population dependence of SBF arises because the stars which contribute most strongly to the fluctuations are those that lie at the tip of the giant branch. Tonry and coworkers parameterize this effect in terms of "effective fluctuation magnitudes" $\overline{M_I}$ (absolute) and $\overline{m_I}$ (apparent). The quantity $\overline{M_I}$ may be thought of as the absolute magnitude of the giant branch stars which dominate the fluctuations; $\overline{m_I}$ is an apparent magnitude obtained from the observed fluctuations. If all galaxies had identical stellar populations, they would all have the same value of $\overline{M_I}$, and their distance moduli would be given simply by $\overline{m_I} - \overline{M_I}$.

Because galaxies do not have identical stellar populations, it is necessary to determine an empirical correction to $\overline{M_I}$ as a function of a distance-independent galaxian property. Tonry and coworkers use $(V - I)$ color for this purpose. Their most recent calibration is

$$\overline{M_I} = -1.74(\pm 0.05) + 4.5(\pm 0.25)\,[(V - I) - 1.15]\,, \tag{6.10}$$

which is valid for $1.0 \leq (V - I) \leq 1.3$ (Tonry et al. 1997). We discuss the zero point of this relation in §6.5.2. The color-dependence indicated by equation (6.10) is readily seen in the $\overline{m_I}$ versus $(V - I)$ diagrams of several tight groups and clusters, as shown in the upper panel of Fig. 6.5. A line of slope 4.5–the solid lines drawn throught the data points–fits the fluctuation magnitude–color data in each group well. (The solid points, as well as the small open squares, are thought to be non-members of the groups.) The different intercepts of the solid lines reflect the different distances to the groups.

The correlation of fluctuation magnitude with color is quite strong. Accurate colors are therefore required in order to minimize systematic effects. The possibility that the slope or zero point of this correlation may not be universal, but instead depend on some as yet undetermined galaxy properties as suggested by Tammann (1992), merits further attention. However, Tonry et al. (1997) show that the most likely manifestation of such a problem, a trend with metallicity of residuals from the $\overline{M_I}$–$(V - I)$ relation, does not exist. It is reassuring, moreover, that theoretical stellar population synthesis models predict a trend of fluctuation magnitude with color that is very similar to the empirical one. This is shown in the lower panel of Fig. 6.5, in which the population

Measurement of galaxy distances

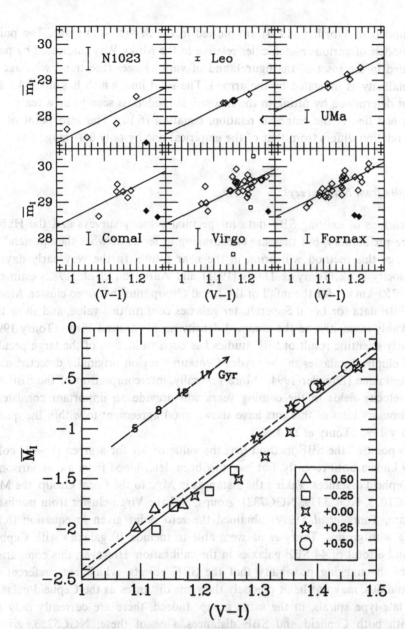

Figure 6.5. *Top panel:* fluctuation apparent magnitudes $\overline{m}_I$ versus $(V-I)$ color for several nearby groups and clusters. The solid lines drawn through the data points all have slope 4.5. *Bottom panel:* a plot of the theoretical $\overline{M}_I$ versus $(V-I)$ relation, from the stellar population synthesis models of Worthey (1994). The different point types indicate different metallicities relative to the Milky Way, as coded in the inset. For each point type, there are several distinct points, corresponding to different stellar population ages, as indicated by the arrow. The solid line is a fit to the theoretical models with slope fixed at 4.5. The dashed line is the empirical relation, equation (6.10). Adapted from Tonry et al. (1997).

synthesis models of Worthey (1994) are plotted in the $\overline{M_I}$–$(V-I)$ plane. The points represent models of various metallicities relative to the Milky Way (indicated by point type as coded in the inset of the figure), and of various ages (the trend with age, at a given metallicity, is indicated by the arrow). The solid line, which has slope 4.5 and an intercept determined by fitting to the theoretical models, is seen to be a reasonable fit. The dashed line is the emprical relation, equation (6.10). The zero point of the theoretical relation differs from that of the empirical one by only 0.07 mag.

6.5.2 Results from SBF surveys

The ramifications of existing SBF data for peculiar-velocity surveys and the Hubble constant are preliminary, but they are encouraging in terms of what they portend for the knowledge this method will bring in the near future. In the very early days of peculiar-velocity work, Tonry & Davis (1981) and Aaronson et al. (1982) estimated values of ~ 250 km s^{-1} for the infall of the Local Group into the Virgo cluster. Model-fits to the SBF data for Local Supercluster galaxies confirm this value, and show that it is remarkably insensitive to the assumed density profile around Virgo (Tonry 1995). Another early scientific result of SBF studies has been validation of the large peculiar motions of elliptical galaxies in the Hydra-Centaurus region originally detected using the D_n–σ technique (Dressler 1994). More generally, intercomparison of the SBF and TF/D_n–σ velocity fields in the coming years will provide an important consistency check. Preliminary tests of this sort have shown good agreement to within the quoted errors (Tonry 1995; Tonry et al. 1997).

The zero point of the SBF method [i.e., the value of $\overline{M_I}$ for a given $(V-I)$ color] was poorly known until recently, but has now been determined from a comparison of SBF and Cepheid distances. Taking the distances in Mpc to the Local Group, the M81, CenA, NGC1023, NGC3379, NGC7331 groups, and the Virgo cluster from published Cepheid data, Tonry et al. (1997) obtained the zero point given in equation (6.10). By working with groups, Tonry et al. were able to include 10 galaxies with Cepheid distances and a total of 44 SBF galaxies in the calibration. However, this comparison suffers from the nagging possibility that the SBF objects, which are preferentially ellipticals and S0s, may not lie at precisely the same distances as the Cepheid galaxies, which are late-type spirals, in the same group. Indeed, there are currently only five galaxies with both Cepheid and SBF distances. One of these, NGC5253, gives a discordant result. If the remaining four are used, Tonry et al. (1997) find an SBF zero point -1.82 ± 0.06, in reasonable agreement with the preferred value of -1.74 ± 0.05 found from the group comparison.

Thus calibrated, the SBF technique can be used as a temporary bridge between Cepheid distances, still too few in number to be reliable calibrators, and the secondary DIs that probe the far-field of the Hubble flow. Tonry et al. (1997) used SBF distances for groups and individual galaxies to provide absolute calibrations for TF, D_n–σ, and Type Ia supernovae (§6.6). In so doing, they obtained distances in Mpc for relatively distant galaxies, and thus estimates of the Hubble constant. The mean value was found to be $H_0 = 81 \pm 6$ km s^{-1} Mpc^{-1} from SBF-calibrated secondary DIs. Such a large value of H_0, if it holds up, may prove problematic for Big Bang cosmology, as discussed in §6.2. However, it should be kept in mind that the absolute calibration of SBF is tied to the Cepheid distance scale, and that the latter might change in coming years as the HST Key Project (§6.2) continues.

6.6 Supernovae

The use of supernovae as distance indicators has grown dramatically in the last few years. Supernovae have been applied to the Hubble Constant problem, to measurement of the cosmological parameters Ω_0 and Λ, and even, in a preliminary way, to constraining bulk peculiar motions. There is every reason to believe that in the next decade supernovae will become still more important as distance indicators. It is certain that many more will be discovered, especially at high redshift.

Supernovae come in two main varieties. Type Ia supernovae (SNe Ia) are thought to result from the nuclear detonation of a white dwarf star that has been overloaded by mass transferred from an evolved (Population II) companion. (Recall that a white dwarf cannot have a mass above the Chandrasekhar limit, 1.4 M$_\odot$. When mass transfer causes the white dwarf to surpass this limit, it explodes.) Type II supernovae result from the imploding cores of high-mass, young (Population I) stars that have exhausted their nuclear fuel.[1] Of the two, it is the SNe of Type Ia that have received the most attention lately. Type II SNe have shown somewhat less promise as distance indicators. They are considerably fainter (~ 2 mag), and thus are detected less often in magnitude-limited surveys (although their intrinsic frequency of occurrence is in fact greater than that of Type Ia). The discussion to follow will be restricted to SNe Ia.

Because SNe Ia result (in all likelihood) from detonating white dwarfs, and because the latter tend to have very similar masses, SNe Ia tend to have very similar luminosities. That is, they are very nearly standard candles, so comparison of their apparent and abolute magnitudes yields a distance. Recent work suggests that Type Ia SNe are not quite standard candles, in that their peak luminosities correlate with the shape of their light curves (Phillips 1994; Hamuy *et al.* 1995; Riess, Press & Kirshner 1995a,b; Perlmutter *et al.* 1997). Basically, broad light curves correspond to brighter, and narrow light curves to fainter, supernovae. When this effect is accounted for, the scatter in SNe Ia-predicted peak magnitudes might be as small as 0.1 mag, as found by Riess, Press & Kirshner (1995b). Hamuy *et al.* (1995) and Perlmutter *et al.* (1997) find that the scatter drops from $\lesssim 0.3$ mag when SNe Ia are treated as standard candles to 0.17 mag when the light-curve shape is taken into account. The precise scatter of SNe Ia remains a subject for further study.

---

[1] It is inconvenient that *Type I* supernovae occur in *Type II* stellar populations, while *Type II* supernovae occur in *Type I* populations. Inconvenient nomenclature is, of course, nothing new in astronomy – and must be tolerated as usual.

The wealth of new SNe data in recent years is due to the advent of large-scale, systematic search techniques. To understand this, it may be worth stating the obvious. It is not possible to pick an arbitrary galaxy and get a supernova distance for it because most galaxies, at a given time, do not have a supernova in them. Thus, it is necessary to search many galaxies at random and somehow identify the small fraction ($\sim 10^{-4}$) in which a supernova is going off at any given time. Methods for doing this have been pioneered by Perlmutter and collaborators (Goobar & Perlmutter 1995; Perlmutter et al. 1995, 1996, 1997). Deep images are taken of the same region of the sky 2–3 weeks apart. Stellar objects which appear in the second image but not in the first are candidate supernovae to be confirmed by spectroscopy. By means of such an approach, of order 30 high-redshift ($z = 0.35$–0.65) are now known. Related approaches for finding moderate- and high-redshift supernovae have been developed by other groups as well (Adams et al. 1995; Hamuy et al. 1995; Schmidt et al. 1994).

Search techniques such as those of the Perlmutter group survey many faint galaxies in limited regions of the sky, and are not very good at finding low-redshift ($z \lesssim 0.03$) supernovae. Thus, they are not particularly relevant to peculiar-velocity studies (but see below). However, precisely because they detect intermediate- to high-redshift supernovae, such techniques will be useful for measuring H_0 (with supernovae found at $z \lesssim 0.2$, where cosmological effects are relatively unimportant), and are among the best existing methods for determining the cosmological parameters Ω_0 and Λ (with supernovae at $z \gtrsim 0.3$, which probe spatial curvature). To see how this works, one can plot Hubble diagrams for recently discovered supernovae both at moderate and high redshift. This is done in Fig. 6.6, which has been adapted from the 1996 San Antonio AAS meeting contribution by the Perlmutter group. The low-redshift data ($\log(cz) < 4.5$) are from Hamuy et al. (1995), and the high-redshift data are from Perlmutter et al. (1996).

Figure 6.6 contains several important features. First, the observed peak apparent magnitudes are plotted versus log redshift in the top panel. To the degree that SNe Ia are standard candles, one expects these apparent magnitudes to go as $\mathrm{const.} + 5\log(cz)$, the straight line plotted through the points at low redshift. Correcting the SNe Ia magnitudes for the light-curve widths (i.e., going from the top to the bottom panel) significantly improves the agreement with this low-redshift prediction. This is the main reason that the light-curve width correction is thought to greatly reduce the SNe Ia scatter. Whether or not the correction is made, however, the data provide unequivocal proof of the linearity of the Hubble law at low ($z \simeq 0.1$) redshift. Second, one expects that that at higher redshifts the m_B–$\log(cz)$ relation will depart from linearity because of spacetime curvature. The departure from linearity is, to first order in z, a function only of the deceleration parameter q_0 – or equivalently, if the universe has vanishing cosmological constant Λ (see below), by the density parameter Ω_0, which in that case is exactly twice q_0. Figure 6.6 assumes $\Lambda \equiv 0$ and thus labels the curves by Ω_0. There is a hint in the behavior of the light-curve-shape corrected magnitudes that this departure from linearity has been detected, and in particular that $\Omega_0 \simeq 1$ is a better fit to the data than $\Omega_0 \simeq 0$ (Perlmutter et al. 1996).

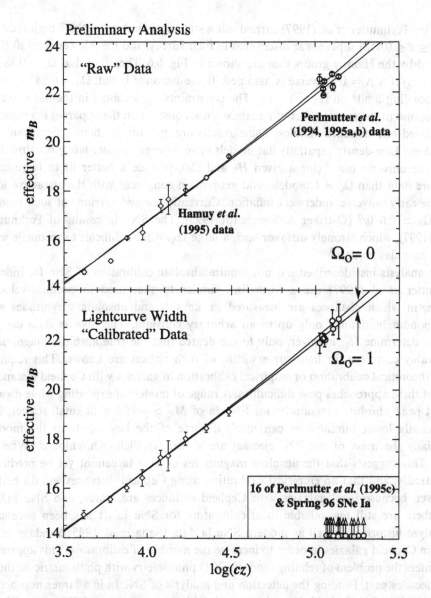

Figure 6.6. Hubble diagrams using SNe Ia. The *upper panel* shows the observed peak apparent magnitudes; in the *lower panel* the magnitudes are corrected for the light-curve width effect (see main text for details). The inset in the lower panel shows the redshifts of 16 additional SNe Ia recently discovered but not yet analyzed by the Perlmutter group.

More generally, the behavior of the Hubble diagram at high redshift depends on the cosmological parameters Ω_0 and $\Omega_\Lambda \equiv \Lambda/3H_0^2$. Perlmutter *et al.* (1997) suggest that the SNe Ia data should be interpreted for now in the context of two cosmological paradigms: a $\Lambda = 0$ universe, and a spatially flat ($\Omega_0 + \Omega_\Lambda = 1$)

universe.[2] Perlmutter et al. (1997) carried out a statistical analysis of the 7 high-redshift ($0.354 \leq z \leq 0.458$) supernovae discovered in their survey, and the 9 lower-redshift SNe Ia found by the Hamuy group, that are shown in Fig. 6.6. They find that $\Omega_0 = 0.88^{+.69}_{-.60}$ (1σ errors) if a $\Lambda = 0$ universe is assumed. If the universe is flat, $\Omega_0 = 0.94^{+.34}_{-.28}$, with corresponding limits on $\Omega_\Lambda = 1 - \Omega_0$. The constraints are stronger in the flat universe case because of the strong effect of a cosmological constant on the apparent magnitudes of high-redshift standard candles. These results are, potentially, highly significant for cosmology. Low-density, spatially flat models have become popular lately because they make the universe older (for a given H_0 and Ω_0), provide a better fit to large-scale structure data than $\Omega_0 = 1$ models, and yet remain consistent with the attractive idea that the early universe underwent inflation. Currently favored versions of such models have $\Omega_\Lambda \simeq 0.6$–$0.7$ (Ostriker & Steinhardt 1995). The SNe Ia results of Perlmutter et al. (1997), which strongly disfavor such a large Ω_Λ, will be difficult to reconcile with low-density flat models.[3]

The analysis just described did not require absolute calibration of SNe Ia. Indeed, Perlmutter et al. (1997) use a formalism similar to that used in peculiar-velocity studies, in which distances are measured in km s^{-1}, and absolute magnitudes are, correspondingly, defined only up to an arbitrary constant. The SNe Ia data can be used to determine H_0, however, only to the degree that the true absolute magnitudes (preferably corrected for light-curve width) of such objects are known. This requires either theoretical calibration or empirical calibration in galaxies with Cepheid distances. Both of these approaches pose difficulties. A range of models of exploding white dwarfs predict peak absolute magnitudes for SNe Ia of $M_V \simeq -19.5$ with small scatter, but significantly lower luminosities can result if some of the key inputs to the models (especially the mass of the ^{56}Ni ejectae) are varied (Höflich, Khokhlov & Wheeler 1995). This suggests that the absolute magnitudes of SNe Ia cannot yet be predicted theoretically, and that an empirical calibration using Cepheid distances will do better. However, because local galaxies with Cepheid distances are scarce, and SNe Ia are rare, there are still few reliable local calibrators for SNe Ia. It has been necessary to analyze historical as well as modern SNe Ia data (Saha et al. 1995; Sandage et al. 1996) in Cepheid galaxies in order to increase the number of calibrators. This approach encounters the problem of relating modern CCD photometry with photometric methods from decades past. Pending the detection and analysis of SNe Ia in a larger number of local galaxies with Cepheid distances, one should view estimates of H_0 inferred from supernovae as preliminary.

Being rare events, SNe Ia are unlikely to provide a detailed map of the local peculiar-velocity field. However, because of their small scatter (see above), a few well-observed SNe Ia distributed on the sky may lead to useful constraints on amplitude and scale of

---

[2] With a large sample of SNe Ia that spans a large redshift range, it may be possible to constrain Ω_0 and Ω_Λ separately, without assuming either a flat universe or a vanishing cosmological constant (Goobar & Perlmutter 1995). The present data are not adequate for this purpose.

[3] More recent SNe Ia data, analyzed in 1997–98 by both the Perlmutter group and the one led by Brian Schmidt, Adam Riess and Robert Kirshner, appear to favor low-density, flat models.

bulk flows. A first attempt at this was carried out by Riess, Press & Kirshner (1995b), who used 13 SNe Ia with peak magnitudes corrected by light-curve widths to place limits on the bulk flow within ~ 7000 km s^{-1}. They found the data to be consistent with at most a small ($\lesssim 400$ km s^{-1}) bulk streaming, and to be inconsistent with the large bulk flow found by Lauer & Postman (1994) using an independent method (cf. §6.7 below). However, one must be cautious in interpreting such results because small-scale power in the velocity field can obscure large-scale motions (Watkins & Feldman 1995). Constraints on bulk flows using SNe Ia are likely to improve in the coming years.

6.7 Brightest cluster galaxies

Another "classical" distance indicator method that has been reborn in modern guise is photometry of brightest cluster galaxies (BCGs). As originally treated by Sandage and coworkers (Sandage 1972; Sandage & Hardy 1973), BCGs were considered to be good standard candles. As such, they were used to demonstrate the linearity of the Hubble diagram to relatively large distances and estimate H_0. Any such estimate was and remains highly suspect, however, because of the difficulty of obtaining a good absolute calibration of the method. The scatter of BCGs as standard candles is around 0.30–0.35 mag, which compares favorably with methods such as TF or D_n–σ.

A dubious assumption in the early work was that BCGs are true standard candles. Gunn & Oke (1975) first suggested that the luminosities of BCGs might correlate with their surface brightness profiles. Following this suggestion, Hoessel (1980) defined a metric radius $r_m = 10h^{-1}$ kpc, and showed that the metric luminosity $L(r_m) \equiv L_m$ varied roughly linearly with a shape parameter α defined by

$$\alpha \equiv \left.\frac{d\log L}{d\log r}\right|_{r_m}. \qquad (6.11)$$

More recently, Lauer & Postman (1992) have shown that the correlation between L_m and α is better modelled by a quadratic relation. The Lauer & Postman (1992) data, along with their quadratic fit, are shown in the upper panel of Fig. 6.7. Thus modelled, the typical distance error incurred by the BCG L_m-α relation is $\sim 16\%$.

A slight hitch in applying the BCG L_m-α relation is the requirement of defining a metric radius r_m for evaluating both L_m and α. This means that the assumed peculiar velocity of a BCG must be factored in to convert redshift to distance, and thus angular to linear diameter. In practice this is not a very serious issue. At the typically large distances ($\gtrsim 7000$ km s^{-1}) at which the relation is applied, peculiar-velocity corrections have a small effect on L_m and α. Iterative techniques in which a peculiar-velocity solution is obtained and then used to modify the r_ms, converge quickly (Lauer & Postman 1994).

Modern scientific results based on BCGs are due to the pioneering work of Lauer and Postman (Lauer & Postman 1992; Lauer & Postman 1994, hereafter LP94; Postman & Lauer 1995). One important – and uncontroversial – such result has been confirmation,

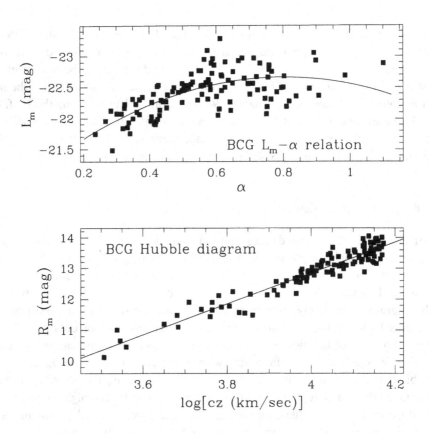

Figure 6.7. *Top panel:* the BCG L_m–α relation exhibited by the sample of Lauer & Postman (1992). Absolute magnitude within the metric radius r_m is plotted against the logarithmic surface brightness slope at r_m. The solid curve shows the quadratic fit to the data. *Bottom panel:* the Hubble diagram for the Lauer & Postman (1992) BCG sample. Apparent magnitude within r_m is plotted against log redshift. The straight line plotted through the points has slope 5, the relation expected for a linear Hubble flow. The data used to make this figure were kindly provided by Marc Postman.

with unprecedented accuracy, of the linearity of the Hubble diagram to redshifts $z \simeq 0.05$ over the entire sky. (The Hubble diagrams using SNe Ia (§6.6), by contrast, are not derived from full-sky samples.) This is shown in the lower panel of Fig. 6.7. However, another result has been considerably more controversial, namely, the detection of a very large-scale bulk peculiar velocity by LP94. The linearity of the BCG Hubble diagram manifests itself with the smallest scatter when the velocities are referred to a local frame that differs significantly from that defined by the CMB dipole. Or, stated another way, the LP94 data indicate that the frame of Abell clusters out to 15,000 km s^{-1} redshift is moving with respect to the CMB frame at a velocity of ~ 700 km s^{-1} toward Galactic coordinates $l \simeq 350°$, $b \simeq 50°$. A reanalysis of the LP94 data by Colless (1995) produced a very similar result for the bulk motion.

The global Hubble flow linearity demonstrated by Lauer & Postman (1992) suggests that the BCG L_m-α relation is an excellent DI out to substantial redshifts. However, the indicated bulk motion is of sufficient amplitude and scale as to appear inconsistent with other indicators of large-scale homogeneity. For example, Strauss et al. (1995) showed that none of the leading models of structure formation that are consistent with other measures of large-scale power can reproduce an LP94-like result in more than a small fraction of realizations. Furthermore, two recent studies, one using the TF relation (Giovanelli et al. 1996) and one using Type Ia SNe (Riess, Press & Kirshner 1995b), suggest that the bulk motion on smaller scales than that probed by the BCGs is inconsistent with the LP94 bulk flow at high-significance levels.

For the above reasons, the status of BCGs as DIs is controversial. However, one should not prejudge the outcome. Velocity studies have yielded a number of surprises in the last 15 years, and it is not inconceivable that the LP94 bulk flow – or something like it – will be vindicated in the long term. Lauer, Postman and Strauss are extending BCG observations to a complete sample with $z \leq 0.1$. Whether or not it confirms LP94, this extended study is likely to greatly clarify the nature of the BCG L_m-α relation.

6.8 Redshift-distance catalogs

As redshift measurements accumulated in the 1970s and 1980s, it was widely recognized that there was a need to assemble these data into comprehensive catalogs. Beginning with the publication of the CfA redshift survey in 1983 (Huchra et al. 1983), all major redshift surveys (see Ch. 5) led to electronically available databases in fairly short order.

Comparable efforts involving redshift-independent distance measurements have been slower in coming. This is largely due to the issue of *uniformity*. Whereas redshift measurements by different observers rarely exhibit major differences, redshift-independent distances obtained by different observers can, and generally do, differ systematically for any number of reasons. In some cases the origin of such differences is different calibrations of the DI. In others, the calibrations are the same but the input data differ in subtle ways. Finally, the way statistical bias effects are treated (§6.9) often differs among those involved in galaxy distance measurements. For all these reasons, it is not possible simply to go to the published literature, find all papers in which galaxy distances are reported, and lump them together in a single database. Instead, individual datasets must be assembled, their input data and selection criteria characterized, their DI relations recalibrated if necessary, and the final distances brought to a uniform system. Only then can the resultant catalog be relied upon – and even then, caution is required.

The first steps toward assembling homogeneous redshift-distance catalogs were taken in the late 1980s by David Burstein. His goal was to combine the then newly-acquired D_n-σ data from the 7-Samurai group (§6.4) with the extant data on spiral galaxy distances, especially the infrared TF data obtained by the Aaronson group (§6.3). Burstein's efforts produced two electronic catalogs, the Mark I (in 1987) and Mark II (in 1989) catalogs of galaxy peculiar velocities.[4] Burstein's chief concern was matching the TF and D_n–σ distance scales. As there are, by definition, no galaxies that have both kinds of distances, this matching could only be carried out indirectly. The approach decided upon by Burstein, in consultation with the other 7 Samurai, was to require the Coma cluster spirals and ellipticals to have the same mean distances. Although this procedure was imperfect, the Mark II catalog was considered reliable enough to be used in the first major effort to constrain the density parameter Ω_0 by comparing velocities and densities (Dekel et al. 1993).

With the publication of a number of large, new TF datasets in the early 1990s, the need for a greatly expanded redshift-distance catalog became apparent. An important development was the superseding of the majority of the older infrared TF data, obtained by the Aaronson group, by CCD-based (R-and I-band) TF data. Han, Mould and coworkers (Han 1992; Han & Mould 1992; Mould et al. 1991) obtained a full-sky cluster TF sample, based on I-band magnitudes and 21-cm velocity widths, comprising over 400 galaxies. Willick (1990, 1991) and Courteau (1992; Courteau et al. 1993) gathered R-band TF data in the Northern sky for over 800 galaxies in total. The largest single contribution was that of Mathewson, Ford & Buchorn (1992) who published an I-band TF sample of 1355 galaxies in the Southern sky. Despite the influx of the new CCD data, one portion of the infrared TF database of the Aaronson group was not rendered obsolete: the sample of over 300 local ($cz \lesssim 3000$ km s^{-1}) galaxies first observed in the late 1970s and early 1980s (Aaronson et al. 1982). This local sample was, however, subjected to a careful reanalysis by Tormen & Burstein (1995), who rederived the H-band magnitudes using a more homogeneous set of galaxy diameters and inclinations than was available to the original researchers a decade earlier.

In 1993, a group of astronomers (David Burstein, Avishai Dekel, Sandra Faber, Stéphane Courteau, and myself) began the process of integrating these TF data and the existing D_n–σ data into a new redshift-distance catalog. Our methodology is described in detail in Willick et al. (1995, 1996), and portions of the catalog are presented in Willick et al. (1997). The full catalog, known as the *Mark III Catalog of Galaxy Peculiar Velocities*, is quite large (nearly 3000 spirals and over 500 ellipticals, although this includes several hundred overlaps between datasets) and is available only electronically, as described in Willick et al. (1997).

Building upon the foundation laid by Burstein in the Mark I and II catalogs, the Mark III catalog was assembled with special emphasis placed on achieving uniform distances among the separate samples it comprises. Four specific steps were taken toward this goal. First, the raw data in all of the TF samples underwent a uniform set of corrections for inclination and extinction (cf. §6.3.2). Second, the TF relations for each sample were recalibrated using a self-consistent procedure that included correction for selection bias (§6.9). Third, final TF zero points were assigned by requiring that the

---

[4] Although these are referred to as "peculiar-velocity" catalogs, they are, first and foremost, *redshift-distance* catalogs, consisting of redshifts and redshift-independent distances. The peculiar velocities follow from these more basic data, although not necessarily in a simple way, given the statistical bias effects studied in §6.9.

TF distances of objects common to two or more samples agree in the mean. This step ensures that the different samples are on similar relative distance scales. The global TF zero point was determined by the full-sky Han–Mould cluster TF sample. (As explained in §6.3, this zero point was such that the distances are given in units of km s^{-1}, not Mpc.) Fourth, the spiral and elliptical distance scales were matched by applying the POTENT algorithm (see §§7.2.3, 7.3.3) to each separately, and requiring that they produce statistically consistent velocity fields.

In parallel with the efforts of the Mark III group, similar enterprises have been undertaken by two other groups. Brent Tully has also assembled and recalibrated much of the extant TF data. Riccardo Giovanelli, Martha Haynes, Wolfram Freudling, Luiz da Costa, and coworkers have acquired new I-band TF data for ~ 2000 galaxies, and have combined it with the sample of Mathewson et al. (1992). Initial scientific results from each of these efforts have been published (Shaya, Peebles & Tully 1995; Giovanelli et al. 1996; da Costa et al. 1996), and the catalogs themselves will soon become publically available.

New distances for elliptical galaxies, now mostly from the FP-relation rather than the D_n–σ relation (§6.4), continue to be obtained as well. Jorgensen, Franx & Kjaergaard (1995a,b) have published distances for E and S0 galaxies in 10 clusters out to 10,000 km s^{-1}. The EFAR group (Burstein, Colless, Davies, Wegner, and colleagues) are now finishing an FP survey of over 80 groups and clusters at distances between 7000 and 16,000 km s^{-1} (Colless et al. 1993; Wegner et al. 1993,1996).

Implicit in all this ongoing work is that the Mark III catalog, like its predecessors, is just one step along a path still being traveled. Just as the Mark III catalog consists in part of recalibrated data already present in the Mark II, so will future catalogs incorporate, partially recalibrate, and expand upon the Mark III. Of particular note are the distances coming from the SBF survey of Tonry and coworkers (Tonry et al. 1997; cf. §6.5). The SBF distances are much more accurate than either TF or D_n–σ and can provide important checks on them. Tonry et al. (1997) have taken initial steps toward such an intercomparison, and the preliminary results, which suggest mutually consistent results among SBF, D_n–σ, and TF, are encouraging. Little comparison of SNe and BCG distances with other DIs has yet been carried out, but will be in the coming years. It is reasonable to hope that, by the turn of the century at the latest, the available redshift-distance catalogs will be superior, in terms of sky coverage, accuracy, and homogeneity, to the best we have today.

6.9 Malmquist and other biases

Distance scale and peculiar-velocity work have long been plagued by statistical biases. These biases are sufficiently confusing and multifaceted that their effects are often misunderstood or misrepresented. It is worth taking a moment to go over a few of the main issues.

The root problem is that our distance indicators contain scatter: a galaxy with distance d inferred from the DI really lies within some range of distances, approximately (but not exactly) centered on d. This range is characterized by a non-Gaussian distribution of characteristic width $d\Delta$, where Δ is the fractional distance error characteristic of the DI. (If σ is the DI scatter in magnitudes, $\Delta \simeq 0.46\sigma$.) Thus, the farther away the object is the bigger the distance error. For most DIs, a good approximation is that the distribution of distance errors is log-normal: if the true distance is r, then the distance estimate d has a probability distribution given by

$$P(d|r) = \frac{1}{\sqrt{2\pi}(d\Delta)} \exp\left[-\frac{[\ln(d/r)]^2}{2\Delta^2}\right]. \qquad (6.12)$$

Two distinct kinds of statistical bias effects can arise when DIs with the above properties are used. Which of the two occurs depends on which of two basic analytic approaches one adopts for treating the DI data. In the first approach, known as *Method I*, one assumes that the DI-inferred distance d is the best a priori estimate of true distance. Any subsequent averaging or modelling of the datapoints assumes galaxies with similar values of d to be neighbors in real space as well. The second approach, known as *Method II*, takes proximity in redshift space as tantamount to real-space proximity; the DI-inferred distances are then treated only in a statistical sense, averaged over objects with similar redshift-space positions. The Method I/Method II terminology originated with Faber & Burstein (1988); a detailed discussion is provided by Strauss & Willick (1995, §6.4; see also §7.2.2).

Let us consider this distinction in relation to peculiar velocity or Hubble constant studies. In a Method I approach, one would take objects whose DI-inferred distances are within a narrow range of some value d, and average their redshifts. Subtracting d from the resulting mean redshift yields a peculiar-velocity estimate; dividing the mean redshift by d gives an estimate of H_0. However, these estimates will be biased, because the distance estimate d itself is biased: *It is not the mean true distance of the objects in question.* To see this, we reason as follows: if $P(d|r)$ is given by equation (6.12) above, then the distribution of true distances of our objects is given, according to Bayes' Theorem, by

$$P(r|d) = \frac{P(d|r)P(r)}{\int_0^\infty P(d|r)P(r)\,dr} = \frac{r^2 n(r)\exp\left(-\frac{[\ln r/d]^2}{2\Delta^2}\right)}{\int_0^\infty r^2 n(r)\exp\left(-\frac{[\ln r/d]^2}{2\Delta^2}\right)dr}, \qquad (6.13)$$

where we have taken $P(r) \propto r^2 n(r)$, where $n(r)$ is the underlying galaxy number density along the line of sight. To obtain the expectation value of the true distance r for a given d, we multiply equation (6.13) by r and integrate over all r. In general, this integral requires knowledge of the density field $n(r)$ and will have to be done numerically. However, in the simplest case that the density field is constant, the integral can be done analytically. The result is that the expected true distance is $de^{7\Delta^2/2}$ (Lynden-Bell et al. 1988; Willick 1991). This effect is called *homogeneous Malmquist bias*. It tells us that, typically, objects lie further away than their DI-inferred distances. The physical cause is more objects "scatter in" from larger true distances (where there is more volume) than "scatter out" from smaller ones. In general, however, variations in the number density cannot be neglected. When this is the case, there is *inhomogeneous Malmquist bias* (IHM). IHM can be computed numerically if one has a model of the density field. Further discussion of this issue may be found in Willick et al. (1997).

The biases which arise in a Method II analysis are quite different. They may be rigorously understood in terms of the probability distribution of the DI-inferred distance d given the redshift cz, $P(d|cz)$ (contrast with equation (6.13), which underlies Method I). In general, this distribution is quite complicated (cf. Strauss & Willick 1995, §8.1.2), and its details are beyond the scope of this chapter. However, under the assumption of a "cold" velocity field – an assumption that appears adequate in ordinary environments – redshifts complemented by a flow model give a good approximation of true distance. Thus, it really is the probability distribution $P(d|r)$ (equation [6.12]), or one similar to it, that counts for a Method II analysis. However, that equation as written does not represent the full story. If severe selection effects such as a magnitude or diameter limit are present, then the log-normal distribution does not apply exactly. Some galaxies are too faint or small to be in the sample; in effect, the large-distance tail of $P(d|r)$ is cut off. It follows that the typical inferred distances are *smaller* than those expected at a given true distance r. As a result, the peculiar-velocity model that allows true distance to be estimated as a function of redshift is tricked into returning shorter distances. This bias goes in the same sense as Malmquist bias, but is fundamentally different. It results not from volume/density effects, but from *sample selection* effects, and is called *selection bias*.

Selection bias can be avoided, or at least minimized, by working in the so-called "inverse direction." What that means is most easily illustrated using the TF relation. When viewed in its "forward" sense, the TF relation is conceived as a prediction of absolute magnitude given a value of the velocity width parameter, $M(\eta)$. However, it is equally valid to view the relation as a prediction of η given a value of M, i.e., as a function $\eta^0(M)$ (the superscript ensures that there is no confusion between the observed width parameter η and the TF prediction). When one uses the forward relation, one imagines fitting a line $m_i = M(\eta_i) + \mu$ by regressing the apparent magnitudes m_i on the velocity widths η_i; the distance modulus μ is the free parameter solved for. Selection bias then occurs because apparent magnitudes fainter than the magnitude limit are "missing" from the sample, so the fitted line is not the same as the true line. However, if one instead fits a line $\eta^0(m_i - \mu)$ by regressing the widths on the magnitudes, the same effect does not occur, provided the sample selection procedure does not exclude large or small velocity widths. In general, this last caveat is more or less valid. Consequently, working in the inverse direction does in fact avoid or at least minimize selection bias.

This fact, first clearly stated by Schechter (1980) and then reiterated in various forms by Aaronson et al. (1982), Tully (1988), Willick (1994), Dekel (1994), and Davis, Nusser & Willick (1996), among others, remains an obscure one, not universally appreciated. It is often heard, for example, that the TF relation applied to relatively

Table 6.1. *"Method Matrix" of distance indicator biases*

DI type/Method	Method I DI-inferred distance best indicator of true distance	Method II Redshift-space data best indicator of true distance
Forward dist-dep (e.g. mag) predicted by dist-indep (e.g. η) quantity	**Malmquist bias** (*selection-independent*)	**Strong selection bias** (*depends on observational selection criteria*)
Inverse dist-indep predicted by dist-dep quantity	**Malmquist bias** (*selection-dependent*)	**Weak or no selection bias** (*bias present if selection related to dist-indep quantity*)

distant galaxies will necessarily result in a Hubble constant that is biased high, because the distances are biased low due to selection bias. The clear conclusion of the previous paragraph, however, is that provided the analysis is done using redshift-space information to assign a priori distances – that is, provided that a Method II approach is taken – working in the inverse direction can render selection bias unimportant. It is also the case that careful analytical methods (Willick 1994) permit a correction for selection bias even when working in the forward direction. It should be borne in mind, however, that both of these approaches (using the inverse relation or correction for forward selection bias) necessitate a careful characterization of sample selection criteria.

Another wrinkle in this complicated subject is that the relatively bias-free character of inverse distance indicators does not carry over to a Method I analysis. It is beyond the scope of this chapter to discuss this issue in full detail; the interested reader is referred to Strauss & Willick (1995, § 6.5). The main point is that a Method I inverse DI analysis is subject to Malmquist bias in much the same way as a Method I forward analysis; indeed, the inverse Malmquist bias is in some ways considerably more complex, as it depends (unlike forward Malmquist bias) on sample selection criteria. So while it is correct to emphasize the bias-free (or nearly so) nature of working in the inverse direction, it is essential to remember that this property holds only for Method II analyses.

Much of the confusion surrounding the relative bias properties of forward versus inverse DIs stems from neglecting the distinction between Method I and Method II analyses. Recognizing this, Strauss & Willick (1995) summarized the issue with what they called the "Method Matrix" (a more memorable term might be the "magic square") of peculiar-velocity analysis. Their table is reproduced as Table 6.1, in a slightly simpler form (the original alluded to several complications that are unecessary here). Reference to this simple diagram might allay some of the controversies surrounding Malmquist and related biases.

6.10 Summary

The measurement of galaxy distances is crucial for some of the basic problems in astronomy and cosmology. In this chapter I have emphasized the role such measurements play in two of the most important: Hubble constant determination and peculiar-velocity analysis. An important distinction between these two efforts, which has been reiterated throughout, is that for peculiar velocities one only needs distances in $km\,s^{-1}$, which are independent of an absolute distance scale, whereas for determination of H_0 distances in Mpc are required. In practice, this means that peculiar-velocity studies may be carried out using distance indicators such as TF or D_n–σ calibrated only relative to the distant Hubble flow. To obtain H_0 the same DIs must be calibrated relative to local galaxies with Cepheid distances. Because the program of Cepheid measurements in local calibrators using HST (Kennicutt *et al.* 1995) is ongoing, reliable far-field measurements of H_0 are still several years away.

The discussion has been organized around the principal distance indicators currently in use. These are:

(i) *Cepheid variables.* The Period–Luminosity relation for these pulsating stars may be calibrated in the Milky Way and in the Magellanic Clouds. However, they are detectable with HST out to ~ 20 Mpc. As such, they will yield accurate absolute distances for ~ 20 local galaxies over the next several years. These local galaxies will in turn provide absolute calibrations for the secondary distances indicators such as TF or SNe Ia that will be used to measure H_0 in the "far field" ($\gtrsim 7000\ km\,s^{-1}$), where peculiar velocities and depth effects are relatively unimportant.

(ii) *The TF relation.* This method has been the workhorse of peculiar-velocity studies, for it applies to the ordinary spiral galaxies that best trace the peculiar-velocity field. When calibrated using HST Cepheid distances, it promises also to yield a value of H_0 accurate to $\sim 10\%$. The TF relation has been shown to apply to spiral galaxies at high redshift (Vogt *et al.* 1996), although evolutionary effects appear to be significant at $z \simeq 0.5$.

(iii) *The D_n–σ relation.* This is a variant of the Fundamental Plane relations for elliptical galaxies. It is comparable to TF in accuracy, and gives similar global results for the large-scale peculiar-velocity field (Kolatt & Dekel 1994). Its best chance for absolute calibration comes from a comparison with SBF distances. Like TF, D_n–σ has been applied to relatively high-redshift galaxies (Bender, Ziegler & Bruzual 1996), again with evidence of evolutionary changes.

(iv) *Surface brightness fluctuations.* This method may be the most accurate DI known for galaxies beyond the range of HST Cepheid measurements, with distance errors as small as 5% under the best conditions and median errors of $\sim 8\%$. Its application is most straightforward for early-type systems, although with care it may be applied to spirals as well. It holds the promise of giving a high-resolution picture of the peculiar-velocity field. It will also provide a crucial check of the reliability of TF and D_n–σ. Its direct application to the H_0 problem remains uncertain because of the great technical challenge involved in extending it to distances $\gtrsim 5000$ km s^{-1}.

(v) *Type Ia supernovae.* SNe are, in principle, excellent DIs, but suffer from the obvious problem that one cannot, in general, be found in a given galaxy at a given time. In recent years, improved search techniques have vastly increased the number of well-observed SNe Ia, both at relatively low (Hamuy et al. 1995) and high (Perlmutter et al. 1996) redshifts. The results of these studies have included beautiful Hubble diagrams that demonstrate the linearity of the Hubble expansion to $z \simeq 0.1$, with tantalizing hints of curvature that hold the promise of constraining the cosmological parameters Ω_0 and Λ. Sandage and coworkers (Sandage et al. 1996; Saha et al. 1995) have calibrated SNe Ia in galaxies with Cepheid distances to obtain Hubble constant estimates of $H_0 \simeq 57$ km s^{-1} Mpc^{-1}. However, considerable uncertainty attaches to these results at present. The quest for a reliable absolute calibration of SNe Ia continues.

(vi) *The BCG L_m–α relation.* The pioneering work of Lauer and Postman (Lauer & Postman 1992, 1994; Postman & Lauer 1995) has demonstrated the potential of BCGs in distance scale and peculiar-velocity work. The detection of very large-scale bulk streaming using BCGs has caused some to question the global validity of the L_m–α relation (e.g., Riess, Press & Kirshner 1995b), but the verdict is not in yet. Ongoing work by Lauer and Postman, in collaboration with Strauss, will greatly clarify the situation.

I conclude by reiterating a point made at the outset of this chapter. The DIs discussed here are empirical relations whose physical origins are only partially understood at best. There is a class of distance indicators that are based on fairly rigorous physics, and whose absolute calibration may be obtained from first principles. Gravitational lensing of time-variable quasars and the Sunyaev–Zeldovich effect in clusters are perhaps the most noteworthy of these. It is conceivable that these methods will mature and add greatly to what we have learned from the empirical DIs about the distance scale and the peculiar-velocity field. However, this additional information will most likely reinforce, rather than supplant, the knowledge obtained from the DIs discussed here.

Acknowledgments

JAW thanks David Burstein, Tod Lauer, Marc Postman, Saul Perlmutter, and John Tonry for enlightening discussions about the distance indicator relations in which they are leading experts, and for providing data or postscript files for several of the figures presented here.

References

Aaronson, M., Huchra, J. & Mould, J. 1979, *ApJ*, **229**, 1

Aaronson, M., Huchra, J., Mould, J., Sullivan, W., Schommer, R. & Bothun, G. 1980, *ApJ*, **239**, 12

Aaronson, M., Huchra, J., Mould, J., Schechter, P. L. & Tully, R. B. 1982, *ApJ*, **258**, 64

Aaronson, M., Bothun, G., Mould, J., Huchra, J., Schommer, R. A. & Cornell, M. E. 1986, *ApJ*, **302**, 536

Adams, M. T., Wheeler, J. C., Ward, M., Wren, W. R. & Schmidt, B. P. 1995, *BAAS*, **187**, 1711

Bender, R., Burstein, D. & Faber, S. M. 1992, *ApJ*, **399**, 462

Bender, R., Burstein, D. & Faber, S. M. 1993, *ApJ*, **411**, 153

Bender, R., Ziegler, B. & Bruzual, G. 1996, *ApJ*, **463**, L51

Baum, W. 1955, *PASP*, **67**, 328

Bernstein, G. M., et al. 1994, *AJ*, **107**, 1962

Bothun, G. D. & Mould, J. R. 1987, *ApJ*, **313**, 629

Bottinelli, L., Gouguenheim, L., Paturel, G. & de Vaucouleurs, G. 1983, *A&A*, **118**, 4

Burstein, D., Willick, J. & Courteau, S. 1995, in *Opacity of Spiral Disks*, NATO Ser., eds. J. A. Davies & D. Burstein (Dordrecht:Kluwer)

Colless, M., et al. 1993, *MNRAS*, 262, 475

Colless, M. M. 1995, *AJ*, **109**, 1946

Courteau, S. 1992, Ph.D. Thesis, University of California, Santa Cruz

Courteau, S., Faber, S. M., Dressler, A. & Willick, J. A. 1993, *ApJ*, **412**, L51

da Costa, L. N., Freudling, W., Wegner, G., Giovanelli, R., Haynes, M. & Salzer, J. J. 1996, *ApJ*, **468**, L5

Davis, M., Nusser, A. & Willick, J. A. 1996, *ApJ*, **473**, 22

Dekel, A. 1994, *ARA&A*, **32**, 371

Dekel, A., Bertschinger, E., Yahil, A., Strauss, M. A., Davis, M. & Huchra J. P. 1993, *ApJ*, **412**, 1

Djorgovski, S. & Davis, M. 1987, *ApJ*, **313**, 59

Dressler, A. 1994, in *Cosmic Velocity Fields*, eds. F. Bouchet & M. Lachièze-Rey, (Gif-sur-Yvette: Editions Frontières), p. 9

Dressler, A., et al. 1987, *ApJ*, **313**, 42

Eastman, R. G., Schmidt, B. P. & Kirsher, R. 1996, *ApJ*, **466**, 911

Eisenstein, D. J. & Loeb, A. 1996, *ApJ*, **459**, 432

Faber, S. M. & Burstein, D. 1988, in *Large Scale Motions in the Universe*, ed. V. C. Rubin & G. V. Coyne (Princeton: Princeton University Press), p. 115

Faber, S. M., et al. 1987, in *Nearly Normal Galaxies*, ed. S. M. Faber (New York: Springer), p. 175

Faber, S. M. & Jackson, R. E. 1976, *ApJ*, **204**, 668

Federspiel, M., Sandage, A. & Tammann, G. A. 1994, *ApJ*, **430**, 29

Ferrarese, L., et al. 1996, *ApJ*, **464**, 568

Fisher, A., Branch, D. & Nugent, P. 1993, *BAAS*, **183**, #38.03

Freedman, W. L. & Madore, B. F. 1990, *ApJ*, **365**, 186

Freedman, W. L., Wilson, C. D. & Madore, B. F. 1991, *ApJ*, **372**, 455

Freedman, W. L., et al. 1994, *Nature*, **371**, 757

Freudling, W., da Costa, L. N., Wegner, G., Giovanelli, R., Haynes, M. P. & Salzer, J. J. 1995, *AJ*, **110**, 920

Giovanelli, R., Haynes, M. P., Salzer, J. J., Wegner, G., da Costa, L. N. & Freudling, W. 1995, *AJ*, **110**, 1059

Giovanelli, R., Haynes, M. P., Wegner, G., da Costa, L. N., Freudling, W. & Salzer, J. J. 1996, *ApJ*, **464**, L99

Goobar, A. & Perlmutter, S. 1995, *ApJ*, **450**, 14

Gunn, J. E. & Oke, J. B. 1975, *ApJ*, **195**, 255

Hamuy, M., Phillips, M. M., Maza, J., Suntzeff, N. B., Schommer, R. A. & Aviles, R. 1995, *AJ*, **109**, 1

Han, M.-S. 1992, *ApJ*, **391**, 617

Han, M.-S. & Mould, J. R. 1992, *ApJ*, **396**, 453

Hoessel, J. G. 1980, *ApJ*, **241**, 493

Höflich, P., Khokhlov, A. M. & Wheeler, J. C. 1995, *ApJ*, **444**, 831

Huchra, J., Davis, M., Latham, D. & Tonry, J. 1983, *ApJS*, **52**, 89

Jorgensen, I., Franx, M. & Kjaergaard, P. 1996, *MNRAS*, **280**, 167

Kennicutt, R. C., Freedman, W. L. & Mould, J. R. 1995, *AJ*, **110**, 1476

Kolatt, T. & Dekel, A. 1994, *ApJ*, **428**, 35

Lauer, T. R. & Postman, M. 1992, *ApJ*, **400**, L47

Lauer, T. R. & Postman, M. 1994, *ApJ*, **425**, 418

Lynden-Bell, D., et al. 1988, *ApJ*, **326**, 19

Mathewson, D. S., Ford, V. L. & Buchhorn, M. 1992, *ApJS*, **81**, 413

Montes, M. J. & Wagoner, R. V. 1995, *ApJ*, **445**, 828

Mould, J. R., et al. 1991, *ApJ*, **383**, 467

Mould, J. R., et al. 1995, *ApJ*, **449**, 413

Ostriker, J. P. & Steinhardt, P. J. 1995, *Nature*, **377**, 600

Pahre, M. A., Djorgovski, S. G. & de Carvalho, R. R. 1995, *ApJ*, **453**, L17

Perlmutter, S., et al. 1995, *ApJ*, **440**, L41

Perlmutter, S., et al. 1996, to appear in *Thermonuclear Supernovae*, (NATO ASI), eds. R. Canal, P. Ruiz-LaPuente & J. Isern

Perlmutter, S., et al. 1997, *ApJ*, **483**, 565

Phillips, M. M. 1994, *ApJ*, **413**, L105

Pierce, M. J. & Tully, R. B. 1988, *ApJ*, **330**, 579

Pierce, M. J. & Tully, R. B. 1992, *ApJ*, **387**, 47

Postman, M. & Lauer, T. R. 1995, *ApJ*, **440**, 28

Rephaeli, Y. 1995, *ARA&A*, **33**, 541

Riess, A. G., Press, W. H. & Kirshner, R. P. 1995a, *ApJ*, **438**, L17

Riess, A. G., Press, W. H. & Kirshner, R. P. 1995b, *ApJ*, **445**, L91

Rix, H.-W., Guhathakurta, P., Colless, M. & Ing, K. 1996, *MNRAS*, **285**, 779

Saglia, R. P., et al. 1993, *MNRAS*, **264**, 971

Saha, A., et al. 1995, *ApJ*, **438**, 8

Sandage, A. 1972, *ApJ*, **178**, 1

Sandage, A. & Hardy, E. 1973, *ApJ*, **183**, 743

Sandage, A. 1994, *ApJ*, **430**, 13

Sandage, A., Saha, A., Tammann, G. A., Labhardt, L., Panagia, N. & Macchetto, F. D. 1996, *ApJ*, **460**, L15

Schechter, P. L. 1980, *AJ*, **85**, 801

Schlegel, D. 1996, Ph.D. Thesis, University of California, Berkeley

Schmidt, B.P., et al. 1994, *ApJ*, **432**, 42

Shaya, E. J., Peebles, P. J. E. & Tully, R. B. 1995, *ApJ*, **454**, 15

Strauss, M. A. & Willick, J. A. 1995, *Physics Reports*, **261**, 271

Strauss, M. A., Cen, R., Ostriker, J. P., Lauer, T. R. & Postman, M. 1995, *ApJ*, **465**, 534

Tammann, G. A. 1992, *Phys. Scr.*, T43, 31

Tammann, G. A. & Sandage, A. 1995, *ApJ*, **452**, 16

Tonry, J. L. 1991, *ApJ*, 373, L1

Tonry, J. L. 1995, Lecture delivered at the Heron Island Workshop on Peculiar Velocities in the Universe (http://qso.lanl.gov/heron)

Tonry, J. & Davis, M. 1981, *ApJ*, **246**, 680

Tonry, J. L. & Schneider, D. P. 1988, *AJ*, **96**, 807

Tonry, J. L., Ajhar, E. A. & Luppino, G. A. 1989, *ApJ*, **346**, L57

Tonry, J. L., Ajhar, E. A. & Luppino, G. A. 1990, *AJ*, **100**, 1416

Tonry, J. L. & Schechter, P. L. 1990, *AJ*, **100**, 1794

Tonry, J. L., Blakeslee, J. P., Ajhar, E. A. & Dressler, A. 1997, *ApJ*, **475**, 399

Tormen, B. & Burstein, D. 1995, *ApJS*, **96**, 123

Tully, R. B. & Fisher, J. R. 1977, *A&A*, **54**, 661

Tully, R. B. 1988, *Nature*, 334, 209

Vogt, N. P., et al. 1996, *ApJ*, **465**, L15

Watkins, R. & Feldman, H. A. 1995, *ApJ*, **453**, L73

Wegner, G., et al. 1996, *ApJS*, **106**, 1

Willick, J. A. 1990, *ApJ*, **351**, L5

Willick, J. A. 1991, Ph.D. Thesis, University of California, Berkeley

Willick, J. A. 1994, *ApJS*, **92**, 1

Willick, J. A., Courteau, S., Faber, S. M., Burstein, D. & Dekel, A. 1995, *ApJ*, **446**, 12

Willick, J. A., Courteau, S., Faber, S. M., Burstein, D., Dekel, A. & Kolatt, T. 1996, *ApJ*, **457**, 460

Willick, J. A., Courteau, S., Faber, S. M., Burstein, D., Dekel, A. & Strauss, M. A. 1997, *ApJS*, **109**, 333

Worthey, G. 1994, *ApJS*, **95**, 107

7 Large-scale flows and cosmological implications

Avishai Dekel

Abstract

The analysis of large-scale peculiar velocities and their cosmological implications are reviewed. Reconstruction methods of the underlying 3-D velocity and mass-density fields from the observed peculiar velocities are discussed. They include POTENT and its offsprings, inverse TF methods, Wiener methods, simultaneous analyses of velocity data and redshift surveys, and methods for solving the mixed-boundary-conditions problem back in time. The resultant dynamical fields in our cosmological neighborhood are displayed in maps and analyzed via statistics such as bulk flow and mass power spectrum. Implications of the velocity data are discussed, alone or combined with redshift surveys and CMB data. The focus is on the value of the universal density parameter Ω, the parameter β and the galaxy "biasing" scheme, the initial probability distribution of fluctuations, their power spectrum index n, the nature of the dark matter, and the growth of fluctuations via gravitational instability.

7.1 Introduction

This chapter addresses central issues in the rapidly developing field of large-scale dynamics, which has become a mature scientific field where observation and theory are confronted in a quantitative way. The focus of the present discussion is on the analysis of peculiar velocities and their theoretical implications, either by themselves or in combination with other data. Closely related aspects of this field are discussed in other chapters of this book: distance measurements by Willick (Ch. 6) and redshift surveys and their analysis by Strauss (Ch. 5). The field of cosmic flows has been reviewed earlier by Dekel (1994) and by Strauss & Willick (1995). A relevant discussion of approximations to gravitational clustering is provided by Yahil (Ch. 2) and in Sahni & Coles (1996). The cosmic microwave background (CMB, Silk §3.5) and gravitational lensing (Narayan & Bartelmann Ch. 10) provide complementary dynamical data of relevance. Additional background material can be found in the books by Kolb & Turner (1990), Peebles (1993), Padmanabhan (1993), and Coles & Lucchin (1995).

The standard theory of the formation of large-scale structure (LSS) involves several ingredients, consisting of working hypotheses that one should try to falsify by the observations. When an hypothesis is found tentatively consistent with the observations, the goal is commonly to determine the associated model characteristics and parameters. The main theoretical ingredients can be classified as follows:

- *Cosmology.* The background cosmology is the standard homogeneous Friedman–Robertson–Walker (FRW) model, possibly following an Inflation phase, where the CMB defines a cosmological "rest frame". If so, then one wishes to determine the cosmological parameters such as the density parameter Ω (or Ω_m), the cosmological constant Λ, and the Hubble constant H (§§1.3, 1.4; Ch. 11).
- *Fluctuations.* The structure has originated from a random field of small-amplitude initial density fluctuations. If so, the goal is to characterize their statistical properties, e.g., to find out whether they were Gaussian and whether the power spectrum ($P(k)$) was scale-invariant ($n = 1$). Also of interest is whether the fluctuations had a tensor component and whether the energy density was perturbed adiabatically or in an isocurvature manner (§7.4).
- *Dark matter.* The spectrum of fluctuations was filtered during the radiation–plasma era in a way characteristic of the nature of the dark matter (DM) which dominates the mass density. The DM could be baryonic or nonbaryonic. If nonbaryonic, it could be "hot" or "cold" depending on when it became nonrelativistic (§1.5).
- *Gravitational engine.* The fluctuations grew by gravitational instability (GI) into the present LSS (§§2, 11.1). This is a sufficient (but not necessary) condition for:
  - *Potential flow.* The mildly-nonlinear-velocity field smoothed over a sufficiently large scale is irrotational.
  - *Velocity field.* The galaxies trace a unique underlying velocity field, apart from possible small "velocity bias" on small scales.
- *Galaxy formation – biasing.* The density fluctuations of visible galaxies are correlated with the underlying mass fluctuations on certain large scales. If this relation can be approximated by a linear function, then the linearized continuity equation in GI implies a linear relation between velocity divergence and galaxy density contrast. The density biasing factor b is then a characteristic parameter of interest (§7.6).

Another working hypothesis specific to the study of peculiar motions is that the distance indicators such as Tully–Fisher (TF) and Fundamental Plane (FP) (Ch. 6) indeed measure *distances*, which then allow the reconstruction of a large-scale velocity field, with random and systematic errors that are well understood.

252 A. Dekel

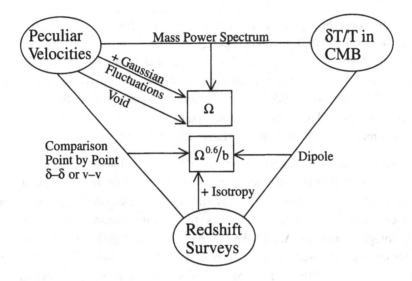

Figure 7.1. Methods for measuring Ω and β from large-scale flows.

The accumulating observations which confront the theories can be classified into the following four categories:

- *Peculiar velocities.* Deviations of galaxies from Hubble flow along the lines of sight (Ch. 6).
- *CMB fluctuations.* Anisotropies in temperature on large angular scales and on sub-degree scales (§3.5).
- *Gravitational lensing.* The lenses are stars, galaxies, and in particular clusters and large-scale fluctuations (§10.4).
- *Galaxy density.* The distribution of luminous objects on the sky and in redshift space (Ch. 5).

Note that the first three types of data are directly related to the first four theoretical ingredients, which are dynamical. The galaxy-density data are the easiest to acquire observationally in large quantities, but their theoretical interpretation is indirect. It involves the uncertain nature of the galaxy-density biasing relation, which requires further hypotheses regarding the process of galaxy formation.

The data from lensing is promising, but their application to LSS is still premature. We therefore focus here on the other three classes of data.

A major goal of the analysis of cosmic flows is measuring the cosmological parameters. As illustrated in Fig. 7.1, the data can be combined to constrain Ω in several different ways. Methods that are based on the peculiar-velocity data alone (§7.5) are independent of galaxy-density biasing. They refer to the present-day LSS which is insensitive to Λ. They can thus serve to measure Ω directly, but they involve relatively large errors. These methods largely rely on the assumption (supported by observations, §7.4.5) that the initial fluctuations were drawn from a *Gaussian* random field.

All the methods that use the spatial distribution of galaxies must depend on the biasing relation between the densities of galaxies and mass. In the linear approximation to GI, which is roughly valid for the fields when they are smoothed on very large scales, what is actually being measured by these methods is the degenerate parameter $\beta \equiv \Omega^{0.6}/b$ rather than Ω itself. The methods include measurements of redshift-space distortions from redshift surveys under the assumption of global isotropy (§5.5.3, §7.6.6), comparisons of the galaxy distribution with the CMB dipole (§7.6.5), and comparisons of the observed peculiar velocities with the galaxy distribution (or the predicted velocities) deduced from redshift surveys (§7.6). The best estimates of β for IRAS galaxies lie in the range $0.5 \leq \beta \leq 1$. As argued below, the contamination by biasing introduces a significant uncertainty in the translation of the various β estimates to a reliable measurement of Ω.

Another way of determining Ω is by comparing the mass power spectrum from observed peculiar velocities with the spectrum of fluctuations in the CMB (§7.4.3, §7.4.4). This comparison involves a more detailed model for the formation of structure, and the current studies are limited to the CDM family of models, allowing for a non-zero cosmological constant, a possible tilt in the spectrum with or without tensor fluctuations, and a mixture of dark-matter species.

The outline of this chapter is as follows: §7.2 describes in some detail reconstruction methods from peculiar velocities; §7.3 brings tests of the basic hypotheses; §7.4 discusses the statistics of mass-density fluctuations, including bulk flow, power spectrum and probability distribution function (PDF), as determined by the velocity data alone or combined with CMB data; §7.5 focuses on direct estimates of Ω from peculiar velocities alone; §7.6 presents estimates of β by comparing velocity and galaxy density data, and addresses the issue of biasing; §7.7 reviews other measures of the cosmological parameters; and §7.8 concludes by addressing again the theoretical ingredients and working hypotheses in light of the observed cosmic flows.

7.2 Reconstruction from peculiar velocities

7.2.1 Data for velocity analysis

The key to velocity analysis is provided by the distance indicators, which are reviewed in detail by Willick (Ch. 6). Only a few important points are stressed here.

So far, the most useful distance indicators for LSS have been of the TF kind, based on intrinsic relations between a distance-dependent quantity such as the flux $\propto L/r^2$ (L is the intrinsic luminosity), and a distance-independent quantity such as the maximum rotation velocity of spirals, v (originally Tully & Fisher 1977). The intrinsic relations are power laws, $L \propto v^\beta$, i.e., $M(\eta) = a - b\eta$ (where $M \equiv -2.5\log L + $ const. is the absolute magnitude and $\eta \equiv \log v - $ const.). The TF slope b can be calibrated in clusters, where all galaxies lie at roughly the same distance. For any other galaxy with observed η and apparent magnitude $m \equiv -2.5\log(L/r^2) + \text{const.}$, one can determine a *relative* distance via $5\log r = m - M(\eta)$. There exists a fundamental freedom in determining the *zero point*, a, which fixes the distances at absolute values (in km s^{-1}, not to be confused with H_0, which is needed for expressing distances in Mpc). Changing a, i.e., multiplying the distances by a factor $(1 + \epsilon)$ while the redshifts are fixed, is equivalent to adding a monopole Hubble-like component, $-\epsilon r$, to the peculiar velocity v, and an offset 3ϵ to the density fluctuation δ (equation [7.7]). The zero point is typically determined by assuming that a set of clusters provides a fair sample of the Hubble flow, or by minimizing the variance of the recovered peculiar-velocity field at grid points inside a large "fair" volume. The current typical intrinsic TF scatter in CCD R and I bands is at best $\sigma_{\text{mag}} \sim 0.33$ mag, corresponding to a relative distance error of $\Delta = (\ln 10/5)\sigma_{\text{mag}} \approx 0.15$.

The physical origins of the scaling relations (which apparently range from globular star clusters to clusters of galaxies) are understood only in vague terms, based on general considerations such as the virial theorem. The details, and in particular the tightness of the relations, are not fully understood, reflecting our limited understanding of galaxy formation. However, the only thing that matters for the purpose of distance measurements is the existence of a mean empirical relation and the measured scatter about it.

The most comprehensive catalog of peculiar-velocity data available today is the Mark III catalog (§6.8; Willick et al. 1995; 1996; 1997a), which is a careful compilation of several datasets under the assumption that all galaxies trace the same underlying velocity field. The merger was non-trivial because the observers differ in their selection procedure, the quantities they measure, the method of measurement and the TF calibration techniques. The careful calibration and merger procedure is crucial for reliable results – in several cases it produced TF distances substantially different from those quoted by the original authors.

The original Mark II catalog, which was used in the first application of POTENT (Dekel, Bertschinger & Faber 1990; Bertschinger et al. 1990), consisted of about 1000 galaxies (mostly Lynden-Bell et al. 1988 and Aaronson et al. 1982a). The extended Mark III catalog consists of ~ 3400 galaxies (dominated by Mathewson, Ford & Buchhorn 1992). This sample enables a reasonable recovery of the dynamical fields with $12 h^{-1}$ Mpc smoothing in a sphere of radius $\sim 60 h^{-1}$ Mpc about the Local Group (LG), extending to $\sim 80 h^{-1}$ Mpc in certain regions (§7.2.3). Part of the data are shown in Fig. 7.2.

More uniformly sampled data of about 1000 spiral galaxies in the Northern Hemisphere (SFI, Giovanelli et al. 1997a; also clusters, SCI, Giovanelli et al. 1997b), combined with the Mathewson, Ford & Buchhorn (1992) data in the Southern Hemisphere, are in preparation for publication, and were already subject to certain analyses (SFI, da Costa et al. 1996). The SFI sample is now in the process of being analyzed using the methods described below. The results will be compared with those obtained already from the Mark III catalog, and eventually all the data will be analyzed together.

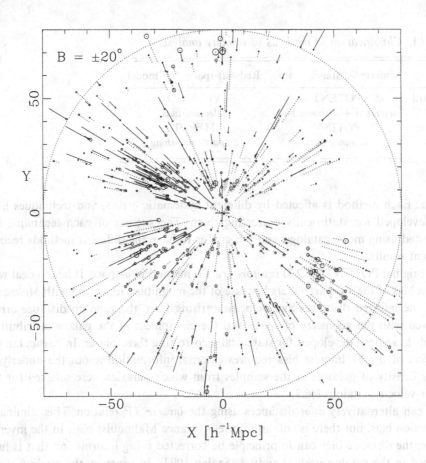

Figure 7.2. Mark III data. Inferred radial peculiar velocities of grouped galaxies from the Mark III catalog in a $\pm 20°$ slice about the Supergalactic Plane. Distances and velocities are in $1000\,\mathrm{km\,s^{-1}}$. The areas of circles at the object positions are proportional to the number of galaxies grouped. Solid and dashed lines distinguish between outgoing and incoming velocities. Converging flows are indicated at the Great Attractor (left) and at Perseus–Pisces (bottom-right) (Willick et al. 1997a).

7.2.2 Methods of velocity analysis

One way of classifying the methods of velocity analysis is summarized in Table 7.1 (see also §6.9).

The "forward" and "inverse" methods refer to whether the TF relation is interpreted as $M(\eta)$ or $\eta(M)$. The difference is crucial because the apparent magnitude depends on distance while η does not, and because the selection depends on magnitude and is independent of η. On the other hand, the velocity field can be computed either from peculiar velocities that were evaluated at each galaxy's TF-inferred position, d, or by fitting a parametric model for the potential (and thus the velocity) field in redshift

Table 7.1. *Classification of methods of velocity analysis*

	Inferred-distance space	Redshift-space + V model
Forward	POTENT	VELMOD
TF	selection + distance bias	selection bias
Inverse	POTINV	MFPOT
TF	distance bias	z-space smoothing

space, z. Each method is affected by different systematic errors, and techniques have been developed for statistically correcting them. The success of each technique has been tested using mock catalogs, and the goal is to have the different methods recover consistent results.

The original POTENT (§7.2.3) is a *forward* TF method in d-space. It has to deal with *selection* bias of the TF parameters because of the magnitude limit, and with *Malmquist* bias in the inferred distances and velocities arising from the random distance errors convolved with the geometry of space and the clumpiness of the galaxy distribution. Methods have been developed for statistically correcting these biases. In particular, the correction of the Malmquist bias requires external information about the underlying number density of galaxies in the samples from which galaxies were selected for the peculiar-velocity catalogs (§6.9).

One can alternatively infer distances using the *inverse* TF relation. This eliminates the selection bias, but there is still an inferred-distance Malmquist bias. In the inverse-TF case, the distance bias can in principle be corrected using information that is fully contained in the catalog itself (Landy & Szalay 1992). In practice, the quality of the correction is limited by the sparseness of the sampling. The POTENT analysis of the inverse data corrected this way is termed POTINV (Eldar, Dekel & Willick 1999).

If the selection does not explicitly depend on η, Malmquist bias can be eliminated by minimizing η residuals in redshift space without ever inferring actual distances to individual galaxies. The distance is replaced by $r = z - u_\alpha(z)$, where u_α is a parametric model for the radial peculiar-velocity field (§7.2.5). If the forward TF relation is used, as in VELMOD (Willick et al. 1997b), the method still has to correct for selection bias. The use of the *inverse* TF relation (Schechter 1980) guarantees in this case that both the selection bias and the distance bias are eliminated, at the expense of over-smoothing due to the representation of the fields in redshift space. Recent implementations of such methods were termed MFPOT (§7.2.5; Blumenthal, Dekel & Yahil 1999) and ITF (§7.6.4; Davis, Nusser & Willick 1996).

Another way of distinguishing between the methods is by their goals. The methods working in d-space can serve for reconstruction of 3-D maps of the velocity and mass-density fields, unbiased and uniformly smoothed with equal-volume weighting throughout the volume. These fields can then be straightforwardly compared to other data and to theory in order to obtain cosmological implications (e.g., §7.6.1). Alternatively, one may direct the method to estimating certain parameters of the model (e.g., β) without ever reconstructing uniform maps. The redshift-space methods serve this purpose well (§7.6.3, §7.6.4).

Yet another characteristic of some of the methods is the usage of a whole-sky redshift survey (such as IRAS 1.2 Jy) as an intrinsic part of the reconstruction from peculiar velocities. This is the case in the SIMPOT, VELMOD and ITF methods (§7.6). These methods are geared towards determining β, with SIMPOT also providing uniform reconstruction maps.

Finally, one can focus on optimal formal treatment of the random errors, which are in fact the main obstacle. A method based on Wiener filtering has been developed for recovering the most probable mean field from the noisy peculiar-velocity data in d-space (§7.2.4). This can serve as a basis for constrained realizations of uniform smoothing, each of which being an equally good guess for the structure in our real cosmological neighborhood.

7.2.3 POTENT analysis

7.2.3.1 Correcting Malmquist bias

The calibration of the forward TF relation is affected by a *selection* bias. A magnitude limit in the selection of the sample used for calibration at a fixed *true* distance (e.g., in a cluster) tilts the forward TF regression line of M on η towards bright M at small η values. This bias is inevitable when the dependent quantity is explicitly involved in the selection process, and it occurs to a certain extent even in the inverse TF relation $\eta(M)$ due to existing dependences of the selection on η. Fortunately, the selection bias can be corrected once the selection function is known (§6.9; Willick 1994).

The TF inferred distance, d, and the mean peculiar-velocity at a given d, suffer from a *Malmquist* or *inferred-distance* bias. The distances, either forward or inverse, are corrected for Malmquist bias in a statistical way before being fed as input to POTENT-like procedures.

If M is distributed normally for a given η, with standard deviation σ_{mag}, then the forward inferred distance d of a galaxy at a true distance r is distributed log-normally about r, with relative error $\Delta \approx 0.46\sigma_{mag}$. Given d, the expectation value of r is

$$E(r|d) = \frac{\int_0^\infty rP(r|d)\,dr}{\int_0^\infty P(r|d)\,dr} = \frac{\int_0^\infty r^3 n(r)\exp\left(-\frac{[\ln(r/d)]^2}{2\Delta^2}\right)dr}{\int_0^\infty r^2 n(r)\exp\left(-\frac{[\ln(r/d)]^2}{2\Delta^2}\right)dr}, \qquad (7.1)$$

where $n(r)$ is the number density in the underlying distribution from which galaxies were selected (by quantities that do not explicitly depend on r). The deviation of $E(r|d)$ from d reflects the bias. The homogeneous part (HM) arises from the geometry of space – the inferred distance d underestimates r because it is more likely to have been scattered by errors from $r > d$ than from $r < d$, the volume being $\propto r^2$. If $n=$const., equation (7.1) reduces to $E(r|d) = de^{3.5\Delta^2}$, in which the distances should simply be multiplied by a factor, 8% for $\Delta = 0.15$, equivalent to changing the zero point of the TF relation.

Fluctuations in $n(r)$ are responsible for the inhomogeneous bias (IM), which systematically enhances the inferred density perturbations and thus, for example, the value of Ω inferred from them. If $n(r)$ is varying slowly with r, and if $\Delta \ll 1$, then equation (7.1) reduces to $E(r|d) = d[1 + 3.5\Delta^2 + \Delta^2(d\ln n/d\ln r)_{r=d}]$, showing the dependence on Δ and on the gradients of $n(r)$. To illustrate, consider a lump of galaxies at one point r with $u=0$. Their inferred distances are randomly scattered to the foreground and background of r. With all galaxies having the same z, the inferred u on either side of r mimic a spurious infall towards r, which is interpreted dynamically as a spurious overdensity at r.

In one version of the Mark III data for POTENT analysis, the forward IM bias is corrected in two steps. First, the galaxies are grouped in z-space (Willick et al. 1995), reducing the distance error of each group of N members to $\Delta/\sqrt{N}$ and thus significantly weakening the bias. With or without grouping, the noisy inferred distance of each object, d, is replaced by $E(r|d)$, with an assumed $n(r)$ properly corrected for grouping if necessary. This procedure has been tested using realistic mock data from N-body simulations, showing that IM bias can be reduced to the level of a few percent. The practical uncertainty is in $n(r)$, which can be approximated for example by the high-resolution density field of IRAS or optical galaxies, or by the recovered mass-density itself in an iterative procedure under certain assumptions about how galaxies trace mass. The resultant correction to the density recovered by POTENT is $< 20\%$ even at the highest peaks.

When the selection of the sample is complicated (e.g., involving different diameter limits in different redshift shells, as in the SFI catalog), the analytic correction may be invalid. In this case one can apply instead an empirical correction based entirely on mock catalogs (e.g., Freudling et al. 1995).

Distances are alternatively inferred via the *inverse* TF relation between internal velocity parameter η and magnitude m, $\eta = \eta^0(m - 5\log d)$. Under the assumption that the selection was independent of η and was not an explicit function of distance, the expectation value of the true distance r, given d, is

$$E(r|d) = d\, e^{3\Delta^2/2} f(de^{\Delta^2})/f(d), \qquad (7.2)$$

where $\Delta \equiv (\ln 10/5)\sigma_\eta/\eta^0$. In this case, the required density function, $f(d)$, is in d-space, and is derivable from the sample itself (Landy & Szalay 1992). Eldar, Dekel & Willick (1998) have applied this correction to the inverse distances in the Mark III catalog, to serve as input for a POTENT analysis (POTINV). The agreement between the forward POTENT and POTINV results are well within the level of the random errors.

7.2.3.2 Smoothing the radial velocities

The goal of the POTENT analysis is to recover from the collection of Malmquist-corrected, radial peculiar velocities u_i at inferred positions d_i the underlying 3-D velocity field $v(x)$ and the associated mass-density fluctuation field $\delta(x)$, smoothed with a Gaussian of radius R_s (we denote hereafter a 3-D Gaussian window of radius

$12\,h^{-1}$ Mpc by G12, etc.). The first, most difficult step is the smoothing, or interpolation, into a radial velocity field with minimum bias, $u(x)$. The desire is to reproduce the $u(x)$ that would have been obtained had the true $v(x)$ been sampled densely and uniformly and smoothed with a spherical Gaussian window of radius R_s. With the data as available, $u(x_c)$ is taken to be the value at $x = x_c$ of an appropriate *local* velocity model $v(\alpha_k, x - x_c)$. The model parameters α_k are obtained by minimizing the weighted sum of residuals,

$$S = \sum_i W_i [u_i - \hat{x}_i \cdot v(\alpha_k, x_i)]^2, \tag{7.3}$$

within an appropriate local window $W_i = W(x_i, x_c)$. The window is a Gaussian, modified such that it minimizes the combined effect of the following three types of errors.

1. *Tensor window bias*. Unless $R_s \ll r$, the u_is cannot be averaged as scalars because the directions $\hat{x}_i$ differ from $\hat{x}_c$, so $u(x_c)$ requires a fit of a local 3-D model as in equation (7.3). The original POTENT used the simplest local model, $v(x) = B$ of 3 parameters, for which the solution can be expressed explicitly in terms of a tensor window function (Dekel, Bertschinger & Faber 1990). However, a bias occurs because the tensorial correction to the spherical window has conical symmetry, weighting more heavily objects of large $\hat{x}_i \cdot \hat{x}_c$, and because variations of v within the window may be interpreted wrongly as a bulk velocity. For example, the bias in the case of a true infall transverse to the line-of-sight is an artificial flow towards the observer, e.g., of $\sim 300\,\mathrm{km\,s^{-1}}$ at the Great Attractor (GA). This bias, if neglected, has the effect of artificially reducing the density contrast between the GA and the LG by a factor of two or more, and of artificially expanding the void between the LG and Perseus–Pisces (PP) (e.g., as seen in da Costa *et al.* 1996). A way to reduce this bias is by generalizing the zeroth-order B into a 9-parameter first-order velocity model, $v(x) = B + \bar{\bar{L}} \cdot (x - x_c)$, with $\bar{\bar{L}}$ a symmetric tensor that automatically ensures local irrotationality. The linear terms tend to "absorb" most of the bias, leaving $v(x_c) = B$ less biased. Unfortunately, a high-order model tends to pick undesired small-scale noise. The optimal compromise for the Mark III data was found to be a 9-parameter model fit out to $r = 40\,h^{-1}$ Mpc, smoothly changing to a 3-parameter fit beyond $60\,h^{-1}$ Mpc (Dekel *et al.* 1999).

2. *Sampling-gradient bias* (SG). If the true velocity field is varying within the effective window, the non-uniform sampling introduces a bias because the smoothing is galaxy-weighted whereas the aim is equal-volume weighting. The simplest way to correct this bias is by weighting each object with the local volume that it "occupies", or the inverse of the local density. A crude estimate of this volume is $V_i \propto R_n^3$, where R_n is the distance to the nth neighboring object (e.g., $n = 4$). This procedure is found via simulations to reduce the SG bias in Mark III to negligible levels typically out to $60\,h^{-1}$ Mpc as long as one keeps out of the Galactic Zone of Avoidance. The $R_n(x)$ field can serve later as a flag for poorly sampled regions, to be excluded from any quantitative analysis.

3. *Reducing random errors*. The ideal weighting for reducing the effect of Gaussian noise has weights $W_i \propto \sigma_i^{-2}$, where σ_i are the distance errors. Unfortunately, this weighting spoils the carefully designed equal-volume weighting, biasing u towards its

values at smaller r_i and at nearby clusters where the errors are small. A successful compromise is to weight by both, i.e.

$$W(x_i, x_c) \propto V_i \sigma_i^{-2} \exp[-(x_i - x_c)^2/2R_s^2] \,. \tag{7.4}$$

The success of this smoothing procedure has been evaluated using carefully designed mock catalogs "observed" from simulations (Kolatt et al. 1996). Note that POTENT could alternatively vary R_s to keep the random errors at a constant level, but only at the expense of producing fields that are not directly comparable to theoretical models of uniform smoothing. An additional way to reduce noise is by eliminating "bad" galaxies with large residuals $|u_i - u(x_i)|$, where $u(x_i)$ is obtained in a preliminary smoothing stage.

The resultant errors in the recovered fields are assessed by Monte-Carlo simulations. The poor-man way to generate random realizations of noisy data is by perturbing the input distances with a Gaussian of standard deviation σ_i (and then dealing with the resulting numerical Malmquist bias, Dekel et al. 1993). The better way to generate noisy data is via full, realistic Monte-Carlo mock catalogs, where the noise is added as scatter in the TF quantities (Kolatt et al. 1996). The error in δ at a grid point is estimated by the standard deviation of the recovered δ over the Monte-Carlo simulations, σ_δ (and similarly σ_v). In the well-sampled regions, which extend in Mark III out to 40–60 h^{-1} Mpc, the errors are $\sigma_\delta \approx 0.1$–$0.3$, but they may blow up in certain regions at large distances. To exclude noisy regions, any quantitative analysis could be limited to points where σ_v and σ_δ are within certain bounds.

7.2.3.3 Potential analysis

If the LSS evolved according to GI, then the large-scale velocity field is expected to be *irrotational*, $\nabla \times v = 0$. Any vorticity mode would have decayed during the linear regime as the universe expanded and, based on Kelvin's circulation theorem, the flow remains vorticity-free in the mildly-nonlinear regime as long as it is laminar. Bertschinger & Dekel (1989) have demonstrated that irrotationality is valid to a good approximation when a nonlinear velocity field is properly smoothed over. Irrotationality implies that the velocity field can be derived from a scalar potential, $v(x) = -\nabla \Phi(x)$, so the radial velocity field $u(x)$ should contain enough information for a full 3-D reconstruction. In the POTENT procedure, the potential is computed by integration along radial rays from the observer,

$$\Phi(x) = -\int_0^r u(r', \theta, \phi) dr' \,. \tag{7.5}$$

The two missing transverse velocity components are then recovered by differentiation.

A few comments on possible variants of POTENT. First, although the integration in equation (7.5) is naturally done along radial paths because the data are radial velocities, note that the smoothing procedure itself determines a local 3-D velocity using the finite opening angle of the window ($\sim R_s/r$). The transverse components typically carry larger uncertainty, but the non-uniform sampling may actually make the minimum-error path be non-radial, especially in nearby regions where $R_s/r \sim 1$. For example, it might be better to compute the potential at the far side of a void by integration along its populated periphery rather than through its empty center. The optimal path can be determined by a *max-flow* algorithm (Simmons, Newsam & Hendry 1995). It turns out in practice that only little can be gained by allowing single non-radial paths because large empty regions usually occur at large distances where the transverse components are very noisy, but averaging over many paths may be useful.

Second, the use of the opening angle to determine transverse velocities can be carried one step further by fitting the data in a wider local window with a model of more parameters, e.g., a power series generalizing the linear model, $v_i(x) = B_i + L_{ij}\tilde{x}_j + Q_{ijk}\tilde{x}_j\tilde{x}_k + ...$, where $\tilde{x} = x - x_c$. If the matrices are all symmetric then the velocity model is automatically irrotational making the computation of the potential unnecessary. The density can be approximated by $\delta_c(x) = \|I - L_{ij}\| - 1$ (equation [7.9] below), so that numerical differentiation is avoided. This fit must be local because the model tends to blow up at large distances. The expansion can be truncated at any order, limited by the tendency of the high-order terms to pick up small-scale noise. The smoothing here is not a separate preceding step and the SG bias is reduced, but the effective smoothing is not straightforwardly related to theoretical models of uniform smoothing.

Finally, another method of potential interest without a preliminary smoothing step is based on wavelet analysis, which enables a natural isolation of the structure on different scales (Rauzy et al. 1993). The effective smoothing involves no loss of information and no specific scale or shape for the wavelet, and the analysis is global and done in one step. How successful this method could be still remains to be seen.

POTENT and its variants apply potential analysis. It is important to note that the existence of *irrotationality* is impossible to prove solely from observations of radial velocities. Irrotationality is first of all assumed based on theoretical considerations. One way to test this assumption is against another hypothesis, that of global isotropy, by evaluating the degree of isotropy of the 3-D velocity field that has been derived by potential analysis inside a volume that is big enough to be a fair sample. The POTENT velocity field from Mark III data is indeed found to be isotropic within the random errors.

Another consistency check of irrotationality is via the degree of similarity of the density field by POTENT and the galaxy-density field, under the assumption that galaxies trace mass (§7.3.2). An attempt to apply POTENT to the radial components of a rotational flow could have resulted in a very wrong density field that would not have resembled the galaxy-density field at all. For example, the POTENT output of a purely rotating density enhancement is an artificial density peak on one side of the true center of the object and a mirror density trough on the other side.

7.2.3.4 *From velocity to density*

The final step of the POTENT procedure is the derivation of the mass-density fluctuation field associated with the peculiar velocity field. This requires a solution to the equations of gravitational instability in the mildly-nonlinear regime with mixed boundary conditions (e.g., §2.3). Useful approximations are discussed below.

Let x, v, and Φ_g be the position, peculiar-velocity and peculiar-gravitational potential in comoving distance units, corresponding to ax, av, and $a^2\Phi_g$ in physical units, with $a(t)$ the universal expansion factor. Let $\delta \equiv (\rho - \bar{\rho})/\rho$ be the mass-density fluctuation. The equations governing the evolution of fluctuations of a pressureless gravitating fluid in a standard cosmological background during the matter era are the *Continuity* equation, the *Euler* equation of motion, and the *Poisson* field equation:

$$\dot{\delta} + \nabla \cdot v + \nabla \cdot (v\delta) = 0 ,$$
$$\dot{v} + 2Hv + (v \cdot \nabla)v = -\nabla \Phi_g ,$$
$$\nabla^2 \Phi_g = (3/2)H^2\Omega\delta , \qquad (7.6)$$

with H and Ω varying in time. (The dynamics do not depend on the value of the Hubble constant H; it is set to unity by measuring distances in km s^{-1}, $1 h^{-1}$ Mpc $= 100$ km s^{-1}).

In the *linear* approximation, the GI equations can be combined into a time evolution equation, $\ddot{\delta} + 2H\dot{\delta} = (3/2)H^2\Omega\delta$. The growing mode of the solution, $D(t)$, is irrotational and can be expressed in terms of $f(\Omega) \equiv H^{-1}\dot{D}/D \approx \Omega^{0.6}$. The linear relation between density and velocity is

$$\delta_1 = -f^{-1}\nabla \cdot v . \qquad (7.7)$$

The use of δ_1 is limited to the small dynamical range between a few tens of megaparsecs and the $\sim 100 h^{-1}$ Mpc extent of the current samples. However, the sampling of galaxies enables reliable dynamical analysis with a smoothing radius as small as $\sim 10 h^{-1}$ Mpc, where $|\nabla \cdot v|$ obtains values larger than unity and therefore nonlinear effects play a role. Even reconstruction with $\sim 5 h^{-1}$ Mpc smoothing may be feasible in well-sampled regions nearby. Unlike the strong nonlinear effects in virialized systems which erase any memory of the initial conditions, mildly-nonlinear effects carry crucial information about the formation of LSS, and should therefore be treated properly. Figure 7.3 shows that δ_1 becomes a severe underestimate at large $|\delta|$. This explains, by the way, why equation (7.7) is invalid in the nonlinear epoch even where $\delta = 0$. the integral requirements that $\int \delta d^3x = 0$ by definition and $\int \nabla \cdot v d^3x = 0$ by isotropy then imply $-\nabla \cdot v > \delta$ at $|\delta| \ll 1$. Fortunately, the small variance of $\nabla \cdot v$ given δ promises that some function of the velocity derivatives may be a good local approximation to δ.

A basis for useful *mildly-nonlinear* relations is provided by the Zel'dovich approximation (Zel'dovich 1970). The displacements of particles from their initial, Lagrangian positions q to their Eulerian positions x at time t are assumed to have a universal time dependence,

$$x(q,t) - q = D(t)\psi(q) = f^{-1}v(q,t) . \qquad (7.8)$$

For the purpose of approximating GI, the Lagrangian Zel'dovich approximation can be interpreted in *Eulerian* space, $q(x) = x - f^{-1}v(x)$, provided that the flow is *laminar* with no orbit mixing. This remains a good approximation when multi-streams are appropriately smoothed over. The solution of the continuity equation then yields (Nusser et al. 1991)

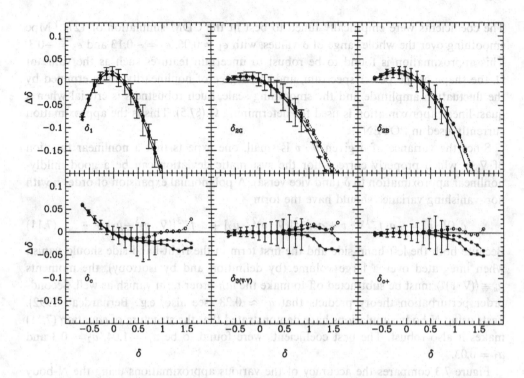

Figure 7.3. Quasi-linear velocity-to-density approximations. $\Delta\delta \equiv \delta_{\text{approx}}(v) - \delta_{\text{true}}$. The mean and standard deviation are from large CDM N-body simulations normalized to $\sigma_8 = 1$, smoothed G12. The three curves correspond to different models: standard CDM (squares, solid line), tilted CDM ($n = 0.6$, stars, dashed line), and open CDM ($\Omega = 0.2$, open circles, dotted line) (Ganon et al. 1999).

$$\delta_c(x) = \|I - f^{-1}\partial v/\partial x\| - 1, \tag{7.9}$$

where the bars denote the Jacobian determinant and I is the unit matrix. The Zel'dovich displacement is first order in f^{-1} and v and therefore the determinant in δ_c includes second- and third-order terms as well, involving sums of double and triple products of partial derivatives which we term $\Delta_2(x)$ and $\Delta_3(x)$ respectively.

The approximations were tested using CDM N-body simulations with a varying degree of tilt in the power spectrum on large scales ($0.6 \le n \le 1$), with varying Ω values in the range $0.1-1$, and with $\sigma_8 = 1$, under Gaussian smoothing radii in the range $(5-20)\,h^{-1}$ Mpc (Mancinelli et al. 1994; Ganon et al. 1999). The simple and physically motivated δ_c was found to have a typical scatter of about 10% of the standard deviation of δ. It is an excellent approximation for $\delta \ge 1$, and is a slight overestimate at the negative tail (Fig. 7.3). The approximation δ_c can be improved by slight adjustments of the coefficients of the nth-order terms,

$$\delta_{c+} = -(1+\epsilon_1)f^{-1}\nabla\cdot v + (1+\epsilon_2)f^{-2}\Delta_2 + (1+\epsilon_3)f^{-3}\Delta_3. \tag{7.10}$$

The coefficients were empirically tuned to best fit the CDM simulation of $12\,h^{-1}$ Mpc smoothing over the whole range of δ values, with $\epsilon_1 = 0.06$, $\epsilon_2 = -0.13$ and $\epsilon_3 = -0.3$. This approximation is found to be robust to uncertain features such as the value of Ω, the shape of the power spectrum, and the degree of nonlinearity as determined by the fluctuation amplitude and the smoothing scale. Such robustness is crucial when a quasi-linear approximation is used for determining Ω (§7.5). This is the approximation currently used in POTENT.

Since the variance of δ given $\nabla \cdot v$ is small, one expects that a nonlinear function of $\nabla \cdot v$ which properly corrects for the systematic deviation can be a good mildly-nonlinear approximation to δ (and vice versa). A polynomial expansion of order n with non-vanishing variance should have the form

$$\delta_n(\nabla \cdot v) = -a_1 f^{-1} \nabla \cdot v + a_2 f^{-2}[(\nabla \cdot v)^2 - \mu_2] + a_3 f^{-3}[(\nabla \cdot v)^3 - \mu_3] + \cdots . \quad (7.11)$$

Because both the left-hand side and the first term in the right-hand side should vanish when integrated over a large volume (by definition and by isotropy), the moments $\mu_n \equiv \langle (\nabla \cdot v)^n \rangle$ must be subtracted off to make the nth-order term vanish as well. Second-order perturbation theory predicts that $a_2 \approx 0.28$ (see also, e.g., Bernardeau 1992). Tests using N-body simulations have demonstrated that the structure of equation (7.11) makes it also robust. The best coefficients were found to be $a_1 = 1.04$, $a_2 = 0.3$ and $a_3 = 0.03$.

Figure 7.3 compares the accuracy of the various approximations using the N-body simulations. δ_c is the best among the physically motivated approximations, which also include two second-order approximations (Bernardeau 1992; Gramman 1993a). The latter do somewhat better at the negative tail, but they provide severe underestimates in the positive tail. δ_{c+} and $\delta_3(\nabla \cdot v)$ are excellent robust fits over the whole mildly-nonlinear regime.

We note in passing that equation (7.9) is not easily invertible to solve for $\nabla \cdot v$ or v when δ is given, e.g., from redshift surveys, but a useful approximation that has been derived from simulations is $\nabla \cdot v = -f\delta/(1 + 0.18\delta)$.

7.2.3.5 Testing with mock catalogs

The way to optimize POTENT and other reconstruction methods is by minimizing the systematic errors when applied to mock catalogs. It is important that these mock catalogs mimic the real data as closely as possible. Such mock catalogs have been produced, for example, to mimic the Mark III and the IRAS 1.2 Jy catalogs (Kolatt et al. 1996). They are publically available and serve as standard "benchmarks" for the competing methods.

The procedure for making these mock catalogs involves two main steps: a dynamical N-body simulation that mimics our actual cosmological neighborhood; and the generation of galaxy catalogs from it.

In step one, the present-day large-scale density field is assumed to be traced, for example, by IRAS galaxies. The real-space density of galaxies, smoothed with a Gaussian of radius $5\,h^{-1}$ Mpc, is derived from the IRAS 1.2 Jy redshift survey (Fisher et al.

1995) via an iterative procedure and a power-preserving filter (PPF) that ensures a fixed variance in space (Sigad et al. 1998; §5.2). This fluctuation field is traced back in time into the linear regime using the Zel'dovich–Bernoulli equation (Nusser & Dekel 1992) and is slightly Gaussianized to correct for imperfections in the PPF procedure. Small-scale fluctuations that obey the IRAS power spectrum (Fisher et al. 1993) is filled in by the method of constrained realizations (Hoffman & Ribak 1991), to provide seeds for galaxy formation and the appropriate small-scale noise. The resultant fluctuation field is fed as initial conditions for a full PM N-body simulation, which is run forward until the present time, marked by $\sigma_8 = 0.7$. The simulation box is of side $256\,h^{-1}$ Mpc centered on the Local Group. Figure 7.4 shows a slice of the simulation mass particles about the Supergalactic Plane.

In step two, galaxies, spirals and ellipticals, are identified assuming a "biasing" scheme which obeys the morphology–density relation (Dressler 1980). Absolute magnitudes and log line widths are assigned by assuming random Gaussian scatter about a given TF relation. Galaxies are selected to mimic the various datasets by following the true selection procedures, and their apparent magnitudes and line widths are "observed" to make the desired Monte-Carlo mock Mark III catalogs for the method testing.

Figure 7.5 demonstrate the quality of the POTENT reconstruction from the Mark III catalog by comparing the recovered density field to the true G12-smoothed field This comparison is done both via maps in the Supergalactic Plane and via a point-by-point comparison at the points of a uniform grid inside a volume of effective radius $40\,h^{-1}$ Mpc. The input to POTENT in the different panels are (a) true smoothed radial velocities $u(r)$ at the points of a spherical grid; (b) true galaxy velocities sampled sparsely and non-uniformly as in the Mark III data; and (c) the average of ten Monte-Carlo mock catalogs of noisy galaxy velocities sampled sparsely and non-uniformly. One can see that the remaining systematic errors are small in most of the sampled volume out to $\sim 60\,h^{-1}$ Mpc. The final systematic error is not correlated with the signal (slope $\sim$ unity in the scatter diagram) and is on the order of $\Delta\delta \sim 0.13$. The random errors are not a major obstacle in certain well-sampled regions (such as the Great Attractor on the left), but they become severe in poorly-sampled regions (such as parts of the Perseus–Pisces region near the Galactic plane, on the right). The errors derived from the noisy mock catalogs are used to eliminate poorly-recovered regions from quantitative analyses.

7.2.3.6 Maps of velocity and density fields

Figure 7.6 shows Supergalactic-Plane maps of the velocity field in the CMB frame and the associated δ_{c+} field (for $\Omega=1$) as recovered by POTENT from the Mark III catalog. The recovery is reliable out to $\sim 60\,h^{-1}$ Mpc in most directions outside the Galactic plane ($Y=0$). Both large-scale ($\sim 100\,h^{-1}$ Mpc) and small-scale ($\sim 10\,h^{-1}$ Mpc) features are important; e.g., the bulk velocity reflects properties of the initial fluctuation power spectrum (§7.4), while the small-scale variations indicate the value of Ω (§7.5).

The velocity map shows a clear tendency for motion from right to left, in the general

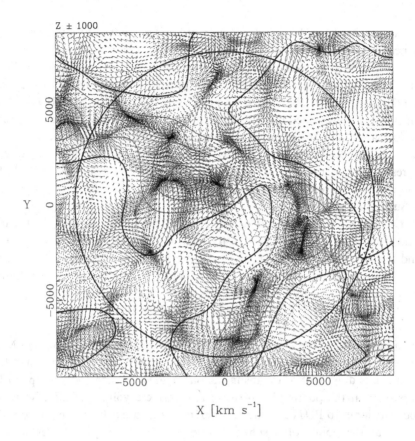

Figure 7.4. Mock universe. The mass distribution in a slice of thickness $\pm 10\,h^{-1}$ Mpc about the Supergalactic Plane from N-body simulation that mimics the structure in the real universe. The initial conditions were inspired by the IRAS 1.2 Jy redshift survey, traced back in time and complemented by constrained small-scale structure. The contours refer to the G12 smoothed density field, with the heavy contour marking $\delta = 0$ and contour spacing $\Delta\delta = 0.2$. Note the "Great Attractor" on the left and "Perseus–Pisces" on the right. This simulation is the basis for detailed mock catalogs that serve as benchmarks for reconstruction methods (Kolatt *et al.* 1996).

direction of the LG motion in the CMB frame ($L, B = 139°, -31°$ in Supergalactic coordinates). The bulk velocity within $60\,h^{-1}$ Mpc is 300–350 km s^{-1} towards ($L, B \approx 166°, -20°$) (§7.4.1) but the flow is not coherent over the whole volume sampled, e.g., there are regions in front of PP (bottom right) and behind the GA (far left) where the XY velocity components vanish, i.e., the streaming relative to the LG is opposite to the bulk flow direction. The velocity field shows local convergences and divergences which indicate strong density variations on scales about twice as large as the smoothing scale.

The bottom panel of Fig. 7.6 shows the POTENT mass-density field in the Supergalactic Plane as a landscape plot. The Great Attractor (with $12\,h^{-1}$ Mpc smoothing

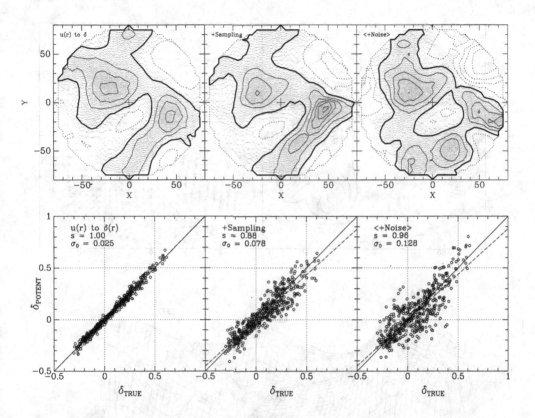

Figure 7.5. Systematic errors in the POTENT analysis. The density field recovered by POTENT from the mock data is compared with the "true" G12 density. The observational problems are introduced gradually from left to right. *Top:* density maps in the Supergalactic Plane (compare with Fig. 7.4). *Bottom:* point-by-point comparison inside the comparison volume (Dekel et al. 1999).

and $\Omega = 1$) is a broad density ramp of maximum height $\delta = 1.4 \pm 0.3$ located near the Galactic plane $Y = 0$ at $X \approx -40\,h^{-1}$ Mpc. The GA extends towards Virgo near $Y \approx 10$ (the "Local Supercluster"), towards Pavo–Indus–Telescopium (PIT) across the Galactic plane to the south ($Y < 0$), and towards the Shapley concentration behind the GA ($Y > 0, X < 0$). The structure at the top is related to the "Great Wall" of Coma, with $\delta \approx 0.6$. The Perseus–Pisces peak which dominates the right-bottom is peaked near Perseus with $\delta = 1.0 \pm 0.4$. PP extends towards the Southern Galactic Hemisphere (Aquarius, Cetus), coinciding with the "Southern Wall" as seen in redshift surveys. Underdense regions separate the GA and PP, extending from bottom-left to top-right. The deepest region in the Supergalactic Plane, with $\delta = -0.8 \pm 0.2$, roughly coincides with the galaxy-void of Sculptor.

One can still find in the literature statements questioning the very existence of the GA (e.g., Rowan-Robinson 1993), which simply reflect ambiguous definitions for this

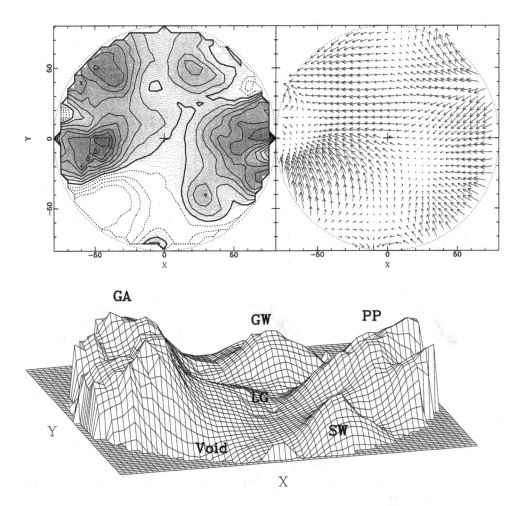

Figure 7.6. POTENT dark-matter maps. The G12 fluctuation fields of peculiar velocity and mass-density in the Supergalactic Plane as recovered by POTENT from the Mark III peculiar velocities. (*Top*) The vectors are projections of the 3-D velocity field in the CMB frame, dominated by large derivatives embedded in a coherent bulk flow. Distances and velocities are in $100\,\mathrm{km\,s^{-1}}$. Contour spacing is 0.2 in δ, with the heavy contour marking $\delta = 0$ and dashed contours $\delta < 0$. The height in the surface plot (*bottom*) is proportional to the total mass-density contrast δ. The LG is at the center, GA on the left, PP and the Southern Wall on the right, Coma Great Wall at the top, and the Sculptor void in between (Dekel *et al.* 1999).

phenomenon. A GA clearly exists in the sense that the dominant feature in the local peculiar-velocity field is a coherent convergence, centered near $X \approx -40$. Whether or not the associated density peak has a counterpart in the galaxy distribution is another question. The GA is ambiguous only in the sense that the good correlation observed between the mass-density inferred from the velocities and the galaxy-density in redshift surveys is perhaps not perfect (§7.3.2, §7.6.7).

Other cosmographic issues of debate are whether there exists a back-flow behind the GA in the CMB frame, and whether PP and the LG are approaching each other. These effects are detected by the current POTENT analysis only at the 1.5σ level in terms of the random uncertainty. Furthermore, the freedom in the zero point of the distance indicators permits adding a Hubble-like peculiar velocity which can balance the GA back-flow and make PP move away from the LG. Thus, these issues remain debatable.

To what extent should one believe the recovery in the Galactic Zone of Avoidance (ZOA) which is empty of tracers? The velocities observed at medium galactic latitudes on the two sides of the ZOA are used as probes of the mass in the ZOA. The interpolation is based on the assumed irrotationality, where the recovered transverse components enable a reconstruction of the mass-density. However, while the SG bias can be corrected where the width of the ZOA is smaller than the smoothing length, the result could be severely biased where the unsampled region is larger. With $12\,h^{-1}$ Mpc smoothing in Mark III the interpolation is suspected of being severely biased in $\sim 50\%$ of the ZOA at $r = 40\,h^{-1}$ Mpc (where $R_4 > R_s$), but the interpolation is pretty safe in the closer, highly populated GA region, for example. Indeed, a deep survey of optical galaxies at small Galactic latitudes and the ROSAT X-ray satellite revealed A3627 as a very rich Abell cluster centered at $(l, b, z) = (325°, -7°, 45\,h^{-1}$ Mpc) (Kraan-Korteweg et al. 1996; Boehringer et al. 1996), near the central peak of the GA as predicted by POTENT at $(320°, 0°, 40\,h^{-1}$ Mpc) (Kolatt, Dekel & Lahav 1995).

The application of a simple version of POTENT to the SFI data lead to an apparent mass distribution in which the gross features roughly agree with the reconstruction from Mark III but the amplitude of the perturbations and other details differ (da Costa et al. 1996). These differences, which are partly due to tensor window bias (§7.2.3.2), go away when the up-to-date version of POTENT is used. In fact, the density fields out of Mark III and SFI are consistent within the errors, with a slight tendency for the SFI density field to be of higher contrast (in preparation).

7.2.4 Wiener reconstruction

The main obstacle in the POTENT reconstruction effort arises from the random distance errors. Still within the family of methods that work in inferred-distance space, the noise can be better dealt with by implementing an alternative analysis based on Wiener Filter (WF, see Press et al. 1995) and Constrained Realizations (CR, Hoffman & Ribak 1991). This analysis is currently limited to the linear regime. The general method is described in Zaroubi et al. (1995). An application to the Mark III data has been attempted by Zaroubi, Hoffman & Dekel (1999).

For optimal extraction of signal from noisy data, consider a set of noisy data $d = \{d_i\}_{i=1}^{M}$ that are related to a true signal $s = \{s_j\}_{j=1}^{N}$ via a linear mapping $d = Rs + \epsilon$, where R is an $M \times N$ response matrix and ϵ is the vector of statistical errors. Assume a prior model for the statistics of the true signal and the errors, e.g., that the signal is a Gaussian random field with some specific power spectrum, and that the errors are also Gaussian. The minimum-variance estimator of the underlying field, i.e., the linear combination of the data that minimizes the variance of the difference between the estimator and all possible realizations of the underlying field, can be proven to be $s_{WF} = F\,d$, where $F \equiv \langle s\,d^\dagger\rangle\langle d\,d^\dagger\rangle^{-1}$ is the Wiener Filter. The assumed power spectrum enters the computation of the covariance matrices that involve the signal. The key feature of the resultant mean field is that it is dominated by the data where the errors are small and by the model where the errors are large.

It is worth knowing that if the assumed prior is Gaussian, the WF estimator is also the most probable field given the data (which is a generalization of the conditional mean given one constraint, Dekel 1981). For a Gaussian field it is also closely related to maximum-entropy reconstruction of noisy pictures (Lahav et al. 1994; Gull & Daniell 1978)

In our case, given a set of observed radial peculiar velocities u_i with errors ϵ_i about an underlying field $u(x_i)$, all in inferred-distance space (forward or inverse) properly corrected for Malmquist bias, the WF density fluctuation field is thus given by

$$\delta_{WF}(x) = \langle \delta(x)u(x_i)^\dagger\rangle \, \langle u(x_i)u(x_j)^\dagger + \epsilon_i\epsilon_j^\dagger\rangle^{-1} u_j \,. \tag{7.12}$$

The covariance matrices $\langle \delta(x)u(x_i)^\dagger\rangle$ and $\langle u(x_i)u(x_j)^\dagger\rangle$ are derived based on the prior power spectrum. The relation between velocity and density assumes linear theory. The noise covariance matrix $\langle \epsilon_i\epsilon_j^\dagger\rangle$ is assumed to be diagonal.

The Wiener reconstruction automatically assigns high weight to good data and suppresses noisy data. It enables reconstruction with higher resolution in good regions, and is therefore ideal for high-resolution study in the local neighborhood. It allows an extrapolation into poorly sampled regions, and dynamical reconstruction of density from observed velocities (and vice versa), provided that the analysis is limited to the linear regime.

A general undesired feature of maximum probability solutions is that they tend to be oversmoothed in regions of poor data, relaxing to $\delta = 0$ in the extreme case of no data. This is unfortunate because the signal of true density is modulated by the density and quality of sampling – an effective sampling bias which replaces the SG bias of POTENT. The effectively varying smoothing length can affect any dynamical use of the reconstructed field (e.g., deriving a gravitational acceleration from a density field), as well as prevent a straightforward comparison with other data or with uniformly smoothed theoretical fields.

A modified filtering method termed PPF can partly cure this problem by forcing the recovered field to have a spatially constant variance at the expense of a slight deviation from the minimal variance solution and amplification of noisy features where the noise is large. In k-space, the PPF is simply the square-root of the WF (Sigad et al. 1998).

A way of generating proper realizations with constant variance is by applying a CR algorithm to the WF mean field. One generates a mock random realization of the signal drawn from the prior model, $\tilde{s}$, and "observes" it at the same places where the

Large-scale flows and cosmological implications

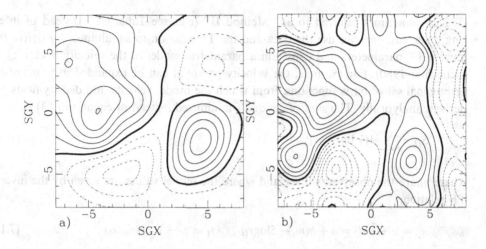

Figure 7.7. Wiener fields. The G12 density and velocity fields as recovered by Wiener Filter from the Mark III velocities. *Left:* the Wiener, most probable, mean field, with oversmoothing in poorly sampled regions. *Right:* One of many random constrained realization of uniform G12 smoothing (Zaroubi, Hoffman & Dekel 1999).

real data is observed to obtain a mock dataset $\tilde{d}$. A constrained realization is then obtained by $s_{CR} = \tilde{s} + F(d - \tilde{d})$, provided that the signal is Gaussian. In our case,

$$\delta_{CR}(x) = \tilde{\delta}(x) + F_j(x)(u_j - \tilde{u}_j) . \tag{7.13}$$

These constrained realizations are uniformly smoothed – any one of them could be a good representation of the real density field. A collection of random constrained realizations is therefore a very useful result of a reconstruction process, which can serve for proper quantitative comparison with theory or other observations.

Earlier attempts to recover a maximum probability mean field are reported in Kaiser & Stebbins (1991), and Stebbins (1994). They modelled the density field via Fourier expansion and recovered the most probable Fourier coefficients subject to the data and an assumed prior. Wiener reconstruction and constrained realizations from the Mark III data has been attempted by Zaroubi, Hoffman & Dekel (1999). Figure 7.7 shows the recovered mean field in the Supergalactic Plane, and one random constrained realization of uniform G12 smoothing.

7.2.5 *Malmquist-free analysis*

As mentioned already, the selection bias can be practically eliminated from the calibration of the *inverse* TF relation, $\eta(M)$, as long as the internal velocity parameter η does not explicitly enter the selection process. The distance Malmquist bias can also be eliminated if the TF relation is not used to directly infer distances to individual galaxies. This is the main feature of the methods listed in the right column of Table 7.1

(§7.2.2), sometimes referred to as "Method II" (e.g., see Table 6.1). Instead of inferring distances, one can use the data and the TF relation to simultaneously derive the inverse TF parameters and constrain a parametric model of the velocity field, $v(\alpha_k, x)$ (Schechter 1980). For example, the velocity potential can be expanded in terms of an orthogonal set of basis functions, from which the model velocity and density fields are derived analytically. The sum of residuals minimized instead of equation (7.3) is

$$S_{II} = \sum_i W_i [\eta_i - \eta_i^0]^2 , \qquad (7.14)$$

where η_i are the observed values, and where the model values are given by the inverse TF relation,

$$\eta_i^0 = \tilde{a} + \tilde{b} M_i = \tilde{a} + \tilde{b}(m_i - 5\log r_i) , \qquad r_i = z_i - \hat{x}_i \cdot v(\alpha_k, x_i) . \qquad (7.15)$$

The varying parameters are the inverse TF parameters $\tilde{a}$ and $\tilde{b}$ of the different datasets and the α_k that characterize the velocity model. Spatial correlations of these residuals may also need to be minimized in order to ensure a robust fit. The "distance" r_i is derived from the redshift z_i, not from the TF relation, and is therefore free of inferred-distance Malmquist bias. In fact, the whole analysis can be carried out in redshift space.

An inverse method was first used by Aaronson et al. (1982b) to fit a few-parameter Virgo-centric toy model to a local sample of spirals. Two recent attempts have been made to implement the inverse z-space method to the Mark III catalog, using a multi-parameter potential model. This is a non-trivial problem of nonlinear multi-parameter minimization with several possible routes.

One attempt, termed ITF, is geared towards a direct comparison with galaxy redshift data rather than a uniform recovery of the mass-density field (Nusser & Davis 1995; §7.6.4). The velocity model is expanded in terms of spherical harmonics and Bessel functions, such that the resolution is decreasing with radius following the trend in the data sampling. The whole analysis is done in redshift space, which inevitably introduces oversmoothing in collapsing regions. In order to enable semi-analytic minimization in this scheme, the inverse TF parameters of the different datasets in the Mark III catalog had to be forced into a common system – a somewhat risky process that carries errors.

The other attempt, termed MFPOT, aims at a uniform reconstruction in z-space or in r-space, as in POTENT (Blumenthal et al. 1998, first described in Dekel 1994). The potential model in this case is a Fourier expansion, to ensure uniform resolution. The minimization is numerical, which allows the simultaneous analysis of heterogeneous data. If the velocity model is expressed in z-space, then r_i is given explicitly by z_i. If the model is expressed in r-space, then r_i is implicit in the second equation of (7.15), requiring iterative minimization, e.g., by carrying r_i from one iteration to the next one. The velocity model could be either global or local, with the former enabling a simultaneous minimization of the TF and velocity parameters and the latter requiring a sequential minimization of global TF parameters and local velocity parameters.

The MFPOT results are free of Malmquist bias and of uncertainties in matching the TF parameters of the different datasets. It provides an excellent way to determine the (inverse) TF parameters of each of the different datasets using all the data, including field and cluster galaxies. It does not require inverse density weighting to avoid SG bias as in POTENT because the velocity converges to the correct global function even when spatial sampling is non-uniform. Another bonus is that the use of Fourier expansion provides a direct measure of the power spectrum. However, the results in redshift space suffer from other biases which have to be carefully diagnosed and corrected for.

It is clear that the Malmquist-free results will have to be consistent with the Malmquist-corrected forward and inverse results in inferred-distance space before one can safely put to rest the issue of Malmquist bias and its effect on our LSS results.

The other attempt to apply inverse TF analysis in redshift space is geared towards a direct estimation of β via a mode-by-mode comparison of the TF data with galaxy redshift data, bypassing the uniform recovery of the velocity field and the mass-density distribution (ITF by Nusser & Davis 1995; Davis, Nusser & Willick 1996). This method is discussed in §7.6.4.

7.3 Testing basic hypotheses

Before embarking on cosmological implications regarding the cosmological parameters and the initial fluctuations, it is important to try to verify that the peculiar-velocity data and the recovered dynamical field are consistent with the basic hypotheses laid out in the Introduction. In particular, we provide three examples here, which test the hypotheses that (a) GI is the engine driving the fluctuation growth; (b) galaxies roughly trace mass; and (c) the TF analysis indeed measures peculiar velocities and the POTENT procedure recovers the mass-density field.

7.3.1 *CMB fluctuations versus local motions: gravitational instability*

If the CMB defines a standard cosmological frame, then the established helio-centric dipole pattern of $\delta T/T = (1.23 \pm 0.01) \times 10^{-3}$ is a direct measurement of peculiar velocity of the LG: $V(0) = 627 \pm 22 \,\mathrm{km\,s^{-1}}$ towards $(l,b) = (276° \pm 3°, +30° \pm 3°)$ (Kogut et al. 1993; Lineweaver et al. 1996). The Copernican hypothesis then implies that large peculiar velocities *exist* in general, and the question is only how large and *coherent* they are. A more esoteric interpretation tried to explain the CMB dipole by a horizon-scale gradient in entropy, relic of bubbly Inflation, which can be made consistent with the smallness of the quadrupole and the achromaticity of the dipole (e.g., Gunn 1989; Paczynski & Piran 1990). In addition to the general objection to global anisotropy based on the simplicity principle of Occam's razor, a non-velocity interpretation fails to explain the fact that the gravitational acceleration vector at the LG, $g(0)$, as inferred from the galaxy distribution in our cosmological neighborhood, is within 20° of $V(0)$ and of a similar amplitude of several hundred $\mathrm{km\,s^{-1}}$ (for any reasonable choice of Ω). This argument in favor of the reality of $V(0)$ and the standard GI picture is strengthened by the fact that a similar $g(0)$ is obtained by inegrating over the POTENT mass fluctuation field as derived from velocity divergences in our cosmological neighborhood (Kolatt, Dekel & Lahav 1995).

The measurements of CMB fluctuations on scales $\leq 90°$ are independent of the local streaming motions, but GI predicts an intimate relation between their amplitudes. The CMB fluctuations are associated with fluctuations in gravitational potential, velocity and density in the surface of last-scattering at $z \sim 10^3$, while similar fluctuations in our neighborhood have grown by gravity to produce the dynamical structure observed locally. The comparison between the two is therefore a crucial test for GI.

Before COBE, the local streaming velocities served to predict the expected level of CMB fluctuations. The local surveyed region of $\sim 100\,h^{-1}$ Mpc corresponds to a $\sim 1°$ patch on the last-scattering surface. An important effect on scales $\geq 1°$ is the Sachs–Wolfe effect (Sachs & Wolfe 1967), where potential fluctuations $\Delta\Phi_g$ induce temperature fluctuations via gravitational redshift, $\delta T/T = \Delta\Phi_g/(3c^2)$. Since the velocity potential is proportional to Φ_g in the linear and mildly-nonlinear regimes, $\Delta\Phi_g$ is $\sim Vx$, where x is the scale over which the bulk velocity is V. Thus $\delta T/T \geq Vx/(3c^2)$. A typical bulk velocity of $\sim 300\,\mathrm{km\,s^{-1}}$ across $\sim 100\,h^{-1}$ Mpc (§7.4.1) corresponds to $\delta T/T \geq 10^{-5}$ at $\sim 1°$. If the fluctuations are roughly scale-invariant ($n = 1$), then $\delta T/T \geq 10^{-5}$ is expected on all scales $> 1°$. Bertschinger, Gorski & Dekel (1990) produced a crude $\delta T/T$ map of the local region as seen by a hypothetical distant observer, and predicted $\delta T/T \geq 10^{-5}$ from the local potential well associated with the GA. An up-to-date version of the $\delta T/T$ maps is provided by Zaroubi et al. (1997b), who added a proper treatment of the acoustic effects on sub-degree scales for various cosmological models. This map is shown in Fig. 7.8.

Now that CMB fluctuations of $\sim 10^{-5}$ have been detected practically on all the relevant angular scales (§3.5.3), the argument can be reversed: if one assumes GI, then the *expected* bulk velocity in the surveyed volume is $\sim 300\,\mathrm{km\,s^{-1}}$, i.e., the inferred motions of §7.2 are most likely real. If, alternatively, one accepts the peculiar velocities as real for other reasons, then their consistency with the CMB fluctuations is a relatively sensitive and robust test of the validity of GI. This test is unique in the sense that it addresses the specific fluctuation growth rate as predicted by GI theory (§7.3.2). It is robust in the sense that it is quite insensitive to the values of the cosmological parameters and is independent of the complex issues involved in the process of galaxy formation. This is therefore the strongest consistency test for GI.

7.3.2 *Galaxies versus dynamical mass: biasing*

The theory of GI combined with the assumption of linear biasing for galaxies predict a correlation between the dynamical density field and the galaxy-density field, which can be addressed quantitatively based on the mock catalogs and the estimated errors in the two datasets. Figure 7.9 compares density maps in the Supergalactic Plane for IRAS

Large-scale flows and cosmological implications

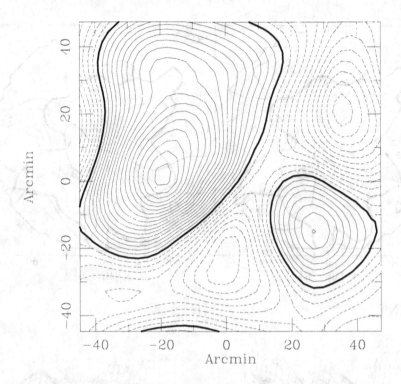

Figure 7.8. CMB map. Temperature fluctuations in a 1.5° CMB patch on the sky of a hypothetical distant observer viewing our local neighborhood at his last-scattering surface from above the Supergalactic Plane. The resolution is 10 arcmin, which is comparable to the smoothing imposed along the line-of-sight by the finite thickness of the last-scattering surface. Contour spacing is 5×10^{-6} in $\Delta T/T$. The universe is assumed to be Einstein–deSitter, with $h = 0.5$ and $\Omega_b h^2 = 0.025$. In general, the temperature is correlated with baryonic density (rather than potential, compare to Fig. 7.7) due to the acoustic effects that dominate on these scales (Zaroubi et al. 1997b).

1.2 Jy galaxies (δ_g) and POTENT Mark III mass (δ), both G12 smoothed. The general correlation is evident – the GA, PP, Coma and the voids all exist both as dynamical entities and as structures of galaxies. To evaluate goodness of fit, Fig. 7.10 shows the statistic $\chi^2 = N^{-1} \sum^N (\delta_g - b\delta)^2/\sigma^2$ as computed from the data in comparison with its distribution over pairs of Mark III and IRAS 1.2 Jy mock catalogs. The fact that the data lies near the center of this distribution indicates that the two datasets are consistent with being noisy versions of an underlying fluctuation field and that the data are in agreement with the hypotheses of GI plus linear biasing (Dekel et al. 1993; Sigad et al. 1998; more in §7.6.1).

What is it exactly that one can learn from the observed v–δ_g correlation (Babul et al. 1994)? First, it argues that the velocities are real because it is hard to invoke any other reasonable way to make the galaxy distribution and the TF measurements agree so

276   A. Dekel

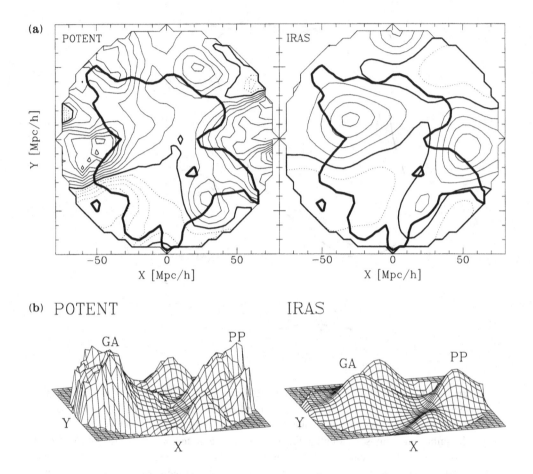

Figure 7.9. Mass versus galaxies. POTENT mass-density field ($\Omega = 1$) versus IRAS galaxy-density fields in the Supergalactic plane, both G12 smoothed. Contour spacing is 0.2. (a) The heavy contour marks the boundary of the comparison volume of effective radius $46\,h^{-1}$ Mpc. (b) The height in the surface plot is proportional to δ. The LG is at the center, GA on the left, PP on the right, and the Sculptor void in between (Sigad *et al.* 1998).

well (§7.2.3.1). On the other hand, although it is true that gravity is the only long-range force that could attract galaxies to stream toward density concentrations, the fact that a v–δ_g correlation is predicted by GI plus linear biasing does not necessarily mean that the observation can serve as a sensitive test for either. Recall that converging (or diverging) flows tend to generate overdensities (or underdensities) simply as a result of mass conservation, independent of the source of the motions.

Let us assume for a moment that galaxies trace mass, i.e., that the linearized continuity equation, $\dot{\delta} = -\nabla \cdot v$, is valid for the galaxies as well. The observed correlation (in the linear approximation) is then $\delta \propto -\nabla \cdot v$, and together they imply that $\dot{\delta} \propto \delta$, or

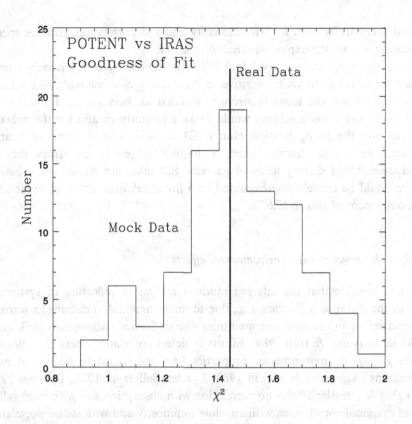

Figure 7.10. Goodness of fit in the comparison of G12 density fields of POTENT mass and IRAS galaxies, as tested by a χ^2 statistic that is an error-weighted sum of differences between the two fields (Sigad et al. 1998).

equivalently that $\nabla \cdot v$ is proportional to its time average. This property is not exclusive to GI; one can construct counter-examples where the velocities are produced by a non-GI impulse.

Even irrotationality does not follow from $\delta \propto -\nabla \cdot v$; it has to be adopted based on theoretical arguments in order to enable reconstruction from radial velocities or from observed densities. Once continuity and irrotationality are assumed, the observed $\delta \propto -\nabla \cdot v$ implies a system of equations which is identical in all its *spatial* properties to the equations of GI, but can differ in the constants of proportionality and their temporal behavior. It is therefore impossible to distinguish between GI and a non-GI model which obeys continuity plus irrotationality based only on snapshots of *present-day* linear fluctuation fields. This makes the relation between CMB fluctuations and velocities an especially important test for GI.

On the other hand, the fact that the constant of proportionality in $\delta \propto -\nabla \cdot v$ is indeed the same everywhere is a non-trivial requirement from a non-GI model. For example, a version of the explosion scenario (Ostriker & Cowie 1981; Ikeuchi 1981), which tested successfully both for irrotationality and v–δ correlation, requires special synchro ization among the explosions (Babul et al. 1994).

So, what is the v–δ_g relation good for? While its sensitivity to GI is only partial, this relation turns out to be quite sensitive to the validity of a *continuity*-like relation for the *galaxies*. When the latter is strongly violated all bets are off for the $v-\delta_g$ relation. A nonlinear biasing scheme would make continuity invalid for the galaxies, which would ruin the $v-\delta_g$ relation even if GI is valid. The observed correlation is thus a sensitive test for density *biasing*. It implies, subject to the errors, that the $\sim 12\,h^{-1}$ Mpc-smoothed density fields of galaxies and mass are related via a biasing relation that could be crudely approximated by a linear relation with b of order unity (but see a refinement of this in §7.6.7).

7.3.3 Ellipticals versus spirals: environmental effects

A priori, it is possible that the inferred motions are just a reflection of systematic variations in the distance indicators, e.g., due to environmental variations in intrinsic galaxy properties, or in the apparent quantities due to Galactic absorption (Silk 1989; Djorgovski, de Carvalho & Han 1989). Efforts to detect correlations between velocities and certain galaxy or environmental properties have led so far to null or at most minor detections (Aaronson & Mould 1983; Lynden-Bell et al. 1988; Burstein 1990; Burstein, Faber & Dressler 1990) – no correlation with absorption and with local galaxy density, and marginal correlations with absolute luminosity and with stellar population (Gregg 1995). Admittedly, these null results are only indicative as one cannot rule out a correlation with some other property not yet tested for.

Qualitative comparisons of the velocities of ellipticals (Es) and spirals (Ss) indicated general agreement (Burstein, Faber & Dressler 1990). Earlier attempts to compare distances to clusters as inferred by TF distances to Ss and FP distances to Es were unsatisfactory because the clusters as defined by the different observers of the two types of galaxies are rarely the same clusters. A recent attempt to measure E distances in ten clusters that were pre-defined by their S population shows consistency of cluster distances at the expected 1σ level (Scodeggio 1997).

The POTENT analysis of the Mark III data enable a quantitative comparison of the E and S fields at the same spatial positions (Kolatt & Dekel 1994). Figure 7.11 compares the two fields as derived independently by POTENT, showing a general resemblance. The radial velocities of each type were interpolated into a smoothed field on a grid, $u(x)$, using $12\,h^{-1}$ Mpc POTENT smoothing (§7.2.3.2), and then compared within a volume limited by the poor sampling of Es to $\simeq (50\,h^{-1}\,\text{Mpc})^3$, containing ~ 10 independent sub-volumes. The two fields were found to be *consistent* with being noisy versions of the same underlying field, while the opposite hypothesis of complete independence is strongly ruled out, indicating that the consistency is not dominated by the errors. A possible discrepancy indicated earlier by the Mark II data (Bertschinger 1991) is gone with the improved data and bias-correction. The E–S correlation is thus

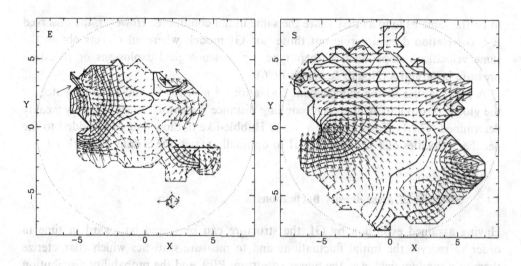

Figure 7.11. Ellipticals versus Spirals. Mass-density and velocity fields in the Supergalactic Plane as recovered by POTENT independently for the $D_n - \sigma$ and TF Mark II data. Smoothing is G12. Contour spacing is 0.2. Distances and velocities are in $1000\,\text{km}\,\text{s}^{-1}$ (Kolatt & Dekel 1994).

consistent with real motions. The strength of this test will improve further as the E sample grows in extent (e.g., project EFAR by Colless et al. 1993) and in accuracy (e.g., the SBF method, §6.5).

The E–S correlation is not a trivial result because of the physical difference between the FP and TF scaling relations for ellipticals and spirals (Gunn 1989; see also §6.3). The FP relation naturally follows from virial equilibrium, $\sigma^2 \propto M/D$ (M mass, D diameter), and a smoothly varying $M/L \propto M^\gamma$, which together yield $DI^{1/(1+\gamma)} \propto \sigma^{2(1-\gamma)/(1+\gamma)}$ (I being the surface brightness out to D). The TF relation, involving only two of the three quantities entering the virial theorem, is more demanding – it requires an additional constraint which is probably imposed at galaxy formation (e.g., the slope of the power spectrum of fluctuations on galactic scales).

Therefore, although the E–S correlation cannot rule out environmental effects, it makes this possibility less plausible. If environmental effects were dominant, then the two distance indicators would have to vary coherently, which would require that the large-scale properties of the environment be different from the local properties determining the galaxy type (E in clusters, S in field). It would also require that the properties affecting the virial equilibrium which determines the FP relation, and those affecting the additional constraint involved in the TF relation at galaxy formation, vary together in space. The one quantity that could plausibly affect the two in a correlated way is M/L, but to test for this one will have to compare the TF and FP velocities to those inferred from an independent indicator, e.g., SBF or supernovae.

The E–S correlation is consistent with the hypothesis of GI because, following Galileo, the velocities of all test bodies in a given gravitational potential are predicted to be the same as long as they share the same initial conditions. However, the observed E-S correlation does not rule out those non-GI models where all objects obtain the same velocities independent of their type, e.g., cosmological explosions or radiation pressure instabilities (Hogan & White 1986).

A practical use of the independent derivation of $u(x)$ for Ss and Es is in matching the global zero points of the corresponding distance indicators, otherwise only weakly determined (§7.2.1). For example, a $\sim 3\%$ Hubble-like outflow had to be added to the peculiar velocities of Es from Mark II to optimally match the S data of Mark III.

7.4 Statistics of mass-density fluctuations

Having assumed evolution by GI, the structure can be traced backward in time in order to recover the initial fluctuations and to measure statistics which characterize them as a random field, e.g., the power spectrum, $P(k)$, and the probability distribution functions (PDF). "Initial" here may refer either to the *linear* regime at $z \sim 10^3$ after the onset of the self-gravitating matter era, or to the origin of fluctuations in the early universe before being filtered on sub-horizon scales during the plasma–radiation era. The spectrum is filtered on scales $\leq 100\,h^{-1}$ Mpc by DM-dominated processes, but its shape on scales $\geq 10\,h^{-1}$ Mpc is not affected much by the mildly-nonlinear effects (because the faster density growth in superclusters roughly balances the slower density depletion in voids at the same wavelength). The shape of the one-point PDF, on the other hand, is expected to survive the plasma era unchanged but it develops strong skewness even in the mildly-nonlinear regime. Thus, the present-day $P(k)$ can be used as is to constrain the origin of fluctuations (on large scales) and the nature of the DM (on small scales), while the PDF needs to be traced back to the linear regime first.

The competing scenarios of LSS formation are reviewed by Primack (§§1.5, 1.7). In summary, if the dark matter is all baryonic, then by nucleosynthesis constraints the universe must be of low density, $\Omega \lesssim 0.1$, and a viable model for LSS is the Primordial Isocurvature Baryonic model (PIB) with several free parameters. With $\Omega \sim 1$, the nonbaryonic DM constituents are either "cold" or "hot", and the main competing models are CDM, HDM, and CHDM – a 7:3 mixture of the two. The main difference in the DM effect on $P(k)$ arises from free-streaming damping of the "hot" component of fluctuations on galactic scales. Currently popular variants of the standard CDM model ($\Omega=1$, $n=1$) include a tilted power spectrum on large scales ($n \lesssim 1$) and a flat, low-Ω universe with a non-zero cosmological constant such that $\Omega_{\rm tot} + \Omega_\Lambda = 1$.

The peculiar velocities of the Mark III catalog enable direct derivations of the mass power spectrum itself, independent of galaxy biasing, roughly in the range $10-100\,h^{-1}$ Mpc. The bulk velocity in spheres of radii up to $60\,h^{-1}$ Mpc is sensitive to even larger wavelengths. In all standard theories, the power spectrum on large scales is expected to be a power law, $P_k \propto k^n$, with n of order unity. It is expected to turn around at $k_{\rm peak} \sim 0.065(\Omega h)^{-1}(h^{-1}\,{\rm Mpc})^{-1}$, corresponding to the horizon scale at the epoch of equal energy densities in matter and radiation. The dark-matter (DM) type mostly affects the shape of the filtered spectrum in the "blue" side of the peak ($k > k_{\text{peak}}$). Once the fluctuation amplitude on very large scales is fixed by COBE's measurments of CMB fluctuations, the bulk velocity is sensitive to n and is insensitive to Ω or the DM type. The steep slope of the CDM-like spectra at $k > k_{\text{peak}}$, where it is best constrained by the data, makes it more sensitive than the bulk velocity to Ωh.

We first describe the bulk velocity (§7.4.1). Then we discuss a model-independent evaluation of $P(k)$ from the velocity field recovered by POTENT (§7.4.2), and a likelihood estimation of $P(k)$ from raw radial peculiar velocities under a prior model (§7.4.3).

7.4.1 Bulk velocity

A simple and robust statistic related to the power spectrum is the bulk velocity - the amplitude of the vector average V of the R_s-smoothed velocity field $v(x)$ over a volume defined by a normalized window function $W_R(r)$ (e.g., top-hat) of a characteristic scale R,

$$V \equiv \int d^3x\, W_R(x)\, v(x)\,, \quad \langle V^2 \rangle = \frac{f(\Omega)^2}{2\pi^2} \int_0^\infty dk\, P(k)\, \widetilde{W}_R^2(k)\,. \tag{7.16}$$

We denote by V_r the bulk velocity in a top-hat sphere of radius $R = r h^{-1}$ Mpc. The ensemble variance $\langle V^2 \rangle$ for a model that is characterized by $P(k)$ is an integral of $P(k)$ in which the wavelengths $\geq R$ are emphasized by $\widetilde{W}_R^2(k)$, the Fourier Transform of $W_R(r)$. The bulk velocity can be obtained from the observed radial velocities by minimizing residuals as in equation (7.3). The first report by Dressler et al. (1987) of $V = 599 \pm 104$ for ellipticals within $\sim 60 h^{-1}$ Mpc was interpreted prematurely as being in severe excess of common predictions, but it quickly became clear that the effective window was much smaller than $\sim 60 h^{-1}$ Mpc due to the non-uniform sampling and weighting (Kaiser 1988). The SG bias can be crudely corrected by volume-weighting as in POTENT (§7.2.3.2), at the expense of larger noise. Courteau et al. (1993) reported, based on an early version of the Mark III data, $V_{60} = 360 \pm 40$ towards $(L, B) = (162°, -36°)$. Alternatively, V_r can be computed from the POTENT v field by simple vector averaging from the grid.

The bulk velocity as a function of R, from several recent sources, is shown in Fig. 7.12. The Mark III POTENT result inside a sphere of radius $R = 50 h^{-1}$ Mpc is $V_{50} = 374 \pm 85\,\text{km s}^{-1}$ towards $(158°, -9°) \pm 10°$. The $\sim 20\%$ error bars are due to distance errors, and one should consider an additional uncertainty of similar magnitude due to the non-uniform sampling. The SFI sample of Sc galaxies yields a similar result inside $R = 50 h^{-1}$ Mpc (contrary to premature rumors), $V_{50} \approx 364\,\text{km s}^{-1}$ towards $(172°, -14°)$ (da Costa et al. 1996). At larger radii the bulk velocity in SFI seems to drop faster than in Mark III. This difference may be related to a difference in matching the zero points of the TF relations between North and South in the two catalogs, but

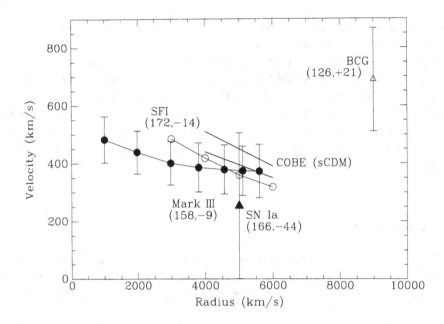

Figure 7.12. Bulk velocity. The amplitude of the bulk velocity relative to the CMB frame in top-hat spheres about the LG, as derived from several datasets. The directions are indicated in Supergalactic coordinates (L, B). The Mark III data and the SFI data yield consistent results at $R \leq 50\,h^{-1}$ Mpc. The new result from Supernovae Type Ia on larger scales is a natural extrapolation. The result from brightest galaxies in clusters (LP) is discrepant at more than the 2σ level.

the fact is that these two samples are not large enough for a reliable estimate of V at larger radii.

Supernovae Type Ia provide more accurate distances, with only $\sim 8\%$ error, and they can be measured at larger distances. The current sample of 44 such SNe by Riess & Kirshner (1997, private communication), which extends out to $\sim 300\,h^{-1}$ Mpc, shows a bulk flow of $V = 253 \pm 252$ km s^{-1} towards $(166°, -44°)$. The effective radius of this dataset for a bulk-flow fit is in fact less than $50\,h^{-1}$ Mpc because the data is weighted inversely by the errors. The SNe bulk flow is consistent with the results from the Mark III and SFI galaxy data. They all make a bulk of sense within the framework of standard isotropic and homogeneous cosmology.

The only apparently discrepant result comes from the velocities measured on a larger scale using brightest cluster galaxies (BCGs) as distance indicators (Lauer & Postman 1993, LP; §6.7). They indicate a large bulk velocity of $V = 689 \pm 178$ towards a very different direction $\sim (126°, 21°)$. An ongoing effort to measure BCGs in a larger sample of clusters and distances to clusters based on other distance indicators will soon tell whether this early result is a $\gtrsim 2\sigma$ statistical fluke (Watkins & Feldman 1995; Strauss et al. 1995), whether the errors were underestimated, or whether something is systematically different between the BCG distances and the distances measured by other indicators.

Shown in comparison are the expected rms bulk velocity in a standard CDM model ($\Omega = 1$, $n = 1$) normalized to COBE, for $h = 0.5$ (bottom curve) and 0.8 (top curve) (Sugiyama 1995). These theoretical curves would not change much if a 20% hot component is mixed with the cold dark matter, or if Ω is lower but still in the range $0.2 - 1.0$, as long as $n \approx 1$. The main effect of Ω and H_0 on $P(k)$ is via k_{peak}. The predicted bulk velocity over $\sim 100\,h^{-1}$ Mpc is effectively an integral of $P(k)$ over $k < k_{\text{peak}}$, and is therefore relatively insensitive to Ω while it is quite sensitive to n. When compared to a theoretical prediction, the error should also include cosmic scatter due to the fact that only one sphere has been sampled from a random field. These errors are typically on the order of the measurement errors.

Mach Number. The bulk velocity is robust but it relates to the normalization of $P(k)$, which is not predicted from first principles by the standard theories; it is, rather, normalized by some other uncertain observations, e.g., the CMB fluctuations, the cluster abundance, or the galaxy distribution with an unknown biasing factor. A statistic which measures the *shape* of $P(k)$ free of its normalization is the cosmic Mach number (Ostriker & Suto 1990, defined as $\mathcal{M} \equiv V/S$, where S is the rms deviation of the local velocity from the bulk velocity,

$$S^2 \equiv \int d^3x\, W_R(x)\,[v(x) - V]^2\,,\quad \langle S^2 \rangle = \frac{f^2}{2\pi^2} \int_0^\infty dk\, P(k)\,[1 - \widetilde{W}_R^2(k)]\,. \quad (7.17)$$

$\mathcal{M}(R, R_s)$ measures the ratio of power on large scales $\gtrsim R$ to power on small scales $\gtrsim R_s$. Strauss, Cen & Ostriker (1993) derived $\mathcal{M} = 1.0$ for the local Ss (Aaronson et al. 1982a) with $R \sim 20\,h^{-1}$ Mpc and $R_s \to 0$, and found $\sim 5\%$ of their standard CDM simulations to have $\mathcal{M}$ as large – a marginal rejection or consistency depending on taste. On larger scales, using POTENT within a top-hat window of $R = 60\,h^{-1}$ Mpc and Gaussian smoothing of $R_s = 12\,h^{-1}$ Mpc, the Mark III data yield $\mathcal{M} \sim 1$, which roughly coincides with the rms value expected from CDM but has only $\sim 5\%$ probability to be that low for a PIB spectrum ($\Omega = 0.1$, $h = 1$, fully ionized, normalized to $\sigma_8 = 1$) (Kolatt & Dekel, unpublished).

7.4.2 Power spectrum from the velocity field via POTENT

One way to compute the power spectrum is via the smoothed mass-density field as recovered by POTENT (Kolatt & Dekel 1997). The key is to correct the result for systematic deviations from the true $P(k)$. The data suffers from distance errors and sparse, non-uniform sampling, and they were heavily smoothed. The $P(k)$ is computed from within a window of effective radius $\sim 50\,h^{-1}$ Mpc, say, where the densities are weighted inversely by the squares of the local errors. The density field is zero-padded in a larger periodic box in order to enable an FFT procedure. The $P(k)$ is computed

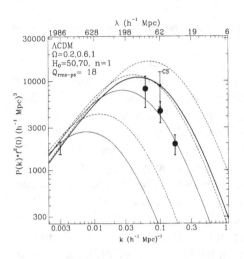 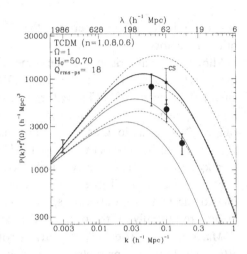

Figure 7.13. The mass power spectrum $[\times f(\Omega)^2]$ from POTENT Mark III velocities (filled symbols), with 1σ random errors. The curves are COBE-normalized theoretical predictions for flat CDM with $h = 0.5$ (solid) and $h = 0.7$ (dashed). *Left:* ΛCDM with $n = 1$ and Ω growing upwards. *Right:* TCDM with $\Omega = 1$ and n growing upwards. Typical cosmic scatter (CS) for $\Omega = 1$ and $n = 1$ is indicated (Kolatt & Dekel 1997).

by averaging the amplitudes of the Fourier Transforms in bins of k. This procedure yields an "observed" $P(k)$, which we term $O(k)$.

The systematic errors in the above procedure are then modelled by $O(k) = M(k)[S(k) + N(k)]$, where $S(k)$ is the true signal $P(k)$, $N(k)$ is the noise, and $M(k)$ represents the effects of sampling, smoothing, applying a window etc. The correction functions $M(k)$ and $N(k)$ can be derived from Monte Carlo mock catalogs (§7.2.3.5). The factor $M(k)$ is derived first by $M^{-1} = S/\langle O \rangle_{\text{no-noise}}$, where S here is the known power spectrum built into the simulations, and the averaging is over mock catalogs not perturbed by noise. Then $N(k)$ is computed by $N = M^{-1}\langle O \rangle_{\text{noise}} - S$, where the averaging is over noisy mock catalogs.

Equipped with the correction functions $M(k)$ and $N(k)$, the $P(k)$ observed from the real universe, $O(k)$, is corrected to yield the true $P(k)$ by $S(k) = M(k)^{-1}O(k) - N(k)$. The recovered mass-density $P(k)$ is shown in Fig. 7.13 in three thick logarithmic bins covering the range $0.04 \leq k \leq 0.2\,(h^{-1}\,\text{Mpc})^{-1}$, within which the results are reliable. The robust result is $P(k)f(\Omega)^2 = (4.6 \pm 1.4) \times 10^3 (h^{-1}\,\text{Mpc})^3$ at $k = 0.1(h^{-1}\,\text{Mpc})^{-1}$ (using the convention where the Fourier Transform is defined with no 2π factors in its coefficient). The logarithmic slope at $k = 0.1$ is -1.45 ± 0.5. This translates to $\sigma_8 \Omega^{0.6} \simeq 0.7$–$0.8$, depending on where the peak in $P(k)$ is (see §7.4.3).

The observed $P(k)$ is compared in Fig. 7.13 to the linear predictions of a family of inflation-motivated flat CDM models ($\Omega_m + \Omega_\Lambda = 1$) with the Hubble constant arbitrarily fixed at $h = 0.5$. For $n = 1$, maximum likelihood is obtained at $\Omega \simeq 0.7h_{50}^{-1.3} \pm 0.1$. For $\Omega = 1$, assuming no tensor fluctuations, the linear power index is $n \simeq 0.75h_{50}^{-0.8} \pm 0.1$.

7.4.3 Power spectrum from velocities and COBE via likelihood

The power spectrum, in a parametric form including Ω, h and n among the parameters, has alternatively been determined from the velocity data via a Baysian likelihood analysis (Zaroubi et al. 1997a; see also Jaffe & Kaiser 1995). According to Bayse, the probability of the model parameters (m) given the data (d), which is the function one wishes to maximize, can be expressed as $P(m|d) = P(d|m) P(m)/P(d)$. The probability $P(d)$ serves here as a normalization constant. Without any external constraints on the model parameters, one assumes that $P(m)$ is a constant in a given range. The remaining task is to maximize the likelihood $\mathscr{L} = P(d|m)$ as a function of the model parameters. This function can be written down explicitly.

Under the assumption that the velocities and the errors are both Gaussian random fields with no mutual correlations, the likelihood can be written as $\mathscr{L} = (2\pi|D|)^{-1/2} \exp(-d_i D_{ij}^{-1} d_j/2)$, where d_i are the data at points $i = 1,..., N$, and D_{ij} is the covariance matrix, which can be split into covariance of signal (s) and covariance of noise (n), $D_{ij} \equiv \langle d_i d_j \rangle = \langle s_i s_j \rangle + \langle n_i n_j \rangle$. If the errors are uncorrelated, the noise matrix is diagonal. The signal matrix is computed from the model $P(k)$ as a a function of the model parameters.

Zaroubi et al. (1997a) used a parametric model for the power spectrum of the general form $P_k = A k^n T(\Gamma_i; k)$, where $T(k)$ is a small-scale filter of an assumed shape characterized by free parameters Γ_i, k^n is the initial $P(k)$ which is still valid on large scales today, and A is a normalization factor. The normalization can either be determined by COBE's data (for given Ω, Λ, h, n and tensor/scalar fluctuations), or be left as a free parameter to be fixed by the velocity data alone. The filter $T(k)$ can either be taken from a specific physical model (e.g., CDM, where $\Gamma = \Omega h$), or be an arbitrary function with enough flexibility to fit the data.

The robust result for all the models is a relatively high amplitude, with $P(k) f(\Omega)^2 = (4.8 \pm 1.5) \times 10^3 (h^{-1} \text{Mpc})^3$ at $k = 0.1 (h^{-1} \text{Mpc})^{-1}$. An extrapolation to smaller scales using the different CDM models yields $\sigma_8 \Omega^{0.6} = 0.88 \pm 0.15$ (for the dispersion in top-hat spheres of radius $8 h^{-1}$ Mpc).

Within the general family of CDM models, allowing for a cosmological constant in a flat universe and a tilt in the spectrum, the parameters are confined by a 90% likelihood contour of the sort $\Omega h_{50}^\mu n^\nu = 0.8 \pm 0.2$, where $\mu = 1.3$ and $\nu = 3.4$ or 2.0 for models with and without tensor fluctuations, respectively. Figure 7.14 displays the likelihood map in the $\Omega - n$ plane for these models. For open CDM the powers are $\mu = 0.95$ and $\nu = 1.4$ (no tensor fluctuations). A Γ-shape model free of COBE normalization yields only a weak constraint: $\Gamma = 0.4 \pm 0.2$ (where Γ is not necessarily Ωh).

Both Ω and n obtained by the likelihood analysis from the raw peculiar velocities tend to be slightly higher ($\sim 20\%$) than their estimates based on the $P(k)$ recovered from the POTENT output. This difference may arise from the different relative weighting assigned to the different wavelengths in the two analyses. The difference between the

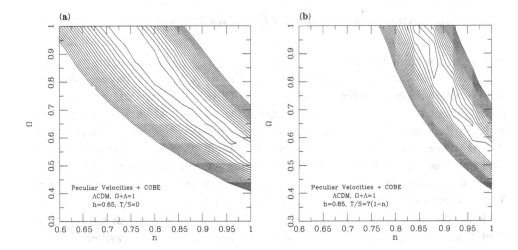

Figure 7.14. Likelihood contour maps in the $\Omega - n$ plane for flat CDM with and without tensor fluctuations and $h = 0.65$. Contour spacing is unity in log-likelihood. Under the assumption of a χ^2 distribution, the two-dimensional 90 percentile corresponds to 2.3 contours, and the one-dimensional 90 percentile corresponds to 1.35 contours (Zaroubi et al. 1997a).

results obtained in the two different ways is on the order of the errors in each analysis and the cosmic scatter.

Very similar estimates of $P(k)$, and the parameters Ω and n, are obtained from a preliminary analysis of the SFI sample of Sc galaxies (in preparation).

In summary: The "standard" CDM model is marginally rejected at the $\sim 2\sigma$ level, while each of the following modifications leads to a good fit to the peculiar velocities and large-scale CMB data: $n \lesssim 1$, $\Omega_\nu \sim 0.3$, or $\Omega \lesssim 1$. The strong implication on the dark matter issue is that values of Ω as low as ~ 0.2 are ruled out with high confidence (independent of Λ), leaving, in particular, no room for the baryonic PIB model.

7.4.4 Peculiar velocities vs. small-scale CMB fluctuations

Sub-degree angular scales at the last-scattering surface correspond to the $\leq 100 h^{-1}$ Mpc comoving scales explored by peculiar velocities today. Thus, under the assumption that the local neighborhood is typical, the power spectrum on these scales is simultaneously constrained by the mass-density fluctuations in our cosmological neighborhood and by the CMB fluctuations.

The sub-degree CMB fluctuations are being explored by many balloon-borne experiments, and in the early 2000s we expect accurate results from the CMB satellites MAP and PLANCK. These measurements will eventually allow a simultaneous likelihood analysis of the two kinds of data. At this point, however, although there are already

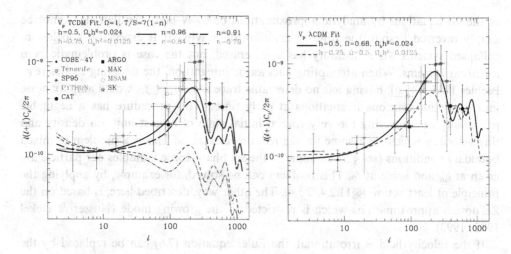

Figure 7.15. Angular power spectrum in the CMB. The current data (symbols) are compared to certain CDM models that fit the peculiar-velocity data at 90% likelihood. The extreme models shown are either with low h and high $\Omega_b h^2$ or vice versa. *Left:* tilted CDM with $\Omega = 1$, including tensor fluctuations. *Right:* Flat CDM with $n = 1$.

preliminary detections of the first acoustic peak in the angular power spectrum, the uncertainties are still large. Any current comparison is therefore limited to the semi-quantitative level. The range of parameters permitted by the peculiar-velocity data for power spectra of the CDM family (Zaroubi *et al.* 1997a) can be translated to a range of angular power spectra, C_l. This range is plotted against current observations in Fig. 7.15.

The immediate conclusions from a visual inspection of the figure are that a wide range of CDM models can simultaneously obey the constraints from the two datasets, but there is a subset of models that fits the velocities well but seems to fail to reproduce a high enough accoustic peak in the CMB spectrum.

The acoustic peak in the CMB is sensitive to Ω_b and the observations prefer a high baryon content, $\Omega_b h^2 \sim 0.025$ (similar to the value measured by Tytler *et al.* 1996; Burles & Tytler 1996), while the peculiar velocities have little to add because $P(k)$ is hardly affected by Ω_b.

The power index n is important in both cases; the peculiar velocities allow values of n significantly lower than unity, but the current CMB data seem not to tolerate values of n below 0.9 or so.

The peculiar-velocity data prefers $\Omega \geq 0.4$, and the location of the first acoustic peak in the sub-degree CMB data indicates in agreement a high value of $\Omega_m + \Omega_\Lambda$.

7.4.5 Back in time: the initial probability distribution

In order to learn more about the initial fluctuations, beyond their power spectrum on large scales, one needs to trace the structure back in time. The forward integration of the GI equations by analytic approximations or by N-body simulations cannot be simply reversed despite the time reversibility of gravity. It is especially hopeless in collapsed systems where memory has been erased, but the case is problematic even for linear systems. When attempting backwards integration, the decaying modes (e.g., Peebles 1993, Ch. 5), having left no detectable trace at present, t_0, would amplify noise into dominant spurious fluctuations at early times. This procedure has a negligible probability of recovering the very special initial state of almost uniform density and tiny velocities which we assume for the real universe at t_{in}. This is a problem of mixed boundary conditions (see §2.3.1): some of the six phase-space variables per particle are given at t_{in} and some at t_0. This problem can be solved, for example, by applying the principle of least action (§§11.2.4, 2.3.4). The other way, described here, is based on the Zel'dovich approximations which is restricted to the growing mode (Nusser & Dekel 1992; 1993).

If the velocity field is irrotational, the Euler equation (7.6) can be replaced by the Bernoulli equation for the potentials, $\dot{\Phi}_v - (\nabla \Phi_v)^2/2 = -2H\Phi_v + \Phi_g$. The Zel'dovich approximation requires that each side vanishes: one side relates linearly Φ_v to Φ_g and the other we term "the Zel'dovich–Bernoulli equation" (ZB), $\dot{\varphi}_v - (\dot{D}/2)(\nabla \varphi_v)^2 = 0$, in which the potentials are expressed in units of $a^2\dot{D}$. The ZB equation can be integrated back in time with a guaranteed uniform solution at t_{in}. The input velocity potential φ_v at t_0 is extractable from observations of peculiar velocities or galaxy density, and the desired fields v and δ at t_{in} can be derived from φ_v of that time using linear theory. [While the ZB approximation conserves momentum, like δ_c of equation (7.9), one can alternatively satisfy continuity under the Zel'dovich approximation (as in $\delta_0 = -f^{-1}(\Omega)\nabla \cdot v$), and obtain instead a second-order equation for φ_g which can be more accurate than ZB for certain purposes (Gramman 1993b, eqs. 2.24, 2.25).] The recovered initial density fluctuation field δ_{in} has deeper valleys and shallower hills compared to naive recovery using linear theory, which makes structures such as the Great Attractor less eccentric as initial fluctuations than they appear today.

An issue of great interest is whether or not the initial fluctuations were statistically a Gaussian random field. A Gaussian field is characterized by the joint probability distributions (PDFs) of all order being generalized Gaussians (Bardeen et al. 1986), and in particular the one-point probability of δ is $P(\delta) \propto \exp[-\delta^2/(2\sigma^2)]$. Most Inflation scenarios naturally predict Gaussian fluctuations, but non-Gaussian fluctuations are allowed by certain versions of Inflation (e.g., Kofman et al. 1990) and by models where the perturbations are seeded by topological defects such as cosmic strings and textures, or by explosions (see Peebles 1993, Ch. 16). The present-day density PDF develops a log-normal-like shape due to nonlinear effects (Coles & Jones 1991; Kofman et al. 1994). The tails become positively skewed because peaks collapse to large densities while the density in voids cannot become negative, and the middle develops negative skewness as density hills contract and valleys expand. On the other hand, the PDF of present-day velocity components is insensitive to mildly-nonlinear effects (Kofman et al. 1994).

The observed PDFs today agree with N-body simulations of Gaussian initial conditions (Bouchet et al. 1993), but they have only limited discriminatory power against initial non-Gaussianities; the development of a density PDF with a general log-normal shape may occur even in certain cases of non-Gaussian initial fluctuations (e.g., Weinberg & Cole 1992), and the velocity PDF becomes Gaussian under general conditions due to the Central Limit Theorem whenever the velocity is generated by several independent density structures. A more effective strategy for addressing the Gaussianity of the initial conditions seems to be to take advantage of the actual dynamical fields at t_0 (as opposed to reducing the information content to the PDF right from the begining), trace the fields back in time, and use the linear fields to discriminate between theories. The Eulerian Zel'dovich approximation can be used to *directly* recover the initial PDF (IPDF) as follows (Nusser & Dekel 1993). The tensor $\partial v_i/\partial x_j$ derived from $v(x)$ is transformed to Lagrangian variables $q(x)$. The corresponding eigenvalues $\mu_i \equiv \partial v_i/\partial x_i$ and $\lambda_i \equiv \partial v_i/\partial q_i$ are related via the key relation $\lambda_i = \mu_i/(1 - f^{-1}\mu_i)$. In the Zel'dovich approximation $v \propto \dot{D}$ so the Lagrangian derivatives λ_i are traced back in time by simple scaling $\propto \dot{D}^{-1}$. The initial densities at $q(x)$ can then be computed using linear theory, $\delta_{in} \propto -(\lambda_1 + \lambda_2 + \lambda_3)$, and the IPDF is computed by bin counting of δ_{in} values across the Eulerian grid, weighted by the present-day densities at the grid points.

A key feature of the recovered IPDF is that it is sensitive to the assumed value of Ω when the input data is peculiar velocities (§7.5.2), and is Ω-independent when the input is density, so it could be used to robustly recover the IPDF from the density field of the 1.2 Jy IRAS survey (Nusser, Dekel & Yahil 1995). The IPDF so determined turns out to be only weakly dependent on galaxy biasing in the range $0.5 \leq b \leq 2$, at least for the power-law biasing relation tested. Errors were evaluated using mock IRAS-like catalogs. The IPDF derived from IRAS for $b = 1$ is shown in Fig. 7.16. It is *consistent with Gaussian*: the initial skewness and kurtosis are limited at the 3σ level to $-0.65 < S < 0.36$ and $-0.82 < K < 0.62$. Constraints of similar nature were derived from the spatial distribution of rich clsuters of galaxies (Kolatt, Dekel & Primack 1997). These limits may be useful for evaluating specific non-Gaussian models.

The COBE measurements are consistent with Gaussian fluctuations, but the large smoothing scale and the measurement errors limit their discriminatory power to strongly non-Gaussian models (Smoot et al. 1994). Future CMB data with sub-degree resolution will be much more useful in addressing the issue of non-Gaussianities. For example, scenarios seeded by topological defects predict many-sigma peaks that should be detectable at high probability by the planned experiments.

7.5 Direct measurements of Ω from peculiar velocities

Assuming that the inferred motions are real and generated by GI, they can be used to estimate Ω in several different ways. Most evidence from virialized systems on scales $\leq 10 h^{-1}$ Mpc suggest a low mean density of $\Omega \sim 0.2$ (see Dekel, Burstein & White 1997). The spatial *variations* of the large-scale velocity field provide ways to measure

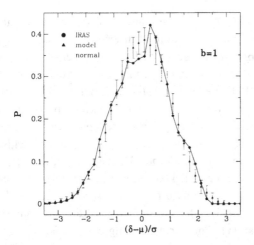

Figure 7.16. The density IPDF recovered from the density field of 1.2 Jy IRAS galaxies (solid) compared to a normal distribution (dotted) and to the IPDF recovered from the density field of Gaussian CDM simulations (triangles). $\mu \equiv \langle \delta \rangle$ and $\sigma^2 \equiv \langle (\delta - \mu)^2 \rangle$ (Nusser, Dekel & Yahil 1995).

the mass density in a volume that may be closer to a "fair" sample. One family of such methods is based on comparing the dynamical fields derived from velocities to the fields derived from galaxy redshifts (§7.6). These methods can be applied in the linear regime but they always rely on the assumed biasing relation between galaxies and mass often parameterized by b, so they actually provide an estimate of $\beta \equiv f(\Omega)/b$. Another family of methods measures β from redshift surveys alone, based on z-space deviations from isotropy (§§5.5.3, 7.6.6). In the present section, we focus on methods that rely on nonlinear effects in the peculiar-velocity data *alone*, and they thus provide estimates of Ω independent of galaxy-density biasing. These methods are based on the assumption that the initial fluctuations were Gaussian.

7.5.1 Outflow in voids

A diverging flow in an extended low-density region can provide a robust dynamical lower bound on Ω, based on the fact that large gravitating outflows are *not* expected in a low-Ω universe (Dekel & Rees 1994). In practice, for any assumed value of Ω, the partial derivatives of the smoothed observed velocity field are used to infer a nonlinear approximation for the mass density via the approximation δ_c, equation (7.9). A key point is that this approximation is typically an overestimate, $\delta_c > \delta$ (when the true value of Ω is used). For fluctuations that started Gaussian, the probability that δ_c is an overestimate, in the range $\delta < -0.5$, is well over 99%. Analogously to $\delta_0 \approx -\Omega^{-0.6} \nabla \cdot v$, the δ_c inferred from a given diverging velocity field is more negative when a smaller Ω is assumed, so for a small enough Ω one may obtain $\delta < -1$ in certain void regions.

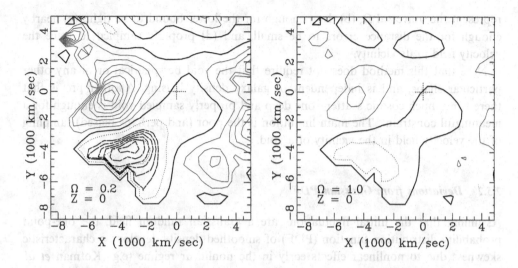

Figure 7.17. Maps of δ_c inferred from the observed velocities near the Sculptor void in the Supergalactic Plane, for two values of Ω. The LG is marked by '+' and the void is confined by the Pavo part of the GA (left) and the Aquarius extension of PP (right). Contour spacing is 0.5, with $\delta_c = 0$ heavy, $\delta_c > 0$ solid, and $\delta_c < 0$ dotted. The heavy-dashed contours mark the illegitimate downward deviation of δ_c below -1 in units of σ_δ, starting from zero (i.e., $\delta_c = -1$) and decreasing with spacing $-0.5\sigma_\delta$. The value $\Omega = 0.2$ is ruled out at the 2.9σ level (Dekel & Rees 1994).

Such values of δ are forbidden because mass is never negative, so this provides a lower bound on Ω.

The inferred δ_c field, smoothed at $12 h^{-1}$ Mpc, and the associated error field σ_δ, were derived by POTENT at grid points from the observed radial velocities of Mark III. Focusing on the deepest density wells, the input Ω was lowered until δ_c became significantly smaller than -1. The most promising "test case" provided by the Mark III data is a broad diverging region centered near the Supergalactic plane at the vicinity of $(X, Y) = (-25, -40)$ in h^{-1} Mpc – the "Sculptor void" of galaxies between the GA and the "Southern Wall" extension of PP (Fig. 7.17, compare to Fig. 7.6).

Values of $\Omega \approx 1$ are perfectly consistent with the data, but δ_c becomes smaller than -1 already for $\Omega = 0.6$. The values $\Omega = 0.3$ and 0.2 are ruled out at the 2.4σ and 2.9σ levels in terms of the random error σ_δ.

A preliminary application of this method to the SFI data, in which the Mathewson, Ford & Buchhorn (1992) data (which is dominant in the Sculptor void) were calibrated differently than in Mark III, suggests that the lower bound on Ω is even tighter.

This result is still to be improved. The systematic errors have been partially corrected in POTENT, but a more specific investigation of the biases affecting the smoothed velocity field in density wells is still in progress. For the method to be effective one needs to find a void that is (a) bigger than the correlation length for its vicinity to represent the universal Ω, (b) deep enough for the lower bound to be tight, (c) nearby enough for the distance errors to be small, and (d) properly sampled to trace the velocity field in its vicinity.

Note that this method does not require that the void be spherical or of any other particular shape, and is independent of galaxy density biasing. Another pro is that there is no much cosmic scatter – one deep and properly sampled void is enough for a meaningful constraint. The main limitation is the poor (and perhaps biased) sampling of the velocity field in the vicinity of a void.

7.5.2 Deviations from Gaussian PDF

Assuming that the initial fluctuations are a Gaussian random field, the one-point probability distribution function (PDF) of smoothed density develops a characteristic skewness due to nonlinear effects early in the nonlinear regime (e.g., Kofman et al. 1994; §7.4.5). The skewness of δ is given according to second-order perturbation theory by

$$\langle \delta^3 \rangle / \langle \delta^2 \rangle^2 \approx (34/7 - 3 - n), \tag{7.18}$$

with n the effective power index of the power spectrum near the (top-hat) smoothing scale (Bouchet et al. 1992). Since this ratio of moments for δ is practically independent of Ω, and since $\nabla \cdot v \sim -f\delta$, the corresponding ratio for $\nabla \cdot v$ must strongly depend on Ω, and indeed in second-order it is (Bernardeau et al. 1995)

$$T_3 \equiv \langle (\nabla \cdot v)^3 \rangle / \langle (\nabla \cdot v)^2 \rangle^2 \approx -f(\Omega)^{-1}(26/7 - 3 - n). \tag{7.19}$$

Using N-body simulations and $12\,h^{-1}$ Mpc smoothing one indeed finds $T_3 = -1.8 \pm 0.7$ for $\Omega = 1$ and $T_3 = -4.1 \pm 1.3$ for $\Omega = 0.3$, where the error is the cosmic scatter for a sphere of radius $40\,h^{-1}$ Mpc in a CDM universe ($H_0 = 75$, $b = 1$). An estimate of T_3 in the current POTENT velocity field within $40\,h^{-1}$ Mpc is -1.1 ± 0.8, where the errors this time represent distance errors. With the two errors added in quadrature, $\Omega = 0.3$ is rejected at the $\sim 2\sigma$ level (somewhat sensitive to the assumed $P(k)$).

Since the present-day PDF contains only part of the information stored in the data and is in some cases not that sensitive to the initial PDF (IPDF), a more powerful bound can be obtained by using the detailed present-day velocity field $v(x)$ to recover the IPDF, and using the latter to constrain Ω by measuring its Ω-dependent deviation from the assumed normal distribution (Nusser & Dekel 1993). The necessary "time machine" is provided by the Eulerian interpretation of the Zel'dovich approximation (§7.4.5).

The velocity out of POTENT Mark II, within a conservatively selected volume, was fed into the IPDF recovery procedure with Ω either 1 or 0.3, and the errors due to distance errors and cosmic scatter were estimated. Figure 7.18 shows the recovered IPDFs. The IPDF recovered for $\Omega = 1$ is marginally consistent with Gaussian, while the one recovered for $\Omega = 0.3$ shows significant deviations. The largest deviation, bin

Large-scale flows and cosmological implications

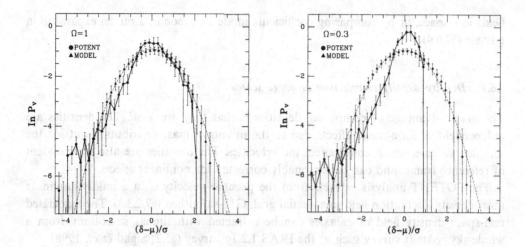

Figure 7.18. Ω from IPDF. The density IPDF recovered from the G2 POTENT peculiar-velocity field (solid), compared to a normal distribution (short dash) and with the IPDF recovered from the velocity field of Gaussian CDM simulations (triangles). The assumed Ω is 1.0 (*left*) or 0.3 (*right*). The simulations are of $\Omega_0 = \Omega$ accordingly (Nusser & Dekel 1993).

by bin in the IPDF, is $\lesssim 2\sigma$ for $\Omega = 1$ and $> 4\sigma$ for $\Omega = 0.3$, and a similar rejection of $\Omega = 0.3$ is obtained with a χ^2-type statistic. The skewness and kurtosis are poorly determined because of noisy tails but the replacements $\langle x|x|\rangle$ and $\langle |x|\rangle$ allow a rejection of $\Omega = 0.3$ at the $(5-6)\sigma$ levels.

The main advantage of the methods based on the PDF is their insensitivity to galaxy-density biasing. The main weakness is the need for a "fair" sample; the cosmic scatter is large due to the large smoothing scale within the limited volume.

7.6 Measurements of β from galaxy density and velocities

In linear GI theory, the density and velocity fluctuations are related via $\delta = -\Omega^{-0.6}\nabla \cdot v$. As a first, crude approximation the density of galaxies can be related to the local mass density by $\delta_g = b\delta$, where b is the linear biasing factor for the galaxies of that type and on the given scale (§7.6.7). Thus, the data of peculiar velocity and the data of galaxy density, which seem to be compatible with the models of GI and linear biasing (§7.3.2), can be combined to constrain the degenerate parameter $\beta \equiv \Omega^{0.6}/b$.

The comparison between the two datasets can be done in several different ways. In particular, it can be done by comparing density fields derived locally from the two datasets (§7.6.1), or by comparing velocities derived more globally from the two datasets (§7.6.4, §7.6.3). It can be done successively by first recovering fields from each dataset and then combining them to obtain β, or by a simultaneous recovery of fields and β determination from the two datasets (§7.6.2). It can be done by direct comparison of

7.6.1 Density–density comparison on large scales

The main advantage of comparing densities is that they are *local*. The densities are independent of long-range effects due to the unknown mass distribution outside the sampled volume, which could affect the velocities. The densities are also independent of reference frame, and can be reasonably corrected for nonlinear effects.

The POTENT analysis extracts from the peculiar-velocity data a mildly-nonlinear mass-density fluctuation field in a spatial grid, G12-smoothed (§7.2.3.4). The associated real-space density field of galaxies can be extracted with similar smoothing from a whole-sky redshift survey such as the IRAS 1.2 Jy survey (§5.2; Sigad et al. 1998).

A brief summary of the recovery of the IRAS density field is as follows. The solution to the linearized GI equation $\nabla \cdot v = -f\delta$ for an irrotational field is

$$v(x) = \frac{f}{4\pi} \int_{\text{all space}} d^3x' \, \delta(x') \frac{x' - x}{|x' - x|^3} \,. \tag{7.20}$$

The velocity is proportional to the gravitational acceleration, which ideally requires full knowledge of the distribution of mass in space. In practice, one is provided with a flux-limited, discrete redshift survey, obeying some radial selection function $\phi(r)$. The galaxy density is estimated by $1 + \delta_g(x) = \sum n^{-1} \phi(r_i)^{-1} \delta^3_{\text{Dirac}}(x - x_i)$, where $n \equiv V^{-1} \sum \phi(r_i)^{-1}$ is the mean galaxy density, and the inverse weighting by ϕ restores the equal-volume weighting. Equation (7.20) is then replaced by

$$v(x) = \frac{\beta}{4\pi} \int_{r<R_{\text{max}}} d^3x' \, \delta_g(x') \, S(|x' - x|) \frac{x' - x}{|x' - x|^3} \,. \tag{7.21}$$

Under the assumption of linear biasing, the cosmological dependence enters through β. The integration is limited to $r<R_{\text{max}}$ where the signal dominates over shot-noise. $S(y)$ is a small-scale smoothing window ($\geq 500 \, \text{km s}^{-1}$) essential for reducing the effects of nonlinear gravity, shot-noise, distance uncertainty, and triple-value zones.

The distances are estimated from the redshifts in the LG frame by

$$r_i = z_i - \hat{x}_i \cdot [v(x_i) - v(0)] \,. \tag{7.22}$$

Equations (7.21)–(7.22) can be solved iteratively: make a first guess for the x_i, compute the v_i by equation (7.21), correct the x_i by equation (7.22), and so on until convergence. The convergence can be improved by increasing β gradually during the iterations from zero to its desired value.

Even under $12 \, h^{-1}$ Mpc smoothing, δ is of order unity in places, necessitating a mildly-nonlinear treatment. Local approximations from v to δ were discussed in §7.2.3.4, but the non-local nature of the inverse problem makes it less straightforward. A possible solution is to find an inverse relation of the sort $\nabla \cdot v = F(\Omega, \delta_g)$, including nonlinear gravity and nonlinear biasing. This is a Poisson-like equation in which

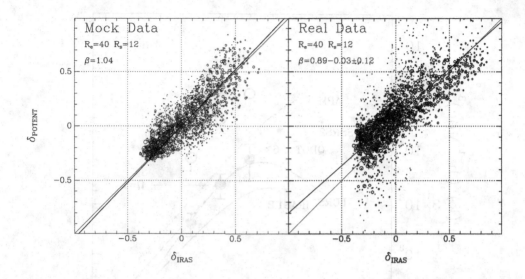

Figure 7.19. β from POTENT vs. IRAS density comparison. The smoothing is G12 and the comparison volume is of effective radius $40\,h^{-1}$ Mpc. Two-dimensional regression lines are marked. *Left:* Regression between the averages of 20 mock catalogs of each type, showing a bias as small as 4%. *Right:* Real data.

$-\beta \delta_g(x)$ is replaced by $F(x)$, and since the smoothed velocity field is still irrotational for mildly-nonlinear perturbations, it can be integrated analogously to equation (7.20). With smoothing of $10\,h^{-1}$ Mpc and $\beta = 1$, the approximation based on an empirical inverse to δ_c, equation (7.9), has an rms error $< 50\,\mathrm{km\,s^{-1}}$.

In recent reconstructions from redshift surveys, the galaxy-density field is recovered from the noisy IRAS data via a Power-Preserving Filter (PPF, by A. Yahil, described in Sigad *et al.* 1998, see our comment in §7.2.4) – a modification of the Wiener Filter. The PPF returns a field that is not far from the Wiener, most probable field, but it mimics the desired fixed smoothing by forcing the variance to be a spatial constant despite the fact that the errors vary. Another recent improvement in the reconstruction from redshift surveys is the implementaion of a statistical method to obtain the distance of galaxies in triple-valued zones (see below: §7.6.3, equation [7.26]).

The simplest way of comparing the POTENT and IRAS density fields is via a two-dimensional linear regression using the values of the fields at grid points within a local comparison volume. The errors of both fields enter the regression. The comparison volume is determined by equal-error contours.

The latest comparison of POTENT Mark III and IRAS 1.2 Jy data at $12\,h^{-1}$ Mpc smoothing within a volume of $(65\,h^{-1}\,\mathrm{Mpc})^3$ yields $\beta_I = 0.86 \pm 0.12$ (Sigad *et al.* 1998). The corresponding scatter diagram is shown in Fig. 7.19. The systematic error in this derivation, of only 4%, is deduced from the analogous scatter diagram for the averages of 20 random mock catalogs of each type. This is an update of the higher estimate

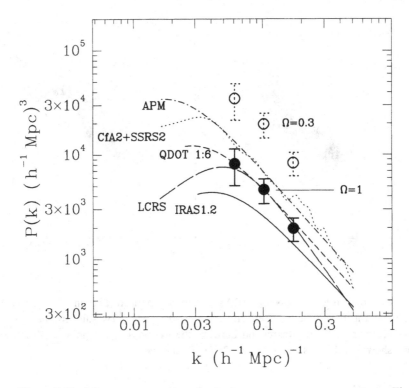

Figure 7.20. β from power spectra of galaxies versus mass power spectrum. The estimates from various galaxy-density samples were all translated from redshift space to real space using Kaiser's approximation and the best-fit value of β. The $P(k)$ from peculiar velocities (Fig. 7.13) is marked by solid symbols for $\Omega = 1$ (open symbols for $\Omega = 0.3$). The values of β (for any Ω) can be read directly from the vertical offset of the solid symbols and the corresponding curves (Kolatt & Dekel 1997).

$\beta_I = 1.3 \pm 0.3$ (Dekel et al. 1993) obtained based on an earlier version of POTENT with the Mark II velocities and the IRAS 1.9 Jy redshifts.

Similar comparisons of the mass-density field with the density of optical galaxies indicate a similar correlation and a $\sim 30\%$ lower estimate for β_O (Hudson et al. 1995), in agreement with the ratio of biasing factors, $b_O/b_I \approx 1.3$, obtained by direct comparison of optical and IRAS galaxy densities.

A comparison of similar nature of the POTENT Mark III data with the density distribution of Abell/ACO $R \geq 0$ clusters and the corresponding predicted velocities at G15 smoothing yields similar consistency out to distances of $\sim 60 h^{-1}$ Mpc, and an estimate of $\beta_C = 0.26 \pm 0.11$ (Branchini et al. 1999). This is consistent with a linear biasing factor for the clusters that is about 4 times larger than that of galaxies, in accordance with the observed ratio of about 4^2 for the corresponding correlation functions (see §4.10).

A direct comparison of the mass power spectrum as derived from peculiar velocities (§7.4.2) with the galaxy power spectra as derived from different redshift and angular surveys is shown in Fig. 7.20 (Kolatt & Dekel 1997). It demonstrates a similarity in shape and yields for the various galaxy types β values in the range 0.77–1.21, with a typical error of ± 0.1. For IRAS galaxies typically $\beta_I \gtrsim 1$, and for optical galaxies $\beta_O \lesssim 1$. These estimates do not directly address the value of Ω, but it is clear from the figure that if Ω is as small as ~ 0.3, then all galaxy types must be severely antibiased.

In principle, the degeneracy of Ω and b is broken in the mildly-nonlinear regime, where $\delta(v)$ is no longer $\propto f^{-1}$. Compatible mildly-nonlinear corrections in POTENT and in the IRAS analysis allowed a preliminary attempt to separate these parameters using the Mark II and IRAS 1.9 Jy data (Dekel et al. 1993). Unfortunately, nonlinear biasing effects are hard to distinguish from nonlinear gravitational effects, so a specific nonlinear biasing scheme is a prerequisite for such an analysis.

7.6.2 Simultaneous fit of velocity and density: SIMPOT

The dynamical fields and β can be recovered simultaneously by a fit of a parametric model for the potential field to the combined data of the observed radial peculiar velocities and the distribution of galaxies in redshift space. This procedure takes advantage of the complementary features of the data in the recovery of the fields, it enforces the same effective smoothing on the data without preliminary reconstruction procedures such as POTENT, and it obtains a more reliable best fit by simultaneous rather than successive minimization. It has been implemented so far to the forward TF data, but it can be generalized in principle to minimize inverse TF residuals.

In the SIMPOT procedure by Nusser & Dekel (1999), the model for the potential field is taken to be an expansion in spherical harmonics Y_{lm} and Bessel functions j_l where the coefficients Φ_{lmn} are the free parameters,

$$\Phi(r) = \sum_{l=1}^{l_{max}} \sum_{m=-l}^{l} \sum_{n=1}^{n_{max}} \Phi_{lmn} j_l(k_n r) Y_{lm}(\hat{r}). \tag{7.23}$$

The model radial velocity is derived from this potential by $u = -\partial \Phi/\partial r$, and the model density in redshift space is derived using linear theory via $\delta = f^{-1}\nabla^2 \Phi + s^{-2}(\partial/\partial s)(s^2 \partial \Phi/\partial s)$. The second term reflects redshift distortions, where s is the radial variable in redshift space. The resulting models for $u(x)$ and $\delta(s)$ are expansions in certain functions A_{lmn} and B_{lmn} that are appropriate combinations of the original base functions.

The combined χ^2 to minimize as a function of the parameters Φ_{lmn} and β is the sum of

$$\chi_u^2 = \sum_i \sigma_{ui}^{-2} \left[u_i^{obs} - \sum_{lmn} A_{lmn}(r_i)\Phi_{lmn} \right]^2 \tag{7.24}$$

and

$$\chi_\delta^2 = \int d^3s \, \sigma_\delta^{-2}(s) \left[\delta^{\text{obs}}(s) - \sum_{lmn} B_{lmn}(s) \Phi_{lmn} \right]^2. \tag{7.25}$$

The observations are the peculiar velocities u_i^{obs} and a continuous density field in redshift space $\delta^{\text{obs}}(s)$, that is somewhat more tricky to obtain. The β dependence enters only in χ_δ^2, both via the velocity–density relation and the redshift distortions.

A SIMPOT fit to the Mark III peculiar-velocity data and the IRAS 1.2 Jy redshift survey yields $\beta_I = 1.0 \pm 0.15$ for a smoothing window equivalent to a Gaussian of $12\,h^{-1}$ Mpc. This analysis provides a marginal hint for scale-dependent biasing, with β_I smaller than unity at smaller smoothing scales.

7.6.3 Velocity–velocity comparison on small scales: VELMOD

Earlier comparisons of the peculiar velocities from the Mark II catalog and the velocities predicted from the IRAS redshift surveys (QDOT, 1.9 Jy, and 1.2 Jy) yielded estimates of β_I in the range 0.4–1.0 (Kaiser et al. 1991; Nusser & Davis 1994).

The more sophisticated VELMOD method of comparison (Willick et al. 1997b) compares the raw peculiar-velocity data with a "model" velocity field that is predicted from the IRAS 1.2 Jy redshift survey. It is done without attempting to reconstruct a velocity field from the data. The key feature of VELMOD is that it explicitly allows for a non-unique mapping between real space and redshift space. Triple valuedness in the redshift field as well as non-negligible small-scale velocity "temperature" are treated in a unified way. This is done by expressing the probability that an object at distance r has a redshift z by

$$P(z|r) = \frac{1}{\sqrt{2\pi}\sigma_u} \exp\left[-\frac{1}{2} \frac{[z - r - u(r)]^2}{2\sigma_u^2} \right] \tag{7.26}$$

where $u(r)$ is the radial component of the model velocity field and σ_u is the small-scale velocity noise (which can in principle be a function of position). The above probability is then multiplied by the TF probability factor, $P(m, \eta, r)$, and integrated over the entire line-of-sight to obtain the probability of the observable quantities (m, η, z). One then maximizes that probability over the entire dataset.

The method is computer-intensive because numerical integrals are required for each galaxy, and for each fit parameter (TF parameters, σ_u, velocity field model parameters, etc.). This effort is worthwhile to the degree that the velocity field is triple-valued or the small-scale noise σ_u is comparable to the TF error. In particular, VELMOD is more rigorous in an analysis of the very local ($z \leq 3000\,\text{km s}^{-1}$) region.

VELMOD has been applied with a Gaussian smoothing of $3\,h^{-1}$ Mpc to the IRAS 1.2 Jy redshift survey and a subset of 838 spiral galaxies from the Mark III catalog within $z \leq 3000\,\text{km s}^{-1}$ of the Local Group. The method was tested successfully using mock catalogs drawn from the N-body simulation of Kolatt et al. (1996). When applied to the real data it yielded consistency with the model of linear GI and linear biasing once an artificial quadrupole was allowed, with $\beta_I = 0.5 \pm 0.1$ at $3h^{-1}$ Mpc. The catch is that it is not at all clear why linear GI and the simplified deterministic biasing should be valid for the densities and velocities at such high resolution. The estimated value of β should therefore be interpreted with caution.

It is interesting to note that Shaya, Peebles & Tully (1995) obtained a similarly low value for β_O from roughly the same local neighborhood. They apply the least-action reconstruction method (§§11.2.4, 2.3) to a composite redshift sample consisting of galaxies within ~3000 km s^{-1}, and obtain predicted distances which they then compare to TF distances of a subsample of their galaxies. Their result is obtained in terms of M/L and t_0, and they end up quoting a value for Ω, but they in fact measure a quantity related to β for a specific model of biasing. This method is likely to underestimate β because it assumes that the mass is all concentrated in the centers of galaxies and groups, which tends to overestimate the gravitational forces between them. This problem must be severe especially at earlier times, when the mass was distributed more uniformly. Preliminary tests using simple N-body simulations indicate that the method underestimates Ω by a factor of ~ 2.5, which Shaya et al. now crudely fix by softening the force on scales $\leq 4h^{-1}$ Mpc. Their latest result is $\Omega \simeq 0.25 \pm 0.1$ (private communication), but the systematic errors are yet to be properly quantified using detailed mock catalogs.

7.6.4 Inverse-TF fit in redshift space

As explained in §7.2.2 and §7.2.5, an inverse TF analysis in redshift space, which uses a parametric model for the velocity field, can be practically free of Malmquist bias. Davis, Nusser & Willick (1996, ITF) have used such a method to estimate β via a mode by mode comparison of the Mark III TF data and the IRAS 1.2 Jy redshift survey, without trying to recover a homogeneous velocity or density field.

The model for the peculiar-velocity field in this application is a set of orthogonal, smooth modes, reducing the data to a set of 56 coefficients fitted to a maximum redshift of 6000 km s^{-1} and maximum spherical harmonic of $l = 3$. The radial resolution of the modes degrades steeply with redshift, from 800 km s^{-1} locally to 3000 km s^{-1} at 4000 km s^{-1} redshift. Equivalent mode coefficients are computed for the gravity field derived from the whole-sky IRAS 1.2 Jy survey. Given the coefficients of the expansions, one compares the velocity and gravity vector fields on a mode by mode basis (or a galaxy by galaxy basis). The comparison shows the two independent fields to be remarkably aligned in general. There seem to be, however, apparent systematic discrepancies in the fields that lead to considerable coherence in the residuals between them. These residuals take the form of a dipole field in the LG frame that grows with distance, which is not consistent with a bulk flow residual alone.

A likelihood analysis in the mode by mode comparison, which takes into account the errors and covariance in both the velocity and gravity coefficients, yields β_I values in the range 0.4–0.6. However, in contrast with results obtained using simulated galaxy catalogs, the χ^2 per degree of freedom for the fit is well in excess of unity, primarily due to the coherent dipole residuals at $cz \gtrsim 3000\,\text{km}\,\text{s}^{-1}$. Thus, despite the general alignment of the velocity fields, they do not agree in full detail. The method is capable of measuring β to an accuracy of 10%, but without understanding these systematic discrepancies, one cannot infer a solid value of β from these data by this analysis.

One possible source of error in this analysis is in a procedure which forced the inverse TF parameters of the different Mark III datasets into a common system in order to enable a semi-analytic minimization. Another difficulty arises from the severe smoothing that is imposed in collapsing regions when analyzed in redshift space. A possible error in the determination of the appropriate Local Group rest frame may contribute an error to the acceleration field predicted from IRAS. Other uncertainties in the IRAS velocity predictions arise from the lack of proper information about the mass distribution at large distances, which may exert unknown quadrupole and higher moments into the velocity field. It is important to note that the density–density comparison (§7.6.1) may be free of such errors because the density field is determined locally by derivatives of the velocity field.

Finally, the high χ^2 may indicate a hidden source of scatter that is intrinsic to the biasing relation between galaxies and mass (§7.6.7).

7.6.5 Dipoles of galaxies versus CMB

Earlier attempts for measuring β involved the velocity of the Local Group in the CMB frame, $v(0)$, measured from the dipole in the CMB to be of amplitude $627\pm22\,\text{km}\,\text{s}^{-1}$. In the linear approximation, this velocity is expected to be proportional to the gravitational acceleration $g(0)$ that is exerted by the mass fluctuations around the LG, and can be computed by integrating the Poisson equation $\nabla \cdot v = -f^{-1}\delta$ under appropriate boundary conditions. One way to estimate $g(0)$ is from a whole-sky galaxy survey where only the angular positions and the fluxes (or diameters) are observed, exploiting the coincidence of nature that both the apparent flux and the gravitational force vary as r^{-2}. If $L \propto M$, then the vector sum of the fluxes in a volume-limited sample is proportional to $g(0)$ due to the mass in that volume. This idea can be modified to deal with a flux-limited sample once the luminosity function is known, and applications to the combined UGC/ESO diameter-limited catalog of optical galaxies yield β_O values in the range 0.3–0.5 (Lahav 1987; Lynden-Bell, Lahav & Burstein 1989). These estimates suffer from limited sky coverage, uncertain corrections for Galactic extinction, and different selection procedures defining the North and South Hemisphere samples. The IRAS catalog provides a superior sky coverage of 96% of the sky, with negligible Galactic extinction and with fluxes observed by one telescope, but with possible undersampling of cluster cores (Kaiser & Lahav 1989). A typical estimate from the angular IRAS catalog is $\beta_I = 0.9 \pm 0.2$ (Yahil, Walker & Rowan-Robinson 1986).

The redshift surveys provide the third dimension which could help deriving $g(0)$ by integrating the Poisson equation, subject to the difficulties associated with discrete, flux-limited sampling. The question is whether $g(0)$ is indeed predominantly due to the mass within the volume sampled, i.e., whether $g(0)$ as computed from successive concentric spheres converges interior to R_{max}. This is an issue of fundamental uncertainty (e.g., Lahav, Kaiser & Hoffman 1990; Juszkiewicz, Vittorio & Wyse 1990; Strauss et al. 1992b). The $r-z$ mapping could either compress or rarefy the z-space volume elements depending on the sign of u in the sense that an outflow makes the z-space density δ_z smaller than the true density δ: $\delta_z(x) \approx \delta(x) - 2[v(x) - v(0)] \cdot \hat{r}/r$. The varying selection function adds to this geometrical effect [in analogy to $n(r)$ in the IM bias (§7.2.3.1) and there is contribution from dv/dr as well (Kaiser 1987). It is thus clear that the redshifts must be corrected to distances and that any uncertainty in $v(x)$ at large x or at $x = 0$ would confuse the derived $g(0)$. The latter is the Kaiser "rocket effect": if $v(0)$ originates from a finite volume $r < r_0$ and the density outside r_0 is uniform with $v = 0$, then the measurements in z-space introduce a fake $g(0)$ in the direction of $v(0)$ due to the matter outside r_0, and this $g(0)$ is logarithmically diverging with r. The value of $v(0)$ is uncertain because it is derived like the rest of $v(x)$ from the density distribution – not from the CMB dipole. These difficulties in identifying convergence limit the effectiveness of this method in determining β and the power spectrum on large scales. The hopes for improvement by increasing the depth are not high because the signal according to conventional $P(k)$ models drops with distance faster than the shot-noise.

Attempting to measure β_I from the IRAS data, Strauss et al. (1992b) computed the probability distribution of $g(0)$ under several models for the statistics of fluctuations, via a self-consistent solution for the velocities and an ad hoc fix to the rocket effect, which enabled partial corrections for shot-noise, finite volume, and small-scale nonlinear effects. They confirmed that the direction of $g(0)$ converges to a direction only $\sim 20°$ away from the CMB dipole, but were unable to determine unambiguously whether $|g(0)|$ converges even within $100\,h^{-1}$ Mpc. A maximum likelihood fit and careful error analysis constrained β_I to the range 0.4–0.85 with little sensitivity to the $P(k)$ assumed. Rowan-Robinson et al. (1991) obtained from the QDOT dipole $\beta_I = 0.8^{+0.2}_{-0.15}$. Hudson's (1993b) best estimate from the optical dipole is $\beta_O = 0.72^{+0.37}_{-0.18}$.

The volume-limited Abell/ACO catalog of clusters with redshifts within $300\,h^{-1}$ Mpc was used to compute $g(0)$ in a similar way under the assumption that clusters trace mass linearly (Scaramella, Vettolani & Zamorani 1991). An apparent convergence was found by $\sim 180\,h^{-1}$ Mpc to the value $g(0) \approx 4860 \beta_C$ km s^{-1}. A comparison with the LG-CMB motion of 600 km s^{-1} yields $\beta_C \approx 0.123$, which corresponds to $\beta_O \approx 0.44$ and $\beta_I \approx 0.56$ if the ratios of biasing factors are 4.5 : 1.3 : 1. A similar analysis by Plionis & Valdarnini (1991) yielded convergence by $\sim 150\,h^{-1}$ Mpc and β values larger by $\sim 30-80\%$.

7.6.6 Redshift distortions

Redshift samples, which contain hidden information about velocities, can be used on their own to measure β (see also §5.5.3). The clustering, assumed isotropic in real space,

x, is anisotropic in z-space, z, where $z = r + v \cdot \hat{x}$ displaces galaxies along the preferred direction $\hat{x}$. While virial velocities on small scales stretch clusters into "Fingers of God" along the line-of-sight, systematic infall motions enhance large-scale structures by artificially squashing them along the line-of-sight. The linear approximation $-\nabla \cdot v = \beta \delta_g$ indicates that the effect is β-dependent because $-\nabla \cdot v$ is related to the anisotropy in z-space while δ_g is isotropic, so the statistical deviations from isotropy can yield β (e.g., Sargent & Turner 1977).

Kaiser (1987) showed in linear theory that the anisotropic Fourier $P(k)$ in z-space is related to the r-space $P(k)$ of mass density via

$$P^z(k, \mu) = P(k)(1 + \beta \mu^2)^2 , \qquad (7.27)$$

where $\mu \equiv \hat{k} \cdot \hat{x}$. This relation is valid only for a fixed μ, i.e., in a distant volume of small solid angle (Zaroubi & Hoffman 1996), but there are ways to apply it more generally. The redshift $P(k)$ can be decomposed into Legendre polynomials, $\mathscr{P}_l(\mu)$, with even multipole moments $P_l^z(k)$,

$$P^z(k, \mu) = \sum_{l=0}^{\infty} P_l^z(k) \mathscr{P}_l(\mu) , \quad P_l^z(k) = \frac{2l+1}{2} \int_{-1}^{+1} d\mu P^z(k, \mu) \mathscr{P}_l(\mu) . \qquad (7.28)$$

Based on equation (7.27) the first two non-vanishing moments are

$$P_0^z(k) = (1 + \frac{2}{3}\beta + \frac{1}{5}\beta^2)P(k) , \quad P_2^z(k) = (\frac{4}{3}\beta + \frac{4}{7}\beta^2)P(k) , \qquad (7.29)$$

so the observable ratio of quadrupole to monopole is a function of β independent of $P(k)$. An application to the 1.2 Jy and QDOT IRAS surveys yield $\beta_I = 0.52 \pm 0.13$ and 0.54 ± 0.3 respectively, at wavelength $30 - 40 h^{-1}$ Mpc. This result is suspected of being an underestimate because of nonlinear effects out to $\sim 50 h^{-1}$ Mpc (Cole, Fisher & Weinberg 1995). Peacock & Dodds (1994) developed a method for reconstructing the linear $P(k)$ and they obtain $\beta_I = 1.0 \pm 0.2$. Fisher & Nusser (1996) obtain $\beta_I = 0.6 \pm 0.2$.

The distortions should be apparent in the z-space two-point correlation function, $\xi_z(r_p, \pi)$, which is the excess of pairs with separation π along the line-of-sight and r_p transversely (Davis & Peebles 1983). The contours of equal ξ, assumed round in r-space, appear in z-space elongated along the line-of-sight at small separations and squashed on large scales depending on β. Hamilton (1992; 1993, 1995) used the multiple moments of ξ_z, in analogy to equations (7.28) and (7.29), and his various estimates from the 1.2 Jy and QDOT IRAS surveys are $\beta_I = 0.7 \pm 0.2$, with some indication that this might be an underestimate. Fisher et al. (1994b) computed $\xi_z(r_p, \pi)$ from the 1.2 Jy IRAS survey, and derived the first two pair-velocity moments. Their attempt to use the velocity dispersion via the Cosmic Virial Theorem led to the conclusion that this is a bad method for estimating Ω, but the mean, $\langle v_{12} \rangle = 109^{+64}_{-47}$ at $10 h^{-1}$ Mpc, yielded $\beta_I = 0.45^{+0.27}_{-0.18}$. The drawbacks of using ξ versus $P(k)$ are that (a) the uncertainty in the mean density affects all scales in ξ whereas it is limited to the $k = 0$ mode of the $P(k)$, (b) the errors on different scales in ξ are correlated whereas they are independent in a linear $P(k)$ for a Gaussian field, and (c) ξ mixes different physical scales, complicating the transition between the linear and nonlinear regimes. Nonlinear effects tend to make all the above results underestimates.

A promising method that is tailored to deal with a realistic redshift survey of a selection function $\phi(r)$ and does not rely on the subtleties of equation (7.27) is based on a weighted spherical harmonic decomposition of $\delta_z(z)$ (Fisher, Scharf & Lahav 1994),

$$a_{lm}^z = \int d^3z \, \phi(r) f(z) \left[1 + \delta_z(z)\right] Y_{lm}(\hat{s}), \tag{7.30}$$

$$\langle |a_{lm}^z|^2 \rangle = \frac{2}{\pi} \int_0^\infty dk \, k^2 P(k) \, |\psi_l^r(k) + \beta \psi_l^c(k)|^2 \,. \tag{7.31}$$

The arbitrary weighting function $f(z)$ is vanishing at infinity to eliminate surface terms. The mean-square of the harmonics is derived in linear theory assuming that the survey is a "fair" sample, and ψ^r and ψ^c are explicit integrals over r of certain expressions involving $\phi(r)$, $f(r)$, Bessel functions and their derivatives. The first term represents real structure and the second is the correction embodying the z-space distortions. The harmonic $P(k)$ in z-space, averaged over m, is thus determined by $P(k)$ and β, where the z-space distortions appear as a β-dependent excess at small l. The harmonic $P(k)$ derived from the 1.2 Jy IRAS survey yield $\beta_I = 1.0 \pm 0.3$ for assumed $\sigma_8 = 0.7$ (motivated by the IRAS ξ, Fisher et al. 1994a), with an additional systematic uncertainty of ± 0.2 arising from the unknown shape of the $P(k)$.

So far, the efforts to deduce β from redshift distortions lead to results that are uncertain, in the range $0.5 \leq \beta_I \leq 1.2$, and are sometimes puzzling (e.g., Hamilton 1995). The main limitations are noise and cosmic scatter – the volume sampled does not contain enough independent large-scale flow patterns. However, these methods have promising future because measuring redshifts is relatively inexpensive and can be done in large quantities. Already available are large surveys such as PSCZ from IRAS to a flux limit of 0.6 Jy and the Las Campanas Redshift Survey of optical galaxies in the south. Even larger surveys, of order one million galaxies, are in progress, such as the Two-Degree Field (2DF) and the Sloan Digital Sky Survey (SDSS) (see §5.7). These surveys will drastically reduce the cosmic scatter. With a sufficiently large redshift survey, one can even hope to be able to use the nonlinear effects to determine Ω and b separately. However, one should bear in mind that the β estimated from redshift distortions may be systematically different from the β estimated by other methods because of the stochastic nature of the galaxy biasing process (Dekel & Lahav 1998; §7.6.7).

Table 7.2 summarizes the estimates of β and Ω from cosmic flows.

7.6.7 Galaxy biasing as a stochastic process

In all the methods described in §7.6, the cosmological parameter of interest, Ω, is contaminated by the uncertain relation between galaxy density and mass density, the

Table 7.2. Ω and β from cosmic flows

Peculiar velocities alone	Gaussian IPDF	Nusser & Dekel 93	$\Omega > 0.3$ ($> 4\sigma$)
	Skewness($\nabla \cdot v$)	Bernardeau et al. 94	$\Omega > 0.3$ (2σ)
	Void	Dekel & Rees 94	$\Omega > 0.3$ (2.4σ)
	Power spectrum	Kolatt & Dekel 97	$\sigma_8\Omega^{0.6} = 0.7 \pm 0.15$
	+COBE	Zaroubi et al. 97a	$\sigma_8\Omega^{0.6} = 0.8 \pm 0.15$
Galaxy density vs. velocities	M2-QDOT v	Kaiser et al. 91	$\beta_\mathrm{I} = 0.9^{+0.2}_{-0.15}$
	M3-I1.2 v-dipole	Nusser & Davis 94	$\beta_\mathrm{I} = 0.6 \pm 0.2$
	M3-I1.2 v-inverse	Davis et al. 96	$\beta_\mathrm{I} = 0.6 \pm 0.2(?)$
	M3-I1.2 v G3	Willick et al. 96	$\beta_\mathrm{I} = 0.5 \pm 0.1$
	M3-I1.2 δ G12	Sigad et al. 98	$\beta_\mathrm{I} = 0.86 \pm 0.15$
	M2-Optical v	Hudson 94	$\beta_\mathrm{O} = 0.5 \pm 0.1$
	TF-Optical	Shaya et al. 97	$\beta_\mathrm{O} \simeq 0.44 \pm 0.15$
	M3-Optical δ G12	Hudson et al. 95	$\beta_\mathrm{O} = 0.75 \pm 0.2$
	M3-clusters G15	Branchini et al. 99	$\beta_\mathrm{C} = 0.26 \pm 0.11$
Redshift distortions	ξ I1.2	Peacock & Dodds 94	$\beta_\mathrm{I} = 1.0 \pm 0.2$
	ξ I1.2	Fisher et al. 94a	$\beta_\mathrm{I} = 0.45^{+0.3}_{-0.2}$
	Y_{lm} I1.2	Fisher et al. 94b	$\beta_\mathrm{I} = 1.0 \pm 0.3$
	P_k I1.2, QDOT	Cole et al. 95	$\beta_\mathrm{I} = 0.5 \pm 0.15$
	ξ I1.2, QDOT	Hamilton 95	$\beta_\mathrm{I} = 0.7 \pm 0.2$
	Y_{lm} I1.2	Heavens & Taylor 95	$\beta_\mathrm{I} = 1.1 \pm 0.3$
	P_k I1.2	Fisher & Nusser 96	$\beta_\mathrm{I} = 0.6 \pm 0.2$
CMB dipole	vs. galaxies angular	Yahil et al. 86	$\beta_\mathrm{I} = 0.9 \pm 0.2$
	vs. galaxies redshift	Strauss et al. 92b	$\beta_\mathrm{I} = 0.4 - 0.85$
		Rowan-Rob. et al. 91	$\beta_\mathrm{I} = 0.8^{+0.2}_{-0.15}$
	vs. galaxies angular	Lynden-Bell et al. 89	$\beta_\mathrm{O} = 0.3 - 0.5$
	vs. galaxies redshift	Hudson 93b	$\beta_\mathrm{O} = 0.7^{+0.4}_{-0.2}$
	clusters	Scaramella et al. 91	$\beta_\mathrm{C} \sim 0.13$
		Plionis et al. 91	$\beta_\mathrm{C} \sim 0.17 - 0.22$

$\beta \equiv \Omega^{0.6}/b$, $b_\mathrm{C} : b_\mathrm{O} : b_\mathrm{I} \approx 4.5 : 1.3 : 1.0$, $\sigma_8\Omega^{0.6} = (0.69 \pm 0.05)\beta_\mathrm{I}$, M3= Mark III, I1.2 = IRAS 1.2 Jy, G12 = Gaussian smoothing $12\,h^{-1}$ Mpc

so called "galaxy biasing". Non-trivial galaxy biasing clearly exists. The fact that galaxies of different types cluster differently (Dressler 1980) implies that at least some do not trace the underlying mass. This is hardly surprising because any reasonable physical theory would predict non-trivial biasing (Kaiser 1984; Davis et al. 1985; Bardeen et al. 1986; Dekel & Silk 1986; Dekel & Rees 1987; Braun, Dekel & Shapiro 1988; Weinberg 1995). In particular, simulations of galaxy formation in a cosmological context (e.g., Cen & Ostriker 1992; 1993; Lemson et al. 1998) indicate a biasing relation that is nonlinear in density, is varying with scale, and has a statistical scatter reflecting dependencies on factors other than density.

One should therefore not be surprised by the fact that the various estimates of β span a large range, from less than one-half to more than unity. Some of this scatter is due to the different types of galaxies involved, and some may be due to remaining effects of nonlinear gravity or other systematic errors, but a significant fraction of the scatter in β is likely to reflect non-trivial properties of the biasing scheme. This means that translating a measured β into Ω is non-trivial; it requires a detailed knowledge of the relevant biasing scheme.

In order to strengthen this point, we demonstrate below that an obvious source of systematic variations in β is the inevitable *statistical* scatter in the biasing process (Dekel & Lahav 1998). This scatter in the relation between densities can be interpreted as reflecting the dependence of galaxy formation efficiency, or galaxy density, on physical properties of the protogalaxy environment other than density. These could be local properties such as the potential field, the deformation tensor, tidal effects, and angular momentum, or long-range effects carried by radiation or particles from neighboring sources. The shot noise can also be an important source of scatter if not properly removed. In the simple example below, we assume that this scatter in the biasing is local and neglect possible spatial correlations.

Let $\delta(x)$ be the field of mass-density fluctuations smoothed with a given window, and let $\delta_g(x)$ be the corresponding field for galaxies of a given type. We treat them as random fields, both with probability densities of zero mean by definition. Denote $\langle \delta^2 \rangle \equiv \sigma^2$ and $\langle \delta^3 \rangle \equiv S$. Consider the "biasing" relation between galaxies and mass to be a *random* process, specified by the *conditional probability* function $P(\delta_g|\delta)$. The common deterministic biasing relation, $\delta_g = b(\delta)\delta$, is replaced by the conditional mean,

$$\langle \delta_g|\delta \rangle \equiv b(\delta)\delta. \tag{7.32}$$

The statistical character of the relation is expressed by the conditional moments of higher order about the mean, such as

$$\langle (\delta_g - b\delta)^2|\delta \rangle \equiv \sigma_b^2 \sigma^2, \quad \text{and} \quad \langle (\delta_g - b\delta)^3|\delta \rangle \equiv S_b S. \tag{7.33}$$

This statistical nature of biasing leads to a different "biasing parameter" for each specific application.

Take for example the ratio of variances, $b_{\text{var}}^2 \equiv \langle \delta_g^2 \rangle / \langle \delta^2 \rangle$, such as being obtained by a ratio of power spectra or two-point correlation functions, or by comparing the mass function of clusters to the variance of δ_g at $8\,h^{-1}$ Mpc (White et al. 1993). One can prove in general that $\langle \delta_g^m \rangle = \langle \langle \delta_g^m|\delta \rangle_{\delta_g} \rangle_\delta$, and therefore, $\langle \delta_g^2 \rangle = \langle b^2(\delta)\delta^2 \rangle + \langle \sigma_b^2(\delta) \rangle \sigma^2$. Thus, in the simple case where $b(\delta)$ is constant, b_{var} is an overestimate of b by

$$b_{\text{var}} = b(1 + \sigma_b^2/b^2)^{1/2}. \tag{7.34}$$

Another common way of estimating β is via linear regression of the noisy field $-\nabla \cdot v(x)$ [$\approx f(\Omega)\delta(x)$] on $\delta_g(x)$ (§7.6.1), or via a regression of the corresponding velocities. The slope of the forward regression of δ_g on δ is $b_{\text{for}} = \langle \delta_g \delta \rangle / \langle \delta^2 \rangle$, and the slope of the inverse regression of δ on δ_g is $b_{\text{inv}}^{-1} = \langle \delta \delta_g \rangle / \langle \delta_g^2 \rangle$. In the case where b is constant, $b_{\text{for}} = b$, and b_{inv} is an overestimate,

$$b_{\text{inv}} = b(1 + \sigma_b^2/b^2). \tag{7.35}$$

The promising method of estimating β from large-scale redshift distortions (§7.6.6) measures yet a different quantity. This measure is relatively insensitive to the stochasticity in the biasing scheme, if local, but it is affected by the nonlinear biasing features. In general, it turns out that most methods for determining β lead similarly to an underestimate.

The magnitude of the effect depends on the actual values of σ_b and similar parameters. One way to estimate the natural biasing scatter at a given smoothing scale is by investigating goodness of fit of the density fields of mass and light and the model of deterministic biasing. By requiring that $\chi^2 = 1$ per degree of freedom one can estimate the scatter needed in addition to the known errors. For example, Hudson et al. (1995) estimated for optical galaxies versus POTENT Mark III mass at $12\,h^{-1}$ Mpc smoothing $\sigma_b^2/b^2 \sim 0.25$. Alternatively, one can estimate σ_b from theoretical simulations. For example, preliminary hydro simulations (Cen & Ostriker 1993) yield under $10\,h^{-1}$ Mpc Gaussian smoothing $\sigma_b^2/b^2 \sim 0.4$. If $b = \Omega = 1$, then the β values derived by the various methods are expected to span the range $0.7 \leq \beta \leq 1$, and this is solely due to the dispersion in the biasing relation. A large skewness in $P(\delta_g|\delta)$ may stretch this range even further.

The discussion above referred to one fixed smoothing length. When the bias relation at a given smoothing is nonlinear (i.e., b is a function of δ), or when it is non-deterministic (e.g., $\sigma_b \neq 0$), further smoothing to a larger scale could significantly alter the biasing relation and its moments. A certain scale-dependence is thus inevitable (Somerville et al. 1999).

A relevant moral from the biasing uncertainty of β is that methods for measuring Ω independent of density biasing (§7.5) are desirable. However, it has to be borne in mind that the galaxies may also be biased tracers of the *velocity* field of the matter (see §11.2.5). Such a "velocity biasing" would affect any attempt to extract dynamical information from large-scale velocities. The expected magnitude of the velocity biasing in the standard scenarios of structure formation is a matter of debate, and even its sign is unclear (e.g., Summers, Davis & Evrard 1995). Based on recent simulations it seems likely to be limited to a ~ 10–20% effect.

7.7 Cosmological parameters

The previous sections discussed measurements of Ω (or β) from large-scale data of peculiar velocities on scales 10–$100\,h^{-1}$ Mpc, alone or combined with redshift surveys or microwave background data. In this section we try to put these estimates in a wider perspective (see Dekel, Burstein & White 1997 for a review). Complementary discussions are provided in Chs. 1, 11, and 8.

One very interesting large-scale constraint that has not been discussed here is based on cluster abundance, that can be predicted for a Gaussian field via the Press–Schechter formalism, and is quite insensitive to the shape of the power spectrum. The current estimates are $\sigma_8 \Omega_m^{0.6} \simeq 0.5$–$0.6$ (White, Cole & Frenk 1993; Eke et al. 1996; Mo, Ying

& White 1996). This is only slightly lower than the estimates of $\sigma_8 \Omega_m^{0.6} \simeq 0.7$–$0.8$ from the power spectrum of the peculiar-velocity data (§§7.4.2, 7.4.3). Note that this quantity is related to β_I via $\sigma_8 \Omega_m^{0.6} = \sigma_{8I} \beta_I$, where σ_{8I} is the rms fluctuation of IRAS galaxies in a top hat window of $8h^{-1}$ Mpc. With the estimate from the IRAS 1.2 Jy survey of $\sigma_{8I} = 0.69 \pm 0.05$ (Fisher et al. 1994a), the results from cluster abundance and from the δ–δ POTENT–IRAS comparison (§7.6.1) are in pleasant agreement.

Constraints from virialized systems such as galaxies and clusters on smaller comoving scales of $1 - 10\,h^{-1}$ Mpc are discussed in §1.4, Chs. 4 and 11; they typically yield low values of $\Omega_m \sim 0.2$–0.4, but with several loopholes. Most interesting among these is the constraint involving the baryonic fraction in clusters from X-ray or Sunyaev–Zel'dovich (SZ) data, and the estimates of Ω_b from the observed deuterium abundance and the theory of Big-Bang Nucleosyntheis. With the baryonic fraction $f_b \simeq 0.1 h_{60}^{-1}$ as recently observed from SZ in five clusters (Myers et al. 1997; see also White & Fabian 1995 for X-ray measurements), and with the recent estimates of $\Omega_b \simeq 0.067 h_{60}^{-2}$ (Tytler, Fan & Burles 1996; Burles & Tytler 1996), the current estimate is $\Omega_m \simeq 0.67 h_{60}^{-1}$. This result is compatible with the estimates from large-scale flows.

In this section we review in some more detail the constraints from global measures of Ω, which commonly involve the cosmological constant Λ. We adopt as our basic working hypothesis the standard cosmological model of Friedmann–Robertson–Walker (FRW), where we assume homogeneity and isotropy and describe gravity by general relativity. We limit the discussion to the matter-dominated era in the 'dust' approximation.

The Friedmann equation that governs the universal expansion can be written in terms of the different contributions to the energy density (e.g., Carroll et al. 1992):

$$\Omega_m + \Omega_\Lambda + \Omega_k = 1,$$

$$\Omega_m \equiv \frac{\rho_m}{(3H^2/8\pi G)}, \quad \Omega_\Lambda \equiv \frac{\Lambda c^2}{3H^2}, \quad \Omega_k \equiv \frac{-kc^2}{a^2 H^2}. \tag{7.36}$$

Here, $a(t)$ is the expansion factor of the universe, $H(t) \equiv \dot{a}/a$ is the Hubble constant, $\rho_m(t)$ is the the mass density, Λ is the cosmological constant, and k is the curvature parameter. Hereafter, the above symbols for the cosmological parameters refer to their values at the present time, t_0.

We denote $\Omega_{tot} \equiv \Omega_m + \Omega_\Lambda$, which by equation (7.36) equals $1 - \Omega_k$; its value relative to unity determines whether the universe is open ($k = -1$), flat ($k = 0$), or closed ($k = +1$). Another quantity of interest is the deceleration parameter, $q_0 \equiv -a\ddot{a}/\dot{a}^2$, which by equation (7.36) is related to the other parameters via $q_0 = \Omega_m/2 - \Omega_\Lambda$.

The FRW model also predicts a relation between the dimensionless product $H_0 t_0$ and the parameters Ω_m and Ω_Λ. For $\Omega_\Lambda = 0$, this product ranges between 1 and $2/3$ for Ω_m in the range 0 to 1 respectively, and it is computable for any values of Ω_m and Ω_Λ (§7.7.5).

The global measures commonly involve combinations of the cosmological parameters. Constraints in the Ω_m–Ω_Λ plane are displayed in Fig. 7.21 (see p. 311).

7.7.1 Occam's Razor

The above working hypotheses, and the order by which more specific models should be considered against observations, are guided by the principle of Occam's Razor, i.e., by simplicity and robustness to initial conditions. The caveat is that different researchers might disagree on the evaluation of "simplicity".

It is commonly assumed that the simplest model is the Einstein–deSitter model, $\Omega_m = 1$ and $\Omega_\Lambda = 0$. One property that makes it robust is the fact that Ω_m remains constant at all times with no need for fine tuning at the initial conditions.

The most natural extension according to the generic model of Inflation is a flat universe, $\Omega_{tot} = 1$, where Ω_m can be smaller than unity but only at the expense of a non-zero cosmological constant.

These simple models could serve as useful references, and even guide the interpretation of the results, but they should not bias the measurements.

7.7.2 Classical tests of geometry

The parameter-dependent large-scale geometry of spacetime is reflected in the volume–redshift relation. There are two classical versions of the tests that utilize this dependence: magnitude versus redshift (or "Hubble diagram") and number density versus redshift. The luminosity distance d_l to a redshift z, which enters the Hubble diagram test, depends on Ω_m and Ω_Λ via the integral (e.g., Carroll, Press & Turner 1992)

$$d_l(z) = \frac{c(1+z)}{H_0 |\Omega_k|^{1/2}} S_k \left[|\Omega_k|^{1/2} \int_0^z F(\Omega_m, \Omega_\Lambda, z') \, dz' \right] ,$$

$$F(\Omega_m, \Omega_\Lambda, z) \equiv [(1+z)^2(1+\Omega_m z) - z(2+z)\Omega_\Lambda]^{-1/2} , \qquad (7.37)$$

where $S_0(x) \equiv x$, $S_{+1} \equiv \sin$ and $S_{-1} \equiv \sinh$. At $z \sim 0.4$, d_l happens to be (to a good approximation) a function of the combination $\Omega_m - \Omega_\Lambda$ (not q_0) (Perlmutter et al. 1996). The angular-diameter distance, which enters the tests based on number density, is simply $d_a = d_l/(1+z)^2$.

The main advantage of such tests is that they are direct measures of global geometry and thus independent of assumptions regarding the mass type and distribution, the statistical nature of the fluctuations, the growth by GI and galaxy biasing. The galaxy-type "standard candles" that were used over the years clearly suffer from severe evolution complications. Supernovae Type Ia are the popular current candidate for a standard candle, based on the assumption that stellar processes are not likely to vary much in time. However, some caution is in place as long as we lack a complete theory for supernovae.

The first seven supernovae analyzed by Perlmutter et al. (1996) at $z \sim 0.4$ yielded $-0.3 < \Omega_m - \Omega_\Lambda < 2.5$ as the 90% two-parameter likelihood contour (see Fig. 7.21). However, improved results from tens of supernovae seem to favor a non-zero cosmological constant, $\Omega_\Lambda - \Omega_m > 0$. The preliminary success of the method may indicate that it will be able to separate the dependences on Ω_m and Ω_Λ within a few years, once several supernovae are measured at $z \sim 1$ (Goobar et al. 1995).

7.7.3 Number count of quasar lensing

This is a promising new version of the classical number density test. When Ω_Λ is positive and comparable to Ω_m, the universe should have gone through a phase of slower expansion in the recent cosmological past, which should be observed as an accumulation of objects at a specific redshift of order unity. In particular, it should be reflected in the observed rate of lensing of high-redshift quasars by foreground galaxies (Fukugita et al. 1990). The probability of lensing of a source at redshift z_s by a population of isothermal spheres of constant comoving density as a function of the cosmological parameters is (Carroll, Press & Turner 1992):

$$P_{lens} \propto \int_0^{z_s} (1+z)^2 F(\Omega_m, \Omega_\Lambda, z) \, [d_a(0,z) d_a(z, z_s) / d_a(0, z_s)]^2 \, dz \,, \tag{7.38}$$

where $d_a(z_1, z_2)$ is the angular diameter distance from z_1 to z_2. The contours of constant lensing probability in the $\Omega_m - \Omega_\Lambda$ plane for $z_s \sim 2$ happen to almost coincide with the lines $\Omega_m - \Omega_\Lambda = \text{const}$. The limits from lensing are thus similar in nature to the limits from SN Ia.

This test shares all the advantages of direct geometrical measures. The high redshifts involved bring about a unique sensitivity to Ω_Λ, compared to the negligible effect that Ω_Λ has on the structure observed at $z \ll 1$. However, the constraint is weakened if the lensed images are obscured by dust in the early-type galaxies responsible for the lensing, or if these galaxies had rapid evolution between $z \sim 1$ and the present.

From the failure to detect the accumulation of lenses, the current limit for a *flat* model is $\Omega_\Lambda < 0.66$ (or $\Omega_m > 0.36$) at the 95% confidence level (Kochanek 1996) (see Fig. 7.21).

7.7.4 Microwave-background acoustic peaks

This is the most promising test, which is expected to provide the most stringent constraints on the cosmological parameters within a decade. The test uses the effect of the background cosmology on the geodesics of photons. Current ground-based and balloon-borne experiments already provide preliminary constraints on the *location* of the first acoustic peak on sub-degree scales in the angular power spectrum of CMB temperature fluctuations, $l(l+1)C_l$. The dependence of the peak location on the cosmological parameters enters via the combined effect of (a) the physical scale of the "sound horizon" that is proportional to the cosmological horizon at recombination, and (b) the geometry of spacetime via the angular-diameter distance. In the vicinity of a flat model, the first peak is predicted at approximately the multipole (e.g., (White & Scott 1996)

$$l_{peak} \simeq 220(\Omega_m + \Omega_\Lambda)^{-1/2}. \tag{7.39}$$

The next generation of CMB satellites (MAP, to be launched by NASA in 2001, and in particular Plank, scheduled by ESA for 2004) are planned to obtain a precision at ~ 10 arcmin resolution that will either rule out the current framework of GI for structure formation or will measure the cosmological parameters to high precision. Detailed evaluation of Plank shows that nominal performance and expected foreground subtraction noise will allow parameter estimation with the following accuracy (ignoring systematics): $H_0 \pm 1\%$, $\Omega_{tot} \pm 0.005$, $\Omega_\Lambda \pm 0.02$, $\Omega_b \pm 2\%$.

The precision hoped for is much better than attainable with any other known method. If the observations fit the model, the precision is such that the model will be confirmed beyond reasonable doubt. The constraints on Ω_{tot} come mostly from geometrical effects. The interpretation is based on well-understood physics of sound waves in the linear regime, and on the assumption of absence of any relevant preferred scale (in the Mpc to Gpc range) in the physics which generated the initial structure. The latter assumption can be checked directly by the observations themselves. On the other hand, the measurements might be messed up by unexpected foreground contamination (e.g., by diffuse matter in galaxy groups).

Balloon-borne and ground-based observations have already confirmed the existence of the first acoustic peak, and have constrained its location to the vicinity $l \sim 200$. The results of COBE's DMR ($l \sim 10$) provide an upper bound of $\Omega_m + \Omega_\Lambda < 1.5$ at the 95% confidence level for a scale-invariant initial spectrum (and the constraint becomes tighter for any "redder" spectrum, $n < 1$) (White & Scott 1996). Several balloon-borne experiments ($l \sim 50 - 200$) strengthen this upper bound (e.g., the Saskatoon experiment, Scott et al. 1996, see §3.5). The Saskatoon experiment and the CAT experiment ($l \sim 350 - 700$) yield a preliminary lower bound of $\Omega_m + \Omega_\Lambda > 0.3$ (Hancock et al. 1998) (Fig. 7.21).

7.7.5 The age of the universe

Measured independent lower bounds on the Hubble constant and on the age of the oldest globular clusters provide a lower bound on $H_0 t_0$ ($= 1.05ht$, where $H_0 \equiv 100h \mathrm{km\,s}^{-1}\,\mathrm{Mpc}^{-1}$ and $t_0 \equiv 10t$ Gyr), and thus an interesting constraint in the $\Omega_m - \Omega_\Lambda$ plane. The exact expressions are computable in the various regions of parameter space. For example, a useful approximation in the presence of a cosmological constant, that is an exact solution for a flat universe, is (Carroll, Press & Turner 1992)

$$H_0 t_0 = \frac{2}{3} \frac{1}{|1-\Omega_a|^{1/2}} S_a^{-1}\left(\frac{|1-\Omega_a|^{1/2}}{\Omega_a^{1/2}}\right), \quad \Omega_a \equiv 0.7\Omega_m - 0.3\Omega_\Lambda + 0.3, \tag{7.40}$$

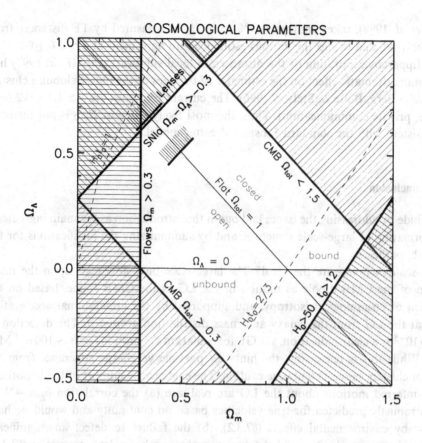

Figure 7.21. Current limits ($\sim 2\sigma$) on the cosmological parameters Ω_m and Ω_Λ from global measures: luminosity distance of SN Ia, lens count, the location of the CMB peak, and the age versus Hubble constant. The short marks are the one-parameter 95% limits from SN Ia and lenses for a flat universe. Also shown (vertical line) is the 95% lower bound on Ω_m from cosmic flows. The most likely value of Ω_m lies in the range 0.5 to 1. The Einstein–deSitter model is permitted. An open model with $\Omega_m \simeq 0.2$ and $\Omega_\Lambda = 0$, or a flat model with $\Omega_m \simeq 0.3$ and $\Omega_\Lambda \simeq 0.7$, are ruled out (Dekel, Burstein & White 1997).

where $S^{-1}_{\Omega_a \leq 1} \equiv \sinh^{-1}$ and $S^{-1}_{\Omega_a > 1} \equiv \sin^{-1}$. A useful crude approximation near $H_0 t_0 \sim 2/3$ is

$$\Omega_m - 0.7\Omega_\Lambda \simeq 5.8(1 - 1.3ht). \tag{7.41}$$

Progress has been made in measuring H_0 via the HST key project detecting Cepheids in nearby clusters for calibration of TF distances, and via accurate distances to SNe Type Ia. The new calibration of local Cepheids by the Hipparcos astrometric satellite (Feast & Catchpole 1997) has reduced the estimates of H_0 by $\sim 10\%$. The indications from SNe velocities for a local void of radius $\sim 75 h^{-1}$ Mpc out to the Great Wall

(Zehavi et al. 1998), takes another $\sim 5\%$ from H_0 as measured by TF distances from within the void, and brings the various estimates into agreement at $h \simeq 0.6 \pm 0.1$.

The Hipparcos calibration of the distances to local subdwarf stars (Reid 1997) had an even more dramatic effect on the estimates of the ages of the oldest globular clusters (e.g., VandenBerg, Bolte & Stetson 1996). The current estimates are $t \simeq 1.1 \pm 0.2$ (e.g., M. Bolte, private communication). Thus, the most likely value of $H_0 t_0$ is not far from 2/3, consistent with the standard Einstein–deSitter model.

7.8 Conclusion

We conclude by addressing the issues laid out in the introduction as the main ingredients of the formation of large-scale structure, and by summarizing the justifications for the working hypotheses.

Large-scale motions – are they real? The large-scale bulk flow rests upon the interpretation of the CMB dipole as motion of the LG in the CMB frame, based on the *assumption* of cosmological isotropy, and supported by the gravitational acceleration derived at the LG from the galaxy and mass distribution around it. The detection of $\delta T/T \sim 10^{-5}$ is a clear indication, via GI, for $\sim 300 \, \mathrm{km \, s^{-1}}$ motions over $\sim 100 \, h^{-1}$ Mpc (§7.3.1). While this is reassuring, the hints for possible very-large coherence from the motion of clusters suggests tentative caution. The evidence in support of the hypothesis that TF-inferred motions about the LG are real, are (a) the correlation $\delta_g \propto -\nabla \cdot v$, which is robustly predicted for true velocities based on continuity and would be hard to mimic by environmental effects (§7.3.2), (b) the failure to detect any significant correlation between velocities and the environment or other galaxy properties (§7.3.3), and (c) the similarity between the velocity fields traced by spirals and by ellipticals (§7.3.3).

Linear biasing – is it a good approximation? The galaxy velocity–density correlation is sensitive to the biasing scheme, and the observed correlation on each of the scales tested, given the errors, is consistent with linear biasing (properly modified in the tails). Generally speaking, the galaxies do trace the mass. However, it is difficult to distinguish nonlinear biasing from nonlinear gravitational effects, and the range of different estimates of β (§7.6) argues that the biasing scheme is not trivial when inspected in detail, e.g., it involves intrinsic scatter, scale-dependence, and nonlinear features.

Gravity – is it the dominant source of LSS? The observed velocity–density correlation is fully consistent with GI (§7.3.2), but it is sensitive to continuity more than to the specific time-dependence implied by gravity. Any non-GI process followed by a gravitating phase would end up consistent with this observation, and certain non-GI models may show a similar spatial behavior even if gravity never plays any role. The elliptical–spiral correlation is also consistent with gravity where galaxies of all types trace the same velocity field, but, admittedly, any model where all galaxies are set into motion by the same mechanism could pass this test. The strongest evidence for gravitational origin comes from the statistical agreement between the fluctuations observed today in our local neighborhood and those implied by the CMB at the time of recombination. A marginal warning signal for GI is provided by the $\sim 700 \,\mathrm{km\, s^{-1}}$ bulk velocity indicated for rich clusters across $\sim 200\, h^{-1}$ Mpc. Such a velocity at face value would be in conflict with the gravitational acceleration implied by the cluster distribution and with the $\delta T/T \sim 10^{-5}$ at $\sim 2°$, but the errors in the cluster measurements are large.

Potential flow. This property, which is used in the reconstruction from either velocities or densities, is impossible to deduce solely from observations of velocities along the lines of sight from one origin. Irrotationality is assumed based on the theory of GI, or it can be tested against the assumption of isotropy by measuring the isotropy of the velocity field derived by potential analysis for a fair sample.

Initial fluctuations. The observed CMB fluctuations provide evidence for initial fluctuations that are consistent with a *power spectrum* near scale invariance, $n \sim 1$. The observed flows confirm the general consistency with $n \sim 1$ on very large scales (§7.3.1). A CDM power spectrum with $\Omega = 1$ can simultaneously fit the CMB and flows data with $n \sim 0.8$–0.9 (§7.4). There are preliminary indications for an excess of power on 100–200 h^{-1} Mpc scale (Broadhurst et al. 1990; S. Landy based on LCRS, private communication; Cohen et al. 1996; Einasto 1997). The question of whether the fluctuations were *Gaussian* cannot be answered by observed velocities alone. The PDF of $\nabla \cdot v$ is consistent with Gaussian initial fluctuations skewed by nonlinear gravity, but this is not a discriminatory test. On the other hand, the galaxy spatial distribution does indicate Gaussian initial fluctuations fairly convincingly (§7.4.5)

Dark matter – what is it made of? In view of the nucleosynthesis constraints on baryonic density, the high Ω indicated by the motions requires nonbaryonic DM. The observed mass-density power spectrum on scales $10-100\, h^{-1}$ Mpc does not yet allow a clear distinction between the competing models involving baryonic, cold and hot DM and possibly a cosmological constant. Within the family of CDM models, those most successful in matching the current LSS data are the following variants: (a) $\Omega = 1$ with a tilted spectrum $n \sim 0.9$, (b) $\Omega \sim 0.5$ with or without a cosmological constant and $n = 1$, and (c) $\Omega = 1$ with 20% hot dark matter. A high acoustic peak in the CMB angular power spectrum, and a peak in the mass-density power spectrum at $\sim 125\, h^{-1}$ Mpc, if confirmed, may indicate a high baryonic content. I do not think that any of the front-runner models can be significantly ruled out based on current tests, contrary to occasional premature statements in the literature about the "death" of a certain model. I predict that were the DM constituent(s) to be securely detected in the lab, the corresponding scenario of LSS would find a way to overcome the $\sim 2\sigma$ obstacles it is facing now.

Background cosmology. All the observations so far are consistent with large-scale homogeneity and isotropy (with the exception of the $\sim 2\sigma$ discrepancy indicated by cluster velocities). The methods based on virialized objects tend to favor low values of Ω (~ 0.2) but with plausible loopholes. The flows provide a unique opportunity to constrain Ω independent of Λ and H_0; current velocity data provide in several different ways a significant ($> 2\sigma$) lower bound of $\Omega > 0.3$. This is consistent with $\Omega = 1$ but "ugly" values near $\Omega \approx 0.4$–0.5 are not ruled out either. The range of β values obtained on different scales may be partly due to underestimated errors and partly due to a non-trivial biasing relation between galaxies and mass. The latest global measures of geometry seem to favor $\Omega_\Lambda - \Omega_m > 0$. Preliminary constraints from CMB anisotropy on sub-degree scales seem to indicate a near-flat universe, $\Omega_\Lambda + \Omega_m \sim 1$. The age constraints are not too demanding; they allow $\Omega_m - \Omega_\Lambda \leq 1$. The data are thus consistent with the predictions of *Inflation*: flat geometry and Gaussian, almost scale-invariant initial fluctuations.

The rapid progress in this field guarantees that many of the results and uncertainties discussed above will soon become obsolete, but we hope that the discussion of concepts will be of lasting value, and that the methods discussed can be useful, either as are or as a basis for improvement.

Acknowledgments

This review is based on work with several close collaborators (see references), supported by grants from the US–Israel Binational Science Foundation, the Israel Science Foundation, the NSF and NASA.

References

Aaronson, M., Huchra, J., Mould, J., Schechter, P. L. & Tully, R. B. 1982b, *ApJ*, **258**, 64

Aaronson, M., et al. 1982a, *ApJS*, **50**, 241

Aaronson, M. & Mould, J. 1983, *ApJ*, **265**, 1

Babul, A., Weinberg, D., Dekel, A. & Ostriker, J. P. 1994, *ApJ*, **427**, 1

Bardeen, J., Bond, J. R., Kaiser, N. & Szalay, A. 1986, *ApJ*, **304**, 1

Bernardeau, F. 1992, *ApJ*, **390**, L61

Bernardeau, F., Juszkiewicz, R., Dekel, A. & Bouchet, F. 1995, *MNRAS*, **274**, 20

Bertschinger, E. 1991, in *Physical Cosmology*, ed. M. Lachieze-Rey (Editions Frontieres)

Bertschinger, E. & Dekel, A. 1989, *ApJ*, **336**, L5

Bertschinger, E., Dekel, A., Faber, S. M., Dressler, A. & Burstein, D. 1990, *ApJ*, **364**, 370

Bertschinger, E., Gorski, K. & Dekel, A. 1990, *Nature*, **345**, 507

Blumenthal, G. R., Dekel, A. & Yahil, A. 1999, in preparation

Boehringer, H., Neumann, D. M., Schindler, S. & Kraan-Korteweg, R. C. 1996, *ApJ*, **467**, 168

Bouchet, F., Juszkiewicz, R., Colombi, S. & Pellat, R. 1992, *ApJ*, **394**, L5

Bouchet, F., Strauss, M. A., Davis, M., Fisher, K. B., Yahil, A. & Huchra, J. P. 1993, *ApJ*, **417**, 3

Branchini, E., Plionis, M., Zehavi, I. & Dekel, A. 1999, *MNRAS*, submitted

Braun, E., Dekel, A. & Shapiro, P. 1988, *ApJ*, **328**, 34

Broadhurst, T. J., Ellis, R. S., Koo, D. C. & Szalay, A. S. 1990, *Nature*, **343**, 726

Burles, S. & Tytler, D. 1996, *ApJ*, **460**, 584

Burstein, D. 1990, *Rep. Prog. Phys.*, **53**, 421

Burstein, D., Faber, S. M. & Dressler, A. 1990, *ApJ*, **354**, 1

Carroll, S. M, Press, W. H. & Turner, E. L. 1992, *ARA&A*, **30**, 499

Cen, R. & Ostriker, J. P. 1992, *ApJ*, **399**, L113

Cen, R. & Ostriker, J. P. 1993, *ApJ*, **417**, 415

Cohen, J. G., et al. 1996, *ApJ*, **471**, 5

Cole, S., Fisher, K. B. & Weinberg, D. 1995, *MNRAS*, **275**, 515

Coles, P. & Jones, B. 1991, *MNRAS*, **248**, 1

Coles, P. & Lucchin, F. 1995, *Cosmology: The Origin and Evolution of Cosmic Structure* (John Wiley & Sons)

Colless, M. et al. 1993, *MNRAS*, **262**, 475

Courteau, S., Faber, S. M., Dressler, A. & Willick J. A. 1993, *ApJ*, **412**, L51

da Costa, L. N., Freudling, W., Wegner, G., Giovanelli, R., Haynes, M. & Salzer, J. J. 1996, *ApJ*, **468**, L5

Davis, M., Efstathieou, G., Frenk, C. S. & White, S. D. M. 1985, *ApJ*, **292**, 371

Davis, M., Nusser, A. & Willick, J. A. 1996, *ApJ*, **473**, 22

Davis, M. & Peebles, P. J. E. 1983, *ARA&A*, **21**, 109

Dekel, A. 1981, *A&A*, **101**, 79

Dekel, A. 1994, *ARA&A*, **32**, 371

Dekel, A., Bertschinger, E., Faber, S. M. 1990, *ApJ*, **364**, 349

Dekel, A., Bertschinger, E., Yahil, A., Strauss, M. A., Davis, M. & Huchra, J. 1993, *ApJ*, **412**, 1

Dekel, A., Burstein, D. & White, S. D. M. 1997, in *Critical Dialogs in Cosmology*, ed. N. Turok (Singapore: World Scientific), p. 175

Dekel, A., et al. 1999, in preparation

Dekel, A. & Lahav, O. 1998, astro-ph/9806193

Dekel, A. & Rees, M. J. 1987, *Nature*, **326**, 455

Dekel, A. & Rees, M. J. 1994, *ApJ*, **422**, L1

Dekel, A., Silk, J. 1986, *ApJ*, **303**, 39

Djorgovski, S., de Carvalho, R. & Han, M. S. 1989, in *The Extragalactic Distance Scale*, eds. S. van den Bergh & C. J. Pritchet (Provo: ASP), p. 3

Dressler, A. 1980, *ApJ*, **236**, 351

Dressler, A., et al. 1987, *ApJ*, **313**, 42

Einasto, J., et al. 1997, *Nature*, **385**, 139

Eke, V. R., Cole, S. & Frenk, C. S. 1996, *MNRAS*, **282**, 263

Eldar, A., Dekel, A. & Willick, J. A. 1999, in preparation

Feast, M. W. & Catchpole, R. M. 1997, *MNRAS*, **286** L1

Fisher, K. B., Davis, M., Strauss, M. A., Yahil, A. & Huchra, J. P. 1993, *ApJ*, **402**, 42

Fisher, K. B., Davis, M., Strauss, M. A., Yahil, A. & Huchra, J. P. 1994a, *MNRAS*, **266**, 50

Fisher, K. B., Davis, M., Strauss, M. A., Yahil, A. & Huchra, J. P. 1994b, *MNRAS*, **267**, 92

Fisher, K. B., Huchra, J. P., Strauss, M. A., Davis, M., Yahil, A. & Schlegel, D. 1995, *ApJS*, **100**, 69

Fisher, K. B. & Nusser, A. 1996, *MNRAS*, **279**, L1

Fisher, K. B., Scharf, C. A. & Lahav, O. 1994, *MNRAS*, **266**, 21

Freudling, W. F., da Costa, L. N., Wegner, G., Giovanelli, R., Haynes, M. P. & Salzer, J. J. 1995, *AJ*, **110**, 920

Fukugita, M., Yamashita, K., Takahara, F. & Yoshii, Y. 1990, *ApJ*, **361** L1

Ganon, G., Dekel, A., Mancinelli, P. & Yahil, A. 1999, in preparation

Giovanelli, R., et al. 1997a, *AJ*, **113**, 22

Giovanelli, R., et al. 1997b, *AJ*, **113**, 53

Goobar, A. & Perlmutter, S. 1995, *ApJ*, **450**, 14

Gramman, M. 1993a, *ApJ*, **405**, L47

Gramman, M. 1993b, *ApJ*, **405**, 449

Gregg, M. D. 1995, *ApJ*, **443**, 527

Gull, S. & Daniell, X. 1978, *Nature*, **272**, 686

Gunn, J. E. 1989, in *the Extragalactic Distance Scale*, eds. S. van den Berg & C. J. Pritchet (Provo: ASP), p. 344

Hamilton, A. J. S. 1992, *ApJ*, **385**, L5

Hamilton, A. J. S. 1993, *ApJ*, **406**, L47

Hamilton, A. J. S. 1995, in *Clustering in the Universe*, eds. S. Maurogordato et al. (Gif-sur-Yvette: Editions Frontieres), p. 143

Hancock, S., Rocha, G., Lasenby, A. N. & Cutierrez, C. M. 1998, *MNRAS*, , in press

Heavens, A. F. & Taylor, A. N. 1995, *MNRAS*, **275**, 483

Hudson, M. J. 1993a, *MNRAS*, **265**, 43

Hudson, M. J. 1993b, *MNRAS*, **265**, 72

Hudson, M. J. 1994, *MNRAS*, **266**, 475

Hudson, M. J., Dekel, A., Courteau, S., Faber, S. M. & Willick, J. A. 1995, *MNRAS*, **274**, 305

Hoffman, Y. & Ribak, E. 1991, *ApJ*, **380**, L5

Hogan, C. J. & White, S. D. M. 1986, *Nature*, **321**, 575

Ikeuchi, S. 1981, *PASJ*, **33**, 211

Jaffe, A. H. & Kaiser, N. 1995, *ApJ*, **455**, 26

Juszkiewicz, R., Vittorio, N. & Wyse, R. F. G. 1990, *ApJ*, **349**, 408

Kaiser, N. 1987, *MNRAS*, **227**, 1

Kaiser, N. 1988, *MNRAS*, **231**, 149

Kaiser, N. 1984, *ApJ*, **284**, L9

Kaiser, N., et al. 1991, *MNRAS*, **252**, 1

Kaiser, N. & Lahav, O. 1989, *MNRAS*, **237**, 12

Kaiser, N. & Stebbins, A. 1991, in *Large Scale Structure and Peculiar Motions in the Universe*, eds. D. W. Latham & L. N. da Costa (San Francisco: ASP), p. 111

Kochanek, C. S. 1996, *ApJ*, **466** 638

Kofman, L., Bertschinger, E., Gelb, J., Nusser, A. & Dekel, A. 1994, *ApJ*, **420**, 44

Kofman, L., Blumenthal, G. R., Hodges, H. & Primack, J. R. 1990, in *Large Scale Structure and Peculiar Motions in the Universe*, eds. D. W. Latham & L. N. da Costa (San Francisco: ASP), p. 251

Kogut, A., et al. 1993, *ApJ*, **419**, 1

Kolatt, T. & Dekel, A. 1994, *ApJ*, **428**, 35

Kolatt, T. & Dekel, A. 1997, *ApJ*, **479**, 592

Kolatt, T., Dekel, A., Ganon, G. & Willick, J. 1996, *ApJ*, **458**, 419

Kolatt, T., Dekel, A. & Lahav, O. 1995, *MNRAS*, **275**, 797

Kolb, E.W. & Turner, M. S. 1990, *The Early Universe* (Addison-Wesley)

Kraan-Korteweg, R. C., Woudt, A. P., Cayatte, V., Fairall, A. P., Balkowski, C. & Henning, P. A. 1996, *Nature*, **379**, 51

Lahav, O. 1987, *MNRAS*, **225**, 213

Lahav, O., Fisher, K. B., Hoffman, Y., Scharf, C. A. & Zaroubi, S. 1994, *ApJ*, **423**, L93

Lahav, O., Kaiser, N. & Hoffman, Y. 1990, *ApJ*, **352**, 448

Landy, S. & Szalay, A. 1992, *ApJ*, **391**, 494

Lauer, T. R. & Postman, M. 1993, *ApJ*, **425**, 41

Lineweaver, C. H., Tenorio, L., Smoot G. F., Keegstra, P., Banday, A. J. & Lubin, P. 1996, *ApJ*, **470**, 38

Lynden-Bell, D., Faber, S. M., Burstein, D., Davies, R. L., Dressler, A., Terlevich, R. J. & Wegner, G. 1988, *ApJ*, **326**, 19

Lynden-Bell, D., Lahav, O. & Burstein, D. 1989, *MNRAS*, **241**, 325

Mancinelli, P. J., Yahil, A., Ganon, G. & Dekel, A. 1994, in *Cosmic Velocity Fields*, eds. F. Bouchet & M. Lachieze-Rey (Editions Frontieres), p. 215

Mathewson, D. S., Ford, V. L. & Buchhorn, M. 1992, *ApJS*, **81**, 41

Mo, H. J., Jing, Y. P. & White, S. D. M. 1996, *MNRAS*, **282**, 1096

Myers, S. T., Baker, J. E., Readhead, A. C. S.,

Leitch, E. M. & Herbig, T. 1997, *ApJ*, **485** 1

Nusser, A. & Davis, M. 1994, *ApJ*, **421**, L1

Nusser, A. & Davis, M. 1995, *MNRAS*, **276**, 1391

Nusser, A. & Dekel, A. 1992, *ApJ*, **391**, 443

Nusser, A. & Dekel, A. 1993, *ApJ*, **405**, 43

Nusser, A. & Dekel, A. 1999, in preparation

Nusser, A., Dekel, A., Bertschinger, E. & Blumenthal, G. R. 1991, *ApJ*, **379** 6

Nusser, A., Dekel, A. & Yahil, A. 1995, *ApJ*, **449**, 439

Ostriker, J. P. & Cowie, L. L. 1981, *ApJ*, **243**, L127

Ostriker, J. P. & Suto, Y. 1990, *ApJ*, **348**, 378

Paczynski, B. & Piran, T. 1990, *ApJ*, **364**, 341

Padmanabhan, T. 1993, *Structure Formation in the Universe* (Cambridge University Press)

Peacock, J. A. & Dodds, S. J. 1994, *MNRAS*, **267**, 102

Peebles, P. J. E. 1993, *Principles of Physical Cosmology*, (Princeton: Princeton University Press)

Perlmutter, S., et al. 1996, *ApJ*, **483**, 565

Plionis, M. & Valdarnini, R. 1991, *MNRAS*, **249**, 46

Press, W. H., Flannery, B. P., Teukolsky, S. A. & Vetterling, W. T. 1995, *Numerical Recipes* (Cambridge University Press)

Rauzy, S., Lachieze-Rey, M., Henriksen, R. N. 1993, *A&A*, **273**, 357

Reid, I. N. 1997, *ApJ*, **115**, 204

Riess, A. G., Press, W. H. & Kirshner, R. P. 1995, *ApJ*, **438**, L17

Rowan-Robinson, M. 1993, *Proc. Nat. Acad. Sci.*, **90**, 482

Rowan-Robinson, M., Lawrence, A., Saunders, W., Leech, K. 1991, *MNRAS*, **253**, 485

Sachs, R. K. & Wolfe, A. M. 1967, *ApJ*, **147**, 73

Sahni, V. & Coles, P. 1996, *Phys. Rep.*, **262**, 1

Sargent, W. L. W. & Turner, E. L. 1977, *ApJ*, **212**, L3

Scaramella, R., Vettolani, G. & Zamorani, G.

1991, *ApJ*, **376**, L1

Schechter, P. 1980, *AJ*, **85**, 801

Scodeggio, M. 1997, in *Galaxy Scaling Relations: Origins, Evolution and Applications*, ed. L. da Costa (Springer) in press

Scott P. F., et al. 1996, *ApJ*, **461**, L1

Shaya, E. J., Peebles, P. J. E. & Tully, R. B. 1995, *ApJ*, **454** 15

Shaya, E. J., Peebles, P. J. E. & Tully, R. B. 1997, private communication

Sigad, Y., Eldar, A., Dekel, A., Strauss, M. S. & Yahil, A. 1998, *ApJ*, **495**, 516

Silk, J. 1989, *ApJ*, **345**, L1

Simmons, J. F. L., Newsam, A. & Hendry, M. A. 1995, *A&A*, **293**, 13

Smoot, G. F., et al. 1994, *ApJ*, **437**, 1

Somerville, R., Lemson, G., Dekel, A., Kauffmann, G. & White, S. D. M. 1999, in preparation

Stebbins, A. 1994, in *Cosmic Velocity Fields*, eds. F. Bouchet & M. Lachieze-Rey (Editions Frontieres), p. 253

Strauss, M. A., Cen, R. & Ostriker, J. P. 1993, *ApJ*, **408**, 389

Strauss, M. A., Cen, R., Ostriker, J. P., Lauer, T. R. & Postman, M. 1995, *ApJ*, **444**, 507

Strauss, M. A., Yahil, A., Davis, M., Huchra, J. P. & Fisher, K. B. 1992b. *ApJ*, **397**, 395

Strauss, M. A. & Willick, J. A. 1995, *Phys. Rep.*, **261**, 271

Sugiyama, N. 1995, *ApJS*, **100**, 281

Summers, F. J., Davis, M. & Evrard, A. E. 1995, *ApJ*, **454**, 1

Tully, R. B. & Fisher, J. R. 1977, *A&A*, **54**, 661

Tytler, D., Fan, X. M. & Burles, S. 1996, *Nature*, **381**, 207

VandenBerg, D., Bolte, M. & Stetson, P. 1996, *ARA&A*, **34**, 461

Watkins, R. & Feldman, H. A. 1995, *ApJ*, **453** 73

Weinberg, D. 1995, in *Wide-Field Spectroscopy and the Distant Universe*, eds. S. J. Maddox & A. Aragon-Salamanca (World Scientific: Singapore), p. 129

Weinberg, D. H. & Cole, S. 1992, *MNRAS*, **259**, 652

White, M. & Scott, D. 1996, *ApJ*, **459**, 415

White, S. D. M., Efstathiou, G. & Frenk, C. S. 1993, *MNRAS*, 262, 1023

White, S. D. M. & Fabian, A. 1995, *MNRAS*, **273**, 72

White, S. D. M., Navaro, J., Evrard, A. & Frenk, C. S. 1993, *Nature*, **366**, 429

Willick, J. 1994, *ApJS*, **92**, 1

Willick, J. A., Courteau, S., Faber, S. M., Burstein, D. & Dekel, A. 1995, *ApJ*, **446**, 1

Willick, J. A., Courteau, S., Faber, S. M., Burstein, D., Dekel, A. & Kolatt, T. 1996, *ApJ*, **457**, 460

Willick, J. A., Courteau, S., Faber, S. M., Burstein, D., Dekel, A. & Strauss, M. A. 1997a, *ApJS*, **109**, 333

Willick, J. A., Strauss, M. S., Dekel, A. & Kolatt, T. 1997b, *ApJ*, **486**, 629

Yahil, A., Walker, X. & Rowan-Robinson, M. 1986, *ApJ*, **301**, L1

Zaroubi, S. & Hoffman, Y. 1996, *ApJ*, **462**, 25

Zaroubi, S., Hoffman, Y. & Dekel, A. 1999, *ApJ*, submitted

Zaroubi, S., Hoffman, Y., Fisher, K. B. & Lahav, O. 1995, *ApJ*, **449**, 446

Zaroubi, S., Sugiyama, N., Silk, J., Hoffman, Y. & Dekel, A. 1997b, *ApJ*, **490**, 473

Zaroubi, S., Zehavi, I., Dekel, A., Hoffman, Y. & Kolatt, T. 1997a, *ApJ*, **486**, 21

Zehavi, I., Riess, A., Kirshner, R. & Dekel, A. 1998, *ApJ*, **503** (astro-ph/9802252)

Zel'dovich, Ya. B. 1970, *A&A*, **5**, 20

Part three

Structure on galactic scales and lensing

8 Cosmological simulations

Jeremiah P. Ostriker

Abstract

The theories described in earlier chapters of this book for the growth of structure are completely quantitative. Accurate computations, which are fairly simple in the linear domain, are, in principle, possible at all scales and to any degree of nonlinearity. No "unknown" physics is required – only careful physical modeling and powerful computers. Progress has been very rapid, with dark-matter-only calculations reaching in the last decade from 64^3 to beyond 512^3 particles. Hydrodynamic calculations, which treat gas and radiation, have gone from 30^3 to 1024^3 cells, with a far richer set of physical processes being treated. This progress has allowed increasingly precise comparisons between model predictions and observations. At present, the best comparisons can be made for the "Lyman-α clouds", which produce the absorption lines seen between us and distant quasars, and for the X-ray emitting clusters of galaxies. Less accurate, but equally interesting, are attempts to simulate the origin and distribution of galaxies, with many observed properties (including "biasing") seeming to find natural explanations on the basis of numerical simulation. Finally, an *ab initio* approach to gravitational lensing is beginning to be possible using the most advanced dark-matter simulations. The current, highly tentative conclusion of this work is that models having matter density Ω_m significantly less than unity fit the observations best. The very rapid advance of computational power implies that this field has great promise for future work.

8.1 Introduction

Cosmology as a science developed from the observations of galaxies. After the realization, in the early decades of the 20th century, that we live in one of many galaxies rushing away from one another, the subject became, for several decades, a discipline striving to determine two numbers. One, the Hubble constant measures the rate of expansion and implicitly indicates how long that has persisted; the second, Ω_0 (or alternatively q_0), is – for the simplest set of models – a measure of the mass density and, hence, curvature of the universe and indicates whether the universe will expand forever or ultimately collapse after gravitational energy overcomes kinetic energy. Individual galaxies were seen as the "building blocks" and were used alone or in combination with others to provide the standard metersticks or standard candles that enabled one to determine, empirically, the two desired numbers. But, in the 1970s, scientists, led by Beatrice Tinsley, realized that, since these building blocks were not part of the original Big Bang, they were formed at some time and, hence, must evolve on a cosmic time-scale. Understanding their formation and evolution, investigated at first so that they could be better used as tools, became an end in itself, and has become a principle goal of cosmogonic study.

To address these questions, quite definite mathematical models for the growth of structure were developed. These all relied in an essential way on the fundamental gravitational instability of quasi-uniform media: overdense regions will expand slower than average and underdense regions will expand faster than average, accentuating any initial irregularities. In addition, there are a variety of kinematic and hydrodynamic feedback mechanisms which can further increase density fluctuations once these have passed into the nonlinear domain. Studying the origin and growth of fluctuations in an initially quasi-uniform medium is a well-defined computational task, but it was not well suited to address the observational data. The reason is that there is no obvious way to connect the computed mass fluctuations with the observational fluctuations in galaxy numbers or properties. In most contexts, when we say that we observe "galaxies," we mean that we observe "stars in galaxies," so the problem of galaxy formation irreducibly comes back to the unsolved problem of star formation. This serious complication resides, in standard parlance, under the heading of "the uncertainty of bias" (cf. §7.6.7, Ch. 11). But it is so serious a problem, in my view, that it has been necessary to back away from the original goal – and to study instead the formation of other structures in the universe, not the structures apparent in the galaxy distribution.

The structures most amenable to reliable computation are those represented by (a) X-ray clusters, (b) gravitational lenses (with separations of lensed images approximately greater than 3″) and (c) intergalactic Lyman-α absorption clouds. The reason for picking these topics is that their formation and structure can be understood via straightforward physics that does *not* involve the unknown physics of star formation. Understanding the formation and evolution of these structures has progressed rapidly in the last few years, so that comparison between observations and model computations is beginning to effectively weed out the less successful models. Let us now summarize the situations when model calculations can be effectively used to confront observations.

a. Dark-matter-only simulations:

(i) The computed quantity must be *directly* derivable from the potential fluctuations, as these are determined largely by the dark-matter component. Thus, one might examine:

Cosmological simulations

    (a) Gravitational lensing which measures potential fluctuations (integrated along the line-of-sight) directly.

    (b) The velocity field, ranging from galaxy pairwise velocities, to numbers of clusters having differing velocity dispersions, to large-scale velocity fields.

  (ii) The computed quantity must be dominated by mass clumping on scales larger than 10 kpc, since, on smaller scales, the condensed (stellar) baryonic component is gravitationally significant. This eliminates, for instance, small-scale (less than 2 arcsec) gravitational lens splittings or galactic rotation curves.

b. Cosmological hydro simulations:

  (i) Most of the baryonic component should be in the gaseous, not condensed (e.g., stellar) state, so that uncertainties in the conversion of gas to stars or galaxies are not important.

  (ii) The principal processes for heating and cooling the gas are either computable, in a self-consistent fashion (e.g., shock heating), or constrained by observations (e.g., the UV and soft X-ray backgrounds), so that the physical state of the gas can be computed with confidence.

  (iii) The dynamical range for the computations must be sufficient to more than bridge the gap from a resolution scale of ΔL, less than the Jeans' length, to a box size L, large enough to provide a fair sample for the structures being studied.

At the present time, with present hardware and software, the X-ray clusters and the Lyman-α clouds both (marginally) satisfy these requirements for phenomena that can be investigated with profit.

8.2 Simulation methods

8.2.1 Specification of models

A specific simulation must adopt a set of parameters for the global cosmology, current Hubble constant H_0 ($\equiv 100\,h\,\mathrm{km\,s^{-1}\,Mpc^{-1}}$), and current values for the mass density (in units of the critical density) in three forms: baryons, dark matter, and cosmological constant ($\Omega_b, \Omega_d, \Omega_\Lambda$ with $\Omega_{tot} \equiv \Omega_b + \Omega_d + \Omega_\Lambda$). Further, the dark matter must be classified as to whether it was hot or cold (relativistic or nonrelativistic) at decoupling (cf. §1.5). If there was a period during which the dark matter was cold, i.e., nonrelativistic and contained most of the mass density before decoupling (the requirement being that the predominant species had a mass particle above a few eV), then gravitational perturbations can grow in this component, while the baryonic gas is still frozen, mechanically coupled to the much more uniform relativistic photon fluid via the Compton drag process. Since most investigators adopt the value for the baryon density, fixed by conventional light element nucleosynthesis [$\Omega_b h^2 = 0.0125 \pm 0.025$, Walker et al. (1991)], Ω_b is not treated as a separate parameter once h has been specified. Work by Tytler et al. (1996) and Miralda-Escudé et al. (1996) indicates that a value almost twice as large as this would be preferable. Typically, the models are broken into two categories – flat or open – dependent on whether Ω_{tot} equals unity or is less than unity (and typically in the range $0.1 \leq \Omega_{tot} \leq 0.4$).

Then one specifies the nature of the perturbations imposed at early epochs, dividing them into "Gaussian" (in the distribution of amplitudes with phases random) or "non-Gaussian" (either locally non-Gaussian, or globally with phases correlated over a horizon, as with cosmic strings). Then one specifies the nature of the perturbations, which are conventionally divided into "adiabatic" (e.g., a simple, local compression or expansion of all components) or "isocurvature" variants (where matter and radiation are oppositely perturbed to leave a variation in entropy per baryon but maintain uniformity in the total energy density). Finally, the spectrum of perturbations is specified, with power-law initial states $P_k \propto k^n$ being popular and the typical choice of n close to the standard, $n = 1$ (Peebles & Yu 1970; Harrison 1970; Zel'dovich 1972) for which perturbations enter the horizon, at early times, with fixed amplitude.

At this point the reader may be bewildered at the range of choices permitted, and think that, with so many apparently free parameters, the subject is almost like a game without rules, or with rules so loosely specified that anyone can claim to have won at any time. Such an unconstrained game is not worth playing. In fact, the range of models considered is strongly circumscribed by both the commendable taste of scientists for simplicity and the strong constraints provided by observation. Specifically, most current investigation is limited to models that are defined by three parameters which are all variants of the "standard" cold dark-matter scenario.

It is useful to first define that standard scenario precisely, even though it is almost certainly incorrect. Motivated by earlier dimensional analysis (Peebles & Yu 1970; Harrison 1970; Zel'dovich 1972) and following the more recent, simple notions of an early inflationary period (Guth & Pi 1982; Starobinskii 1982; Bardeen, Steinhardt & Turner 1983), one assumes a spectrum of perturbations with $n = 1$ precisely. Only one (cold) type of nonbaryonic dark matter is presumed to exist and the cosmological constant is taken as zero, with the total producing a flat universe: $\Omega_{tot} = \Omega_b + \Omega_c = 1$. Such a model is defined by two parameters, the amplitude, (symbolically) A, and the Hubble constant h. These parameters can be fixed by direct observations: A can be determined from microwave background fluctuations measured by the COBE satellite and h from direct measurements of the Hubble constant. Since most recent estimates of the latter are in the range $h = 0.70 \pm 0.15$, it is difficult to credit very low values of h which are preferred for standard CDM. As a compromise, the "standard" CDM scenario modelers have tended to adopt $h = 0.5$ (but see Bartlett et al. (1995) who present arguments for a much lower value), which yields an age of less than 13 billion years that is marginally consistent with the age determinations for the oldest galactic stars (Bolte & Hogan 1995). The model defined in this fashion fails to match observations on several counts (summarized, with references, in Ostriker (1993),[1] see also Chs. 1, 4, 7, and 11.) These include an inability to match the properties of the clusters of galaxies, too high a small-scale velocity dispersion and the wrong slope for the large scale galaxy-galaxy correlation function.

This failure, the disagreement between calculation and observation, has led to detailed examination of several variants of the CDM scenario that are defined by three parameters (A, h, p). For the "tilted" scenario "TCDM," p represents $n-1$, the slope of the power spectrum; for the open scenario "OCDM," p represents the value of $\Omega_{tot} < 1$; for the Λ model "ΛCDM," p represents Ω_Λ with flatness maintained ($\Omega_{tot} = \Omega_b + \Omega_c + \Omega_\Lambda = 1$); and for the mixed dark-matter model, p represents the mass in the neutrino component Ω_ν (with $\Omega_{tot} = \Omega_b + \Omega_c + \Omega_\nu = 1$). One way to specify these three-parameter models is to use the prescription advocated in Ostriker & Steinhardt (1995), relying on three of the more accurately defined observational constraints:

(i) From cosmic microwave background (CMB) fluctuations (primarily the COBE results): $A \times (1 \pm 8\%)$.
(ii) From the abundance of rich clusters: $\eta \equiv \sigma_8 \Omega_m^{0.56} = 0.56 \times (1 \pm 11\%)$.
(iii) From the slope of the galaxy–galaxy correlation function: $\Gamma \equiv \Omega_m h = 0.25 \times (1 \pm 20\%)$.

The second constraint has been obtained by Bahcall, Cen & Ostriker (1996) and by White, Efstathiou & Frenk (1993), the third by Peacock & Dodds (1994). Using these three constraints, the variant models denoted above are fairly well tied down to give power spectra which are observationally fixed at both short and long wavelengths. All properties of the model then are specified to an accuracy of approximately 15% (the accuracy of the original specification of the model – see (i), (ii), (iii) above), so an accurate calculation of their properties followed by comparison to observations (at least as accurate as $\pm 15\%$) might lead to a statistically significant conflict, i.e., falsification of the model.

Now a few words on technical details. The power spectrum in the linear domain must be computed using a "transfer function" which tracks the growth of the amplitude in waves of each component (Ω_b, Ω_ν, Ω_c, Ω_γ) through decoupling to the point when the numerical simulation is to begin. A clear and well-motivated discussion of transfer functions is presented by Efstathiou (1990). A commonly used source for the CDM transfer function is Bardeen et al. (1986). Also available, by anonymous FTP, are codes for computing transfer functions for a broad variety of models from Bertschinger (1995).

Finally, one must specify the initial mass distribution at some starting epoch. This can be done most simply by displacing particles from a regular grid (Efstathiou

---

[1] No attempt to provide a comprehensive review of the literature will be made in this chapter, which will emphasize the Princeton contribution to the subject (as best evaluated by the reviewer) without prejudice to the preponderance of excellent work done elsewhere.

et al. 1985) with velocities determined by the Zel'dovich approximation. Also, more sophisticated "glass" initial conditions have come into vogue (cf. White (1994) for an example).

8.2.2 Physical processes and numerical methods

8.2.2.1 Hydrodynamics

One must make an early choice between a particle-based, Lagrangian approach (e.g., SPH (Monaghan & Gingold 1983; Hernquist & Katz 1989)) or a mesh based Eulerian model (cf., for example, Cen et al. (1990)). This is not the place for a detailed discussion of the merits and drawbacks of the two approaches; they are addressed (with numerous references to the technical literature) in the context of comparisons between methods addressing a realistic simulation, by Kang et al. (1994). In any case, the continual drive to ever greater dynamic range has led to the development of hybrid schemes (Gnedin 1995; Pen 1995; Xu 1997) which combine the useful attributes of the two approaches.

8.2.2.2 Gravity

If the mass density is defined at N points in space, then a direct determination of the gravitational potential via Poisson's integral equation requires $O(N^2)$ operations. As this places an intolerable (and unnecessary) computational burden on the whole calculation (other parts scale as $O(N)$), many algorithms have been developed which reduce the problem to $O[N\ln(N)]$ operations. For example, if the mass is defined on a regular grid, then one can use efficient Fourier Transform methods (e.g., FFT) to transform to k-space, solve Poisson's differential equation there (using periodic boundary conditions) and then transform back. While this is the fastest method used at present (for given N), the regular grid it requires is clearly not the best way of specifying the mass distribution. Other, less rapid techniques are available for unstructured grids that can be designed to put a finer mesh in regions of higher density. Alternatively, to reach higher resolutions, one can compute forces of particles within a cell directly (P^3M (Hockney & Eastwood 1988; Efstathiou et al. 1985)); here, the cost becomes very high when the number of particles in any cell becomes large. Finally, there are completely unstructured methods, the prototype of which is the Tree algorithm (Barnes & Hut 1986), which take advantage of the fact that, from a distance, the potential at any finite group of particles can be accurately represented by a multipole expansion. An interesting recent hybrid of the particle, mesh and tree algorithms has been developed by Xu (1995), which can be implemented efficiently on a parallel machine architecture. In typical codes, the time spent on gravity is comparable to the time spent on all other aspects of the problem in a given timestep. The reason for this is that, for each mass element, gravity requires $O[\ln(N)]$ operations, which is comparable to the number of other variables.

8.2.2.3 Atomic physics

There exist numerous good atomic physics texts to guide the student (e.g., Osterbrock 1989), as a closely related set of problems has been studied for decades in the context of the interstellar medium. The only word of warning offered is that one must be careful about employing short cuts. For example, one can write, under some circumstances, the cooling rate per unit volume of gas having a specified chemical composition as $n^2 \Lambda(T)$ with the "cooling function," $\Lambda(T)$, tabulated by several groups. But the hidden assumption, that ionization is purely determined by collisional processes, is typically very poor in the low-density intergalactic medium, where background radiation fields can ionize gas efficiently (at a rate $\propto n$, not n^2). Furthermore, the normal assumption of ionization equilibrium can be badly off in low-density environments. As a consequence of these complications, the Princeton group has typically taken the brute force approach of computing for each timestep and each mass element the change in the numbers of each relevant ionized state. Thus, symbolically,

$$dQ_n/dt = -Q_n \int \sigma_\nu J_\nu dV - Q_n N_e I(T) + Q_{n+1} N_e \alpha(T) , \qquad (8.1)$$

where the first term represents photoionization, the second collisional ionization, and the third recombination from the next-higher state of ionization. Since the number of important species is limited, if one is working with gas of a primeval composition, the cost here (aside from the cost of the programmer's time) is less than that of the gravitational computation.

Photoionization is a crucial process, so it is very important to calculate the mean radiation field properly. One must allow for emission of photons, absorption processes, and the cosmological dilution and redshifting processes. If one computes only the mean (frequency-dependent) radiation field, $J_\nu(t)$, then the computational burden is not large. As far as it is known to the author, no one has yet attempted to compute the radiation field as a function of both space and direction, $J_\nu(\theta, r, t)$, but some simpler treatments which allow for shielding are now implemented (Katz et al. 1996; Ostriker & Gnedin 1996).

A detailed presentation of all of the relevant equations for treating the heating and cooling of a hydrogen-helium plasma (with $T > 10^{3.5} K$) is contained in Cen (1992).

8.2.2.4 Magnetic fields

Magnetic fields, even if not primordially existent, will be generated naturally by the "Biermann battery" whenever shocks cause a significant angle between pressure and density gradients:

$$\frac{\partial \boldsymbol{B}}{\partial t} - \nabla \times (\boldsymbol{v} \times \boldsymbol{B}) = \frac{\nabla p \times \nabla \rho}{\rho^2} \left(\frac{cm_e}{e} \right). \qquad (8.2)$$

Recently, Kulsrud et al. (1997) rediscovered a remarkable result implied by Biermann's (1950) original work. If the vector fields for both cyclotron frequency $\omega_{cy} \equiv (eB/m_e c)$ and the vorticity $\boldsymbol{\omega} \equiv \nabla \times \boldsymbol{v}$ are initially zero, and dissipation can be neglected, and, moreover, the magnetic forces are unimportant dynamically, then $\omega_{cy} = -\omega$, at all times and places. Both quantities satisfy the same partial differential equation and initial conditions. Thus, an accurate calculation of the velocity field – neglecting magnetic terms altogether – allows one to compute the magnetic fields, provided the latter are small. A first computation by Kulsrud et al. (1997) shows that average fields of order 10^{-21} Gauss are generated, but that the fields in clusters of galaxies will be orders of magnitude larger, as turbulence leads to a rapid amplification of field strength until equipartition is reached.

8.2.2.5 Galaxy/star formation

As the growth of structure proceeds, we know, empirically, that condensation to ever higher densities leads to the formation of stars and quasars. Given the current state of our knowledge, we cannot model this evolutionary process accurately. But the problem cannot be avoided; some scheme *must* be invented. The reason is that, since any numerical code has finite spatial and temporal resolution, situations will occur when the code simply cannot cope with the physics of collapsing, fragmenting matter. The computations would necessarily halt if no prescription were included in the code to treat regions which were

$$\text{contracting;} \quad \nabla \cdot v < 0, \tag{8.3}$$

$$\text{gravitationally unstable;} \quad M_{\text{Jeans}} < M_{\text{gas}}, \tag{8.4}$$

and

$$\text{cooling rapidly;} \quad t_{\text{cool}} < t_{\text{dynamical}}. \tag{8.5}$$

Such mass elements cannot be followed by the code. In fact, any code, with any defined spatial resolution scale (greater than the size of a star) will hit this barrier and will fail at some point during calculation of gravitational collapse. In the Princeton approach to the problem, we have allowed these mass elements to condense out as subgalactic units treated as particles. Thus, if all three conditions (equations [8.3]–[8.5]) are satisfied (and some minimal overdensity criterion, as well, is adopted for convenience), then we remove from the cell a mass in gas

$$\Delta M_{\text{gas}} = -\frac{M_{\text{gas}}}{\rho} \frac{\Delta t}{t_{\text{dyn}}}, \tag{8.6}$$

where Δt is the timestep t_{dyn}, the free-fall time within the cell ($\propto (G\rho_{\text{cell}})^{-1/2}$). The mass removed from the gas phase is used to create a new particle with mass $= \Delta M_{\text{gas}}$ and a velocity equal to the cells' hydrodynamic velocity. Thus, the operation conserves mass and momentum, but, of course, it does not conserve energy. The energy radiated during the successive collapse and fragmentation phases, which we cannot capture, has been included in the computation as the hidden binding energy of the created particles. We can track these subgalactic units and group them, at any time into "galaxies," with designated "age" (time since creation), mass, and angular momentum.

One would hope that this prescription for forming the condensed component would be robust, i.e., that it would not depend on the spatial or temporal resolution of the code nor on the numerical coefficients that might be included on the right-hand sides of equations (8.4), (8.5) and (8.6). The notion behind this hope is the expectation that once gas starts to collapse, it will not stop but will continue to do so at an accelerated pace. If so, it should not matter when, in the collapsing phase, one steps in and labels the material as designated for the condensed state. Further, the process should be self-regulatory to some extent; as gas is removed from the contracting component via equation (8.6), the remaining gas has lower density and will not as easily satisfy criteria (8.4) and (8.5). Actual numerical tests are only mildly reassuring. As one changes the prescription within reasonable bounds, the results for the amount of condensed (galactic) material formed do change, but by moderate amounts, so that we feel it possible to compute the ρ_{gal} to within approximately a factor of two.

8.2.2.6 Secondary radiation fields

Since the condensed matter presumably is now divided into the familiar categories of stars, interstellar gas, quasars, etc., it will necessarily emit radiation which will affect the remainder of the computation.

Here we have two choices. First, we could simply rely on our observational estimates of the background radiation in fields $J_\nu(z)$, and we could say that, after the onset of galaxy formation, we will put in, by hand, a background with those properties. This option has been chosen by the Weinberg, Hernquist & Katz (1997) collaboration. It has the virtue of simplicity, but ambiguities exist in specifying the time-dependence (should $J_\nu(z)$ be taken to be proportional to the rate of galaxy formation, the lagged rate, or the integrated rate?), and radiation transfer (we do not know observationally the background in just those spectral regions near atomic edges, where absorption is most important but knowledge of J_ν is most critical). The Princeton group has chosen a second alternative: to allow the condensed component to emit radiation with some appropriate source function S_ν and to allow for absorption and cosmological effects:

$$dJ_\nu = (S_\nu - \rho \kappa_\nu J_\nu + \text{cosmological terms})\, dt\,. \tag{8.7}$$

An efficiency of turning condensed baryonic matter into radiation must then be specified,

$$\epsilon_{rad} \equiv \Delta E_{rad}/\Delta M_{condensed} c^2\,, \tag{8.8}$$

which, while picked to reasonably match star-forming regions like Orion (ϵ is approximately equal to $10^{-4.5}$), is essentially arbitrary and must be adjusted to produce approximately the observed background radiation field, J_ν. The spectral slope of the source function is designed to have a soft component appropriate to hot stars and a comparable hard, power-law component designed to mimic AGN spectra.

But, the point cannot be made too strongly that it is essential to utilize *some* method for adding an ionizing background radiation field since (a) it exists in nature, (b) it strongly affects the thermal state of gas at temperatures below 10^5 K, and (c) it cannot be computed *ab initio* until star and quasar formation are understood in detail.

For some purposes, knowledge of $J_\nu(z)$ is vital but can be constrained by observations, as in the computation of the gaseous component that produces the Lyman-α clouds. For some other purposes $J_\nu(z)$ is essentially irrelevant, as in the computation of the properties of the X-ray clusters ($T > 10^7$K). But, for the all important purpose of computing galaxy formation itself, the issue is extremely complex. Recent work by Steinmetz (1995) and by Navarro & Steinmetz (1997) and Weinberg, Hernquist & Katz (1997) indicates that the computed formation of low-mass systems $M_{\text{gal}} < 10^9 M_\odot$ is strongly dependent on the background radiation field, but ordinary ($M_{\text{gal}} \approx 10^{11}$–$10^{12} M_\odot$) systems are virtually unaffected. The situation is further complicated by the fact that a smooth radiation field cannot be assumed; the forming galaxies are themselves local sources of radiation. Much more work will need to be done before the picture becomes clear.

8.3 Results: comparison with observations

The Princeton group has made a large number of hydrodynamical simulations varying both the detailed physical input and the model to be examined. A recent summary of that work is presented in Ostriker & Cen (1996). High-resolution, dark-matter-only simulations relevant to gravitational lensing are discussed in Wambsganss *et al.* (1995). Early phases, during which the first generation of stars recognizes the intergalactic medium (IGM), have recently been treated by Ostriker & Gnedin (1996) and Gnedin & Ostriker (1997). Let us now turn to a discussion of the specific physical components.

8.3.1 Hot components

8.3.1.1 X-ray clusters

Several groups have by now computed the X-ray luminosity emitted by the hot, dense regions observationally identified with clusters of galaxies. These are regions, vertices where Zel'dovich-like pancakes or ribbons intersect, where the potential wells are deepest and where gas, dark matter, and galaxies collect into massive assemblages. The velocity dispersion is typically comparable to or greater than 1000 km s^{-1} and the temperatures are greater than 10^8 K. The fact that such distinct regions, with the appropriate properties, form in all plausible cosmological scenarios is a first triumph of cosmological hydrodynamics. Had these giant X-ray emitting regions not been observed first (Gursky *et al.* 1971), they would have been an early prediction of the simulations. The number of such regions per unit volume as a function of luminosity, $N(L_x)$, predicted by a specific scenario, can now be used to test the model. Also available are the distribution of temperatures, $N(T_x)$, the correlation between these observables, $T_x(L_x)$, the core radii of the X-ray emitting gas, R_x, and numerous other quantities such as, for example, the ratio of gas-to-total mass (as a function of radius) within the X-ray clusters.

It is numerically difficult to compute a large enough volume so that it contains a fair sample of clusters. The requirement is that the volume be greater than $(200 h^{-1}$ Mpc$)^3$

with a resolution length small enough to compute the luminosity of specific clusters accurately (ΔL is less than $20\,h^{-1}$ kpc). Thus the required dynamic range necessary for a secure calculation is 10^4. All errors, if this dynamic range is not reached, conspire to underestimate the number of computed high-luminosity clusters. Thus, a result, such as that by Cen & Ostriker (1994), indicating that the COBE normalized standard CDM model produces too many high-temperature, luminous X-ray clusters, is useful in ruling out some cosmological models. Other work by Bryan et al. (1994) for the CHDM model, and by Cen & Ostriker (1994) for the ΛCDM model indicates that successful models may exist, but numerically better work is needed before either result is confirmed.

8.3.1.2 Background hot gas

There is also a much larger volume and mass of gas in the temperature range 10^6–10^7 K, which would be associated observationally with more moderate density enhancements (groups and weak clusters of galaxies). For this gas, line emission and cooling are important and resolution is still more important than it is for the hotter gas. Preliminary results by Cen et al. (1995) indicate that the emission from this gas corresponds well with one component of the observed soft X-ray background (Gendreau et al. 1995), and that the bulk of the baryonic mass is in this component at $z = 0$.

8.3.2 Warm components

8.3.2.1 Lyman-α forest

The neutral hydrogen detected in gas at roughly 10^4 K, as it produces absorption lines between us and distant quasars, at the rest-frame Lyman-α wavelength, has been identified observationally as the "Lyman-α Forest" and treated, because of its clumpy nature, in heuristic models, as distinct "clouds" of uncertain geometry. The confinement mechanism has remained unknown, with candidates being gravity in mini-halos, external gas pressure or no-confinement (i.e., free expansion). Recent detailed simulation work by at least three groups (Cen et al. 1994; Hernquist et al. 1996; Zhang, Anninos & Norman 1995) has indicated that the "clouds" arise naturally in popular cosmogonic scenarios, that there is some truth in all of the earlier conceptual models, but that the dominant mechanism is ram pressure confinement of infalling gas. Filament-like caustics form from small wavelength perturbations going nonlinear. These caustics become weakly shocked, Zel'dovich pancake-like ribbons of gas, which are overdense by a factor of order 10^1–10^2, with the temperature being maintained by photoheating processes at about $10^{4.5}$ K. Not only are the individual and group statistical properties of the Lyman-α clouds reproduced well by the simulations, but it appears that the spatial correlations (Miralda-Escudé et al. 1996) are well modeled also. Study of the Lyman-α clouds should be a fruitful field for further work, because of the vast body of cosmological information accumulated in the detailed observational absorption line studies now becoming amenable to theoretical analysis.

8.3.2.2 Gas near galaxies

The gaseous components discussed so far are either so hot as to be essentially unaffected by stellar sources (X-ray clusters), or so far from them as to be unreachable except by the dilute metagalactic radiation field (Lyman-α clouds). In addition, there will be the all important overdense components in temperature range $10^{4.0}$–$10^{5.5}$ K within the caustic surfaces from which galaxies will form and which may contain a significant fraction of the total baryonic component of the universe (since galaxy formation is, like star formation, notably inefficient). Here the high-resolution SPH work by Evrard, Summers & Davis (1994), Katz, Weinberg & Hernquist (1996) or Steinmetz (1996) is best. The problem is, however, very difficult, since the detailed output from massive stars must be known (UV, X-rays, SN ejecta, etc.) and spatially dependent radiative transfer is required. This area is wide open to further investigation, but the technical problems are formidable.

8.3.3 Cold condensed components

Perhaps easier to treat is the truly cold component which cools to the opaque neutral atomic or the molecular state or proceeds further into a condensed stellar component. As noted earlier, once gas has started to cool, collapse, and fragment in an accelerated fashion, it must end in one of these phases (neutral atomic, molecular or stellar), although which phase cannot be computed absent a good theory of star formation. Codes should be able to compute the masses, angular momenta, and spatial distributions in these cold components – the principle components of galaxies – even when their information concerning the internal states of the objects is poor. Thus, it is encouraging that even the first computations (cf. Cen & Ostriker (1993)) reproduced a plausible galaxy mass function and the density–morphology relations found in the real universe. Specifically, if one identifies the most gas-poor systems as "ellipticals," they tend, statistically, to have stars with the oldest average age, to live in the regions of highest baryon density, and to be (on average) most massive. The computed gas-poor systems are strikingly like normal elliptical galaxies and gas-rich systems like spirals or irregulars. To be believed, this work must be confirmed by higher-resolution studies and by other investigators using different approximations with regard to the physical modeling.

8.4 Conclusions

8.4.1 X-ray emitting clusters of galaxies

Comparison between simulation and observation has led to certain fairly definite conclusions:

(i) The normalization for the power spectrum on the $8\,h^{-1}$ Mpc scale is quite well pinned down by observations of these systems at $z = 0$.
(ii) Comparison between the normalization thus obtained for the relatively rare systems (containing a few percent of the total mass density) and the overall potential fluctuations measured at the same length scale by the smoothed velocity field argues strongly for a Gaussian origin of the potential field fluctuations. Topological defects such as textures or cosmic strings would produce a much larger relative abundance of clusters than is observed (Chiu, Ostriker & Strauss 1998).
(iii) The low rate of evolution of the X-ray cluster properties (Eke, Cole & Frenk 1996) strongly indicates a universe where the material which can gravitationally cluster has a density substantially less than the critical density. Figure 12 of Cen & Ostriker (1994) shows a much slower evolution of $\langle T_x \rangle$ (the luminosity weighted average cluster gas temperature) in an $\Omega_m = 0.5$ universe than in a standard $\Omega_m = 1$ universe. ROSAT observations strongly favor the former model.

8.4.2 Lyman-α clouds

The recent high-resolution (e.g., Keck) results for the Lyman-α forest are fit well by CDM dominated models (Miralda-Escudé 1996; Katz et al. 1996) with no strong preference for any specific value of Ω indicated. Using the simulations, if normalized to observed values of the ionizing radiation field J_ν at the Lyman Limit (and redshift $z \sim 2$–3), one can derive an estimate for the baryon density which puts it at the high end of the range currently being considered: $\Omega_b h^2$ is approximately equal to 0.025.

Further computations of other scenarios should be important in discriminating amongst models. Since the Lyman-α clouds measure the fluctuations on the relatively small physical scales which are affected by a massive neutrino component, an upper bound on this component can surely be placed by comparing models to observations.

8.4.3 Gravitational lensing

Very high Λ models can be ruled out (cf. Kochanek 1996) as they require so much path length (to a given redshift) as to overproduce lenses, given the known–galactic–lenses candidates. However, standard $\Omega = 1$ CDM at small $\Delta\theta < 2''$ separation overproduces large splitting $\Delta\theta > 5''$ lenses by a large amount. Thus, it appears that low-Ω models will be required to satisfy the lensing observations, and it remains to be seen if these can be combined with a cosmological constant to make a viable model (e.g., $\Omega_m = 0.4$, $\Omega_\Lambda = 0.6$).

This summary touches on only a few of the active areas of research and it emphasizes the Princeton work with which the author is most familiar. It is now very apparent that a combination of improved observations and rapidly advancing computer technology will allow much more extensive comparisons between theory and observation in the next few years and that these comparisons should allow us to greatly narrow the range of viable world models. Current work indicates a preference for Gaussian models containing a significant amount of cold dark matter but having a total matter component which is small compared to the critical value.

References

Bahcall, N. A., Cen, R. Y. & Ostriker, J. P. 1996, *ApJ*, **462**, L49

Bardeen, J. M., Bond, J. R., Kaiser, N. & Szalay, A. S. 1986, *ApJ*, **304**, 15

Bardeen, J., Steinhardt, P. & Turner, M. S. 1983, *Phys. Rev. D.*, **28**, 679

Barnes, J. E. & Hut, P. 1986, *Nature*, **324**, 446

Bartlett, J. G., Blanchard, A., Silk, J. & Turner, M. S. 1995, *Science*, **267**, 980

Bertschinger, E. 1995, astro-ph/9506070

Biermann, L. 1950, *Z. Natureforsch*, **5a**, 65

Bolte, M. & Hogan, C. J. 1995, *Nature*, **376**, 399

Bryan, G. L., Klypin, A., Loken, C., Norman, M. L. & Burns, J. O. 1994, *ApJ*, **437**, L5

Cen, R. 1992, *ApJS*, **78**, 341

Cen, R. Y., Jameson, A., Liu, F. & Ostriker, J. P. 1990, *ApJ*, **362**, L41

Cen, R., Kang, H., Ostriker, J. P. & Ryu, D. 1995, *ApJ*, **451**, 436

Cen, R., Miralda-Escudé, J., Ostriker, J. P. & Rauch, M. 1994, *ApJ*, **437**, L9

Cen, R. & Ostriker, J. P. 1993, *ApJ*, **417**, 415

Cen, R. & Ostriker, J. P. 1994, *ApJ*, **429**, 4

Chiu, W., Ostriker, J. P. & Strauss, M. A. 1998, *ApJ*, **494**, 479

Efstathiou, G. 1990, in *Physics of the Early Universe*, eds. J. A. Peacock, A. F. Heavens & A. T. Davies (Edinburgh: Edinburgh University Press), p. 361

Efstathiou, G., Davis, M., Frenk, C. S. & White, S. D. M. 1985, *ApJS*, **57**, 241

Eke, V. R., Cole, S. & Frenk, C. S. 1996, *MNRAS*, **282**, 263

Evrard, A. E., Summers, F. J. & Davis, M. 1994, *ApJ*, **422**, 11

Gendreau, K. C., et al. 1995, *PASJ*, **47**, L5

Gnedin, N. 1995, *ApJS*, **97**, 231

Gnedin, N. Y. & Ostriker, J. P. 1997, *ApJ*, **486**, 581

Gursky, H., Kellogg, E., Murray, S., Leong, C., Tananbaum, H. & Giacconi, R. 1971, *ApJ*, **167**, L81

Guth, A. H. & Pi, S.-Y. 1982, *PRL*, **49**, 1110

Harrison, E. R. 1970, *Phys. Rev. D.*, **1**, 2726

Hernquist, L. & Katz, N. 1989, *ApJS*, **70**, 419

Hernquist, L., Katz, N., Weinberg, D. H. & Miralda-Escudé, J. 1996, *ApJ*, **457**, L51

Hockney, R. W. & Eastwood, J. W. 1988, in *Computer Simulations Using Particles* (New York: Wiley & Sons)

Kang, H., et al. 1994, *ApJ*, **430**, 83

Katz, N., Weinberg, D. & Hernquist, L. 1996, *ApJS*, **105**, 19

Katz, N., Weinberg, D. H., Hernquist, L. & Miralda-Escudé, J. 1996, *ApJ*, **457**, L57

Kochanek, C. S. 1996, *ApJ*, **466**, 638

Kulsrud, R. M., Cen R., Ryu, D. & Ostriker, J. P. 1997, *ApJ*, **480**, 481

Miralda-Escudé, J., Cen, R., Ostriker, J. P. & Rauch, M. 1996, *ApJ*, **471**, 528

Monaghan, J. J. & Gingold, R. A. 1983, *J. Comp. Phys.*, **52**, 374

Navarro, J. F. & Steinmetz, M. 1997, *ApJ*, **478**, 13

Osterbrock, D. E. 1989, *Astrophysics of*

Gaseous Nebulae and Active Galactic Nuclei (Mill Valley, CA: University Science Books).

Ostriker, J. P. 1993, *ARA&A*, bf 31, 689

Ostriker, J. P. & Cen, R. 1996, *ApJ*, **464**, 27

Ostriker, J. P. & Gnedin, N. Y. 1996, *ApJ*, **472**, L63

Ostriker, J. P. & Steinhardt, P. 1995, *Nature*, 377, 600

Peacock, J. & Dodds, S. J. 1994, *MNRAS*, **267**, 1020

Peebles, P. J. E. & Yu, J. T. 1970, *ApJ*, **162**, 815

Pen, U., 1995, *ApJS*, **100**, 269

Starobinskii, A. A. 1982, *Phys. Lett. B.*, **117**, 175

Steinmetz, M. 1995, *Proc. 17th Texas Symposium on Relativistic Astrophysics*, Annals of the New York Academy of Sciences, **759**, 628

Steinmetz, M. 1996, *MNRAS*, **278**, 1005

Tytler, D., Fan, X. & Burless, S. 1996, *Nature*, 381, 207

Walker, T., Steigman, G., Schramm, D. N., Olive, K. A. & Kang, H. 1991, *ApJ*, **376**, 51

Wambsganss, J., Cen, R., Ostriker, J. P. & Turner, E. L. 1995, *Science*, **268**, 274

Weinberg, D., Hernquist, L. & Katz, N. 1997, *ApJ*, **477**, 8, and references therein.

White, S. 1994, in *Clusters of Galaxies*, Proceedings of the XXIX Rencontres de Moriond, Meribel, Savoie, France, March 12–19, 1994 (Gif Sur Yvette: France)

White, S. D. M., Efstathiou, G. & Frenk, C. S. 1993, *MNRAS*, **262**, 1023

Xu, G. 1995, *ApJS*, **98**, 355

Xu, G. 1997, *MNRAS*, **288**, 903

Zel'dovich, Ya. B. 1972, *MNRAS*, **160**, 1p

Zhang, Y., Anninos, P. & Norman, M. 1995, *ApJ*, **453**, L57

9 Black holes in galaxy centers

S. M. Faber

Abstract

The centers of galaxies are extreme astrophysical environments, with the highest stellar densities and velocity dispersions known. These circumstances put theories of stellar dynamics, gas dynamics, and star formation to the ultimate test. Centers of galaxies also harbor the engines that drive quasars and active galactic nuclei, widely thought to be massive black holes (BHs) of mass $10^{7-9} M_\odot$. Far from being disconnected from the lives of their parent galaxies, centers can also exchange material with the rest of the galaxy in a variety of ways, and their present state was shaped in part by the larger forces of galactic evolution. For all these reasons, galaxy centers are becoming key objects of study in cosmology. The first half of this review describes the latest data on the density distribution of stars in galactic centers from the *Hubble Space Telescope*. The second half describes their dynamics and the current state of the kinematic search for central massive BHs. Finally, certain evidence suggests that BHs may have had substantial gravitational feedback on galaxy centers during galaxy merger events, and thus played a major role in shaping the central structures of galaxies seen today.

9.1 Introduction

The centers of galaxies are fascinating and challenging objects of study. From the standpoint of stellar kinematics, they represent extreme environments in the universe, with higher densities and higher velocity dispersions than found anywhere else. This promotes a wealth of interesting dynamical phenomena, involving stellar relaxation, dynamical friction, and even (if the density is high enough) stellar collisions. Galaxy centers are also the bottoms of deep potential wells. Matter that loses energy eventually finds its way there and, once there, is likely to stay. Such matter can come from stars and gas in the galaxy itself or from outside, brought in by galaxy mergers or cooling flows.

Galactic centers also harbor active galactic nuclei (AGNs) and quasars (QSOs). Popular models for these objects involve matter accretion onto a massive BH. Simple arguments, which we repeat below, suggest that nearly all large galaxies at one time possessed QSOs, and their fossil BH descendents lurk today in nearby galaxies, invisible for lack of fuel. A major goal of present research is to confirm by independent means whether these BHs are present, and if so what their frequency and masses are. New, high-resolution techniques involving the *Hubble Space Telescope* (HST) and very-long baseline radio observations are central to this effort.

The upshot of this activity is a subtle shift in attitude toward galactic centers. Whereas formerly it might have been reasonable to regard studies of centers as rather specialized and disconnected from the mainstream of research on normal galaxies, that view is no longer tenable. Instead, much current research is focusing on newly appreciated intimate connections between centers and their parent galaxies, and seeks to use those connections as an avenue to study the evolution of the whole galaxy.

This chapter focuses on the properties of "normal" galactic centers that (for the most part) today lack AGNs. We are interested in establishing the fundamental parameters of the stellar background, including the luminosity density, mass density, and star and gas kinematics. The emphasis will be on ellipticals and spiral bulges because they are the only types of systems that have as yet been systematically surveyed. They are also presently the best BH candidates (Kormendy & Richstone 1995).

For those unfamiliar with galaxy morphology, ellipticals and spiral bulges are "hot," dense stellar systems at one end of the Hubble morphology sequence. The other end consists of rather amorphous objects of low density called "irregular" galaxies. Irregulars do not possess dense centers and are almost certainly not good candidates for harboring massive BHs. In between are intermediate-type spiral galaxies. These have regular spiral structure and moderately high overall density but lack the centrally concentrated, hot stellar *spheroids* characteristic of ellipticals and spiral bulges. The central properties of these intermediate Hubble types have not yet been characterized systematically, but, based on the incidence of AGNs in some of them, a fair fraction probably contain BHs. For further reference on galaxy morphologies, see *The Hubble Atlas of Galaxies* by Sandage (1961).

The alert reader may have observed the use of the word "center" in this essay, whereas older literature would have used the terms "nucleus" or "core". This reflects recent conventions adopted by the group of collaborators mentioned in the acknowledgments, the so-called "nukers" group. Motivated by structures seen in HST images, we have elected to reserve the words *nucleus* and *core* for two very specific phenomena, as explained below, and utilize the more general word *center* to express the broader notion of "middles of galaxies." The term *hot galaxy* will be further used to encompass both elliptical galaxies and the bulges of spirals. This term reflects the fact that both types of spheroidal system are supported by random stellar motions, not ordered rotation, although, as we shall see, some systems do rotate gently and this proves to be important.

This chapter is in two parts. The first part describes the photometric and kinematic properties of normal galaxy centers, together with some important correlations with global size, shape, and rotation. The second part focuses on the central kinematics and the current state of the BH search. We close with some speculative evidence that BHs are ubiquitous in nearby galaxies and have had a major influence on shaping central structures. As background to this chapter, one can recommend some excellent recent reviews on the global structure and evolution of hot galaxies generally, and the search for BHs in particular: Kormendy & Djorgovski (1989), de Zeeuw & Franx (1991), Barnes & Hernquist (1992), and Kormendy & Richstone (1995). The basic reference on galactic stellar dynamics is the superb textbook *Galactic Dynamics* by Binney & Tremaine (1987).

9.2 Surface-brightness and stellar distributions

9.2.1 Surface brightness profiles

When we image the center of a galaxy, virtually all of the detected light comes from the background sea of stars orbiting in the galactic potential well. Systematic study of the resulting *surface brightness profiles* began with Lauer (1983, 1985) and Kormendy (1985). A collection of 55 inner hot galaxy profiles from the *Hubble Space Telescope* is shown in Fig. 9.1 (Faber *et al.* 1997, based on data in Lauer *et al.* 1995, Forbes, Franx & Illingworth 1995, Jaffe *et al.* 1994, other references, and unpublished profiles in the HST archive as of June 1993). These profiles are plotted in log–log coordinates because the basic surface brightness distributions of hot galaxies are close to being power laws, which would look like straight lines in Fig. 9.1. The actual profiles are not *quite* straight, and the "breaks," or places where the profiles bend, define characteristic radii and surface brightnesses. These important parameters characterize the size and density of the overall structure and, in concert with dynamical information, can be used to measure the mass.

As an example of such features, Lauer (1983, 1985) and Kormendy (1985) detected inner regions in many hot galaxies where the slope of the surface brightness profile turned over, as seen in Fig. 9.1. They measured the size and surface brightness of these *cores*, as they termed them, and demonstrated systematic *parameter relations* linking core properties with one another and with the global properties of galaxies as a whole. In particular, they showed that cores in bright galaxies were much larger and of lower density than cores in small galaxies.

The latest version of the core parameter relations using ground-based data was published by Kormendy & McClure (1993). However, ground data are limited because core features in many galaxies are smaller than a few arc seconds and are badly washed out by atmospheric seeing. HST's angular resolution is 20 times better, which is an enormous improvement. Updated examples of the parameter relations based on HST data are shown below.

The curves in Fig. 9.1 represent projected surface brightness on the sky. With appropriate mathematical techniques and some assumptions, it is possible to invert this information to obtain the *volume luminosity density* of starlight, $v(r)$. Fig. 9.2 shows such

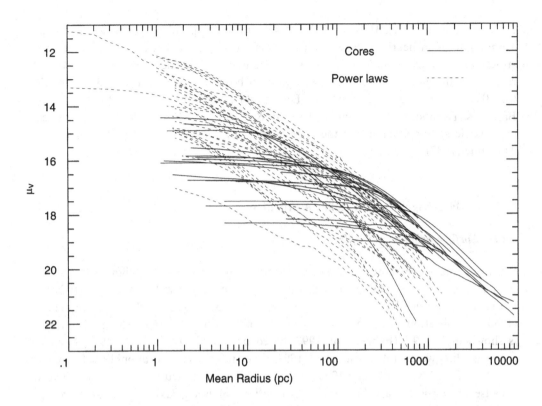

Figure 9.1. Collected V-band surface brightness profiles of 55 ellipticals and spiral bulges observed with HST (from Faber et al. 1997). All were observed in the Planetary Camera through filter F555W in Cycles 1 and 2, when HST suffered from spherical aberration. Deconvolution (Lauer et al. 1992a) recovers the profile accurately into $0\rlap{.}''1$. The abscissa is in parsecs (1 pc = 3×10^{18} cm). The ordinate is the logarithm of surface brightness, I, expressed as V-mag arcsec^{-2}.

inversions for many of the galaxies in Fig. 9.1 (Gebhardt et al. 1996). These have been scaled in both v and r to unit density at the radius of maximum logarithmic curvature in the projected surface brightness profile, i.e., the maximum of $d \log S/d \log r$, where $S \equiv |d \log I/d \log r|$. This point is called the *break radius*, as explained further below.

Both Fig. 9.1 and Fig. 9.2 suggest that brightness profiles fall into two groups. One group is rather straight in log radius–log surface brightness, which means that they do not have a well-defined radial scale length (at least not here in the inner parts; all galaxies possess *global* radii, which are visible as gentle bends in the profile farther out). Because of their relative straightness, we have termed such profiles *power laws*. The second type of profile shows a much more pronounced inner turnover. These are the *cores* previously identified by Lauer and Kormendy.

The appearance of cores in galaxy profiles initially did not surprise astronomers. This was probably because cores were already known in many star clusters, and because

Black holes in galaxy centers

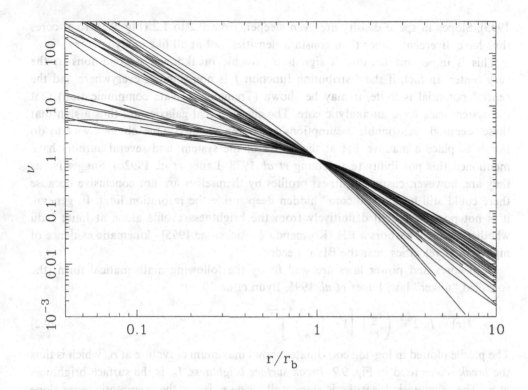

Figure 9.2. Stellar luminosity density profiles of many hot galaxies from Fig. 9.1, scaled to the radius and luminosity at the radius of maximum curvature in the surface brightness. Note the apparent division into cores (flatter) and power laws (steeper). From Gebhardt et al. (1996).

familiar stellar dynamical models, such as the isothermal sphere and King models (King 1966) also show cores. Tremaine (1998) notes that, in the absence of a central compact mass, all physical variables should vary smoothly near the origin of a stellar system, and hence can be expanded in a Taylor series. In particular, the surface brightness can be written

$$I(r) = I_0 + I_1 r^2 + O(r^4), \tag{9.1}$$

where r is projected radius. A quadratic of this sort flattens strongly in log–log coordinates and shows a core. Tremaine suggests the term *analytic core* for systems with cores that arise simply from the smoothness of their central properties. The cores of familiar dynamical models (and most star clusters) are thus analytic.

Figures 9.1 and 9.2 call into question whether real galaxy cores in fact have anything to do with analytic cores. In the latter, the central density flattens completely at small radii, whereas real cores show shallow power-law *cusps* into the resolution limit (about 0".1 with HST). Fits to surface brightness as in Fig. 9.1 (Byun et al. 1996) yield moderate inner cusps with slope $\gamma \equiv |d \log I / d \log r|$ in the range 0.05 to 0.3 (see also Lauer et al.

1995). Slopes in space density are even steeper, from 0.2 to 1.2 (Fig. 9.2). Real cores thus have divergent rather than constant densities, not at all like analytic cores.

This is important because it signals a possible breakdown of conditions at the very center. In fact, if the distribution function f is non-singular everywhere and the central potential is finite, it may be shown (Tremaine, private communication) that the system *must* have an analytic core. The cusps in real galaxies are thus a sign that these seemingly reasonable assumptions are being violated. One obvious way to do this is to place a massive BH at the center of the system, and several authors have mentioned this possibility (e.g., Young et al. 1978, Lauer et al. 1992b). Suggestive as they are, however, cuspy brightness profiles by themselves are not conclusive because there could still be a "mini-core" hidden deep inside the resolution limit. In general, it is not possible to tell definitively from the brightness profile alone at finite radii whether a galaxy harbors a BH (Kormendy & Richstone 1995) – kinematic evidence of high orbital velocities near the BH is needed.

Both cores and power laws are well fit by the following mathematical form (the so-called "nuker" law, Lauer et al. 1995, Byun et al. 1996):

$$I(r) = I_b \, 2^{\frac{\beta-\gamma}{\alpha}} \left(\frac{r_b}{r}\right)^\gamma \left[1 + \left(\frac{r}{r_b}\right)^\alpha\right]^{\frac{\gamma-\beta}{\alpha}}. \tag{9.2}$$

The profile plotted in log–log coordinates shows maximum curvature at r_b, which is thus the *break radius* used in Fig. 9.2. *Break surface brightness*, I_b, is the surface brightness at r_b. The asymptotic logarithmic slope well inside r_b is $-\gamma$, the asymptotic outer slope is $-\beta$, and the parameter α measures the sharpness of the break. Equation (9.2) is intended to fit only over radii $< 10''$ accessible to the HST Planetary Camera. For typical values of β, there must be a further turndown at large radii to ensure that the total luminosity be finite.

A final vocabulary word before proceeding. Close inspection of Fig. 9.1 reveals that some power laws show excess light above the prediction of equation (9.2) within the inner few tenths of an arc second. We term these excesses *nuclei*. Lauer et al. (1995) show examples of nuclei of varying degrees of prominence. Based on nearby Local Group examples, most nuclei are probably massive *star clusters*. However, some are non-thermal sources associated with well-known active galaxies, which we term AGNs to distinguish them from nuclei. Nuclei are interesting star systems in their own right. Some are quite massive yet are centered on BHs that are even more massive than they are. A particularly interesting nucleus in the galaxy NGC3115 is described below.

9.2.2 *Central-parameter relations*

Byun et al. (1996) fitted the profiles of 60 galaxies to equation (9.2) and derived values of r_b and I_b. These have been supplemented with other dynamical information from the literature such as central velocity dispersions to plot updated versions of the central parameter scaling relations (Faber et al. 1997).

Two such examples are shown in Fig. 9.3. The first plots r_b versus total galaxy

Black holes in galaxy centers

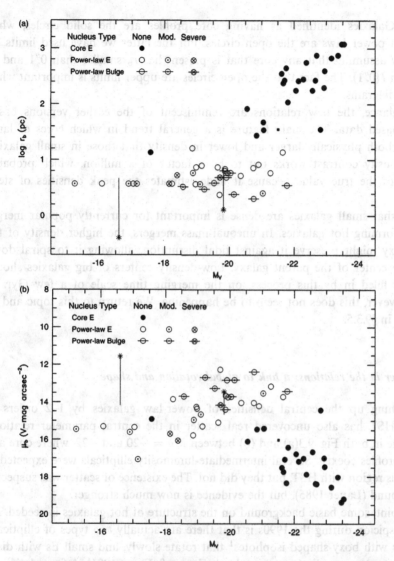

Figure 9.3. (a) HST break radii r_b for hot galaxies vs. galaxy absolute V luminosity. The solid circles are cores using r_b from the fit to equation (9.2). The open circles are power laws, using $r_b = 0\overset{\prime\prime}{.}1$ as an upper limit. Bulges are the hot components of early-type S0–Sb spirals. The asterisks are M31 and M32, as seen nearby in the Local Group and 24 times farther away near the Virgo Cluster. This shows the likely effect of resolution on other power law galaxies. The discrepant core galaxy is the peculiar merger remnant Fornax A. Aside from that, core galaxies show a narrow core size relation vs. absolute magnitude. Note the scatter at middle magnitudes and compare to Fig. 9.4. (b) Same as panel (a) but showing break surface brightness I_b (the limit $I(0\overset{\prime\prime}{.}1)$ is used for power laws). The turndown for small galaxies is probably mostly an artifact of limited resolution, as shown by the motion of M32 (left asterisk) under distance change. The range of a factor of 10^3 in surface brightness translates to a factor of 10^6 in density (Faber et al. 1997).

magnitude. Galaxies identified as having core profiles are the solid circles, while galaxies with power laws are the open circles. For the latter we have used limits on r_b and I_b by assuming that any core that is present has r_b smaller than $0''.1$ and I_b brighter than $I(0''.1)$. The fact that the open circles are upper limits is important when viewing the diagrams.

At first glance, the new relations are reminiscent of the earlier versions based on ground-based data. The main feature is a general trend in which cores of large galaxies are both physically larger and lower in density that those in small galaxies. The total density contrast works out to be a factor of a million, which probably *under*estimates the true value because it underestimates the peak densities of steep power laws.

The fact that small galaxies are dense is important for currently popular merger models for forming hot galaxies. In unequal-mass mergers, the higher density of the smaller galaxy might preserve it against tidal disruption, allowing it to spiral down intact to the center of the parent galaxy. Low-density centers of big galaxies should therefore be filled in by this process, on the merging time scale of a few Gyr. In practice, however, this does not seem to be happening. We return to this topic and its implications in §9.3.5.

9.2.3 Scatter in the relations: a link to global rotation and shape

Besides pushing up the central densities of power-law galaxies by 1–2 orders of magnitude, HST has also uncovered real scatter in the central parameter relations. This is visible in both Fig. 9.3(a) and (b) between $M_V = -20$ and -22, where core and power-law profiles coexist. Several intermediate-luminosity ellipticals were expected to resolve in this region with HST, but they did not. The existence of scatter was suspected from the ground (Lauer 1985), but the evidence is now much stronger.

At this point, some basic background on the structure of hot galaxies is needed. An emerging suspicion during the 1990s is that there are actually *two* types of ellipticals: luminous Es with boxy-shaped isophotes[1] that rotate slowly, and small Es with disky isophotes that rotate rapidly (see review by de Zeeuw & Franx 1991). A formal division of the Hubble sequence for ellipticals into two classes has recently been suggested based on these criteria (Kormendy & Bender 1996). We shall refer to the two subtypes as *boxy* and *disky* respectively.

A major result from HST is that *scatter in the central parameter relations of Fig. 9.3 appears to correlate with boxy/disky subtype*. This is illustrated in Fig. 9.4, which replots r_b vs. M_V from Fig. 9.3(a) but with symbols now indicating galaxy rotation speed. Power-law galaxies are seen to be mainly small rotating ellipticals or bulges, while cores are found in large, slowly rotating ellipticals. At intermediate magnitudes,

---

[1] The isophotal shapes of elliptical galaxies are near-perfect ellipses. The boxy and disky shapes referred to are subtle distortions about the best-fitting ellipse.

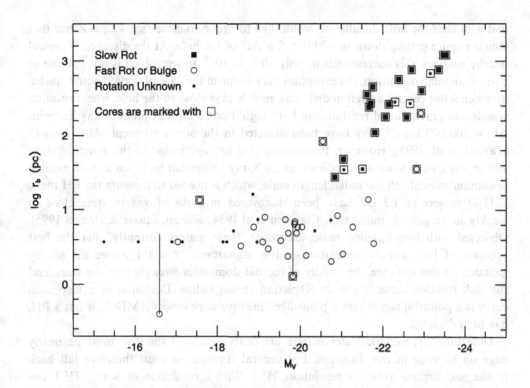

Figure 9.4. Replot of break radii r_b from Fig. 9.3(a) with symbols indicating speed of rotation. Rapidly rotating hot galaxies tend to be disky and have high bulk rotation speeds, while slowly rotating galaxies tend to be boxy and rotate slowly.

rotating galaxies have much smaller cores (if any) than non-rotating galaxies. A figure based on isophote shape rather than rotation speed would look basically similar.

This correlation is actually not new. It was first noticed by Nieto, Bender & Surma (1991) based on a large sample of high-quality ground data, and then pointed out by Jaffe et al. (1994) based on a sample of 14 Virgo cluster galaxies observed with HST. However, the present sample of 61 galaxies is the best and largest yet.

To summarize, it appears likely that disky and boxy ellipticals have different kinds of centers. The key point is the new connection between *central* and *global* properties. Because global properties are involved, whatever process established this link was probably a *major event in the life of the galaxy*. In the final section, we return to this link and explore a possible relation to massive central BHs.

9.3 Kinematic evidence for central massive black holes

9.3.1 Detection criteria

The central question is, how does one search for a *fossil* BH, which by definition is not shining? The only signature is the high orbital speeds of visible components (stars and gas) near the hole. Ideally, we would like to probe so close that $V_{orb} \approx c$, but that would require getting down to within a few AU of the hole. At the distance of typical nearby galaxies, this corresponds to only 10^{-5} to 10^{-6} arcsec, which is not feasible at current angular resolution. An exception may occur in the case of certain active nuclei, in which a hot central accretion disk may emit X-rays close to the hole, where rotation speeds and gravitational redshifts are both high. For example, a broad X-ray line with $\Delta V \approx 100,000$ km s^{-1} may have been detected in the active elliptical MCG-6-30-15 (Tanaka et al. 1995). However, this technique is not applicable to the general galaxy and in any case is somewhat indirect, as no X-ray image can be taken at high enough resolution to establish the radial length scale, which is needed to measure the BH mass.

H_2O masers at 1.3 cm have been discovered in disks of gas rotating close to weakly active galactic nuclei (e.g., Claussen et al 1984; Wilson, Braatz & Henkel 1995). Observed with long-baseline radio telescopes, these masers currently offer the best prospect of high angular resolution–a few milliarcsec, or ≈ 0.1 parsec for nearby galaxies. At this distance, the gravity of the BH dominates strongly over the stars, and the disk rotation curve is closely Keplerian versus radius. Detection of a Keplerian curve is a powerful signal that a point-like "massive dark object" (MDO), if not a BH, has been detected.

Unfortunately, galactic water masers are fairly rare, and the disk must be nearly edge on to mase in our direction. For general surveys, we must therefore fall back to the next highest available resolution, HST. With a resolution of $\sim 0\rlap{.}''1$, HST can probe down to about a parsec in nearby galaxies. At this point, however, V_{orb} is only modestly enhanced above normal, and the question becomes acute as to whether a detected MDO is really a BH.

The elevation of V_{orb} above normal can be quantified using the concept of *mass-to-light ratio*. From rotation and σ measurements, one constructs a dynamical model that yields the mass $M(r)$ in solar masses at any radius r. From optical images, one measures the amount of light within the same radius $L(r)$, in solar luminosities. The quantity

$$M/L = \frac{M(r)}{L(r)} \tag{9.3}$$

in solar units is called the mass-to-light ratio. For a star like the sun, it is unity by definition. For the dimmest M-dwarfs, it can be as high as 1000. For typical, normal stellar populations, it is in the range 1–15, and for non-luminous objects such as planets and neutron stars, it is infinite. Generally speaking, values in excess of $\gtrsim 20$ signal either a highly unusual stellar population or the presence of an MDO.[2]

Non-BH models for MDOs must involve collections of smaller objects. That is because no form of matter is strong enough to support a single *uncollapsed* self-gravitating object of the mass $\gtrsim 10^6 M_\odot$ typical of central MDOs–such an object, if

---

[2] M/L scales with the adopted distance to an object as D^{-1}, and hence with the Hubble constant as H_0. I am using a Hubble constant of $H_0 = 80$ km s^{-1} Mpc^{-1}.

formed, would immediately collapse to form a BH. Thus, to avoid a BH one is forced to distribute the mass among many smaller objects in an equilibrium self-gravitating *cluster*. Objections to the stability and longevity of such clusters focus on three general concerns:

(i) Too-short relaxation time: a cluster may evaporate too quickly as individual objects are scattered into unbound orbits by close gravitational encounters with other cluster members.

(ii) Too-short collision time: objects may physically collide in the cluster and coalesce to make larger objects in a runaway process that could eventually build a BH.

(iii) Implausible constituents and/or formation scenarios: to avoid (i) and (ii), one is forced to make the objects both physically small and light. This points toward elementary particles, planets, brown dwarfs, white dwarfs, neutron stars, and stellar-mass BHs. Formation scenarios for clusters of such bizarre objects do not exist. Even if M/L is not huge, say in the range 20–50, it still requires a very contrived scenario to put *most* of the mass into very dim stars or invisible stellar remnants. For example, a normal stellar population with a typical distribution of masses will have converted only ~10% of its initial mass into remnants by today, and more mass will actually have been ejected (in stellar winds and supernovae) than went into remnants. On the other hand, since the populations in question would likely have formed in the abnormal environment of an active galactic nucleus, arguments based on what would be "impossible" are somewhat weak.

Before discussing recent BH searches, a digression is made to summarize evidence from quasars that massive BHs should in fact be common in nearby galactic centers.

9.3.2 BH demography: predictions from quasars

The following section is adapted from Tremaine (1998). The basic assumptions are that quasars are fueled by infall of matter onto BHs and that this matter was converted into radiation with an efficiency $\epsilon \equiv \Delta E/\Delta Mc^2$. The local mass density in dead quasar BHs is then related to the total local energy density of quasar radiation, as follows.

The local energy density in quasar light is (Chokshi and Turner 1992)

$$u = 1.3 \times 10^{-15} \text{ erg cm}^{-3}. \tag{9.4}$$

The mean local mass density of dead quasars must therefore be at least (Soltan 1982, Chokshi & Turner 1992)

$$\rho_\bullet = \frac{u}{\epsilon c^2} = 2.2 \times 10^5 \left(\frac{0.1}{\epsilon}\right) M_\odot \text{ Mpc}^{-3}, \tag{9.5}$$

assuming that the universe is homogeneous and transparent.

The mass of a dead quasar may be written

$$M_\bullet = \frac{L_Q \tau}{\epsilon c^2} = 7 \times 10^8 \, M_\odot \left(\frac{L_Q}{10^{12} L_\odot}\right)\left(\frac{\tau}{10^9 \, \text{yr}}\right)\left(\frac{0.1}{\epsilon}\right), \qquad (9.6)$$

where L_Q is the quasar luminosity and τ is its lifetime. An upper limit to the lifetime is the evolution time-scale for the quasar population as a whole, $\sim 10^9$ yr; however, upper limits to BH masses in nearby galaxies and direct estimates of BH masses in AGNs both suggest that actual masses and lifetimes are smaller by a factor 10–100 (Haenelt & Rees 1993), which implies $M_\bullet = 10^{7-8} \, M_\odot$.

To focus discussion, we adopt a straw-man model in which a fraction f of all galaxies contains a central BH and BH mass is proportional to galaxy luminosity. Thus $M_\bullet = \Upsilon$, where Υ is the (black hole) to (galaxy) mass-to-light ratio. The luminosity density of galaxies is $j = 1.5 \times 10^8 \, L_\odot \, \text{Mpc}^{-3}$ in the blue band (Efstathiou et al. 1988, adjusted to a Hubble constant of $H_0 = 80 \, \text{km s}^{-1} \, \text{Mpc}^{-1}$). Thus

$$\Upsilon = \frac{\rho_\bullet}{fj} = \frac{0.0015}{f}\left(\frac{0.1}{\epsilon}\right)\frac{M_\odot}{L_\odot}. \qquad (9.7)$$

This value is an average over the light of all local galaxies. However, if we assume that massive BHs are found chiefly in the centers of hot galaxies (i.e., spheroids) as the nearest BH candidates seem to suggest (Kormendy and Richstone 1995), the above number can be converted to the BH mass-to-light ratio *per spheroid* by noting that approximately 30% of the local B-band light is emitted by spheroids (Dressler & Schechter 1987). Correcting for this and converting to the V-band yields

$$\Upsilon_V^s = \frac{0.004}{f_s}\left(\frac{0.1}{\epsilon}\right)\frac{M_\odot}{L_\odot}, \qquad (9.8)$$

where Υ_V^s is now the estimated ratio $M_\bullet/L_V$ per hot component and f_s is the fraction of *spheroids* with BHs.

An independent estimate of Υ_V^s from quasars comes from dividing the typical dead quasar mass derived above, $M_\bullet \approx 10^{7.5} \, M_\odot$, by the typical luminosity of a bright spheroid, $8.5 \times 10^9 \, L_\odot$ (Binggeli, Sandage & Tammann 1988, adjusted to $H_0 = 80 \, \text{km s}^{-1} \, \text{Mpc}^{-1}$). The result is $\Upsilon_V^s \approx 0.004$. Consistency with equation (9.8) then requires $f_s \approx 1$ if $\epsilon \approx 0.1$, or that roughly every spheroid contains a BH.

A further consistency check on Υ_V^s uses individual mass estimates for BH candidates in nearby galaxies (Kormendy & Richstone 1995). It appears that BH mass may scale with galaxy luminosity, so that Υ_V^s is roughly constant with galaxy mass. The mean value from 6 nearby BH candidates is $\Upsilon_V^s = 0.016$ (Faber et al. 1997). This is in reasonable agreement with $\Upsilon_V^s = 0.004$ from quasars above, given that these best BH candidates may be biased somewhat more massive than average.

If every hot galaxy contains a BH, a reasonable estimate for Υ_V^s is the logarithmic mean

$$\Upsilon_V^s = 0.008, \qquad (9.9)$$

for which the corresponding value of $M_\bullet/M_{\text{gal}}$ turns out to be

$$M_\bullet/M_{gal} = 0.002, \qquad (9.10)$$

based on a global average mass-to-light ratio $M_{gal}/L_{gal} = 4$ per galaxy (Faber et al. 1997).

9.3.3 BH searches: a brief early history

The subject began with a pair of seminal papers by Sargent et al. (1978) and Young et al. (1978) on the active elliptical galaxy M87. Pioneering measurements of both the stellar velocity dispersion and the light profile were combined to estimate M/L with radius. A central increase was claimed, based on a substantial rise in σ within the inner 10". The authors inferred a $3 \times 10^9 \, M_\odot$ BH. A shallow cuspy inner core was also detected, which was further noted as additional evidence for a singular central mass distribution, and thus a BH.

This result turned out to hinge on a rather subtle assumption that many workers found objectionable. This can be understood by considering the first moment of the collisionless Boltzmann equation, which expresses the radial momentum balance for a self-gravitating equilibrium stellar system (Binney and Tremaine 1987, p. 204; also Kormendy & Richstone 1995). For a spherical mass distribution and a stellar velocity ellipsoid with one principal axis pointing to the center:

$$M(r) = \frac{Vr^2}{G} + \frac{\sigma_r^2 r}{G}\left[-\frac{d\log v}{d\log r} - \frac{d\log \sigma_r^2}{d\log r} - \left(1 - \frac{\sigma_\theta^2}{\sigma_r^2}\right) - \left(1 - \frac{\sigma_\phi^2}{\sigma_r^2}\right)\right]. \qquad (9.11)$$

Here V is the mean stellar rotation velocity, and σ_r, σ_θ, and σ_ϕ are the radial and tangential components of the velocity dispersion. The density v is not the mass density (that is represented by $M(r)$) but rather the density of the (massless) tracer population whose kinematics (V and σ) we observe. In practice, we assume that v follows the light, i.e., that M/L for the tracer population is constant. With V, σ, and v everywhere measured, we can solve for $M(r)$.

Two problems arise. First, the quantities in equation (9.11) are local, whereas the light and kinematic variables we observe are projected values. Much effort has gone into deprojecting the observations, which turns out to be quite tricky. Assuming this can be done, we encounter the second, more fundamental problem. M87 is not rotating much, so V can be ignored. However, not all three components of σ can be measured independently from the line-of-sight σ alone. Even if one assumes *perfect* spherical symmetry, so that $\sigma_\theta = \sigma_\phi \equiv \sigma_t$, the velocity anisotropy parameter $\beta \equiv [1 - (\sigma_t^2/\sigma_r^2)]$ is still unknown. A variation in β versus radius can lead to an apparent trend in M/L, and hence the appearance of a spurious central mass. The Young et al. (1978) model for M87 made the innocent-seeming but ultimately crucial assumption that $\beta = 0$, i.e., that the velocity dispersion was everywhere isotropic. This key assumption was soon attacked (Duncan & Wheeler 1980, Binney & Mamon 1982), and it was shown that almost *any* reasonable σ gradient can be accounted for without recourse to a BH provided that β is allowed to vary. The sense required is that orbits be more plunging toward the center of the galaxy. In short, without an independent constraint on β, σ profiles by themselves are inconclusive.

Observers nevertheless persevered in trying to find the most pathologically steep σ gradients. Dressler (1984) and Tonry (1984,1987) were the first to spot a steep central σ gradient in the Local Group elliptical M32, which cooperated by having a rotation term that also contributed significantly. Rotation is desirable because it sets an absolute lower limit on the mass (equation (9.11)). Dressler (1984) and Kormendy (1988a) detected high central σ and rotation in the bulge of M31. Other galaxies followed outside the Local Group (e.g., NGC4594, Kormendy 1988b; NGC3115, Kormendy & Richstone 1992; NGC3377, Kormendy 1992). More detailed observations and analysis were done on some of these objects, and ultimately, measured central M/Ls exceeded ~ 30 for a handful of systems.

This was roughly the status as of 1995, when the refurbished HST and very-long-baseline radio interferometry came on line. The early phase had been sufficient to establish the unambiguous existence of central massive dark objects (MDOs) that were either true BHs or some other peculiar structure with high M/L, definitely not normal stars. The next section reviews subsequent progress based on HST and radio interferometry.

9.3.4 BH searches: three illustrative objects

Three recent measurements collectively illustrate the frontiers of current BH research.

9.3.4.1 NGC3115

NGC3115 is a moderate-size E/S0 galaxy about 670 $\mathrm{km\,s^{-1}}$ away (9 Mpc for $H_0 = 80$ $\mathrm{km\,s^{-1}\,Mpc^{-1}}$) with total luminosity $L_V = 1.7 \times 10^{10} L_\odot$. Unlike the objects discussed in the next two sections, it has absolutely no sign of AGN activity – no radio source, no Seyfert nucleus, not even a trace of emission. Its central stellar kinematics were observed recently with HST's Faint Object Spectrograph (Kormendy et al. 1996). NGC3115 had been one of the best BH candidates from the ground (Kormendy & Richstone 1992,1995), and the case substantially improved with HST. Fig. 9.5 compares the new FOS data (filled circles) versus earlier ground-based data at lower resolution. Through the FOS $0''.21$ aperture, central σ increased by 20% and rotation speed *tripled*. Subtracting off projected bulge light from the central spectrum yields a central σ of 600 $\mathrm{km\,s^{-1}}$, and broad wings are clearly visible on the stellar velocity distribution function out to 1200 $\mathrm{km\,s^{-1}}$. These are the most extreme stellar kinematics yet seen.

Four earlier models from Kormendy & Richstone (1992) are shown in Fig. 9.5. The ME models were their best attempt to fit the data with no (ME1) or at most a small ($10^8\,M_\odot$) BH (ME2). Both of these are now excluded. Further experiments indicate that the central dark mass has approximately $M_\bullet = 2 \times 10^9\,M_\odot$.

Is this MDO a black hole? It is surrounded by a compact, bright stellar nucleus with half-light radius $r \leq 0''.05$ (2 pc) (Kormendy et al. 1996). If the stars in this nucleus are

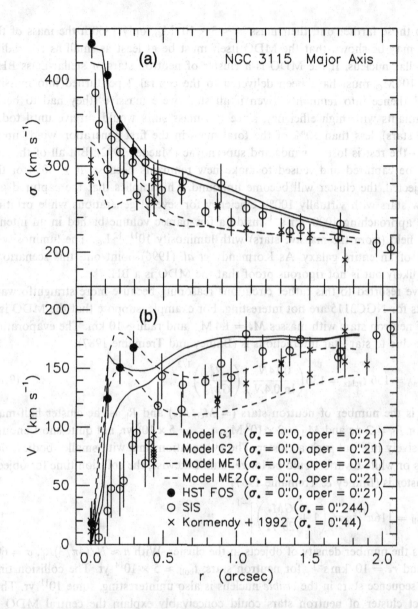

Figure 9.5. Kinematics of the E/S0 galaxy NGC3115 from HST's FOS spectrograph (filled circles) and from earlier Canada–France–Hawaii ground-based data (open circles and crosses). Aperture sizes and rms Gaussian widths of the seeing point-spread-function (the latter is negligible for HST) are indicated. Four BH models from Kormendy & Richstone (1992) have been "re-observed" to match the HST resolution. Rotation and central dispersion both increased with HST, as expected for a central BH of $2 \times 10^9 \, M_\odot$.

similar to those farther out, their mass is $\sim 3 \times 10^7\,M_\odot$, only 1/60th the mass of the MDO. It may be shown that the MDO itself must be at least as small as the radius of this stellar nucleus. If the MDO is a cluster of neutron stars or stellar-mass BHs, then $2 \times 10^9\,M_\odot$ must have been delivered to the central 2 pc, turned into massive stars, and thence into remnants. Even if all stars were massive (they had to be to make remnants with high efficiency, since low-mass stars would survive until today as visible stars), less than 10% of the total mass in the first generation winds up in remnants – the rest is lost in winds and supernovae (Maeder 1992). But all of the lost gas must be captured and reused to make new remnants because, if $\geq 50\%$ of the mass is ejected, the cluster will become unbound. This implies that the captured gas made new stars with virtually 100% efficiency for several generations while orbiting at speeds approaching 1000 km s^{-1} inside a few-parsec volume bathed in an intense radiation field (from the massive stars) with luminosity $10^{11-12}\,L_\odot$. The luminosity is the equal of an entire galaxy. As Kormendy et al. (1996) point out, this scenario is highly unlikely but is not rigorous proof that the MDO is a BH.

We have resorted to this rather circuitous reasoning because more straightforward arguments for NGC3115 are not interesting. For example, suppose that the MDO is a cluster of neutron stars with masses $M_* = 1.4\,M_\odot$ and radii ~ 10 km. The evaporation time-scale due to star–star interactions is (Binney and Tremaine 1987):

$$t_{\text{evap}} = 136\, t_{\text{relax}} = 136 \left(\frac{0.14 N}{\ln 0.4 N} \right) \left(\frac{R_h^3}{GM_{\text{cl}}} \right)^{1/2}, \tag{9.12}$$

where N is the number of neutron stars ($= M_{\text{cl}}/M_*$) and R_h is the cluster half-mass radius. For $R_h = 2$ pc and $M_{\text{cl}} = 2 \times 10^9\,M_\odot$, $t_{\text{evap}} = 5 \times 10^9$ yr, not quite short enough to conclusively rule out such a cluster. The evaporation time with smaller bodies such as planets or white dwarfs would be longer still. Likewise, the collision time for objects in the cluster is (Binney & Tremaine 1987)

$$t_{\text{coll}} = \left[16\pi^{1/2} n\sigma r_* \left(1 + \frac{GM_*}{2\sigma^2 r_*} \right) \right]^{-1}, \tag{9.13}$$

where n is the number density of objects in the cluster. With $n \approx N/(4\pi/3)R_h^3$, $\sigma = 600$ km s^{-1}, and $r_* = 10$ km s^{-1} for neutron stars, $t_{\text{coll}} \approx 5 \times 10^{14}$ yr. The collision time for main sequence stars in the *visible* nucleus is also uninteresting, some 10^{11} yr. Thus a compact cluster of neutron stars could conceivably explain the central MDO in NGC3115, and arguments against it rest on difficulties of formation rather than on fundamental physics.

9.3.4.2 M87

M87 is a well-known, massive elliptical galaxy ($7.8 \times 10^{11}\,L_\odot$) that dominates the main subclump of the Virgo cluster of galaxies ($D \sim 15$ Mpc). It has a double-lobed radio source with a one-sided optical jet plus a central point-like visible non-thermal AGN. Early HST images (Lauer et al. 1992b) confirmed and strengthened the cusp that was found from the ground in the early measurements of Young et al. (1978). In pioneering observations made with HST after refurbishment, Harms et al. (1994) and Ford et al. (1994) discovered an inner, rotating ionized gas disk and measured symmetric rotation speeds of $\pm$ 550 km s^{-1} at 0.''2 radius on either side. The resultant MDO mass is 3×10^9 M$_\odot$, in good agreement with the earlier mass estimate of Sargent et al. (1978).

The mass and radius of measurement are comparable to NGC3115, so similar formation and longevity arguments apply. As before, a cluster of neutron stars would be stable over cosmic time – but also difficult to form. The main things that have been gained over NGC3115 are a significant increase in the rotation velocity V, 550 km s^{-1} vs. 150 km s^{-1}, plus the simpler geometry of a rotating gaseous disk that is probably close to circular motion. This sidesteps the β anisotropy problem of equation (9.11), and the derived MDO mass is more certain.

The good agreement between the new HST mass and the earlier mass estimate of Sargent et al. (1978) is intriguing. The older value was based on the questionable isotropic model with $\beta = 0$, but agreement now suggests that the velocity dispersion is nearly isotropic after all. This is not inconsistent with current merger scenarios for hot-galaxy formation via mergers, which do not easily yield strongly radially anisotropic velocity dispersions. With the pasage of time, isotropic or nearly isotropic models have come to seem more plausible.

It would be good to compare more gaseous BH mass estimates (based on rotating disks) with stellar σ estimates from the same galaxy. Several such pairs could confirm the isotropic picture or yield an average correction factor to the stellar σ values. Calibrating stellar σ estimates would open the way to large BH surveys for the first time.

9.3.4.3 NGC4258

NGC4258 is a nearby, peculiar Sbc spiral about 6.4 Mpc away. It has a rather feeble AGN with a LINER spectrum (Heckman 1980) and very weak, broad emission lines characteristic of Seyferts (Filippenko & Sargent 1985). It has radio synchrotron lobes (van der Kruit, Oort & Mathewson 1972), jets (Cecil, Wilson & de Pree 1995), and anomalous Hα arms between the regular stellar arms (Courtes & Cruvellier 1961, Deharveng & Pellet 1970). On the AGN paradigm it is thus a good candidate to host a BH.

A powerful new probe was provided by the discovery of H$_2$O maser emission at 1.3 cm in a probable circumnuclear disk (Claussen et al. 1984; Claussen & Lo 1986; Nakai, Inoue & Miyoshi 1993). Remarkable observations with the Very Long Baseline Array (VLBA) (Miyoshi et al. 1995) reveal the structure of this disk at very high angular resolution (5–40 microarcsec). Maser emission is detected in three clumps: directly along the line-of-sight to the disk at the systemic velocity of the galaxy, and roughly symmetrically in a zone to either side. The maser emission appears to arise from an exceedingly thin annulus of gas with inner and outer radii 5 to 8 milliarsec (0.16 to 0.26 pc). The emission comes from regions of the disk where the velocity difference along the line-of-sight is less than $\sim$ 1 km s^{-1}, permitting a long path length ($\sim 10^{16}$ cm) for maser amplification. Each masing patch is nearly directly along the line-of-sight or tangent to a circular orbit on either side.

Strong confirmation of this picture comes from the rotation curve, which extends from 5 to 8 milliarcsec and peaks at ± 1000 km s^{-1} at the inner edge. The curve is closely Keplerian to within 3 km s^{-1} (J. Herrnstein, quoted by Maoz 1995), and the derived MDO mass is $3.6 \times 10^7 M_\odot$ at a distance of 6.4 Mpc. Long-term spectroscopic monitoring shows a secular frequency drift for the systemic features at a rate of 9.5 ± 1.1 km s^{-1} yr^{-1} (Haschick, Baan & Peng 1994, Greenhill et al. 1995). This is the expected centripetal acceleration of features as they cross the line-of-sight, $\dot{V} = V^2/r$ and agrees closely with the predicted rate at this distance.

The BH in NGC4258 is not as massive as either of those in NGC3115 or M87, but a very small radius limit and higher density place qualitatively different and tighter constraints on the MDO cluster model. In an elegant analysis, Maoz (1995) shows that the half-mass radius, R_h, of any MDO cluster must be ≤ 0.016 pc (0.5 milliarcsec), otherwise the cluster wings would extend far enough to perturb the Keplerian rotation curve. From equation (9.12) and $M_{cl} = 3.6 \times 10^7 M_\odot$, the evaporation time is

$$t_{evap} \approx 1.5 \times 10^8 \left(\frac{M_*}{1.4 M_\odot}\right)^{-1} \text{yr} . \quad (9.14)$$

A cluster of neutron stars with $M_* = 1.4 M_\odot$ would therefore be very short-lived. For a minimum lifetime of 6 Gyr, the object mass M_* must be less than $0.03 M_\odot$.

This greatly restricts the list of candidate objects – to planets, brown dwarfs, very light BHs, and weakly interacting elementary particles. The first two are extended objects with finite collision cross-sections. Assuming they have zero internal temperature yields a minimum radius and minimum cross-section for any mass (Zapolsky & Salpeter 1969; Stevenson 1991). However, the density factor n is the dominant term, not the cross-section, with the result that the collision time actually *decreases* at lower masses as the number of objects goes up. At $M_* = 0.03 M_\odot$, it is already only 10^5 yr. Thus we have a contradiction: for $M_* > 0.03 M_\odot$, the evaporation time is too short, while for $M_* < 0.03 M_\odot$, the collision time is even shorter. All cluster MDO arguments are ruled out by these arguments unless the objects are *extremely* dense, i.e., small BHs or weakly interacting elementary particles. Once again we have to resort to formation arguments to rule out such clusters, but nuclear clusters comprised purely of tiny BHs or elementary particles strain credulity to the limit.

9.3.5 Core formation by binary BHs?

The above result sets a new standard of stringency for BH searches, but the crucial resolution boost from the maser technique is unfortunately not not available for most galaxies. Central masers are rare, and Miyoshi et al. (1995) further estimate that the probability is only 6% that the disk of NGC4258 is properly oriented to mase in our direction. Even high-resolution kinematic studies using HST are limited to fairly bright, nearby galaxies. In this situation, to broaden the evidence it may be worth appealing to a more speculative photometric argument that, though far from proven in any particular case, has the virtue of applying to a large number of objects.

Recall that the existence of a central BH cannot be inferred from the light distribution alone and that no particular profile shape is in itself rigorous evidence for a BH. While this is true for any individual object, it may be that the *systematics* of galaxy profile shapes, when coupled with plausible scenarios for galaxy formation, can yield important clues to central BHs. A particular puzzle raised by the core systematics in Fig. 9.3 is the low-density cores of very large ellipticals. Many of these are brightest cluster galaxies in high-density environments near cluster centers. Estimates of the accretion rate (Lauer 1988, Merritt 1984, Faber et al. 1997) suggest that such objects should capture two smaller satellite galaxies on average every 5 Gyr. If these satellites survived tidal disruption on account of their higher density (Fig. 9.3(b)), they would spiral down to the center of the parent galaxy and fill in its low density core. Given the existence of low-density cores in all large galaxies so far surveyed, this process is evidently not taking place. Why not?

The contradiction is not yet iron clad because the fate of satellites is not yet certain. N-body experiments are unanimous in suggesting that dense satellites would survive (Barnes 1992, Hernquist 1993, Hernquist, Spergel & Heyl 1993). On the other hand, work in progress using a different technique by Weinberg (1997) suggests they may get torn apart. Although the answer is not clear, the situation is intriguing enough to motivate us to look for a separate explanation for cores, one that might simultaneously explain both their formation and survival.

One possibility (Ebisuzaki, Makino & Okumura 1991 (EMO)) appeals to the argument that essentially *all* hot galaxies contain massive BHs. These holes will be accreted in galaxy–galaxy mergers and will form a binary with any pre-existing BH in the parent galaxy. The binary decays via dynamical friction and stellar scattering in a gradual "death spiral," ending ultimately in the release of gravitational radiation and coalescence (Begelman, Blandford & Rees 1980). EMO considered the dynamical feedback and heating from the binary on the surrounding stellar population. As the binary decays, it transfers energy to the stars, swelling the population and ejecting individual stars from the center. This process can whittle away a previously existing power law, they suggested, and create a core.

Suppose for definiteness that an early generation of QSOs created a nascent population of BHs in the progenitors of modern-day hot galaxies, with mass ratios $M_\bullet/M_{\rm gal} = 0.002$ as given by equation (9.10). As the progenitor population continued to merge hierarchically, BHs would be accreted also, and $M_\bullet/M_{\rm gal}$ would thus remain constant. Quinlan (1996) has constructed high-resolution N-body models that follow this process. He finds that stars are ejected from the centers as predicted, and core profiles are created. A measure of core size is *indicative core mass* $M_{\rm core} \equiv \pi r_b^2 \Sigma_b$ (Faber et al. 1997), where Σ_b is the projected mass surface density at the break radius

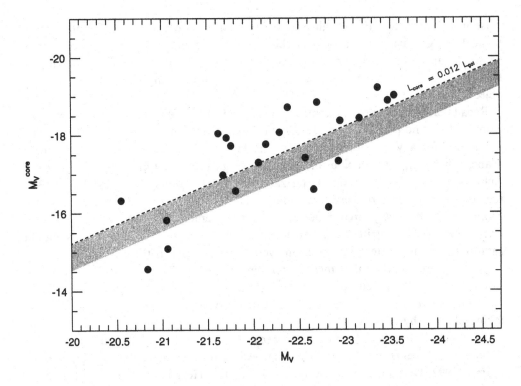

Figure 9.6. Indicative core luminosity vs. total luminosity. The dashed line $L_{core} = 0.012 L_{gal}$ is a fit to the data assuming unit slope. The grey area denotes the predictions of two decaying BH binary models by Quinlan (1996) and Quinlan & Hernquist (1997).

r_b, analogous to I_b. Quinlan's profiles allow one to calculate the ratio $M_{core}/M_\bullet$, and thus predict the core mass and radius for a given value of $M_\bullet/M_{gal}$.

The resulting predictions for $M_\bullet/M_{gal} = 0.002$ are compared to the HST core data in Fig. 9.6. The dashed line shows the mean fit to the data $L_{core} = 0.012 L_{gal}$. A similar relation is also found between core radius and global galaxy radius. Core sizes from a pair of Quinlan's models close in mass to $M_\bullet/M_{gal} = 0.002$ are shown as the grey area. These vary slightly because they start with different initial galaxy profiles. Core size is sensitive to starting profile and is larger in models where the initial profile is less steep.

The agreement with the data is encouraging, and similar agreement is also found for r_b vs. global radius. However, major uncertainties remain. First, the starting conditions of the models are unrealistic in that the two BHs start off in circular orbit far from the center within an already equilibrated potential. These conditions are inconsistent with an ongoing merger and neglect the role that large-scale violent relaxation may play in bringing the holes close to the center without the need for BH heating. Second, the predicted core mass scales directly with the assumed value of $M_\bullet/M_{gal} = 0.002$, which is uncertain. Next, a third BH binary may be accreted before the first can merge, and the resultant three-body interaction may eject one or more of the holes (Hut & Rees

1992, Xu & Ostriker 1994). This would break the assumption that $M_\bullet/M_{gal}$ is constant, which is essential to the relation in Fig. 9.6.

The fourth point brings us back to the beginning. The simple model just outlined assumes that all hot-galaxy progenitors began life with nascent BHs. If they also form by mergers, then all will inherit decaying BH binaries and grow cores. Put another way, where do power-law profiles come from in this picture?

A possibility is that they are formed during *gas-rich* mergers, in which fresh gas is brought to the center, creating a star burst and generating a power-law stellar profile. N-body simulations of merging galaxies show that gas is indeed brought to the center efficiently via tidal torques and cloud–cloud collisions (Barnes & Hernquist 1991), and powerful central star bursts in real mergers are tangible evidence of this process (Sanders *et al.* 1988, Wang, Schweizer & Scoville 1992). To carry the argument further, it has also been suggested that gaseous mergers promote rotating, disky galaxies (Nieto 1988, Kormendy 1989, Nieto, Bender & Surma 1991, Bender, Burstein & Faber 1992), while purely stellar mergers (which occur after all the gas is used up) promote boxy galaxies (Barnes 1988, 1992). This could explain the connection between central profile shape and global shape and rotation in Fig. 9.5 if cores were created in gas-free mergers by the action of binary BHs, while power laws are disky galaxies that had their centers replenished by star formation in gas-rich mergers.

This picture is in some ways attractive, but it ultimately rests on sand for lack of a quantitative theory for how and where stars form in a star burst, and hence what the predicted final stellar profile would be (Mihos & Hernquist 1994). This discussion brings us to the limits of present knowledge. The point seems strong that galaxy centers are closely linked to the global evolution of galaxies as a whole, and that understanding that link is important for a complete picture of galaxy evolution. However, as often happens in galaxy research, the ultimate barrier turns out to be the lack of a rigorous theory of star formation.

Acknowledgments

Much of this review is based on joint work with fellow astronomers studying centers of galaxies with HST, from whom SMF has learned most of what she knows about this subject. SMF would particularly like to acknowledge a longstanding and rewarding collaboration with Ed Ajhar, Yong Ik Byun, Alan Dressler, Karl Gebhardt, Carl Grillmair, John Kormendy, Tod Lauer, Doug Richstone, and Scott Tremaine.

References

Barnes, J. E. 1988, *ApJ*, **331**, 699

Barnes, J. E. 1992, *ApJ*, **393**, 484

Barnes J. E. & Hernquist, L. 1991, *ApJ*, **370**, L65

Barnes J. E. & Hernquist, L. 1992, *ARA&A*, **30**, 705

Begelman, M. C., Blandford, R. D. & Rees, M. J. 1980, *Nature*, **287**, 307

Bender, R., Burstein. D. & Faber, S. M. 1992, *ApJ*, **399**, 462

Binggeli, B., Sandage, A. & Tammann, G. A. 1988, *ARA&A*, **26**, 509

Binney, J. & Mamon, G. A. 1982, *MNRAS*, **200**, 361

Binney, J. & Tremaine, S. 1987, *Galactic Dynamics* (Princeton: Princeton University Press)

Byun, Y.-I., et al. 1996, *AJ*, **111**, 1889

Cecil, G., Wilson, A. S. & de Pree, C. 1995, *ApJ*, **440**, 181

Chokshi, A. & Turner, E. L. 1992, *MNRAS*, **259**, 421

Claussen, M. J., Hellingman, G. M. & Lo, K. Y. 1984, *Nature*, **310**, 298

Claussen, M. J. & Lo, K. Y. 1986, *ApJ*, **308**, 592

Courtes, G. & Cruvellier, P. 1961, *C. R. Acad. Sci. Paris*, **253**, 218

de Zeeuw, T. & Franx, M. 1991, *ARA&A*, **29**, 239

Deharveng, J. M. & Pellet, A. 1970, *A&A*, **9**, 181

Dressler, A. 1984, *ApJ*, **286**, 97

Dressler, A. & Schechter, P. 1987, *AJ*, **94**, 563

Duncan, M. J. & Wheeler, J. C. 1980, *ApJ*, **237**, L27

Ebisuzaki, T., Makino, J. & Okumura, S. K. 1991, *Nature*, **354**, 212

Efstathiou, G., Ellis, R. S. & Peterson, B. A. 1988, *MNRAS*, **232**, 431

Faber, S. M., et al. 1996, *AJ*, **114**, 1771

Filippenko, A. & Sargent, W. L. W. 1985, *ApJS*, **57**, 503

Forbes. D. A., Franx, M. & Illingworth, G. D. 1995, *AJ*, **109**, 1988

Ford, H. C., et al. 1994, *ApJ*, **435**, L27

Gebhardt, K., et al. 1996, *AJ*, **112**, 105

Greenhill, L., et al. 1995, *ApJ*, **440**, 619

Haenelt, M. G. & Rees, M. J. 1993, *MNRAS*, **263**, 168

Harms, R. J., et al. 1994, *ApJ*, **435**, L35

Haschik, A. D., Baan, W. A. & Peng, E. W. 1994, *ApJ*, **437**, L35

Heckman, T. 1980, *A&A*, **87**, 152

Hernquist, L. 1993. *ApJ*, **409**, 548

Hernquist, L., Spergel, D. N. & Heyl, J. S. 1993, *ApJ*, **416**, 415

Hut, P. & Rees, M. J. 1992, *MNRAS*, **259**, 27p

Jaffe, W., Ford, H. C., O'Connell, R. W., van den Bosch, F. C., & Ferrarese, L. 1994, *AJ*, **108**, 1567

King, I. R. 1966, *AJ*, **71**, 64

Kormendy, J. 1985, *ApJ*, **292**, L9

Kormendy, J. 1988a, *ApJ*, **325**, 128

Kormendy, J. 1988b, *ApJ*, **335**, 40

Kormendy, J. 1989, *ApJ*, **342**, L63

Kormendy, J. 1992, in *Testing the AGN Paradigm*, eds. S. S. Holt, S. G. Neff & C. M. Urry (New York: American Institue of Physics), p. 23

Kormendy, J. & Bender, R. 1996, *ApJ*, **464**, L119

Kormendy, J. & Djorgovski, S. 1989, *ARA&A*, **27**, 235

Kormendy, J. & McClure, R. D. 1993, *AJ*, **105**, 1793

Kormendy, J. & Richstone, D. 1992, *ApJ*, **393**, 559

Kormendy, J. & Richstone, D. 1995, *ARA&A*, **33**, 581

Kormendy, J., et al. 1996, *ApJ*, **393**, L559

Lauer, T. R. 1983, Ph.D. Thesis, University of California, Santa Cruz

Lauer, T. R. 1985, *ApJS*, **57**, 473

Lauer, T. R. 1988, *ApJ*, **325**, 49

Lauer, T. R. et al. 1992a, *AJ*, **104**, 552 M32

Lauer, T. R., et al. 1992b, *AJ*, **103**, 703

Lauer, T. R., et al. 1995, *AJ*, **110**, 2622

Maeder, A. 1992, *A&A*, **264**, 105

Maoz, E. 1995, *ApJ*, **447**, L91

Merritt, D. 1984, *ApJ*, **276**, 26

Mihos, J. C. & Hernquist, L. 1994, *ApJ*, **437**, L47

Miyoshi, M. et al. 1995, *Nature*, **373**, 127

Nakai, N., Inoue, M. & Miyoshi, M. 1993, *Nature*, **361**, 45

Nieto, J.-L. 1988 *Bol. Acad. Nac. Cine. Cordoba*, **58**, 239

Nieto, J.-L., Bender, R. & Surma, P. 1991, *A&A*, **244**, 137

Quinlan, G. 1996, *New Astronomy*, **1**, 35

Quinlan, G. & Hernquist, L. 1997, *New Astronomy*, **2**, 533

Sandage, A. R. 1961, *The Hubble Atlas of Galaxies* (Washington: Carnegie Institution of Washington)

Sanders, D. B., et al. 1988, *ApJ*, **325**, 74

Sargent, W. L. W., Young, P. J., Boksenberg, A., Shortridge, K., Lynds, C. R. & Hartwick, F. D. A. 1978, *ApJ*, **221**, 731

Soltan, A. 1982, *MNRAS*, **200**, 115

Stevenson, D. J. 1991, *ARA&A*, **29**, 163

Tanaka, Y., Nandra, K., Fabian, A. C. & Inoue, H. 1995, *Nature*, **375**, 659

Tonry, J. L. 1984, *ApJ*, **283**, L27

Tonry, J. L. 1987, *ApJ*, **322**, 632

Tremaine, S. 1998, in *Some Unsolved Problems in Astrophysics*, (Princeton: Princeton University Press)

van der Kruit, P. C., Oort, J. H. & Mathewson, D. S. 1972, *A&A*, **29**, 249

Wang, Z., Schweizer, F. & Scoville, N. Z. 1992, *ApJ*, **396**, 510

Weinberg, M. 1997, *ApJ*, **478**, 435

Wilson, A. S., Braatz, J. A. & Henkel, C. 1995, *ApJ*, **455**, L127

Xu, G. & Ostriker, J. P. 1994, *ApJ*, **437**, 184

Young, P. J., et al. 1978, *ApJ*, **221**, 721

Young, P. J. 1980, *ApJ*, **242**, 1232

Zapolsky, H. S. & Salpeter, E. E. 1969, *ApJ*, **158**, 809

10 Gravitational lensing

Ramesh Narayan and Matthias Bartelmann

Abstract

This chapter gives an introduction to the phenomenon of gravitational lensing. Topics discussed include lensing by point masses, by galaxies, and by clusters of galaxies and larger-scale structures. The relevant theory is developed and applications to astrophysical problems are discussed.

10.1 Introduction

One of the consequences of Einstein's General Theory of Relativity is that light rays are deflected by gravity. Although this discovery was made only in the 20th century, the possibility that there could be such a deflection had been suspected much earlier, by Newton and Laplace, among others. Soldner (1804) calculated the magnitude of the deflection due to the sun, assuming that light consists of material particles and using Newtonian gravity. Later, Einstein (1911) employed the equivalence principle to calculate the deflection angle and re-derived Soldner's formula. Later yet, Einstein (1915) applied the full field equations of General Relativity and discovered that the deflection angle is actually twice his previous result, the factor of two arising because of the curvature of the metric. According to this formula, a light ray which tangentially grazes the surface of the sun is deflected by 1″.7. Einstein's final result was confirmed in 1919 when the apparent angular shift of stars close to the limb of the sun (see Fig. 10.1) was measured during a total solar eclipse (Dyson, Eddington & Davidson 1920). The quantitative agreement between the measured shift and Einstein's prediction was immediately perceived as compelling evidence in support of the theory of General Relativity. The deflection of light by massive bodies, and the phenomena resulting therefrom, are now referred to as *Gravitational Lensing*.

Eddington (1920) noted that under certain conditions there may be multiple light paths connecting a source and an observer. This implies that gravitational lensing can give rise to multiple images of a single source. Chwolson (1924) considered the creation of fictitious double stars by gravitational lensing of stars by stars, but did not comment

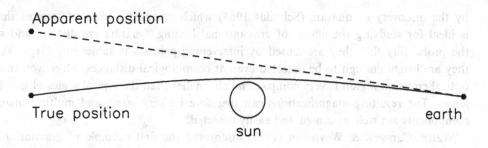

Figure 10.1. Angular deflection of a ray of light passing close to the limb of the sun. Since the light ray is bent toward the sun, the apparent positions of stars move away from the sun.

on whether the phenomenon could actually be observed. Einstein (1936) discussed the same problem and concluded that there is little chance of observing lensing phenomena caused by stellar-mass lenses. His reason was that the angular image splitting caused by a stellar-mass lens is too small to be resolved by an optical telescope.

Zwicky (1937a) elevated gravitational lensing from a curiosity to a field with great potential when he pointed out that galaxies can split images of background sources by a large enough angle to be observed. At that time, galaxies were commonly believed to have masses of $\sim 10^9$ M$_\odot$. However, Zwicky had applied the virial theorem to the Virgo and Coma clusters of galaxies and had derived galaxy masses of $\sim 4 \times 10^{11}$ M$_\odot$. Zwicky argued that the deflection of light by galaxies would not only furnish an additional test of General Relativity, but would also magnify distant galaxies which would otherwise remain undetected, and would allow accurate determination of galaxy masses. Zwicky (1937b) even calculated the probability of lensing by galaxies and concluded that it is on the order of one percent for a source at reasonably large redshift.

Virtually all of Zwicky's predictions have come true. Lensing by galaxies is a major sub-discipline of gravitational lensing today. The most accurate mass determinations of the central regions of galaxies are due to gravitational lensing, and the cosmic telescope effect of gravitational lenses has enabled us to study faint and distant galaxies which happen to be strongly magnified by galaxy clusters. The statistics of gravitational lensing events, whose order of magnitude Zwicky correctly estimated, offers one of the promising ways of inferring cosmological parameters.

In a stimulating paper, Refsdal (1964) described how the Hubble constant H_0 could in principle be measured through gravitational lensing of a variable source. Since the light-travel times for the various images are unequal, intrinsic variations of the source would be observed at different times in the images. The time delay between images is proportional to the difference in the absolute lengths of the light paths, which in turn is proportional to H_0^{-1}. Thus, if the time delay is measured and if an accurate model of a lensed source is developed, the Hubble constant could be measured.

All of these ideas on gravitational lensing remained mere speculation until real examples of gravitational lensing were finally discovered. The stage for this was set by the discovery of quasars (Schmidt 1963) which revealed a class of sources that is ideal for studying the effects of gravitational lensing. Quasars are distant, and so the probability that they are lensed by intervening galaxies is sufficiently large. Yet, they are bright enough to be detected even at cosmological distances. Moreover, their optical emission region is very compact, much smaller than the typical scales of galaxy lenses. The resulting magnifications can therefore be very large, and multiple-image components are well separated and easily detected.

Walsh, Carswell & Weymann (1979) discovered the first example of gravitational lensing, the quasar QSO 0957+561A,B. This source consists of two images, A and B, separated by 6". Evidence that 0957+561A,B does indeed correspond to twin lensed images of a single QSO is provided by (i) the similarity of the spectra of the two images, (ii) the fact that the flux ratio between the images is similar in the optical and radio wavebands, (iii) the presence of a foreground galaxy between the images, and (iv) VLBI observations which show detailed correspondence between various knots of emission in the two radio images. Over a dozen convincing examples of multiple-imaged quasars are known today (Keeton & Kochanek 1996) and the list continues to grow.

Paczyński (1986b) revived the idea of lensing of stars by stars when he showed that at any given time one in a million stars in the Large Magellanic Cloud (LMC) might be measurably magnified by the gravitational lens effect of an intervening star in the halo of our Galaxy. The magnification events, which are called *microlensing* events, have time-scales between two hours and two years for lens masses between $10^{-6}\,M_\odot$ and $10^2\,M_\odot$. Initially, it was believed that the proposed experiment of monitoring the light curves of a million stars would never be feasible, especially since the light curves have to be sampled frequently and need to be distinguished from light curves of intrinsically variable stars. Nevertheless, techniques have advanced so rapidly that today four separate collaborations have successfully detected microlensing events (Alcock et al. 1993; Aubourg et al. 1993; Udalski et al. 1993; Alard 1995), and this field has developed into an exciting method for studying the nature and distribution of mass in our Galaxy.

Einstein rings, a particularly interesting manifestation of gravitational lensing, were discovered first in the radio waveband by Hewitt et al. (1987). About half-a-dozen radio rings are now known and these sources permit the most detailed modelling yet of the mass distributions of lensing galaxies.

Gravitational lensing by galaxy clusters had been considered theoretically even before the discovery of QSO 0957+561. The subject entered the observational realm with the discovery of giant blue luminous *arcs* in the galaxy clusters A370 and Cl2244 (Soucail et al. 1987a,b; Lynds & Petrosian 1986). Paczyński (1987) proposed that the arcs are the images of background galaxies which are strongly distorted and elongated by the gravitational lens effect of the foreground cluster. This explanation was confirmed when the first arc redshifts were measured and found to be significantly greater than that of the clusters.

Apart from the spectacular giant luminous arcs, which require special alignment between the cluster and the background source, clusters also coherently distort the images of other faint background galaxies (Tyson 1988). These distortions are mostly weak, and the corresponding images are referred to as *arclets* (Fort *et al.* 1988; Tyson, Valdes & Wenk 1990). Observations of arclets can be used to reconstruct parameter-free, two-dimensional mass maps of the lensing cluster (Kaiser & Squires 1993). This technique has attracted a great deal of interest, and two-dimensional maps have been obtained of several galaxy clusters (Bonnet *et al.* 1993; Bonnet, Mellier & Fort 1994; Fahlman *et al.* 1994; Broadhurst 1995; Smail *et al.* 1995; Tyson & Fischer 1995; Squires *et al.* 1996a,b; Seitz *et al.* 1996).

As this brief summary indicates, gravitational lensing manifests itself through a very broad and interesting range of phenomena. At the same time, lensing has developed into a powerful tool to study a host of important questions in astrophysics. The applications of gravitational lensing may be broadly classified under three categories.

- The magnification effect enables us to observe objects which are too distant or intrinsically too faint to be observed without lensing. Lenses therefore act as "cosmic telescopes" and allow us to infer source properties far below the resolution limit or sensitivity limit of current observations. However, since we do not have the ability to point this telescope at any particular object of interest but have to work with whatever nature gives us, the results have been only modestly interesting.

- Gravitational lensing depends solely on the projected, two-dimensional mass distribution of the lens, and is independent of the luminosity or composition of the lens. Lensing therefore offers an ideal way to detect and study dark matter, and to explore the growth and structure of mass condensations in the universe.

- Many properties of individual lens systems or samples of lensed objects depend on the age, the scale, and the overall geometry of the universe. The Hubble constant, the cosmological constant, and the density parameter of the universe can be significantly constrained through lensing.

This chapter is divided into three main sections: §10.2 discusses the effects of point-mass lenses; §10.3 considers galaxy-scale lenses; and §10.4 discusses lensing by galaxy clusters and large-scale structure in the universe. References to the original literature are given throughout the text. The following are some general or specialized review articles and monographs.

Monograph

- Schneider, P., Ehlers, J. & Falco, E. E. 1992, *Gravitational Lenses* (Berlin: Springer Verlag)

General reviews

- Blandford, R. D. & Narayan, R. 1992, Cosmological Applications of Gravitational Lensing, *ARA&A*, **30**, 311
- Refsdal, S., & Surdej, J. 1994, Gravitational Lenses, *Rep. Progr. Phys.*, **57**, 117
- Schneider, P. 1996, Cosmological Applications of Gravitational Lensing, in: *The universe at high-z, large-scale structure and the cosmic microwave background, Lecture Notes in Physics*, eds. E. Martínez-González & J.L. Sanz (Berlin: Springer Verlag)
- Wu, X.-P. 1996, Gravitational Lensing in the Universe, *Fundamentals of Cosmic Physics*, **17**, 1

Special reviews

- Fort, B. & Mellier, Y. 1994, Arc(let)s in Clusters of Galaxies, *Astr. Ap. Rev.*, **5**, 239
- Bartelmann, M. & Narayan, R. 1995, Gravitational Lensing and the Mass Distribution of Clusters, in: *Dark Matter*, AIP Conf. Proc. 336, eds. S. S. Holt & C. L. Bennett (New York: AIP Press)
- Keeton II, C.R. & Kochanek, C.S. 1996, Summary of Data on Secure Multiply-Imaged Systems, in: *Cosmological Applications of Gravitational Lensing*, IAU Symp. 173, eds. C. S. Kochanek & J. N. Hewitt
- Paczyński, B. 1996, Gravitational Microlensing in the Local Group, *ARA&A*, **34**, 419
- Roulet, E. & Mollerach, S. 1997, Microlensing, *Physics Reports*, **279**, 67

10.2 Lensing by point masses in the universe

10.2.1 Basics of gravitational lensing

The propagation of light in arbitrary curved spacetimes is in general a complicated theoretical problem. However, for almost all cases of relevance to gravitational lensing, we can assume that the overall geometry of the universe is well described by the Friedmann–Lemaître–Robertson–Walker metric and that the matter inhomogeneities which cause the lensing are no more than local perturbations. Light paths propagating from the source past the lens to the observer can then be broken up into three distinct zones. In the first zone, light travels from the source to a point close to the lens through unperturbed spacetime. In the second zone, near the lens, light is deflected. Finally, in the third zone, light again travels through unperturbed spacetime. To study light deflection close to the lens, we can assume a locally flat, Minkowskian spacetime which is weakly perturbed by the Newtonian gravitational potential of the mass distribution constituting the lens. This approach is legitimate if the Newtonian potential Φ is small, $|\Phi| \ll c^2$, and if the peculiar velocity v of the lens is small, $v \ll c$.

These conditions are satisfied in virtually all cases of astrophysical interest. Consider for instance a galaxy cluster at redshift ~ 0.3 which deflects light from a source at redshift ~ 1. The distances from the source to the lens and from the lens to the observer are ~ 1 Gpc, or about three orders of magnitude larger than the diameter of the cluster. Thus, zone 2 is limited to a small local segment of the total light path. The relative peculiar velocities in a galaxy cluster are $\sim 10^3$ km s$^{-1} \ll c$, and the Newtonian potential is $|\Phi| < 10^{-4} c^2 \ll c^2$, in agreement with the conditions stated above.

10.2.1.1 Effective refractive index of a gravitational field

In view of the simplifications just discussed, we can describe light propagation close to gravitational lenses in a locally Minkowskian spacetime perturbed by the gravitational potential of the lens to first post-Newtonian order. The effect of spacetime curvature on the light paths can then be expressed in terms of an effective index of refraction n, which is given by (e.g., Schneider, Ehlers & Falco 1992)

$$n = 1 - \frac{2}{c^2} \Phi = 1 + \frac{2}{c^2} |\Phi| . \tag{10.1}$$

Note that the Newtonian potential is negative if it is defined such that it approaches zero at infinity. As in normal geometrical optics, a refractive index $n > 1$ implies that light travels slower than in free vacuum. Thus, the effective speed of a ray of light in a gravitational field is

$$v = \frac{c}{n} \simeq c - \frac{2}{c} |\Phi| . \tag{10.2}$$

Figure 10.2 shows the deflection of light by a glass prism. The speed of light is reduced inside the prism. This reduction of speed causes a delay in the arrival time of a signal through the prism relative to another signal traveling at speed c. In addition, it causes wavefronts to tilt as light propagates from one medium to another, leading to a bending of the light ray around the thick end of the prism.

The same effects are seen in gravitational lensing. Because the effective speed of light is reduced in a gravitational field, light rays are delayed relative to propagation in vacuum. The total time delay Δt is obtained by integrating over the light path from the observer to the source:

$$\Delta t = \int_{\text{source}}^{\text{observer}} \frac{2}{c^3} |\Phi| \, dl . \tag{10.3}$$

This is called the Shapiro delay (Shapiro 1964).

As in the case of the prism, light rays are deflected when they pass through a gravitational field. The deflection is the integral along the light path of the gradient of n perpendicular to the light path, i.e.

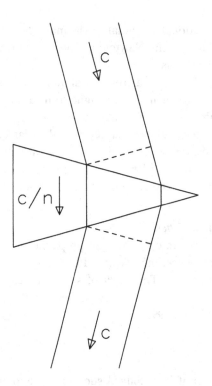

Figure 10.2. Light deflection by a prism. The refractive index $n > 1$ of the glass in the prism reduces the effective speed of light to c/n. This causes light rays to be bent around the thick end of the prism, as indicated. The dashed lines are wavefronts. Although the geometrical distance between the wavefronts along the two rays is different, the travel time is the same because the ray on the left travels through a larger thickness of glass.

$$\hat{\alpha} = -\int \nabla_\perp n \, dl = \frac{2}{c^2} \int \nabla_\perp \Phi \, dl. \tag{10.4}$$

In all cases of interest the deflection angle is very small. We can therefore simplify the computation of the deflection angle considerably if we integrate $\nabla_\perp n$ not along the deflected ray, but along an unperturbed light ray with the same impact parameter. (As an aside we note that while the procedure is straightforward with a single lens, some care is needed in the case of multiple lenses at different distances from the source. With multiple lenses, one takes the unperturbed ray from the source as the reference trajectory for calculating the deflection by the first lens, the deflected ray from the first lens as the reference unperturbed ray for calculating the deflection by the second lens, and so on.)

As an example, we now evaluate the deflection angle of a point mass M (cf. Fig. 10.3). The Newtonian potential of the lens is

$$\Phi(b, z) = -\frac{GM}{(b^2 + z^2)^{1/2}}, \tag{10.5}$$

Gravitational lensing

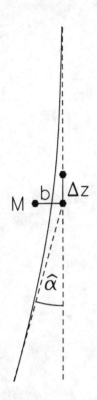

Figure 10.3. Light deflection by a point mass. The unperturbed ray passes the mass at impact parameter b and is deflected by the angle $\hat{\alpha}$. Most of the deflection occurs within $\Delta z \sim \pm b$ of the point of closest approach.

where b is the impact parameter of the unperturbed light ray, and z indicates distance along the unperturbed light ray from the point of closest approach. We therefore have

$$\nabla_\perp \Phi(b,z) = \frac{GM\,b}{(b^2 + z^2)^{3/2}}, \tag{10.6}$$

where $\mathbf{b}$ is orthogonal to the unperturbed ray and points toward the point mass. Equation (10.6) then yields the deflection angle

$$\hat{\alpha} = \frac{2}{c^2} \int \nabla_\perp \Phi\, dz = \frac{4GM}{c^2 b}. \tag{10.7}$$

Note that the Schwarzschild radius of a point mass is

$$R_S = \frac{2GM}{c^2}, \tag{10.8}$$

so that the deflection angle is simply twice the inverse of the impact parameter in units of the Schwarzschild radius. As an example, the Schwarzschild radius of the sun is 2.95 km, and the solar radius is 6.96×10^5 km. A light ray grazing the limb of the sun is therefore deflected by an angle $(5.9/7.0) \times 10^{-5}$ radians $= 1\rlap{.}''7$.

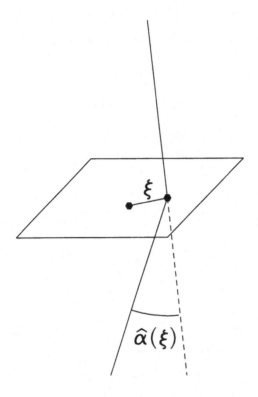

Figure 10.4. A light ray which intersects the lens plane at ξ is deflected by an angle $\hat{\alpha}(\xi)$.

10.2.1.2 Thin screen approximation

Figure 10.3 illustrates that most of the light deflection occurs within $\Delta z \sim \pm b$ of the point of closest encounter between the light ray and the point mass. This Δz is typically much smaller than the distances between observer and lens and between lens and source. The lens can therefore be considered thin compared to the total extent of the light path. The mass distribution of the lens can then be projected along the line-of-sight and be replaced by a mass sheet orthogonal to the line-of-sight. The plane of the mass sheet is commonly called the lens plane. The mass sheet is characterized by its surface-mass density

$$\Sigma(\xi) = \int \rho(\xi, z) \, dz \,, \tag{10.9}$$

where ξ is a two-dimensional vector in the lens plane. The deflection angle at position ξ is the sum of the deflections due to all the mass elements in the plane:

$$\hat{\alpha}(\xi) = \frac{4G}{c^2} \int \frac{(\xi - \xi')\Sigma(\xi')}{|\xi - \xi'|^2} \, d^2\xi' \,. \tag{10.10}$$

Figure 10.4 illustrates the situation.

In general, the deflection angle is a two-component vector. In the special case of a circularly symmetric lens, we can shift the coordinate origin to the center of symmetry and reduce light deflection to a one-dimensional problem. The deflection angle then points toward the center of symmetry, and its modulus is

$$\hat{\alpha}(\xi) = \frac{4GM(\xi)}{c^2\xi}, \qquad (10.11)$$

where ξ is the distance from the lens center and $M(\xi)$ is the mass enclosed within radius ξ,

$$M(\xi) = 2\pi \int_0^\xi \Sigma(\xi')\xi'\,d\xi'. \qquad (10.12)$$

10.2.1.3 Lensing geometry and lens equation

The geometry of a typical gravitational lens system is shown in Fig. 10.5. A light ray from a source S is deflected by the angle $\hat{\alpha}$ at the lens and reaches an observer O. The angle between the (arbitrarily chosen) optic axis and the true source position is β, and the angle between the optic axis and the image I is θ. The (angular diameter) distances between observer and lens, lens and source, and observer and source are D_d, D_{ds}, and D_s, respectively.

It is now convenient to introduce the reduced deflection angle

$$\alpha = \frac{D_{ds}}{D_s}\hat{\alpha}. \qquad (10.13)$$

From Fig. 10.5 we see that $\theta D_s = \beta D_s + \hat{\alpha} D_{ds}$. Therefore, the positions of the source and the image are related through the simple equation

$$\beta = \theta - \alpha(\theta). \qquad (10.14)$$

Equation (10.14) is called the *lens equation*, or ray-tracing equation. It is nonlinear in the general case, and so it is possible to have multiple images θ corresponding to a single source position β. As Fig. 10.5 shows, the lens equation is trivial to derive and requires merely that the following Euclidean relation should exist between the angle enclosed by two lines and their separation,

$$\text{separation} = \text{angle} \times \text{distance}. \qquad (10.15)$$

It is not obvious that the same relation should also hold in curved spacetimes. However, if the distances $D_{d,s,ds}$ are *defined* such that equation (10.15) holds, then the lens equation must obviously be true. Distances so defined are called angular-diameter distances, and equations (10.13), (10.14) are valid only when these distances are used. Note that in general $D_{ds} \neq D_s - D_d$.

As an instructive special case consider a lens with a constant surface-mass density Σ. From equation (10.11), the (reduced) deflection angle is

$$\alpha(\theta) = \frac{D_{ds}}{D_s}\frac{4G}{c^2\xi}(\Sigma\pi\xi^2) = \frac{4\pi G \Sigma}{c^2}\frac{D_d D_{ds}}{D_s}\theta, \qquad (10.16)$$

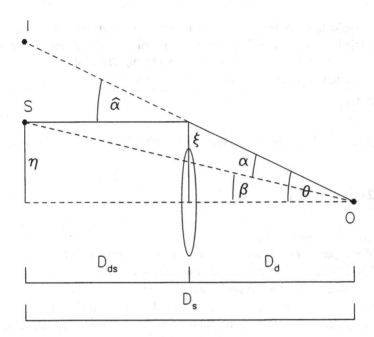

Figure 10.5. Illustration of a gravitational lens system. The light ray propagates from the source S at transverse distance η from the optic axis to the observer O, passing the lens at transverse distance ξ. It is deflected by an angle $\hat{\alpha}$. The angular separations of the source and the image from the optic axis as seen by the observer are β and θ, respectively. The reduced deflection angle α and the actual deflection angle $\hat{\alpha}$ are related by eq. 10.13. The distances between the observer and the source, the observer and the lens, and the lens and the source are D_s, D_d, and D_{ds}, respectively.

where we have set $\xi = D_d\theta$. In this case, the lens equation is linear; that is, $\beta \propto \theta$. Let us define a critical surface-mass density

$$\Sigma_{cr} = \frac{c^2}{4\pi G}\frac{D_s}{D_d D_{ds}} = 0.35\,\mathrm{g\,cm^{-2}}\left(\frac{D}{1\,\mathrm{Gpc}}\right)^{-1}, \qquad (10.17)$$

where the effective distance D is defined as the combination of distances

$$D = \frac{D_d D_{ds}}{D_s}. \qquad (10.18)$$

For a lens with a constant surface-mass density Σ_{cr}, the deflection angle is $\alpha(\theta) = \theta$, and so $\beta = 0$ for all θ. Such a lens focuses perfectly, with a well-defined focal length. A typical gravitational lens, however, behaves quite differently. Light rays which pass the lens at different impact parameters cross the optic axis at different distances behind the lens. Considered as an optical device, a gravitational lens therefore has almost all the aberrations one can think of. However, it does not have any chromatic aberration because the geometry of light paths is independent of wavelength.

Gravitational lensing

A lens which has $\Sigma > \Sigma_{cr}$ somewhere within it is referred to as being *supercritical*. Usually, multiple imaging occurs only if the lens is supercritical, but there are exceptions to this rule (e.g., Subramanian & Cowling 1986).

10.2.1.4 Einstein radius

Consider now a circularly symmetric lens with an arbitrary mass profile. According to equations (10.11) and (10.13), the lens equation reads

$$\beta(\theta) = \theta - \frac{D_{ds}}{D_d D_s} \frac{4GM(\theta)}{c^2 \theta}. \tag{10.19}$$

Due to the rotational symmetry of the lens system, a source which lies exactly on the optic axis ($\beta = 0$) is imaged as a ring if the lens is supercritical. Setting $\beta = 0$ in equation (10.19) we obtain the radius of the ring to be

$$\theta_E = \left[\frac{4GM(\theta_E)}{c^2} \frac{D_{ds}}{D_d D_s} \right]^{1/2}. \tag{10.20}$$

This is referred to as the *Einstein radius*. Figure 10.6 illustrates the situation. Note that the Einstein radius is not just a property of the lens, but depends also on the various distances in the problem.

The Einstein radius provides a natural angular scale to describe the lensing geometry for several reasons. In the case of multiple imaging, the typical angular separation of images is of order $2\theta_E$. Further, sources which are closer than about θ_E to the optic axis experience strong lensing in the sense that they are significantly magnified, whereas sources which are located well outside the Einstein ring are magnified very little. In many lens models, the Einstein ring also represents roughly the boundary between source positions that are multiply imaged and those that are only singly imaged. Finally, by comparing equations (10.17) and (10.20) we see that the mean surface-mass density inside the Einstein radius is just the critical surface density Σ_{cr}.

For a point mass M, the Einstein radius is given by

$$\theta_E = \left(\frac{4GM}{c^2} \frac{D_{ds}}{D_d D_s} \right)^{1/2}. \tag{10.21}$$

To give two illustrative examples, we consider lensing by a star in the Galaxy, for which $M \sim M_\odot$ and $D \sim 10$ kpc, and lensing by a galaxy at a cosmological distance with $M \sim 10^{11} M_\odot$ and $D \sim 1$ Gpc. The corresponding Einstein radii are

$$\theta_E = (0.9 \text{ mas}) \left(\frac{M}{M_\odot} \right)^{1/2} \left(\frac{D}{10 \text{ kpc}} \right)^{-1/2},$$

$$\theta_E = (0''.9) \left(\frac{M}{10^{11} M_\odot} \right)^{1/2} \left(\frac{D}{\text{Gpc}} \right)^{-1/2}. \tag{10.22}$$

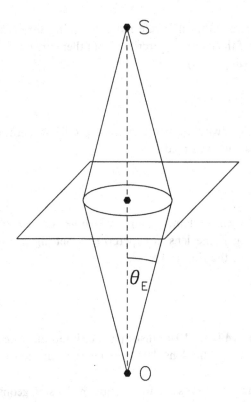

Figure 10.6. A source S on the optic axis of a circularly symmetric lens is imaged as a ring with an angular radius given by the Einstein radius θ_E.

10.2.1.5 Imaging by a point-mass lens

For a point-mass lens, we can use the Einstein radius (10.20) to rewrite the lens equation in the form

$$\beta = \theta - \frac{\theta_E^2}{\theta} \ . \tag{10.23}$$

This equation has two solutions,

$$\theta_\pm = \frac{1}{2}\left(\beta \pm \sqrt{\beta^2 + 4\theta_E^2}\right) \ . \tag{10.24}$$

Any source is imaged twice by a point-mass lens (Fig. 10.7). The two images are on either side of the source, with one image inside the Einstein ring and the other outside. As the source moves away from the lens (i.e., as β increases), one of the images approaches the lens and becomes very faint, while the other image approaches closer and closer to the true position of the source and tends toward a magnification of unity.

Gravitational light deflection preserves surface brightness (because of Liouville's theorem), but gravitational lensing changes the apparent solid angle of a source. The

Gravitational lensing

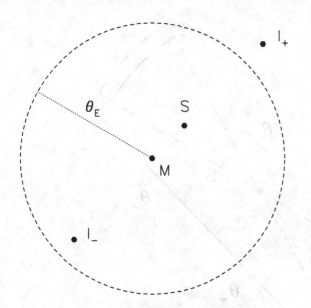

Figure 10.7. Relative locations of the source S and images I_+, I_- lensed by a point mass M. The dashed circle is the Einstein ring with radius θ_E. One image is inside the Einstein ring and the other outside.

total flux received from a gravitationally lensed image of a source is therefore changed in proportion to the ratio between the solid angles of the image and the source,

$$\text{magnification} = \frac{\text{image area}}{\text{source area}}. \qquad (10.25)$$

Figure 10.8 shows the magnified images of a source lensed by a point mass.

For a circularly symmetric lens, the magnification factor μ is given by

$$\mu = \frac{\theta}{\beta} \frac{d\theta}{d\beta}. \qquad (10.26)$$

For a point-mass lens, which is a special case of a circularly symmetric lens, we can substitute for β using the lens equation (10.23) to obtain the magnifications of the two images,

$$\mu_\pm = \left[1 - \left(\frac{\theta_E}{\theta_\pm}\right)^4\right]^{-1} = \frac{u^2+2}{2u\sqrt{u^2+4}} \pm \frac{1}{2}, \qquad (10.27)$$

where u is the angular separation of the source from the point mass in units of the Einstein angle, $u = \beta \theta_E^{-1}$. Since $\theta_- < \theta_E$, $\mu_- < 0$, and hence the magnification of the image which is inside the Einstein ring is negative. This means that this image has its parity flipped with respect to the source. The net magnification of flux in the two images is obtained by adding the absolute magnifications,

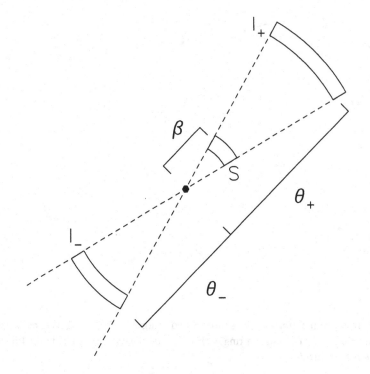

Figure 10.8. Magnified images of a source lensed by a point mass.

$$\mu = |\mu_+| + |\mu_-| = \frac{u^2 + 2}{u\sqrt{u^2 + 4}}. \tag{10.28}$$

When the source lies on the Einstein radius, we have $\beta = \theta_E$, $u = 1$, and the total magnification becomes

$$\mu = 1.17 + 0.17 = 1.34. \tag{10.29}$$

How can lensing by a point mass be detected? Unless the lens is massive ($M > 10^6 \, M_\odot$ for a cosmologically distant source), the angular separation of the two images is too small to be resolved. However, even when it is not possible to see the multiple images, the magnification can still be detected if the lens and source move relative to each other, giving rise to lensing-induced time variability of the source (Chang & Refsdal 1979; Gott 1981). When this kind of variability is induced by stellar-mass lenses it is referred to as *microlensing*. Microlensing was first observed in the multiply imaged QSO 2237+0305 (Irwin et al. 1989), and may also have been seen in QSO 0957+561 (Schild & Smith 1991; see also §10.3.7.4). Paczyński (1986b) had the brilliant idea of using microlensing to search for so-called *Massive Astrophysical Compact Halo Objects* (MACHOs, Griest 1991) in the Galaxy. We discuss this topic in some depth in §10.2.2.

10.2.2 Microlensing in the galaxy

10.2.2.1 Basic relations

If the closest approach between a point-mass lens and a source is $\leq \theta_E$, the peak magnification in the lensing-induced light curve is $\mu_{max} \geq 1.34$. A magnification of 1.34 corresponds to a brightening by 0.32 magnitudes, which is easily detectable. Paczyński (1986b) proposed monitoring millions of stars in the LMC to look for such magnifications in a small fraction of the sources. If enough events are detected, it should be possible to map the distribution of stellar-mass objects in our Galaxy.

Perhaps the biggest problem with Paczyński's proposal is that monitoring a million stars or more primarily leads to the detection of a huge number of variable stars. The intrinsically variable sources must somehow be distinguished from stars whose variability is caused by microlensing. Fortunately, the light curves of lensed stars have certain tell-tale signatures – the light curves are expected to be symmetric in time and the magnification is expected to be achromatic because light deflection does not depend on wavelength (but see the more detailed discussion in §10.2.2.4 below). In contrast, intrinsically variable stars typically have asymmetric light curves and do change their colors.

The expected time-scale for microlensing-induced variations is given in terms of the typical angular scale θ_E, the relative velocity v between source and lens, and the distance to the lens:

$$t_0 = \frac{D_d \theta_E}{v} = 0.214\,\text{yr} \left(\frac{M}{M_\odot}\right)^{1/2} \left(\frac{D_d}{10\,\text{kpc}}\right)^{1/2} \left(\frac{D_{ds}}{D_s}\right)^{1/2} \left(\frac{200\,\text{km s}^{-1}}{v}\right). \tag{10.30}$$

The ratio $D_{ds} D_s^{-1}$ is close to unity if the lenses are located in the Galactic halo and the sources are in the LMC. If light curves are sampled with time intervals between about an hour and a year, MACHOs in the mass range $10^{-6}\,M_\odot$ to $10^2\,M_\odot$ are potentially detectable. Note that the measurement of t_0 in a given microlensing event does not directly give M, but only a combination of M, D_d, D_s, and v. Various ideas to break this degeneracy have been discussed. Figure 10.9 shows microlensing-induced light curves for six different minimum separations $\Delta y = u_{min}$ between the source and the lens. The widths of the peaks are $\sim t_0$, and there is a direct one-to-one mapping between Δy and the maximum magnification at the peak of the light curve. A microlensing light curve therefore gives two observables, t_0 and Δy.

The chance of seeing a microlensing event is usually expressed in terms of the optical depth, which is the probability that at any instant of time a given star is within an angle θ_E of a lens. The optical depth is the integral over the number density $n(D_d)$ of lenses times the area enclosed by the Einstein ring of each lens, i.e.

$$\tau = \frac{1}{\delta\omega} \int dV\, n(D_d)\, \pi \theta_E^2, \tag{10.31}$$

where $dV = \delta\omega\, D_d^2\, dD_d$ is the volume of an infinitesimal spherical shell with radius D_d which covers a solid angle $\delta\omega$. The integral gives the solid angle covered by the Einstein

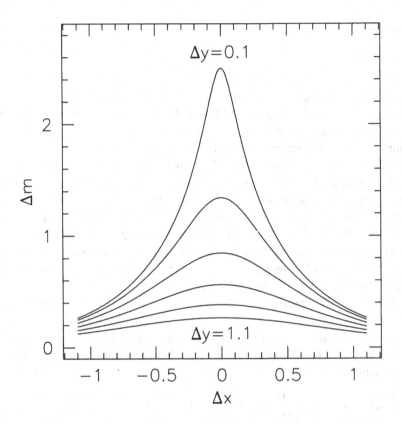

Figure 10.9. Microlensing-induced light curves for six minimum separations between the source and the lens, $\Delta y = 0.1, 0.3, \ldots, 1.1$. The separation is expressed in units of the Einstein radius. Δm is the magnification in magnitudes and Δx is time about the peak.

circles of the lenses, and the probability is obtained upon dividing this quantity by the solid angle $\delta\omega$ which is observed. Inserting equation (10.21) for the Einstein angle, we obtain

$$\tau = \int_0^{D_s} \frac{4\pi G\rho}{c^2} \frac{D_d D_{ds}}{D_s} dD_d = \frac{4\pi G}{c^2} D_s^2 \int_0^1 \rho(x) x(1-x) dx, \qquad (10.32)$$

where $x \equiv D_d D_s^{-1}$ and ρ is the mass density of MACHOs. In writing (10.32), we have made use of the fact that space is locally Euclidean, hence $D_{ds} = D_s - D_d$. If ρ is constant along the line-of-sight, the optical depth simplifies to

$$\tau = \frac{2\pi}{3} \frac{G\rho}{c^2} D_s^2. \qquad (10.33)$$

It is important to note that the optical depth τ depends on the *mass density* of lenses ρ and not on their *mass* M. The time-scale of variability induced by microlensing, however, does depend on the square-root of the mass, as shown by equation (10.30).

Gravitational lensing

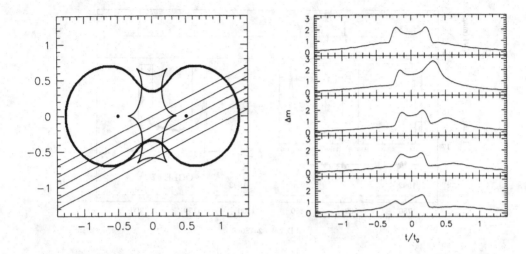

Figure 10.10. *Left panel:* A binary lens composed of two equal point masses. The critical curve is shown by the heavy line, and the corresponding caustic is indicated by the thin line with six cusps. (See §10.3.3.2 for a definition of critical curves and caustics.) Five source trajectories across this lens system are indicated. *Right panel:* Light curves corresponding to an extended source moving along the trajectories indicated in the left panel. Double-peaked features occur when the source comes close to both lenses.

10.2.2.2 Ongoing galactic microlensing searches

Paczyński's suggestion that microlensing by compact objects in the Galactic halo may be detected by monitoring the light curves of stars in the LMC inspired several groups to set up elaborate searches for microlensing events. Four groups, MACHO (Alcock et al. 1993), EROS (Aubourg et al. 1993), OGLE (Udalski et al. 1992), and DUO (Alard 1995), are currently searching for microlensing-induced stellar variability in the LMC (EROS, MACHO) as well as in the Galactic bulge (DUO, MACHO, OGLE).

So far, about 100 microlensing events have been observed, and their number is increasing rapidly. Most events have been seen toward the Galactic bulge. The majority of events have been caused by single lenses, and have light curves similar to those shown in Fig. 10.9, but at least two events so far are due to binary lenses. Strong lensing by binaries (defined as events where the source crosses one or more caustics, see Fig. 10.10) was estimated by Mao & Paczyński (1991) to contribute about 10 percent of all events. Binary lensing is most easily distinguished from single-lens events by characteristic double-peaked or asymmetric light curves; Fig. 10.10 shows some typical examples.

The light curve of the first observed binary microlensing event, OGLE #7, is shown in Fig. 10.11.

10.2.2.3 Early results on optical depths

Both the OGLE and MACHO collaborations have determined the microlensing optical depth toward the Galactic bulge. The results are

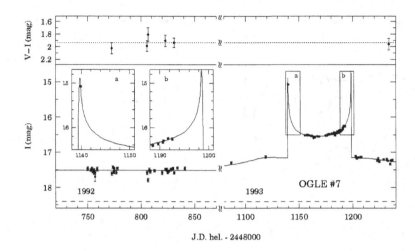

Figure 10.11. Light curve of the first binary microlensing event, OGLE #7 (taken from the OGLE WWW home page at: http://www.astrouw.edu.pl/~ftp/ogle/ogle.html).

$$\tau = \begin{cases} (3.3 \pm 1.2) \times 10^{-6} & \text{(Paczyński et al. 1994)} \\ (3.9^{+1.8}_{-1.2}) \times 10^{-6} & \text{(Alcock et al. 1997)} \end{cases} \quad (10.34)$$

Original theoretical estimates (Paczyński 1991; Griest, Alcock & Axelrod 1991) had predicted an optical depth below 10^{-6}. Even though this value was increased slightly by Kiraga & Paczyński (1994) who realized the importance of lensing of background bulge stars by foreground bulge stars (referred to as self-lensing of the bulge), the measured optical depth is nevertheless very much higher than expected. Paczyński et al. (1994) suggested that a Galactic bar which is approximately aligned with the line-of-sight toward the Galactic bulge might explain the excess optical depth. Self-consistent calculations of the bar by Zhao, Spergel & Rich (1995) and Zhao, Rich & Spergel (1996) give $\tau \sim 2 \times 10^{-6}$, which is within one standard deviation of the observed value. However, using COBE/DIRBE near-infrared data of the inner Galaxy and calibrating the mass-to-light ratio with the terminal velocities of HI and CO clouds, Bissantz et al. (1997) find a significantly lower optical depth, $0.8 \times 10^{-6} \lesssim \tau \lesssim 0.9 \times 10^{-6}$. Zhao & Mao (1996) describe how the shape of the Galactic bar can be inferred from measuring the spatial dependence of the optical depth. Zhao, Spergel & Rich (1995) claim that the duration distribution of the bulge events detected by OGLE is compatible with a roughly normal stellar mass distribution.

In principle, moments of the mass distribution of microlensing objects can be inferred from moments of the duration distribution of microlensing events (De Rújula, Jetzer & Massó 1991). Mao & Paczyński (1996) have shown that a robust determination of mass-function parameters requires ~ 100 microlensing events even if the geometry of the microlens distribution and the kinematics are known.

Based on three events from their first year of data, of which two are of only modest significance, Alcock et al. (1996) estimated the optical depth toward the LMC to be

$$\tau = 9^{+7}_{-5} \times 10^{-8}, \qquad (10.35)$$

in the mass range $10^{-4} M_\odot < M < 10^{-1} M_\odot$. This is too small for the entire halo to be made of MACHOs in this mass range. At the 95% confidence level, the first-year data of the MACHO collaboration rule out a contribution from MACHOs to the halo mass $\gtrsim 40\%$ in the mass range $10^{-3} M_\odot \leq M \leq 10^{-2} M_\odot$, and $\gtrsim 60\%$ within $10^{-4} M_\odot \leq M \leq 10^{-1} M_\odot$. Sahu (1994) argued that all events can be due to objects in the Galactic disk or the LMC itself. The EROS collaboration, having better time resolution, is able to probe smaller masses, $10^{-7} M_\odot \leq M \leq 10^{-1} M_\odot$ (Aubourg et al. 1995). The 95% confidence level from the EROS data excludes a halo fraction $\gtrsim (20-30)\%$ in the mass range $10^{-7} M_\odot \leq M \leq 10^{-2} M_\odot$ (Ansari et al. 1996; Renault et al. 1997; see also Roulet & Mollerach 1997).

More recently, the MACHO group reported results from 2.3 years of data. Based on eight events, they then estimated the optical depth toward the LMC to be

$$\tau = 2.9^{+1.4}_{-0.9} \times 10^{-7}, \qquad (10.36)$$

and the halo fraction to be $0.45 - 1$ in the mass-range $0.2 M_\odot \leq M \leq 0.5 M_\odot$ at 68% confidence. Further, they could not reject, at the 99% confidence level, the hypothesis that the entire halo is made of MACHOs with masses $0.2 M_\odot \leq M \leq 1 M_\odot$ (Sutherland et al. 1996). More data are needed before any definitive conclusion can be reached on the contribution of MACHOs to the halo.

10.2.2.4 Other interesting discoveries

In the simplest scenario of microlensing in the Galaxy, a single point-like source is lensed by a single point mass which moves with constant velocity relative to the source. The light curve observed from such an event is time-symmetric and achromatic. At the low optical depths that we expect in the Galaxy, and ignoring binaries, microlensing events should not repeat since the probability that the same star is lensed more than once is negligibly small.

In practice, the situation is more complicated and detailed interpretations of observed light curves must account for some of the complications listed below. The effects of binary lenses have already been mentioned above. In the so-called *resonant* case, the separation of the two lenses is comparable to their Einstein radii. The light curve of such a lens system can have dramatic features such as the double peaks shown in Figs. 10.10 and 10.11. At least two such events have been observed so far, OGLE #7 (Udalski et al. 1994; Bennett et al. 1995) and DUO #2 (Alard, Mao & Guibert 1995). In the non-resonant case, the lenses are well separated and act as almost independent lenses. Di Stefano & Mao (1996) estimated that a few percent of all microlensing events should "repeat" due to consecutive magnification of a star by the two stars in a wide binary lens.

The sensitivity of microlensing searches to binaries may make this a particularly powerful method to search for planets around distant stars, as emphasized by Mao & Paczyński (1991) and Gould & Loeb (1992).

Multiple sources can give rise to various other complications. Since the optical depth is low, microlensing searches are performed in crowded fields where the number density of sources is high. Multiple-source stars which are closer to each other than $\sim 1''$ appear as single because they are not resolved. The Einstein radius of a solar mass lens, on the other hand, is $\sim 0''.001$ (cf. equation (10.22)). Therefore, if the projected separation of two sources is $< 1''$ but $> 0''.001$, a single lens affects only one of them at a time. Several effects can then occur. First, the microlensing event can apparently recur when the two sources are lensed individually (Griest & Hu 1992). Second, if the sources have different colors, the event is chromatic because the color of the lensed star dominates during the event (Udalski et al. 1994; Kamionkowski 1995; Buchalter, Kamionkowski & Rich 1996). Third, the observed flux is a blend of the magnified flux from the lensed component and the constant flux of the unlensed components, and this leads to various biases (Di Stefano & Esin 1995; see also Alard & Guibert 1997). A systematic method of detecting microlensing in blended data has been proposed and is referred to as "pixel lensing" (Crotts 1992; Baillon et al. 1993; Colley 1995; Gould 1996).

Stars are not truly point-like. If a source is larger than the impact parameter of a single lens or the caustic structure of a binary lens, the finite source size modifies the light curve significantly (Gould 1994a; Nemiroff & Wickramasinghe 1994; Witt & Mao 1994; Witt 1995).

Finally, if the relative transverse velocity of the source, the lens, and the observer is not constant during the event, the light curve becomes time-asymmetric. The parallax effect due to the acceleration of the earth was predicted by Gould (1992b) and detected by Alcock et al. (1995b). The detection of parallax provides an additional observable which helps partially to break the degeneracy among M, v, D_d and D_{ds} mentioned in §10.2.2.1.

Another method of breaking the degeneracy is via observations from space. The idea of space measurements was suggested by Refsdal as early as 1966 as a means to determine distances and masses of lenses in the context of quasar lensing (Refsdal 1966b). Some obvious benefits of space-based telescopes include absence of seeing and access to wavebands like the UV or IR which are absorbed by the earth's atmosphere. The particular advantage of space observations for microlensing in the Galaxy arises from the fact that the Einstein radius of a subsolar-mass microlens in the Galactic halo is of order 10^8 km and thus comparable to the AU, cf. equation (10.22). Telescopes separated by ~ 1 AU would therefore observe different light curves for the same microlensing event. This additional information on the event can break the degeneracy between the parameters defining the time-scale t_0 (Gould 1994b). In the special (and rare) case of very high magnification when the source is resolved during the event (Gould 1994a), all four parameters may be determined.

Interesting discoveries can be expected from the various microlensing "alert systems" which have been recently set up (GMAN, Pratt 1996; PLANET, Sackett 1996, Albrow et al. 1996). The goal of these programs is to monitor ongoing microlensing events in almost real time with very high time resolution. It should be possible to detect anomalies in the microlensing light curves which are expected from the complications listed in this section, or to obtain detailed information (e.g., spectra, Sahu 1996) from objects while they are being microlensed.

Jetzer (1994) showed that the microlensing optical depth toward the Andromeda galaxy M31 is similar to that toward the LMC, $\tau \simeq 10^{-6}$. Experiments to detect microlensing toward M31 have recently been set up (e.g., Gondolo et al. 1997; Crotts & Tomaney 1996), and results are awaited.

10.2.3 Extragalactic microlenses

10.2.3.1 Point masses in the universe

It has been proposed at various times that a significant fraction of the dark matter in the universe may be in the form of compact masses. These masses will induce various lensing phenomena, some of which are very easily observed. The lack of evidence for these phenomena can therefore be used to place useful limits on the fraction of the mass in the universe in such objects (Press & Gunn 1973).

Consider an Einstein–de Sitter universe with a constant comoving number density of point lenses of mass M corresponding to a cosmic density parameter Ω_M. The optical depth for lensing of sources at redshift z_s can be shown to be

$$\begin{aligned} \tau(z_s) &= 3\Omega_M \left[\frac{(z_s + 2 + 2\sqrt{1+z_s})\ln(1+z_s)}{z_s} - 4 \right] \\ &\simeq \Omega_M \frac{z_s^2}{4} \quad \text{for } z_s \ll 1 \\ &\simeq 0.3\Omega_M \quad \text{for } z_s = 2. \end{aligned} \qquad (10.37)$$

We see that the probability for lensing is $\sim \Omega_M$ for high-redshift sources (Press & Gunn 1973). Hence the number of lensing events in a given source sample directly measures the cosmological density in compact objects.

In calculating the probability of lensing it is important to allow for various selection effects. Lenses magnify the observed flux, and therefore sources which are intrinsically too faint to be observed may be lifted over the detection threshold. At the same time, lensing increases the solid angle within which sources are observed so that their number density in the sky is reduced (Narayan 1991). If there is a large reservoir of faint sources, the increase in source number due to the apparent brightening outweighs their spatial dilution, and the observed number of sources is increased due to lensing. This magnification bias (Turner 1980; Turner, Ostriker & Gott 1984; Narayan & Wallington 1993) can substantially increase the probability of lensing for bright optical quasars whose number-count function is steep.

10.2.3.2 Current upper limits on Ω_M in point masses

Various techniques have been proposed and applied to obtain limits on Ω_M over a broad range of lens masses (see Carr 1994 for a review). Lenses with masses in the

Table 10.1. *Summary of techniques to constrain Ω_M in point masses, along with the current best limits*

Technique	References	Mass range $M_\odot$	Limit on Ω_M
Image doubling of bright QSOs	Surdej et al. (1993)	$10^{10} - 10^{12}$	< 0.02
Doubling of VLBI compact sources	Kassiola et al. (1991) Henstock et al. (1995)	$10^6 - 10^8$	< 0.05
Echoes from γ-ray bursts	Nemiroff et al. (1993)	$10^{6.5} - 10^{8.1}$	$\lesssim 1$ excluded
	Nemiroff et al. (1994)	$(10^3 \rightarrow)$	null result
Diff. magnification of QSO continuum vs. broad emission lines	Canizares (1982) Dalcanton et al. (1994)	$10^{-1} - 20$ $10^{-3} - 60$	< 0.1 < 0.2
Quasar variability	Schneider (1993)	$10^{-3} - 10^{-2}$	< 0.1
Femtolensing of γ-ray bursts	Gould (1992a) Stanek et al. (1993)	$10^{-17} - 10^{-13}$	– –

range $10^{10} < M/M_\odot < 10^{12}$ will split images of bright QSOs by $0''.3 - 3''$. Such angular splittings are accessible to optical observations; therefore, it is easy to constrain Ω_M in this mass range. The image splitting of lenses with $10^6 < M/M_\odot < 10^8$ is on the order of milliarcseconds and falls within the resolution domain of VLBI observations of radio quasars (Kassiola, Kovner & Blandford 1991). The best limits presently are due to Henstock et al. (1995). A completely different approach utilizes the differential time delay between multiple images. A cosmological γ-ray burst, which is gravitationally lensed will be seen as multiple repetitions of a single event (Blaes & Webster 1992). By searching the γ-ray burst database for (lack of) evidence of repetitions, Ω_M can be constrained over a range of masses which extends below the VLBI range mentioned above. The region within QSOs where the broad emission lines are emitted is larger than the region emitting the continuum radiation. Lenses with $M \sim 1\,M_\odot$ can magnify the continuum relative to the broad emission lines and thereby reduce the observed emission line widths. Lenses of still smaller masses cause apparent QSO variability, and hence from observations of the variability an upper limit to Ω_M can be derived. Finally, the time delay due to lenses with very small masses can be such that the light beams from multiply imaged γ-ray bursts interfere so that the observed burst spectra should show interference patterns. Table 10.1 summarizes these various techniques and gives the most recent results on Ω_M.

As Table 10.1 shows, we can eliminate $\Omega_M \sim 1$ in virtually all astrophysically plausible mass ranges. The limits are especially tight in the range $10^6 < M/M_\odot < 10^{12}$, where Ω_M is constrained to be less than a few percent.

Gravitational lensing

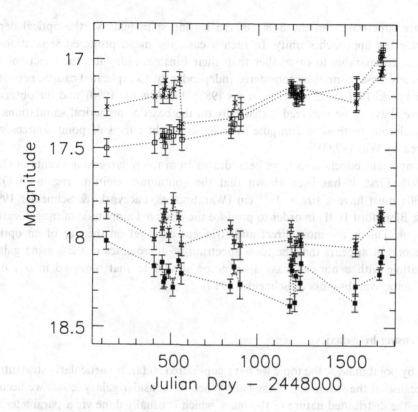

Figure 10.12. Light curves of the four images in the "Einstein Cross" QSO 2237+0305 since August 1990 (from Østensen *et al.* 1996).

10.2.3.3 Microlensing in QSO 2237+0305

Although the Galactic microlensing projects described earlier have developed into one of the most exciting branches of gravitational lensing, the phenomenon of microlensing was, in fact, first detected in a cosmological source, the quadruply imaged QSO 2237+0305 (Irwin et al. 1989; Corrigan *et al.* 1991; Webster *et al.* 1991; Østensen et al. 1996). The lensing galaxy in QSO 2237+0305 is a spiral at a redshift of 0.04 (Huchra *et al.* 1985). The four quasar images are almost symmetrically located in a cross-shaped pattern around the nucleus of the galaxy; hence the system has been named the "Einstein Cross". Uncorrelated flux variations have been observed in QSO 2237+0305, possibly in all four images, and these variations provide evidence for microlensing due to stars in the lensing galaxy. Figure 10.12 shows the light curves of the four images.

The interpretation of the microlensing events in QSO 2237+0305 is much less straightforward than in the case of microlensing in the Galaxy. When a distant galaxy forms multiple images, the surface-mass density at the locations of the images is of order the critical density, Σ_{cr}. If most of the local mass is made of stars or other massive compact objects (as is likely in the case of QSO 2237+0305 since the four

images are superposed on the bulge of the lensing spiral galaxy), the optical depth
to microlensing approaches unity. In such a case, the mean projected separation of
the stars is comparable to or smaller than their Einstein radii, and the effects of the
various microlenses cannot be considered independently. Complicated caustic networks
arise (Paczyński 1986a; Schneider & Weiss 1987; Wambsganss 1990), and the observed
light curves have to be analyzed statistically on the basis of numerical simulations. A
new and elegant method to compute microlensing light curves of point sources was
introduced by Witt (1993).

Two important conclusions have been drawn from the microlensing events in QSO
2237+0305. First, it has been shown that the continuum emitting region in QSO
2237+0305 must have a size $\sim 10^{15}$ cm (Wambsganss, Paczyński & Schneider 1990;
Rauch & Blandford 1991) in order to produce the observed amplitude of magnification
fluctuations. This is the most direct and stringent limit yet on the size of an optical
QSO. Second, it appears that the mass spectrum of microlenses in the lensing galaxy
is compatible with a normal mass distribution similar to that observed in our own
Galaxy (Seitz, Wambsganss & Schneider 1994).

10.3 Lensing by galaxies

Lensing by point masses, the topic we have considered so far, is particularly straightforward because of the simplicity of the lens. When we consider galaxy lenses we need to
allow for the distributed nature of the mass, which is usually done via a parameterized
model. The level of complexity of the model is dictated by the application at hand.

10.3.1 Lensing by a singular isothermal sphere

A simple model for the mass distribution in galaxies assumes that the stars and other
mass components behave like particles of an ideal gas, confined by their combined,
spherically symmetric gravitational potential. The equation of state of the "particles",
henceforth called stars for simplicity, takes the form

$$p = \frac{\rho k T}{m}, \qquad (10.38)$$

where p is the pressure and where ρ and m are the mass density and the mass of
the stars. In thermal equilibrium, the temperature T is related to the one-dimensional
velocity dispersion σ_v of the stars through

$$m\sigma_v^2 = kT. \qquad (10.39)$$

The temperature, or equivalently the velocity dispersion, could in general depend on
radius r, but it is usually assumed that the stellar gas is isothermal, so that σ_v is
constant across the galaxy. The equation of hydrostatic equilibrium then gives

Gravitational lensing

$$\frac{p'}{\rho} = -\frac{GM(r)}{r^2}, \quad M'(r) = 4\pi r^2 \rho,\tag{10.40}$$

where $M(r)$ is the mass interior to radius r, and primes denote derivatives with respect to r. A particularly simple solution of equations (10.38)–(10.40) is

$$\rho(r) = \frac{\sigma_v^2}{2\pi G} \frac{1}{r^2}.\tag{10.41}$$

This mass distribution is called the *singular isothermal sphere*. Since $\rho \propto r^{-2}$, the mass $M(r)$ increases $\propto r$, and therefore the rotational velocity of test particles in circular orbits in the gravitational potential is

$$v_{\rm rot}^2(r) = \frac{GM(r)}{r} = 2\sigma_v^2 = \text{constant}.\tag{10.42}$$

The flat rotation curves of galaxies are naturally reproduced by this model.

Upon projecting along the line-of-sight, we obtain the surface-mass density

$$\Sigma(\xi) = \frac{\sigma_v^2}{2G} \frac{1}{\xi},\tag{10.43}$$

where ξ is the distance from the center of the two-dimensional profile. Referring to equation (10.11), we immediately obtain the deflection angle

$$\hat{\alpha} = 4\pi \frac{\sigma_v^2}{c^2} = (1.''4) \left(\frac{\sigma_v}{220 \,\text{km s}^{-1}}\right)^2,\tag{10.44}$$

which is independent of ξ and points toward the center of the lens. The Einstein radius of the singular isothermal sphere follows from equation (10.20),

$$\theta_E = 4\pi \frac{\sigma_v^2}{c^2} \frac{D_{ds}}{D_s} = \hat{\alpha}\frac{D_{ds}}{D_s} = \alpha.\tag{10.45}$$

Due to circular symmetry, the lens equation is essentially one-dimensional. Multiple images are obtained only if the source lies inside the Einstein ring, i.e., if $\beta < \theta_E$. When this condition is satisfied, the lens equation has the two solutions

$$\theta_\pm = \beta \pm \theta_E.\tag{10.46}$$

The images at $\theta_\pm$, the source, and the lens all lie on a straight line. Technically, a third image with zero flux is located at $\theta = 0$. This third image acquires a finite flux if the singularity at the center of the lens is replaced by a core region with a finite density.

The magnifications of the two images follow from equation (10.26),

$$\mu_\pm = \frac{\theta_\pm}{\beta} = 1 \pm \frac{\theta_E}{\beta} = \left(1 \mp \frac{\theta_E}{\theta_\pm}\right)^{-1}.\tag{10.47}$$

If the source lies outside the Einstein ring, i.e., if $\beta > \theta_E$, there is only one image at $\theta = \theta_+ = \beta + \theta_E$.

10.3.2 Effective lensing potential

Before proceeding to more complicated galaxy-lens models, it is useful to develop the formalism a little further. Let us define a scalar potential $\psi(\theta)$ which is the appropriately scaled, projected Newtonian potential of the lens,

$$\psi(\theta) = \frac{D_{ds}}{D_d D_s} \frac{2}{c^2} \int \Phi(D_d \theta, z)\, dz \,. \tag{10.48}$$

The derivatives of ψ with respect to θ have convenient properties. The gradient of ψ with respect to θ is the deflection angle,

$$\nabla_\theta \psi = D_d \nabla_\xi \psi = \frac{2}{c^2} \frac{D_{ds}}{D_s} \int \nabla_\perp \Phi\, dz = \alpha\,, \tag{10.49}$$

while the Laplacian is proportional to the surface-mass density Σ,

$$\nabla_\theta^2 \psi = \frac{2}{c^2} \frac{D_d D_{ds}}{D_s} \int \nabla_\xi^2 \Phi\, dz = \frac{2}{c^2} \frac{D_d D_{ds}}{D_s} \cdot 4\pi G \Sigma = 2\frac{\Sigma(\theta)}{\Sigma_{cr}} \equiv 2\kappa(\theta)\,, \tag{10.50}$$

where Poisson's equation has been used to relate the Laplacian of Φ to the mass density. The surface-mass density scaled with its critical value Σ_{cr} is called the *convergence* $\kappa(\theta)$. Since ψ satisfies the two-dimensional Poisson equation $\nabla_\theta^2 \psi = 2\kappa$, the effective lensing potential can be written in terms of κ

$$\psi(\theta) = \frac{1}{\pi} \int \kappa(\theta') \ln|\theta - \theta'|\, d^2\theta'\,. \tag{10.51}$$

As mentioned earlier, the deflection angle is the gradient of ψ, hence

$$\alpha(\theta) = \nabla \psi = \frac{1}{\pi} \int \kappa(\theta') \frac{\theta - \theta'}{|\theta - \theta'|^2}\, d^2\theta'\,, \tag{10.52}$$

which is equivalent to equation (10.10) if we account for the definition of Σ_{cr} given in equation (10.17).

The local properties of the lens mapping are described by its Jacobian matrix $\mathcal{A}$,

$$\mathcal{A} \equiv \frac{\partial \beta}{\partial \theta} = \left(\delta_{ij} - \frac{\partial \alpha_i(\theta)}{\partial \theta_j} \right) = \left(\delta_{ij} - \frac{\partial^2 \psi(\theta)}{\partial \theta_i \partial \theta_j} \right) = \mathcal{M}^{-1}\,. \tag{10.53}$$

As we have indicated, $\mathcal{A}$ is nothing but the inverse of the magnification tensor $\mathcal{M}$. The matrix $\mathcal{A}$ is therefore also called the inverse magnification tensor. The local solid-angle distortion due to the lens is given by the determinant of $\mathcal{A}$. A solid-angle element $\delta\beta^2$ of the source is mapped to the solid-angle element of the image $\delta\theta^2$, and so the magnification is given by

$$\frac{\delta\theta^2}{\delta\beta^2} = \det \mathcal{M} = \frac{1}{\det \mathcal{A}}\,. \tag{10.54}$$

This expression is the appropriate generalization of equation (10.26) when there is no symmetry.

Equation (10.53) shows that the matrix of second partial derivatives of the potential ψ (the Hessian matrix of ψ) describes the deviation of the lens mapping from the identity mapping. For convenience, we introduce the abbreviation

$$\frac{\partial^2 \psi}{\partial \theta_i \partial \theta_j} \equiv \psi_{ij} . \tag{10.55}$$

Since the Laplacian of ψ is twice the convergence, we have

$$\kappa = \frac{1}{2}(\psi_{11} + \psi_{22}) = \frac{1}{2} \operatorname{tr} \psi_{ij} . \tag{10.56}$$

Two additional linear combinations of ψ_{ij} are important, and these are the components of the *shear* tensor,

$$\begin{aligned} \gamma_1(\theta) &= \frac{1}{2}(\psi_{11} - \psi_{22}) \equiv \gamma(\theta)\cos[2\phi(\theta)] , \\ \gamma_2(\theta) &= \psi_{12} = \psi_{21} \equiv \gamma(\theta)\sin[2\phi(\theta)] . \end{aligned} \tag{10.57}$$

With these definitions, the Jacobian matrix can be written

$$\begin{aligned} \mathcal{A} &= \begin{pmatrix} 1-\kappa-\gamma_1 & -\gamma_2 \\ -\gamma_2 & 1-\kappa+\gamma_1 \end{pmatrix} \\ &= (1-\kappa)\begin{pmatrix} 1 & 0 \\ 0 & 1 \end{pmatrix} - \gamma\begin{pmatrix} \cos 2\phi & \sin 2\phi \\ \sin 2\phi & -\cos 2\phi \end{pmatrix} . \end{aligned} \tag{10.58}$$

The meaning of the terms convergence and shear now becomes intuitively clear. Convergence acting alone causes an isotropic focusing of light rays, leading to an isotropic magnification of a source. The source is mapped onto an image with the same shape but larger size. Shear introduces anisotropy (or astigmatism) into the lens mapping; the quantity $\gamma = (\gamma_1^2 + \gamma_2^2)^{1/2}$ describes the magnitude of the shear and ϕ describes its orientation. As shown in Fig. 10.13, a circular source of unit radius becomes, in the presence of both κ and γ, an elliptical image with major and minor axes

$$(1-\kappa-\gamma)^{-1}, \quad (1-\kappa+\gamma)^{-1} . \tag{10.59}$$

The magnification is

$$\mu = \det \mathcal{M} = \frac{1}{\det \mathcal{A}} = \frac{1}{[(1-\kappa)^2 - \gamma^2]} . \tag{10.60}$$

Note that the Jacobian $\mathcal{A}$ is in general a function of position θ.

10.3.3 Gravitational lensing via Fermat's Principle

10.3.3.1 The time-delay function
The lensing properties of model gravitational lenses are especially easy to visualize by application of Fermat's Principle of geometrical optics (Nityananda 1984, unpublished; Schneider 1985; Blandford & Narayan 1986; Nityananda & Samuel 1992). From the

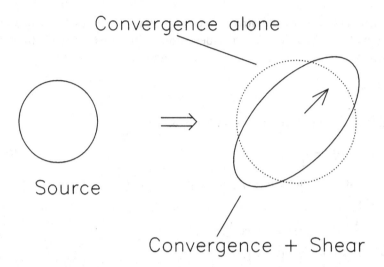

Figure 10.13. Illustration of the effects of convergence and shear on a circular source. Convergence magnifies the image isotropically, and shear deforms it to an ellipse.

lens equation (10.14) and the fact that the deflection angle is the gradient of the effective lensing potential ψ, we obtain

$$(\theta - \beta) - \nabla_\theta \psi = 0. \tag{10.61}$$

This equation can be written as a gradient,

$$\nabla_\theta \left[\frac{1}{2}(\theta - \beta)^2 - \psi \right] = 0. \tag{10.62}$$

The physical meaning of the term in square brackets becomes more obvious by considering the time-delay function,

$$\begin{aligned} t(\theta) &= \frac{(1+z_d)}{c} \frac{D_d D_s}{D_{ds}} \left[\frac{1}{2}(\theta - \beta)^2 - \psi(\theta) \right] \\ &= t_{\text{geom}} + t_{\text{grav}}. \end{aligned} \tag{10.63}$$

The term t_{geom} is proportional to the square of the angular offset between β and θ and is the time delay due to the extra path length of the deflected light ray relative to an unperturbed null geodesic. The coefficient in front of the square brackets ensures that the quantity corresponds to the time delay as measured by the observer. The second term t_{grav} is the time delay due to gravity and is identical to the Shapiro delay introduced in equation (10.3), with an extra factor of $(1+z_d)$ to allow for time stretching. Equations (10.62) and (10.63) imply that images satisfy the condition $\nabla_\theta t(\theta) = 0$ (Fermat's Principle).

In the case of a circularly symmetric deflector, the source, the lens and the images have to lie on a straight line on the sky. Therefore, it is sufficient to consider the section

Gravitational lensing

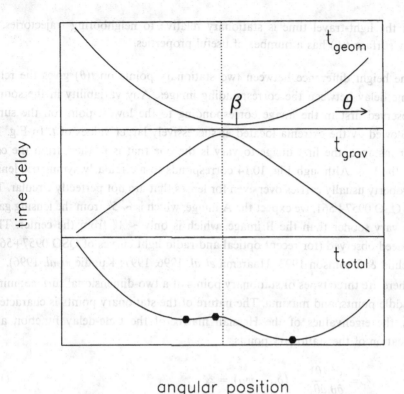

Figure 10.14. Geometric (*top*), gravitational (*middle*), and total (*bottom*) time delay of a circularly symmetric lens for a source that is slightly offset from the symmetry axis. The dotted line shows the location of the center of the lens, and β shows the position of the source. Images are located at points where the total time-delay function is stationary. The image positions are marked with filled hexagons in the bottom panel.

along this line of the time-delay function. Figure 10.14 illustrates the geometrical and gravitational time delays for this case. The top panel shows t_{geom} for a slightly offset source. The curve is a parabola centered on the position of the source. The central panel displays t_{grav} for an isothermal sphere with a softened core. This curve is centered on the lens. The bottom panel shows the total time delay. According to the above discussion images are located at stationary points of $t(\theta)$. For the case shown in Fig. 10.14 there are three stationary points, marked by filled hexagons, and the corresponding values of θ give the image positions.

10.3.3.2 Properties of the time-delay function

In the general case it is necessary to consider image locations in the two-dimensional space of θ, not just on a line. The images are then located at those points θ_i where the two-dimensional time-delay surface $t(\theta)$ is stationary. This is *Fermat's Principle* in geometrical optics, which states that the actual trajectory followed by a light ray is such that the light-travel time is stationary relative to neighboring trajectories. The time-delay surface $t(\theta)$ has a number of useful properties.

1. The height difference between two stationary points on $t(\theta)$ gives the relative time delay between the corresponding images. Any variability in the source is observed first in the image corresponding to the lowest point on the surface, followed by the extrema located at successively larger values of t. In Fig. 10.14 for instance, the first image to vary is the one that is farthest from the center of the lens. Although Fig. 10.14 corresponds to a circularly symmetric lens, this property usually carries over even for lenses that are not perfectly circular. Thus, in QSO 0957+561, we expect the A image, which is $\sim 5''$ from the lensing galaxy, to vary sooner than the B image, which is only $\sim 1''$ from the center. This is indeed observed (for recent optical and radio light curves of QSO 0957+561 see Schild & Thomson 1993; Haarsma et al. 1996, 1997; Kundić et al. 1996).

2. There are three types of stationary points of a two-dimensional surface: minima, saddle points, and maxima. The nature of the stationary points is characterized by the eigenvalues of the Hessian matrix of the time-delay function at the location of the stationary points,

$$\mathcal{T} = \frac{\partial^2 t(\theta)}{\partial \theta_i \partial \theta_j} \propto (\delta_{ij} - \psi_{ij}) = \mathcal{A}. \tag{10.64}$$

The matrix $\mathcal{T}$ describes the local curvature of the time-delay surface. If both eigenvalues of $\mathcal{T}$ are positive, $t(\theta)$ is curved "upward" in both coordinate directions, and the stationary point is a minimum. If the eigenvalues of $\mathcal{T}$ have opposite signs we have a saddle point, and if both eigenvalues of $\mathcal{T}$ are negative, we have a maximum. Correspondingly, we can distinguish three types of images. Images of type I arise at minima of $t(\theta)$ where $\det \mathcal{A} > 0$ and $\operatorname{tr} \mathcal{A} > 0$. Images of type II are formed at saddle points of $t(\theta)$ where the eigenvalues have opposite signs, hence $\det \mathcal{A} < 0$. Images of type III are located at maxima of $t(\theta)$ where both eigenvalues are negative and so $\det \mathcal{A} > 0$ and $\operatorname{tr} \mathcal{A} < 0$.

3. Since the magnification is the inverse of $\det \mathcal{A}$, images of types I and III have positive magnification and images of type II have negative magnification. The interpretation of a negative μ is that the parity of the image is reversed. A little thought shows that the three images shown in Fig. 10.14 correspond, reading from the left, to a saddle point, a maximum and a minimum, respectively. The images A and B in QSO 0957+561 correspond to the images on the right and left in this example, and ought to represent a minimum and a saddle point, respectively, in the time-delay surface. VLBI observations do indeed show the expected reversal of parity between the two images (Gorenstein et al. 1988).

4 The curvature of $t(\theta)$ measures the inverse magnification. When the curvature of $t(\theta)$ along one coordinate direction is small, the image is strongly magnified along that direction, while if $t(\theta)$ has a large curvature the magnification is small. Figure 10.15 displays the time-delay function of a typical circularly symmetric lens and a source on the symmetry axis (top panel), a slightly offset source (central panel), and a source with a large offset (bottom panel). If the separation between source and lens is large, only one image is formed, while if the source is close to the lens three images are formed. Note that, as the source moves, two images approach each other, merge and vanish. It is easy to see that, quite generally, the curvature of $t(\theta)$ goes to zero as the images approach each other; in fact, the curvature varies as $\Delta\theta^{-1}$. Thus, we expect that the brightest image configurations are obtained when a pair of images are close together, just prior to merging. The lines in θ-space on which images merge are referred to as *critical lines*, while the corresponding source positions in β-space are called *caustics*. Critical lines and caustics are important because (i) they highlight regions of high magnification, and (ii) they demarcate regions of different image multiplicity. (The reader is referred to Blandford & Narayan 1986 and Erdl & Schneider 1992 for more details.)

5 When the source is far from the lens, we expect only a single image, corresponding to a minimum of the time-delay surface. New extrema are always created in pairs (e.g., Fig. 10.15). Therefore, the total number of extrema, and thus the number of images of a generic (non-singular) lens, is odd (Burke 1981).

10.3.4 Circularly symmetric lens models

Table 10.2 compiles formulae for the effective lensing potential and deflection angle of four commonly used circularly symmetric lens models; point mass, singular isothermal sphere, isothermal sphere with a softened core, and constant density sheet. In addition, one can have more general models with non-isothermal radial profiles, e.g., density varying as radius to a power other than -2.

The gravitational time-delay functions $t_{\text{grav}}(\theta) \propto -\psi(\theta)$ of the models in Table 10.2 are illustrated in Fig. 10.16. Note that the four potentials listed in Table 10.2 all are divergent for $\theta \to \infty$. (Although the three-dimensional potential of the point mass drops $\propto r^{-1}$, its projection along the line-of-sight diverges logarithmically.) The divergence is, however, not serious since images always occur at finite θ where the functions are well behaved.

The image configurations produced by a circularly symmetric lens are easily discovered by drawing time-delay functions $t(\theta)$ as in Fig. 10.15 corresponding to various offsets of the source with respect to the lens. Figures 10.17 and 10.18 show typical image configurations. The right halves of the figures display the source plane, and the

Table 10.2. *Examples of circularly symmetric lenses. The effective lensing potential $\psi(\theta)$ and the deflection angle $\alpha(\theta)$ are given. The core radius of the softened isothermal sphere is θ_c*

Lens model	$\psi(\theta)$	$\alpha(\theta)$
Point mass	$\dfrac{D_{ds}}{D_s}\dfrac{4GM}{D_d c^2}\ln|\theta|$	$\dfrac{D_{ds}}{D_s}\dfrac{4GM}{c^2 D_d |\theta|}$
Singular isothermal sphere	$\dfrac{D_{ds}}{D_s}\dfrac{4\pi\sigma^2}{c^2}|\theta|$	$\dfrac{D_{ds}}{D_s}\dfrac{4\pi\sigma^2}{c^2}$
Softened isothermal sphere	$\dfrac{D_{ds}}{D_s}\dfrac{4\pi\sigma^2}{c^2}\left(\theta_c^2+\theta^2\right)^{1/2}$	$\dfrac{D_{ds}}{D_s}\dfrac{4\pi\sigma^2}{c^2}\dfrac{\theta}{\left(\theta_c^2+\theta^2\right)^{1/2}}$
Constant density sheet	$\dfrac{\kappa}{2}\theta^2$	$\kappa|\theta|$

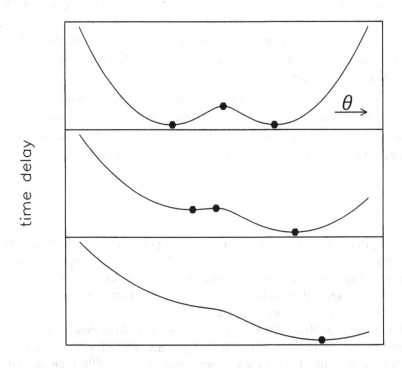

Figure 10.15. The time-delay function of a circularly symmetric lens for a source exactly behind the lens (*top panel*), a source offset from the lens by a moderate angle (*center panel*) and a source offset by a large angle (*bottom panel*).

Gravitational lensing

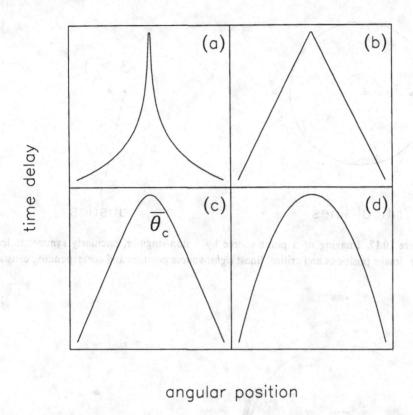

Figure 10.16. Gravitational time-delay functions for the four circularly symmetric effective potentials listed in Table 10.2. (a) Point mass; (b) singular isothermal sphere; (c) softened isothermal sphere with core radius θ_c; (d) constant density sheet.

left halves show the image configuration in the lens plane. Since $\mathscr{A}$ is a 2×2 matrix, a typical circularly symmetric lens has two critical lines where $\det \mathscr{A}$ vanishes, and two corresponding caustics in the source plane. The caustic of the inner critical curve is a circle while the caustic of the outer critical curve degenerates to a critical point because of the circular symmetry of the lens. A source which is located outside the outermost caustic has a single image. Upon each caustic crossing, the image number changes by two, indicated by the numbers in Fig. 10.17. The source shown as a small rectangle in the right panel of Fig. 10.17 has three images as indicated in the left panel. Of the three images, the innermost one is usually very faint; in fact, this image vanishes if the lens has a singular core (the curvature of the time-delay function then becomes infinite) as in the point mass or the singular isothermal sphere.

Figure 10.18 shows the images of two extended sources lensed by the same model as in Fig. 10.17. One source is located close to the point-like caustic in the center of the lens. It is imaged onto the two long, tangentially oriented arcs close to the outer critical curve and the very faint image at the lens center. The other source is located on the

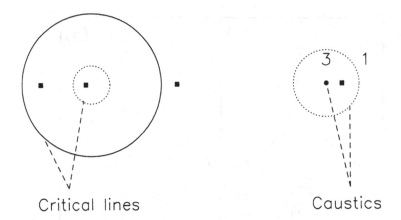

Figure 10.17. Imaging of a point source by a non-singular, circularly symmetric lens. *Left*: image positions and critical lines; *right*: source position and corresponding caustics.

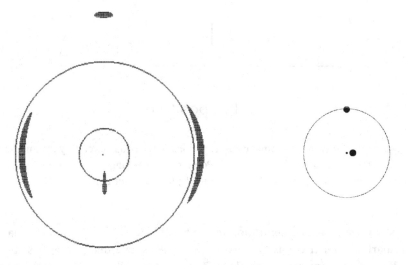

Figure 10.18. Imaging of an extended source by a non-singular circularly symmetric lens. A source close to the point caustic at the lens center produces two tangentially oriented arc-like images close to the outer critical curve, and a faint image at the lens center. A source on the outer caustic produces a radially elongated image on the inner critical curve, and a tangentially oriented image outside the outer critical curve. Because of these image properties, the outer and inner critical curves are called *tangential* and *radial*, respectively.

outer caustic and forms a radially elongated image which is composed of two merging images, and a third tangentially oriented image outside the outer caustic. Because of the image properties, the outer critical curve is called *tangential*, and the inner critical curve is called *radial*.

10.3.5 Non-circularly symmetric lens models

A circularly symmetric lens model is much too idealized and is unlikely to describe real galaxies. Therefore, considerable work has gone into developing non-circularly symmetric models. The breaking of the symmetry leads to qualitatively new image configurations (see Grossman & Narayan 1988; Narayan & Grossman 1989; Blandford et al. 1989).

10.3.5.1 Elliptical galaxy model

To describe an elliptical galaxy lens, we should ideally consider elliptical isodensity contours. A straightforward generalization of the isothermal sphere with finite core gives

$$\Sigma(\theta_1, \theta_2) = \frac{\Sigma_0}{\left[\theta_c^2 + (1-\epsilon)\theta_1^2 + (1+\epsilon)\theta_2^2\right]^{1/2}}, \qquad (10.65)$$

where θ_1, θ_2 are orthogonal coordinates along the major and minor axes of the lens measured from the center. The potential $\psi(\theta_1, \theta_2)$ corresponding to this density distribution has been calculated by Kassiola & Kovner (1993) but is somewhat complicated. For the specific case when the core radius θ_c vanishes, the deflection angle and the magnification take on a simple form,

$$\alpha_1 = \frac{8\pi G \Sigma_0}{\sqrt{2\epsilon}c^2} \tan^{-1}\left[\frac{\sqrt{2\epsilon}\cos\phi}{(1-\epsilon\cos 2\phi)^{1/2}}\right],$$

$$\alpha_2 = \frac{8\pi G \Sigma_0}{\sqrt{2\epsilon}c^2} \tanh^{-1}\left[\frac{\sqrt{2\epsilon}\sin\phi}{(1-\epsilon\cos 2\phi)^{1/2}}\right],$$

$$\mu^{-1} = 1 - \frac{8\pi G \Sigma_0}{c^2(\theta_1^2 + \theta_2^2)^{1/2}(1-\epsilon\cos 2\phi)^{1/2}}, \qquad (10.66)$$

where ϕ is the polar angle corresponding to the vector position $\theta \equiv (\theta_1, \theta_2)$.

Instead of the elliptical density model, it is simpler and often sufficient to model a galaxy by means of an elliptical effective lensing potential (Blandford & Kochanek 1987)

$$\psi(\theta_1, \theta_2) = \frac{D_{ds}}{D_s} 4\pi \frac{\sigma_v^2}{c^2} \left[\theta_c^2 + (1-\epsilon)\theta_1^2 + (1+\epsilon)\theta_2^2\right]^{1/2}, \qquad (10.67)$$

where ϵ measures the ellipticity. The deflection law and magnification tensor corresponding to this potential are easily calculated using the equations given in §10.3.2. When ϵ is large, the elliptical potential model is inaccurate because it gives rise to dumbbell-shaped isodensity contours, but for small ϵ, it is a perfectly viable lens model.

10.3.5.2 External shear

The environment of a galaxy, including any cluster surrounding the primary lens, will in general contribute both convergence and shear. The effective potential due to the local environment then reads

$$\psi(\theta_1, \theta_2) = \frac{\kappa}{2}(\theta_1^2 + \theta_2^2) + \frac{\gamma}{2}(\theta_1^2 - \theta_2^2) \tag{10.68}$$

in the principal axes system of the external shear, where the convergence κ and shear γ are locally independent of θ. An external shear breaks the circular symmetry of a lens and therefore it often has the same effect as introducing ellipticity in the lens (Kovner 1987). It is frequently possible to model the same system either with an elliptical potential or with a circular potential plus an external shear.

10.3.5.3 Image configurations with a non-circularly symmetric lens

In contrast to the circularly symmetric case, for a non-circular lens the source, lens and images are not restricted to lie on a line. Therefore, it is not possible to analyze the problem via sections of the time-delay surface as we did in Figs. 10.14 and 10.15. Fermat's Principle and the time-delay function are still very useful but it is necessary to visualize the full two-dimensional surface $t(\theta)$ (Blandford & Narayan 1986). Those who attended the lectures in Jerusalem (in January 1996) may recall the lecturer demonstrating many of the qualitative features of imaging by elliptical lenses using a Mexican hat to simulate the time-delay surface. In the following, we merely state the results.

Figures 10.19 and 10.20 illustrate the wide variety of image configurations produced by an elliptical galaxy lens (or a circularly symmetric lens with external shear). In each panel, the source plane with caustics is shown on the right, and the image configurations together with the critical curves are shown on the left. Compared to the circularly symmetric case, the first notable difference introduced by ellipticity is that the central caustic which was point-like is now expanded into a diamond shape; it is referred to as the *astroid* caustic (also tangential caustic). Figure 10.19 shows the images of a compact source moving away from the lens center along a symmetry line (right panel) and a line bisecting the two symmetry directions (left panel). A source behind the center of the lens has five images because it is enclosed by two caustics. One image appears at the lens center, and the four others form a cross-shaped pattern. When the source is moved outward, two of the four outer images move toward each other, merge, and disappear as the source approaches and then crosses the astroid (or tangential) caustic. Three images remain until the source crosses the radial caustic, when two more images merge and disappear at the radial critical curve. A single weakly distorted image is finally left when the source has crossed the outer caustic. When the source moves toward a cusp point (right panel of Fig. 10.19), three images merge to form a single image. All the image configurations shown in Fig. 10.19 are exhibited by various observed cases of lensing of QSOs and radio quasars (e.g., Keeton & Kochanek 1996).

Figure 10.20 illustrates what happens when a source with a larger angular size is imaged by the same lens model as in Fig. 10.19. Large arc-like images form which

Gravitational lensing

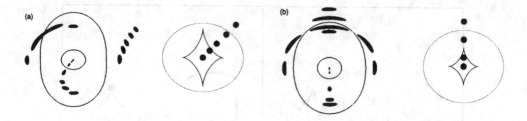

Figure 10.19. Compact source moving away from the center of an elliptical lens. *Left panel:* source crossing a fold caustic; *right panel:* source crossing a cusp caustic. Within each panel, the diagram on the left shows critical lines and image positions and the diagram on the right shows caustics and source positions.

consist either of three or two merging images, depending on whether the source lies on top of a cusp in the tangential caustic (top left panel) or on an inter-cusp segment (a so-called fold caustic, top right panel). If the source is even larger (bottom panels), four images can merge, giving rise to complete or incomplete rings. Radio rings such as MG 1131+0456 (Hewitt *et al.* 1987) correspond to the configuration shown at bottom right in Fig. 10.20.

10.3.6 Studies of galaxy lensing

10.3.6.1 Detailed models of individual cases of lensing

Gravitational lens observations provide a number of constraints which can be used to model the mass distribution of the lens. The angular separation between the images determines the Einstein radius of the lens and therefore gives the mass M (equation (10.22)) or the velocity dispersion σ_v (equation (10.45)) in simple models. The appearance or absence of the central image constrains the core size of the lens. The number of images and their positions relative to the lens determine the ellipticity of the galaxy, or equivalently the magnitude and orientation of an external shear. Since the radial and tangential magnifications of images reflect the local curvatures of the time-delay surface in the corresponding directions, the relative image sizes constrain the slope of the density profile of the lens. This does not work very well if all one has are multiply imaged point images (Kochanek 1991). However, if the images have radio structure which can be resolved with VLBI, matters improve considerably.

Figure 10.21 shows an extended, irregularly shaped source which is mapped into two images which are each linear transformations of the unobservable source. The two transformations are described by symmetric 2×2 magnification matrices $\mathcal{M}_1$ and $\mathcal{M}_2$ (cf. equation (10.53)). These matrices cannot be determined from observations since the original source is not seen. However, the two images are related to each other by a linear transformation described by the relative magnification matrix $\mathcal{M}_{12} = \mathcal{M}_1^{-1}\mathcal{M}_2$ which can be measured via VLBI observations (Gorenstein *et al.* 1988; Falco, Gorenstein

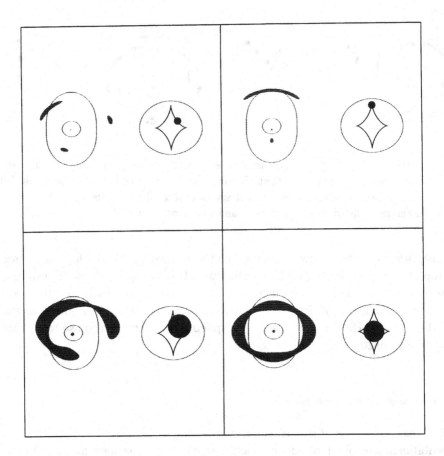

Figure 10.20. Images of resolved sources produced by an elliptical lens. *Top panels:* Large arcs consisting of two or three merging images are formed when the source lies on top of a fold section (*top left panel*) or a cusp point (*top right panel*) of the tangential caustic. *Bottom panels:* A source which covers most of the diamond-shaped caustic produces a ring-like image consisting of four merging images.

& Shapiro 1991). The matrix $\mathcal{M}_{12}$ is in general not symmetric and thus contains four independent components, which are each functions of the parameters of the lens model. In favorable cases, as in QSO 0957+561, it is even possible to measure the spatial gradient of $\mathcal{M}_{12}$ (Garrett *et al.* 1994) which provides additional constraints on the model.

Radio rings with hundreds of independent pixels are particularly good for constraining the lens model. As shown in the bottom panels of Fig. 10.20, ring-shaped images are formed from extended sources which cover a large fraction of the central diamond-shaped caustic. Rings provide large numbers of independent observational constraints and are, in principle, capable of providing the most accurate mass reconstructions

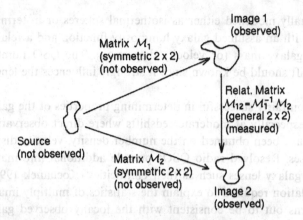

Figure 10.21. Shows an extended source which is mapped into two resolved images. While the source and the individual magnification matrices $\mathcal{M}_1$ and $\mathcal{M}_2$ are not observable, the relative magnification matrix $\mathcal{M}_{12} = \mathcal{M}_1^{-1}\mathcal{M}_2$ can be measured. This matrix provides four independent constraints on the lens model.

of the lens. However, special techniques are needed for analyzing rings. Three such techniques have been developed and applied to radio rings, viz.

1. The Ring Cycle algorithm (Kochanek et al. 1989) makes use of the fact that lensing conserves surface brightness. Surface elements of an extended image which arise from the same source element should therefore share the same surface brightness (to within observational errors). This provides a large number of constraints which can be used to reconstruct the shape of the original source and at the same time optimize a parameterized lens model.
2. The LensClean technique (Kochanek & Narayan 1992) is a generalization of the Ring Cycle algorithm which uses the Clean algorithm to allow for the finite beam of the radio telescope.
3. LensMEM (Wallington, Narayan & Kochanek 1994; Wallington, Kochanek & Narayan 1996) is analogous to LensClean, but uses the Maximum Entropy Method instead of Clean.

10.3.6.2 Statistical modelling of lens populations

The statistics of lensed QSOs can be used to infer statistical properties of the lens population. In this approach, parameterized models of the galaxy and QSO populations in the universe are used to predict the number of lensed QSOs expected to be observed in a given QSO sample and to model the distributions of various observables such as the image separation, flux ratio, lens redshift, source redshift, etc. An important aspect of such studies is the detailed modelling of selection effects in QSO surveys (Kochanek 1993a) and proper allowance for magnification bias (Narayan & Wallington 1993).

The lensing galaxies are usually modelled either as isothermal spheres or in terms of simple elliptical potentials, with an assumed galaxy luminosity function and a relation connecting luminosity and galaxy mass (or velocity dispersion). The QSO number-count as a function of redshift should be known since it strongly influences the lensing probability.

Statistical studies have been fairly successful in determining properties of the galaxy population in the universe, especially at moderate redshifts where direct observations are difficult. Useful results have been obtained on the number density, velocity dispersions, core radii, etc. of lenses. Resolved radio QSOs provide additional information on the internal structure of galaxy lenses such as their ellipticities (Kochanek 1996b). By and large, the lens population required to explain the statistics of multiply imaged optical and radio QSOs turns out to be consistent with the locally observed galaxy population extrapolated to higher redshifts (Kochanek 1993b; Maoz & Rix 1993; Surdej et al. 1993; see below).

So far, statistical studies of galaxy lensing have neglected the contribution from spirals because their velocity dispersions are significantly lower than those of ellipticals. However, most of the lenses found by the CLASS survey (Myers et al. 1995) are classified as S0 or spiral galaxies. This result has recently triggered investigations of lens models that contain disks in addition to halos. While realistic disks increase the multiple-image cross sections of halo-only models only by $\sim 10\%$ (Wang & Turner 1997; Keeton & Kochanek 1998), they allow for much more convincing models of lens systems such as B 1600, where a nearly edge-on disk is observed (Maller, Flores & Primack 1997).

10.3.7 Astrophysical results from galaxy lensing

10.3.7.1 Galaxy structure

The structure of galaxies influences lensing statistics as well as the appearances of individual lensed objects. Gravitational lensing can therefore be used to constrain galaxy models in various ways.

As described earlier, galaxy lens models predict a weak central image whose flux depends on the core radius of the galaxy. The central image is missing in virtually every known multiply-imaged quasar. The lensing galaxies in these cases must therefore have very small core radii, $r_c < 200$ pc (Wallington & Narayan 1993; Kassiola & Kovner 1993; Grogin & Narayan 1996).

Kochanek (1993b) has shown that the observed distribution of image separations in the observed lens sample of quasars requires that most galaxies must have dark halos with characteristic velocity dispersions of $\sigma^*_{\text{dark}} \sim 220 \pm 20$ km s^{-1}. If these dark halos were absent, virtually no image separations larger than 2" would be produced (Maoz & Rix 1993), whereas several wide separation examples are known. Multiply-imaged quasars do not generally constrain the size of the halo because the constraints only extend out to about the Einstein radius, which is ~ 10 kpc at the distance of the lens.

The largest halo inferred from direct modelling of a multiply-imaged quasar is in the lensing galaxy of QSO 0957+561; the halo of this galaxy has been shown to have a radius of at least $15h^{-1}$ kpc, where $h = H_0/100$ km s^{-1} Mpc^{-1} is the reduced Hubble constant (Grogin & Narayan 1996). Brainerd, Blandford & Smail (1996) investigated weak lensing of background galaxies by foreground galaxies and found statistical evidence for halos extending out to radii ~ 100 kpc. At these radii, they determined that an L_* galaxy must have a mass $\sim 10^{12} M_\odot$. Comparable results were obtained by Dell'Antonio & Tyson (1996) and Griffiths et al. (1996). Natarajan & Kneib (1996) show that the sizes of halos around galaxies in clusters could be inferred by measuring the weak lensing effect of these galaxies on background sources.

Only in two cases has it been possible to constrain significantly the radial mass-density variation of the lensing galaxy. Assuming a surface-mass density profile $\Sigma \propto r^{-\alpha}$, the best fitting values of α in these two examples are

$$\alpha = \begin{cases} (0.9 - 1.1) & \text{in MG 1654+134} \quad \text{(Kochanek 1995a)} \\ (1.0 - 1.2) & \text{in QSO 0957+561} \quad \text{(Grogin & Narayan 1996)}. \end{cases} \quad (10.69)$$

Both sources have particularly good data – the first is a radio ring and the second has extensive VLBI observations – and it is this feature that allows a good constraint on α. Note that the density variation is close to isothermal in both cases. Recent observations of QSO 0957+561 with the Hubble Space Telescope (Bernstein et al. 1997) show that the lensing galaxy is shifted by 45 mas from the position assumed by Grogin & Narayan in their model. This will modify the estimate of α for this galaxy, but perhaps by only a fraction of the stated uncertainty.

The observed morphologies of images in lensed quasars are similar to those shown in Fig. 10.19, which means that most lenses are not circularly symmetric. If the non-circularity is entirely due to the ellipticity of the galaxy mass, then typical ellipticities are fairly large, $\sim$ E3–E4 (Kochanek 1996b). However, it is possible that part of the effect comes from external shear. The data are currently not able to distinguish very well between the effects of galaxy ellipticity and external shear. In many well-modelled examples, the mass ellipticity required to fit the images is larger than the ellipticity of the galaxy isophotes, suggesting either that the dark matter is more asymmetric than the luminous matter or that there is a significant contribution from external shear (Kochanek 1996b; Bar-Kana 1996). Keeton, Kochanek & Seljak (1997) find the external cosmic shear insufficient to explain fully the discrepancy between the ellipticity of the galaxy isophotes and the ellipticity of the mass required by lens models. This implies that at least part of the inferred ellipticity of the mass distribution is intrinsic to the galaxies.

10.3.7.2 Galaxy formation and evolution

The angular separations of multiple images depend on the lens mass, and the number of observed multiply imaged quasars with a given separation depends on the number density of galaxies with the corresponding mass. The usual procedure to set limits on the galaxy population starts with the present galaxy population and extrapolates it to higher redshifts assuming some parameterized prescription of evolution. The parameters are then constrained by comparing the observed statistics of lensed sources to that predicted by the model (Kochanek 1993b; Maoz & Rix 1993; Rix et al. 1994; Mao & Kochanek 1994).

If galaxies formed recently, most of the optical depth for multiple imaging will be from low-redshift galaxies. An analysis which uses all the known information on lensed quasars, such as the redshifts of lenses and sources, the observed fraction of lensed quasars, and the distribution of image separations, can be used to set limits on how recently galaxies could have formed. Mao & Kochanek (1994) conclude that most galaxies must have collapsed and formed by $z \sim 0.8$ if the universe is well described by the Einstein–de Sitter model.

If elliptical galaxies are assembled from merging spiral galaxies, then with increasing redshift the present population of ellipticals is gradually replaced by spirals. This does not affect the probability of producing lensed quasars as the increase in the number of lens galaxies at high redshift is compensated by the reduced lensing cross-sections of these galaxies. However, because of their lower velocity dispersion, spirals produce smaller image separations than ellipticals (the image splitting is proportional to σ_v^2, cf. equation (10.45)). Therefore, a merger scenario will predict smaller image separations in high redshift quasars, and the observed image separations can be used to constrain the merger rate (Rix et al. 1994). Assuming that the mass of the galaxies scales with σ_v^4 and is conserved in mergers, Mao & Kochanek (1994) find that no significant mergers could have occurred more recently than $z \sim 0.4$ in an Einstein–de Sitter universe.

If the cosmological constant is large, say $\Lambda_0 > 0.6$, the volume per unit redshift is much larger than in an Einstein–de Sitter universe. For a fixed number density of galaxies, the total number of available lenses then increases steeply (Turner 1990). For such model universes, lens statistics would be consistent with recent rapid evolution of the galaxy population. However, studies of gravitational clustering and structure formation show that galaxies form at high redshifts precisely when Λ is large. When this additional constraint is included it is found that there is no scenario which allows recent galaxy formation or evolution in the universe (see also §10.3.7.5).

Since lensing statistics are fully consistent with the known local galaxy population extrapolated to redshifts $z \sim 1$, the number densities of any dark "failed" galaxies are constrained quite strongly. As a function of velocity dispersion σ_*, the current constraints are (Mao & Kochanek 1994)

$$n_{\text{dark}} < \begin{cases} 0.15\, h^3\, \text{Mpc}^{-3} & \text{for} \quad \sigma_* = 100 \text{ km s}^{-1} \\ 0.032\, h^3\, \text{Mpc}^{-3} & \text{for} \quad \sigma_* = 150 \text{ km s}^{-1} \\ 0.017\, h^3\, \text{Mpc}^{-3} & \text{for} \quad \sigma_* = 200 \text{ km s}^{-1}. \end{cases} \quad (10.70)$$

10.3.7.3 Constraint on CDM

The popular cold dark-matter (CDM) scenario of structure formation in its "standard" variant ($\Omega_0 = 1$, $\Lambda_0 = 0$ and COBE normalized) predicts the formation of large numbers of dark-matter halos in the mass range between galaxies and galaxy clusters.

The implications of these halos for lensing were considered by Narayan & White (1988) and more recently by Cen et al. (1994); Wambsganss et al. (1995); Wambsganss, Cen & Ostriker (1998); and Kochanek (1995b). The latter authors have shown quite convincingly that the standard CDM model produces many more wide-separation quasar pairs than observed. For example, a recent search of a subsample of the HST snapshot survey for multiply imaged QSOs with image separations between 7" and 50" found a null result (Maoz et al. 1997). To save CDM, either the normalization of the model needs to be reduced to $\sigma_8 \sim 0.5 \pm 0.2$, or the long-wavelength slope of the power spectrum needs to be lowered to $n \sim 0.5 \pm 0.2$. Both of these options are inconsistent with the COBE results. The problem of the over-production of wide angle pairs is just a manifestation of the well-known problem that standard COBE-normalized CDM over-produces cluster-scale mass condensations by a large factor. Models which are adjusted to fit the observed number density of clusters also satisfy the gravitational lens constraint.

If the dark halos have large core radii, their central density could drop below the critical value for lensing and this would reduce the predicted number of wide-separation lens systems. Large core radii thus may save standard CDM (Flores & Primack 1996), but there is some danger of fine-tuning in such an explanation. As discussed in §10.3.7.1, galaxy cores are quite small. Therefore, one needs to invoke a rather abrupt increase of core radius with increasing halo mass.

10.3.7.4 Hubble constant

The lens equation is dimensionless, and the positions of images as well as their magnifications are dimensionless numbers. Therefore, information on the image configuration alone does not provide any constraint on the overall scale of the lens geometry or the value of the Hubble constant. Refsdal (1964) realized that the time delay, however, is proportional to the absolute scale of the system and does depend on H_0 (cf. Fig. 10.22).

To see this, we first note that the geometrical time delay is simply proportional to the path lengths of the rays which scale as H_0^{-1}. The potential time delay also scales as H_0^{-1} because the linear size of the lens and its mass have this scaling. Therefore, for any gravitational lens system, the quantity

$$H_0 \Delta \tau \qquad (10.71)$$

depends only on the lens model and the geometry of the system. A good lens model which reproduces the positions and magnifications of the images provides the scaled time delay $H_0 \Delta \tau$ between the images. Therefore, a measurement of the time delay $\Delta \tau$ will yield the Hubble constant H_0 (Refsdal 1964, 1966a).

To measure the time delay, the fluxes of the images need to be monitored over a period of time significantly longer than the time delay in order to achieve reasonable accuracy. In fact, the analysis of the resulting light curves is not straightforward because of uneven data sampling, and careful and sophisticated data analysis techniques have to be applied. QSO 0957+561 has been monitored both in the optical (Vanderriest et al. 1989; Schild & Thomson 1993; Kundić et al. 1996) and radio wavebands (Lehár

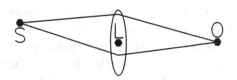

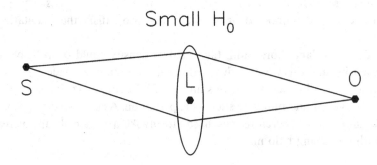

Figure 10.22. Sketch of the dependence of the overall scale of a lens system on the value of the Hubble constant.

et al. 1992; Haarsma et al. 1996, 1997). Unfortunately, analysis of the data has led to two claimed time delays:

$$\Delta\tau = (1.48 \pm 0.03) \text{ yr} \tag{10.72}$$

(Press, Rybicki & Hewitt 1992a,b) and

$$\Delta\tau \simeq 1.14 \text{ yr} \tag{10.73}$$

(Schild & Thomson 1993; Pelt et al. 1994, 1996). The discrepancy appears to have been resolved in favor of the shorter delay. Haarsma et al. (1996) find $\Delta\tau = 1.03$–1.33 yr, and Kundić et al. (1996) derive $\Delta\tau = 417 \pm 3$ d using a variety of statistical techniques.

In addition to a measurement of the time delay, it is also necessary to develop a reliable model to calculate the value of $H_0 \Delta\tau$. QSO 0957+561 has been studied by a number of groups over the years, with recent work incorporating constraints from VLBI imaging (Falco, Gorenstein & Shapiro 1991). Grogin & Narayan (1996) estimate the Hubble constant to be given by

$$H_0 = (82 \pm 6)(1 - \kappa) \left(\frac{\Delta\tau}{1.14 \text{ yr}}\right)^{-1} \text{ km s}^{-1} \text{ Mpc}^{-1} \tag{10.74}$$

where κ refers to the unknown convergence due to the cluster surrounding the lensing galaxy. Since the cluster κ cannot be negative, this result directly gives an upper bound on the Hubble constant ($H_0 < 88$ km s^{-1} Mpc^{-1} for $\Delta\tau = 1.14$ yr). Actually, κ can also be modified by large-scale structure along the line-of-sight. In contrast to the effect of the cluster, this fluctuation can have either sign, but the rms amplitude is estimated to be only a few percent (Seljak 1994; Bar-Kana 1996). Surpi, Harari & Frieman (1996) confirm that large-scale structure does not modify the functional relationship between lens observables, and therefore does not affect the determination of H_0.

To obtain an actual value of H_0 instead of just an upper bound, we need an independent estimate of κ. Studies of weak lensing by the cluster (Fischer et al. 1997) give $\kappa = 0.24 \pm 0.12$ (2σ) at the location of the lens (cf. Kundić et al. 1996). This corresponds to $H_0 = 62^{+12}_{-13}$ km s^{-1} Mpc^{-1}. Another technique is to measure the velocity dispersion σ_{gal} of the lensing galaxy, from which it is possible to estimate κ (Falco, Gorenstein & Shapiro 1991; Grogin & Narayan 1996). Falco et al. (1997) used the Keck telescope to measure $\sigma_{\text{gal}} = 279 \pm 12$ km s^{-1}, which corresponds to $H_0 = 66 \pm 7$ km s^{-1} Mpc^{-1}. Although most models of QSO 0957+561 are based on a spherically symmetric galaxy embedded in an external shear (mostly due to the cluster), introduction of ellipticity in the galaxy, or a point mass at the galaxy core, or substructure in the cluster seem to have little effect on the estimate of H_0 (Grogin & Narayan 1996).

A measurement of the time delay has also been attempted in the Einstein ring system B 0218+ 357. In this case, a single galaxy is responsible for the small image splitting of 0."3. The time delay has been determined to be 12 ± 3 d (1σ confidence limit) which translates to $H_0 \sim 60$ km s^{-1} Mpc^{-1} (Corbett et al. 1996).

Schechter et al. (1997) recently announced a time delay of $\Delta\tau = 23.7 \pm 3.4$ d between images B and C of the quadruple lens PG 1115+080. Using a different statistical technique, Bar-Kana (1997) finds $\Delta\tau = 25.0^{+3.3}_{-3.8}$ d (95% confidence) from the same data. Their best fitting lens model, where the lens galaxy as well as an associated group of galaxies are modelled as singular isothermal spheres, gives $H_0 = 42 \pm 6$ km s^{-1} Mpc^{-1}. Other models give larger values of H_0 but fit the data less well. Keeton & Kochanek (1997) have considered a more general class of models where the lensing galaxy is permitted to be elliptical, and present a family of models which fit the PG 1115+080 data well. They estimate $H_0 = 60 \pm 17$ km s^{-1} Mpc^{-1}. With more accurate data on the position of the lens galaxy, this estimate could be improved to $H_0 = 53^{+10}_{-7}$ km s^{-1} Mpc^{-1} (Courbin et al. 1997).

The determination of H_0 through gravitational lensing has a number of advantages over other techniques.

1. The method works directly with sources at large redshifts, $z \sim 0.5$, whereas most other methods are local (observations within ~ 100 Mpc) where peculiar velocities are still comparable to the Hubble flow.
2. While other determinations of the Hubble constant rely on distance ladders which progressively reach out to increasing distances, the measurement via

gravitational time delay is a one-shot procedure. One measures directly the geometrical scale of the lens system. This means that the lens-based method is absolutely independent of every other method and at the very least provides a valuable test of other determinations.

3 The lens-based method is based on fundamental physics (the theory of light propagation in General Relativity), which is fully tested in the relevant weak-field limit of gravity. Other methods rely on models for variable stars (Cepheids) or supernova explosions (Type II), or empirical calibrations of standard candles (Tully–Fisher distances, Type I supernovae). The lensing method does require some information on the "shapes" of galaxies which is used to guide the choice of a parameterized lens model.

10.3.7.5 Cosmological constant

A large cosmological constant Λ_0 increases the volume per unit redshift of the universe at high redshift. As Turner (1990) realized, this means that the relative number of lensed sources for a fixed comoving number density of galaxies increases rapidly with increasing Λ_0. Turning this around it is possible to use the observed probability of lensing to constrain Λ_0. This method has been applied by various authors (Turner 1990; Fukugita & Turner 1991; Fukugita *et al.* 1992; Maoz & Rix 1993; Kochanek 1996a), and the current limit is $\Lambda_0 < 0.65$ (2σ confidence limit) for a universe with $\Omega_0 + \Lambda_0 = 1$. With a combined sample of optical and radio lenses, this limit is slightly improved to $\Lambda_0 < 0.62$ (2σ; Falco, Kochanek & Muñoz 1998). Malhotra, Rhoads & Turner (1996) claim that there is evidence for considerable amounts of dust in lensing galaxies. They argue that the absorption in dusty lenses can reconcile a large cosmological constant with the observed multiple-image statistics.

A completely independent approach (Kochanek 1992) considers the redshift distribution of lenses. For a given source redshift, the probability distribution of z_d peaks at higher redshift with increasing Λ_0. Once again, by comparing the observations against the predicted distributions one obtains an upper limit on Λ_0. This method is less sensitive than the first, but gives consistent results.

Another technique consists in comparing the observed QSO image separations to those expected from the redshifts of lenses and sources and the magnitudes of the lenses, assuming certain values for Ω_0 and Λ_0. The cosmological parameters are then varied to optimize the agreement with the observations. Applying this approach to a sample of seven lens systems, Im, Griffiths & Ratnatunga (1997) find $\Lambda_0 = 0.64^{+0.15}_{-0.26}$ (1σ confidence limit) assuming $\Omega_0 + \Lambda_0 = 1$.

10.4 Lensing by galaxy clusters and large-scale structure

Two distinct types of lensing phenomena are observed with clusters of galaxies (Fig. 10.23):

Gravitational lensing

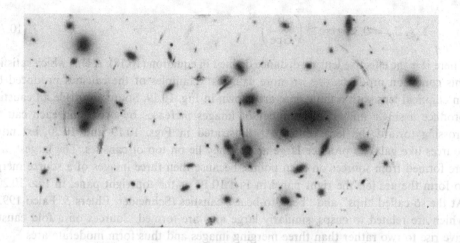

Figure 10.23. *Hubble Space Telescope* image of the cluster Abell 2218, showing a number of arcs and arclets around the two centers of the cluster. (NASA HST Archive).

1. Rich centrally condensed clusters occasionally produce giant arcs when a background galaxy happens to be aligned with one of the cluster caustics. These instances of lensing are usually analyzed with techniques similar to those described in §10.2 for galaxy lenses. In brief, a parameterized lens model is optimized so as to obtain a good fit to the observed image.

2. Every cluster produces weakly distorted images of large numbers of background galaxies. These images are called arclets and the phenomenon is referred to as weak lensing. With the development of the Kaiser & Squires (1993) algorithm and its variants, weak lensing is being used increasingly to derive parameter-free two-dimensional mass maps of lensing clusters.

In addition to these two topics, we also discuss in this section weak lensing by large-scale structure in the universe. This topic promises to develop into an important branch of gravitational lensing, and could in principle provide a direct measurement of the primordial power spectrum $P(k)$ of the density fluctuations in the universe.

10.4.1 Strong lensing by clusters – giant arcs

10.4.1.1 Basic optics
We begin by summarizing a few features of generic lenses which we have already discussed in the previous sections. A lens is fully characterized by its surface-mass density $\Sigma(\theta)$. Strong lensing, which is accompanied by multiple imaging, requires that the surface-mass density somewhere in the lens should be larger than the critical surface-mass density,

$$\Sigma \gtrsim \Sigma_{\mathrm{cr}} = 0.35 \, \mathrm{g\,cm^{-3}} \left(\frac{D}{\mathrm{Gpc}}\right)^{-1}, \qquad (10.75)$$

where D is the effective lensing distance defined in equation (10.18). A lens which satisfies this condition produces one or more caustics. Examples of the caustics produced by an elliptical lens with a finite core are shown in Fig. 10.19. Sources outside all caustics produce a single image; the number of images increases by two upon each caustic crossing toward the lens center. As illustrated in Figs. 10.19 and 10.20, extended sources like galaxies produce large arcs if they lie on top of caustics. The largest arcs are formed from sources on cusp points, because then three images of a source merge to form the arc (cf. the right panel in Fig. 10.19 or the top right panel in Fig. 10.20). At the so-called "lips" and "beak-to-beak" caustics (Schneider, Ehlers & Falco 1992), which are related to cusps, similarly large arcs are formed. Sources on a fold caustic give rise to two rather than three merging images and thus form moderate arcs.

10.4.1.2 Cluster mass inside a giant arc

The location of an arc in a cluster provides a simple way to estimate the projected cluster mass within a circle traced by the arc (cf. Fig. 10.24). For a circularly symmetric lens, the average surface-mass density $\langle \Sigma \rangle$ within the tangential critical curve equals the critical surface-mass density Σ_{cr}. Tangentially oriented large arcs occur approximately at the tangential critical curves. The radius θ_{arc} of the circle traced by the arc therefore gives an estimate of the Einstein radius θ_E of the cluster.

Thus we have

$$\langle \Sigma(\theta_{\mathrm{arc}}) \rangle \approx \langle \Sigma(\theta_E) \rangle = \Sigma_{\mathrm{cr}}, \qquad (10.76)$$

and we obtain for the mass enclosed by $\theta = \theta_{\mathrm{arc}}$

$$M(\theta) = \Sigma_{\mathrm{cr}} \pi (D_\mathrm{d} \theta)^2 \approx 1.1 \times 10^{14} \, M_\odot \left(\frac{\theta}{30''}\right)^2 \left(\frac{D}{1\,\mathrm{Gpc}}\right). \qquad (10.77)$$

Assuming an isothermal model for the mass distribution in the cluster and using equation (10.45), we obtain an estimate for the velocity dispersion of the cluster,

$$\sigma_v \approx 10^3 \, \mathrm{km\,s^{-1}} \left(\frac{\theta}{28''}\right)^{1/2} \left(\frac{D_\mathrm{s}}{D_\mathrm{ds}}\right)^{1/2}. \qquad (10.78)$$

In addition to the lensing technique, two other methods are available to obtain the mass of a cluster: the observed velocity dispersion of the cluster galaxies can be combined with the virial theorem to obtain one estimate, and observations of the X-ray gas combined with the condition of hydrostatic equilibrium provides another. These three quite independent techniques yield masses which agree with one another to within a factor $\sim$ 2–3.

The mass estimate (10.77) is based on very simple assumptions. It can be improved by modelling the arcs with parameterized lens mass distributions and carrying out more detailed fits of the observed arcs. We list in Table 10.3 masses, mass-to-blue-light

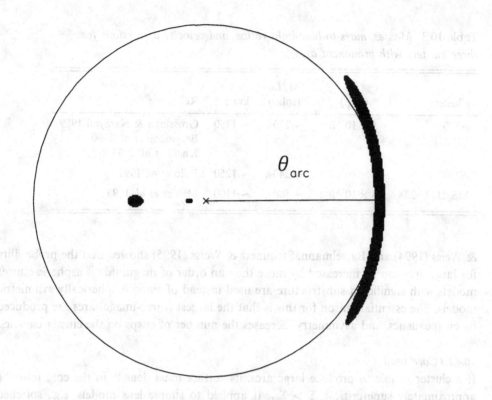

Figure 10.24. Tangential arcs constrain the cluster mass within a circle traced by the arcs.

ratios, and velocity dispersions of three clusters with prominent arcs. Additional results can be found in the review article by Fort & Mellier (1994).

10.4.1.3 Asphericity of cluster mass

The fact that the observed giant arcs never have counter-arcs of comparable brightness, and rarely have even small counter-arcs, implies that the lensing geometry has to be non-spherical (Grossman & Narayan 1988; Kovner 1989; see also Figs. 10.19 and 10.20). Cluster potentials therefore must have substantial quadrupole and perhaps also higher multipole moments. In the case of A370, for example, there are two cD galaxies, and the potential quadrupole estimated from their separation is consistent with the quadrupole required to model the observed giant arc (Grossman & Narayan 1989). The more detailed model of A370 by Kneib *et al.* (1993) shows a remarkable agreement between the lensing potential and the strongly aspheric X-ray emission of the cluster.

Large deviations of the lensing potentials from spherical symmetry also help increase the probability of producing large arcs. Bergmann & Petrosian (1993) argued that the apparent abundance of large arcs relative to small arcs and arclets can be reconciled with theoretical expectations if aspheric lens models are taken into account. Bartelmann

Table 10.3. *Masses, mass-to-blue-light ratios, and velocity dispersions for three clusters with prominent arcs*

Cluster	M ($M_\odot$)	M/L_B (solar)	σ km s^{-1}	Reference
A370	$\sim 5 \times 10^{13} h^{-1}$	$\sim 270 h$	~ 1350	Grossman & Narayan 1989
				Bergmann et al. 1990
				Kneib et al. 1993
A2390	$\sim 8 \times 10^{13} h^{-1}$	$\sim 240 h$	~ 1250	Pelló et al. 1991
MS 2137–23	$\sim 3 \times 10^{13} h^{-1}$	$\sim 500 h$	~ 1100	Mellier et al. 1993

& Weiss (1994) and Bartelmann, Steinmetz & Weiss (1995) showed that the probability for large arcs can be increased by more than an order of magnitude if aspheric cluster models with significant substructure are used instead of smooth spherically symmetric models. The essential reason for this is that the largest (three-image) arcs are produced by cusp caustics, and asymmetry increases the number of cusps on the cluster caustics.

10.4.1.4 Core radii
If a cluster is able to produce large arcs, its surface-mass density in the core must be approximately supercritical, $\Sigma \gtrsim \Sigma_{cr}$. If applied to simple lens models, e.g., softened isothermal spheres, this condition requires

$$\theta_{core} \lesssim 15'' \left(\frac{\sigma_v}{10^3 \text{ km s}^{-1}}\right)^2 \left(\frac{D_{ds}}{D_s}\right). \tag{10.79}$$

Narayan, Blandford & Nityananda (1984) argued that cluster mass distributions need to have smaller core radii than those derived from optical and X-ray observations if they are to produce strong gravitational lens effects. This has been confirmed by many later efforts to model giant arcs. In virtually every case the core radius estimated from lensing is significantly smaller than the estimates from optical and X-ray data. Some representative results on lens-derived core radii are listed in Table 10.4, where the estimates correspond to $H_0 = 50$ km s^{-1} Mpc^{-1}.

Statistical analyses based on spherically symmetric cluster models lead to similar conclusions. Miralda-Escudé (1992, 1993) argued that cluster core radii can hardly be larger than the curvature radii of large arcs. Wu & Hammer (1993) claimed that clusters either have to have singular cores or density profiles much steeper than isothermal in order to reproduce the abundance of large arcs. Although this conclusion can substantially be altered once deviations from spherical symmetry are taken into account (Bartelmanni, Steinmetz & Weiss 1995), it remains true that we require $r_{core} \lesssim 100$ kpc in all observed clusters. Cores of this size can also be reconciled with large-arc statistics.

Interestingly, there are at least two observations which seem to indicate that cluster cores, although small, must be finite. Fort et al. (1992) discovered a radial arc near

Table 10.4. *Limits on cluster core radii from models of large arcs*

Cluster	r_{core} (kpc)	Reference
A370	< 60	Grossman & Narayan (1989)
	< 100	Kneib et al. (1993)
MS 2137−23	∼ 50	Mellier et al. (1993)
Cl0024+1654	< 130	Bonnet et al. (1994)
MS 0440+0204	≪ 90	Luppino et al. (1993)

the center of MS 2137−23, and Smail et al. (1996) found a radial arc in A370. To produce a radial arc with a softened isothermal sphere model, the core radius has to be roughly equal to the distance between the cluster center and the radial arc (cf. Fig. 10.19). Mellier, Fort & Kneib (1993) find $r_{core} \gtrsim 40$ kpc in MS 2137−23, and Smail et al. (1996) infer $r_{core} \sim 50$ kpc in A370. Bergmann & Petrosian (1993) presented a statistical argument in favor of finite cores by showing that lens models with singular cores produce fewer large arcs (relative to small arcs) than observed. The relative abundance increases with a small finite core. These results, however, have to be interpreted with caution because it may well be that the softened isothermal sphere model is inadequate to describe the interiors of galaxy clusters. While this particular model indeed requires core radii on the order of the radial critical radius, other lens models can produce radial arcs without having a flat core, and there are even singular density profiles which can explain radial arcs (Miralda-Escudé 1995; Bartelmann 1996). Such singular profiles for the dark matter are consistent with the fairly large core radii inferred from the X-ray emission of clusters, if the intracluster gas is isothermal and in hydrostatic equilibrium with the dark-matter potential (Navarro, Frenk & White 1996; J.P. Ostriker, private communication).

10.4.1.5 Radial density profile
Many of the observed giant arcs are unresolved in the radial direction, some of them even when observed under excellent seeing conditions or with the *Hubble Space Telescope*. Since the faint blue background galaxies which provide the source population for the arcs seem to be resolved (e.g., Tyson 1995), the giant arcs appear to be demagnified in width. It was realized by Hammer & Rigaut (1989) that spherically symmetric lenses can radially demagnify giant arcs only if their radial density profiles are steeper than isothermal. The maximum demagnification is obtained for a point-mass lens, where it is a factor of two. Kovner (1989) and Hammer (1991) demonstrated that, irrespective of the mass profile and the symmetry of the lens, the thin dimension of an arc is compressed by a factor $\approx 2(1 - \kappa)$, where κ is the convergence at the

position of the arc. Arcs which are thinner than the original source therefore require $\kappa \lesssim 0.5$. Since giant arcs have to be located close to those critical curves in the lens plane along which $1 - \kappa - \gamma = 0$, large and thin arcs additionally require $\gamma \gtrsim 0.5$.

In principle, the radius of curvature of large arcs relative to their distance from the cluster center can be used to constrain the steepness of the radial density profile (Miralda-Escudé 1992, 1993), but results obtained from observed arcs are not yet conclusive (Grossman & Saha 1994). One problem with this method is that substructure in clusters tends to enlarge curvature radii irrespective of the mass profile of the dominant component of the cluster (Miralda-Escudé 1993; Bartelmann, Steinmetz & Weiss 1995).

Wu & Hammer (1993) argued for steep mass profiles on statistical grounds because the observed abundance of large arcs appears to require highly centrally condensed cluster mass profiles in order to increase the central mass density of clusters while keeping their total mass constant. However, their conclusions are based on spherically symmetric lens models and are significantly changed when the symmetry assumption is dropped (Bartelmann, Steinmetz & Weiss 1995).

It should also be kept in mind that not all arcs are thin. Some "thick" and resolved arcs are known (e.g., in A2218, Pelló-Descayre et al. 1988; and in A2390, Pelló et al. 1991), and it is quite possible that thin arcs predominate just because they are more easily detected than thick ones due to observational selection effects. Also, Miralda-Escudé (1992, 1993) has argued that intrinsic source ellipticity can increase the probability of producing thin arcs, while Bartelmann, Steinmetz & Weiss (1995) showed that the condition $\kappa \lesssim 0.5$ which is required for thin arcs can be more frequently fulfilled in clusters with substructure where the shear is larger than in spherically symmetric clusters.

10.4.1.6 Mass sub-condensations

The cluster A370 has two cD galaxies and is a clear example of a cluster with multiple mass centers. A two-component mass model centered on the cD galaxies (Kneib et al. 1993) fits very well the lens data as well as X-ray and deep optical images of the cluster. Abell 2390 is an interesting example because it contains a "straight arc" (Pelló et al. 1991, see also Mathez et al. 1992) which can be produced only with either a lips or a beak-to-beak caustic (Kassiola, Kovner & Blandford 1992). If the arc is modelled with a lips caustic, it requires the mass peak to be close to the location of the arc, but this is not where the cluster light is centered. With a beak-to-beak caustic, the model requires two separate mass condensations, one of which could be at the peak of the luminosity, but then the other has to be a dark condensation. Pierre et al. (1996) find enhanced X-ray emission at a plausible position of the secondary mass clump, and from a weak lensing analysis Squires et al. (1996b) find a mass map which is consistent with a mass condensation at the location of the enhanced X-ray emission.

Abell 370 and A2390 are the most obvious examples of what is probably a widespread phenomenon, namely that clusters are in general not fully relaxed but have substructure as a result of ongoing evolution. If clusters are frequently clumpy, this can lead to systematic effects in the statistics of arcs and in the derived cluster parameters (Bartelmann et al. 1995; Bartelmann 1995b).

10.4.1.7 Lensing results vs. other mass determinations

Enclosed mass Three different methods are currently used to estimate cluster masses. Galaxy velocity dispersions yield a mass estimate from the virial theorem, and hence the galaxies have to be in virial equilibrium for such estimates to be valid. It may also be that the velocities of cluster galaxies are biased relative to the velocities of the dark-matter particles (Carlberg 1994), though current estimates suggest that the bias is no more than about 10%. The X-ray emission of rich galaxy clusters is dominated by free–free emission of thermal electrons and therefore depends on the squared density of the intracluster gas, which in turn traces the gravitational potential of the clusters. Such estimates usually assume that the cluster gas is in thermal hydrostatic equilibrium and that the potential is at least approximately spherically symmetric. Finally, large arcs in clusters provide a mass estimate through equation (10.77) or by more detailed modelling. These three mass estimates are in qualitative agreement with each other up to factors of $\approx$ 2–3.

Miralda-Escudé & Babul (1995) compared X-ray and large-arc mass estimates for the clusters A1689, A2163 and A2218. They took into account deviations from spherical symmetry and obtained lensing masses from individual lens models which reproduce the observed arcs. They arrived at the conclusion that in A1689 and A2218 the mass required for producing the large arcs is higher by a factor of 2–2.5 than the mass required for the X-ray emission, and proposed a variety of reasons for such a discrepancy, among them projection effects and non-thermal pressure support. Loeb & Mao (1994) specifically suggested that strong turbulence and magnetic fields in the intracluster gas may constitute a significant non-thermal pressure component in A2218 and thus render the X-ray mass estimate too low. Bartelmann & Steinmetz (1996) used gas dynamical cluster simulations to compare their X-ray and lensing properties. They found a similar discrepancy as that identified by Miralda-Escudé & Babul (1995) in those clusters that show structure in the distribution of line-of-sight velocities of the cluster particles, indicative of merging or infall along the line-of-sight. The discrepancy is probably due to projection effects.

Bartelmann (1995b) showed that cluster mass estimates obtained from large arcs by straightforward application of equation (10.77) are systematically too high by a factor of $\approx$ 1.6 on average, and by as much as a factor of $\approx$ 2 in 1 out of 5 cases. This discrepancy arises because equation (10.77) assumes a smooth spherically symmetric mass distribution whereas realistic clusters are asymmetric and have substructure. Note that Daines et al. (1996) found evidence for two or more mass condensations along the line of sight toward A1689, while the arclets in A2218 show at least two mass concentrations. It appears that cluster mass estimates from lensing require detailed lens models in order to be accurate to better than $\approx$ 30–50 percent. In the case of MS 1224, Fahlman et al. (1994) and Carlberg, Yee & Ellingson (1994) have obtained masses using the Kaiser & Squires weak-lensing cluster reconstruction method. Their mass estimates are 2 − 3 times higher than the cluster's virial mass. Carlberg et al. find evidence from velocity measurements that there is a second poor cluster in the foreground of MS 1224 which may explain the result. All of these mass discrepancies illustrate that cluster masses must still be considered uncertain to a factor of ≈ 2 in general.

Core radii Lensing estimates of cluster core radii are generally much smaller than the core radii obtained from optical or X-ray data. The upper limits on the core radii from lensing are fairly robust and probably reliable. Many clusters with large arcs have cD galaxies which can steepen the central mass profile of the cluster. However, there are also non-cD clusters with giant arcs, e.g., A1689 and Cl1409 (Tyson 1990), and MS 0440+02 (Luppino et al. 1993). In fact, Tyson (1990) claims that A1689 and Cl1409 have cores smaller than $100\,h^{-1}$ kpc, similar to upper limits for core radii found in other arc clusters with cD galaxies. As mentioned in §10.4.1.4, even the occurrence of radial arcs in clusters does not necessarily require a non-singular core, and so all the lensing data are consistent with singular cores in clusters. The X-ray core radii depend on whether or not the cooling regions of clusters are included in the emissivity profile fits, because the cooling radii are of the same order of magnitude as the core radii. If cooling is included, the best-fit core radii are reduced by a factor of $\simeq 4$ (Gerbal et al. 1992; Durret et al. 1994). Also, isothermal gas in hydrostatic equilibrium in a singular dark-matter distribution develops a flat core with a core radius similar to those observed. Therefore, the strongly peaked mass distributions required for lensing seem to be quite compatible with the extended X-ray cores observed.

Does mass follow light? Leaving the core radius aside, does mass follow light? It is clear that the mass cannot be as concentrated within the galaxies as the optical light is (e.g., Bergmann, Petrosian & Lynds 1990). However, if the optical light is smoothed and assumed to trace the mass, then the resulting mass distribution is probably not very different from the true mass distribution. For instance, in A370, the elongation of the mass distribution required for the giant arc is along the line connecting the two cD galaxies in the cluster (Grossman & Narayan 1989) and in fact Kneib et al. (1993) are able to achieve an excellent fit of the giant arc and several arclets with two mass concentrations surrounding the two cDs. Their model potential also agrees very well with the X-ray emission of the cluster. In MS 2137, the optical halo is elongated in the direction indicated by the arcs for the overall mass asymmetry (Mellier, Fort & Kneib 1993), and in Cl0024, (see p. 419, Fig. 10.25), the mass distribution is elongated in the same direction as the galaxy distribution (Wallington, Kochanek & Koo 1995). Smail et al. (1995) find that the mass maps of two clusters reconstructed from weak lensing agree fairly well with their X-ray emission. An important counter example is the cluster A2390, where the straight arc requires a mass concentration which is completely dark in the optical (Kassiola, Kovner & Blandford 1992). Pierre et al. (1996), however, find excess X-ray emission at a position compatible with the arc.

What kinds of clusters produce giant arcs? Which parameters determine whether or not a galaxy cluster is able to produce large arcs? Clearly, large velocity dispersions and small core radii favor the formation of arcs. As argued earlier, intrinsic asymmetries and substructure also increase the ability of clusters to produce arcs because they increase the shear and the number of cusps in the caustics.

The abundance of arcs in X-ray luminous clusters appears to be higher than in optically selected clusters. At least a quarter, maybe half, of the 38 X-ray bright clusters selected by Le Fèvre et al. (1994) contain large arcs, while Smail et al. (1991) found only one large arc in a sample of 19 distant optically selected clusters. However, some clusters which are prominent lenses (A370, A1689, A2218) are moderate X-ray sources, while other clusters which are very luminous X-ray sources (A2163, Cl1455) are poor lenses. The correlation between X-ray brightness and enhanced occurrence of arcs may suggest that X-ray bright clusters are more massive and/or more centrally condensed than X-ray quiet clusters.

Substructure appears to be at least as important as X-ray brightness for producing giant arcs. For example, A370, A1689 and A2218 all seem to have clumpy mass distributions. Bartelmann & Steinmetz (1996) used numerical cluster simulations to show that the optical depth for arc formation is dominated by clusters with intermediate rather than the highest X-ray luminosities.

Another possibility is that giant arcs preferentially form in clusters with cD galaxies. A370, for instance, even has two cDs. However, non-cD clusters with giant arcs are known, e.g., A1689, Cl1409 (Tyson, Valdes & Wenk 1990), and MS 0440+02 (Luppino et al. 1993).

10.4.2 Weak lensing by clusters – arclets

In addition to the occasional giant arc, which is produced when a source happens to straddle a caustic, a lensing cluster also produces a large number of weakly distorted images of other background sources which are not located near caustics. These are the arclets. There is a population of distant blue galaxies in the universe whose spatial density reaches 50–100 galaxies per square arc minute at faint magnitudes (Tyson 1988). Each cluster therefore has on the order of 50–100 arclets per square arc minute exhibiting a coherent pattern of distortions. Arclets were first detected by Fort et al. (1988).

The separations between arclets, typically $\sim (5-10)''$, are much smaller than the scale over which the gravitational potential of a cluster as a whole changes appreciably. The weak and noisy signals from several individual arclets can therefore be averaged by statistical techniques to get an idea of the mass distribution of a cluster. This technique was first demonstrated by Tyson, Valdes & Wenk (1990). Kochanek (1990) and Miralda-Escudé (1991a) studied how parameterized cluster-lens models can be constrained with arclet data.

The first systematic and parameter-free procedure to convert the observed ellipticities of arclet images to a surface-density map $\Sigma(\theta)$ of the lensing cluster was developed by

Kaiser & Squires (1993). An ambiguity intrinsic to all such inversion methods which are based on shear information alone was identified by Seitz & Schneider (1995a). This ambiguity can be resolved by including information on the convergence of the cluster; methods for this were developed by Broadhurst, Taylor & Peacock (1995) and Bartelmann & Narayan (1995a).

10.4.2.1 The Kaiser & Squires algorithm

The technique of Kaiser & Squires (1993) is based on the fact that both convergence $\kappa(\theta)$ and shear $\gamma_{1,2}(\theta)$ are linear combinations of second derivatives of the effective lensing potential $\psi(\theta)$. There is thus a mathematical relation connecting the two. In the Kaiser & Squires method one first estimates $\gamma_{1,2}(\theta)$ by measuring the weak distortions of background galaxy images, and then uses the relation to infer $\kappa(\theta)$. The surface density of the lens is then obtained from $\Sigma(\theta) = \Sigma_{cr}\kappa(\theta)$ (see equation (10.50)).

As shown in §10.3, κ and $\gamma_{1,2}$ are given by

$$\kappa(\theta) = \frac{1}{2}\left(\frac{\partial^2\psi(\theta)}{\partial\theta_1^2} + \frac{\partial^2\psi(\theta)}{\partial\theta_2^2}\right),$$

$$\gamma_1(\theta) = \frac{1}{2}\left(\frac{\partial^2\psi(\theta)}{\partial\theta_1^2} - \frac{\partial^2\psi(\theta)}{\partial\theta_2^2}\right),$$

$$\gamma_2(\theta) = \frac{\partial^2\psi(\theta)}{\partial\theta_1\partial\theta_2}. \quad (10.80)$$

If we introduce Fourier Transforms of κ, $\gamma_{1,2}$, and ψ (which we denote by hats on the symbols), we have

$$\hat{\kappa}(k) = -\frac{1}{2}(k_1^2 + k_2^2)\hat{\psi}(k),$$

$$\hat{\gamma}_1(k) = -\frac{1}{2}(k_1^2 - k_2^2)\hat{\psi}(k),$$

$$\hat{\gamma}_2(k) = -k_1 k_2 \hat{\psi}(k), \quad (10.81)$$

where k is the two-dimensional wavevector conjugate to θ. The relation between κ and $\gamma_{1,2}$ in Fourier space can then be written

$$\begin{pmatrix}\hat{\gamma}_1 \\ \hat{\gamma}_2\end{pmatrix} = k^{-2}\begin{pmatrix}(k_1^2 - k_2^2) \\ 2k_1 k_2\end{pmatrix}\hat{\kappa},$$

$$\hat{\kappa} = k^{-2}\left[(k_1^2 - k_2^2), (2k_1 k_2)\right]\begin{pmatrix}\hat{\gamma}_1 \\ \hat{\gamma}_2\end{pmatrix}. \quad (10.82)$$

If the shear components $\gamma_{1,2}(\theta)$ have been measured, we can solve for $\hat{\kappa}(k)$ in Fourier space, and this can be back transformed to obtain $\kappa(\theta)$ and thereby $\Sigma(\theta)$. Equivalently, we can write the relationship as a convolution in θ space,

$$\kappa(\theta) = \frac{1}{\pi}\int d^2\theta' \, \mathrm{Re}\left[\mathcal{D}^*(\theta - \theta')\gamma(\theta')\right], \quad (10.83)$$

where $\mathcal{D}$ is the complex convolution kernel,

$$\mathcal{D}(\theta) = \frac{(\theta_2^2 - \theta_1^2) - 2i\theta_1\theta_2}{\theta^4}, \tag{10.84}$$

and $\gamma(\theta)$ is the complex shear,

$$\gamma(\theta) = \gamma_1(\theta) + i\gamma_2(\theta). \tag{10.85}$$

The asterisk denotes complex conjugation.

The key to the Kaiser & Squires method is that the shear field $\gamma(\theta)$ can be measured. (Elaborate techniques to do so were described by Bonnet & Mellier 1995 and Kaiser, Squires & Broadhurst 1995.) If we define the ellipticity of an image as

$$\epsilon = \epsilon_1 + i\epsilon_2 = \frac{1-r}{1+r} e^{2i\phi}, \quad r \equiv \frac{b}{a}, \tag{10.86}$$

where ϕ is the position angle of the ellipse and a and b are its major and minor axes, respectively, we see from equation (10.59) that the average ellipticity induced by lensing is

$$\langle \epsilon \rangle = \left\langle \frac{\gamma}{1-\kappa} \right\rangle, \tag{10.87}$$

where the angle brackets refer to averages over a finite area of the sky. In the limit of weak lensing, $\kappa \ll 1$ and $|\gamma| \ll 1$, and the mean ellipticity directly gives the shear,

$$\langle \gamma_1(\theta) \rangle \approx \langle \epsilon_1(\theta) \rangle, \quad \langle \gamma_2(\theta) \rangle \approx \langle \epsilon_2(\theta) \rangle. \tag{10.88}$$

The $\gamma_1(\theta)$, $\gamma_2(\theta)$ fields so obtained can be transformed using the integral (10.83) to obtain $\kappa(\theta)$ and thereby $\Sigma(\theta)$. The quantities $\langle \epsilon_1(\theta) \rangle$ and $\langle \epsilon_2(\theta) \rangle$ in (10.88) have to be obtained by averaging over sufficient numbers of weakly lensed sources to have a reasonable signal-to-noise ratio.

10.4.2.2 Practical details and subtleties

In practice, several difficulties complicate the application of the elegant inversion technique summarized by equation (10.83). Atmospheric turbulence causes images taken by ground-based telescopes to be blurred. As a result, elliptical images tend to be circularized so that ground-based telescopes measure a lower limit to the actual shear signal. This difficulty is not present for space-based observations.

The point-spread function of the telescope can be anisotropic and can vary across the observed field. An intrinsically circular image can therefore be imaged as an ellipse just because of astigmatism of the telescope. Subtle effects like slight tracking errors of the telescope or wind at the telescope site can also introduce a spurious shear signal.

In principle, all these effects can be corrected for. Given the seeing and the intrinsic brightness distribution of the image, the amount of circularization due to seeing can be estimated and taken into account. The shape of the point-spread function and its variation across the image plane of the telescope can also be calibrated. However, since the shear signal especially in the outskirts of a cluster is weak, the effects have to be determined with high precision, and this is a challenge.

The need to average over several background galaxy images introduces a resolution limit to the cluster reconstruction. Assuming 50 galaxies per square arc minute, the typical separation of two galaxies is $\sim 8''$. If the average is taken over ~ 10 galaxies, the spatial resolution is limited to $\sim 30''$.

We have seen in equation (10.87) that the observed ellipticities strictly do not measure γ, but rather a combination of κ and γ,

$$\langle \epsilon \rangle = \langle g \rangle \equiv \left\langle \frac{\gamma}{1-\kappa} \right\rangle. \tag{10.89}$$

Inserting $\gamma = \langle \epsilon \rangle (1-\kappa)$ into the reconstruction equation (10.83) yields an integral equation for κ which can be solved iteratively. This procedure, however, reveals a weakness of the method. Any reconstruction technique which is based on measurements of image ellipticities alone is insensitive to isotropic expansions of the images. The measured ellipticities are thus invariant against replacing the Jacobian matrix $\mathscr{A}$ by some scalar multiple $\lambda \mathscr{A}$ of it. Putting

$$\mathscr{A}' = \lambda \mathscr{A} = \lambda \begin{pmatrix} 1-\kappa-\gamma_1 & -\gamma_2 \\ -\gamma_2 & 1-\kappa+\gamma_1 \end{pmatrix}, \tag{10.90}$$

we see that scaling $\mathscr{A}$ with λ is equivalent to the following transformations of κ and γ,

$$1-\kappa' = \lambda(1-\kappa), \quad \gamma' = \lambda\gamma. \tag{10.91}$$

Manifestly, this transformation leaves g invariant. We thus have a one-parameter ambiguity in shear-based reconstruction techniques,

$$\kappa \to \lambda\kappa + (1-\lambda), \tag{10.92}$$

with λ an arbitrary scalar constant.

This invariance transformation was highlighted by Schneider & Seitz (1995) and was originally discovered by Falco, Gorenstein & Shapiro (1985) in the context of galaxy lensing. If $\lambda \lesssim 1$, the transformation is equivalent to replacing κ by κ plus a sheet of constant surface-mass density $1-\lambda$. The transformation (10.92) is therefore referred to as the mass-sheet degeneracy.

Another weakness of the Kaiser & Squires method is that the reconstruction equation (10.83) requires a convolution to be performed over the entire θ plane. Observational data however are available only over a finite field. Ignoring everything outside the field and restricting the range of integration to the actual field is equivalent to setting $\gamma = 0$ outside the field. For circularly symmetric mass distributions, this implies vanishing total mass within the field. The influence of the finiteness of the field can therefore be quite severe.

Finally, the reconstruction yields $\kappa(\theta)$, and in order to calculate the surface-mass density $\Sigma(\theta)$ we must know the critical density Σ_{cr}, but since we do not know the redshifts of the sources there is a scaling uncertainty in this quantity. For a lens with given surface-mass density, the distortion increases with increasing source redshift. If the sources are at much higher redshifts than the cluster, the influence of the source redshift becomes weak. Therefore, this uncertainty is less serious for low-redshift clusters.

Nearly all the problems mentioned above have been addressed and solved. The solutions are discussed in the following subsections.

Gravitational lensing

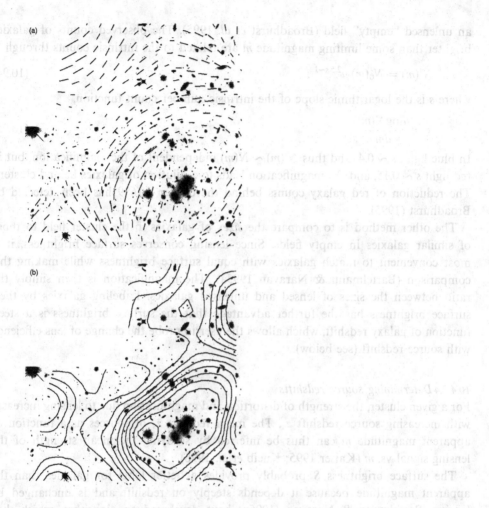

Figure 10.25. HST image of the cluster Cl0024, overlaid (a) with the shear field obtained from an observation of arclets with the *Canada–France Hawaii Telescope* (courtesy of Y. Mellier & B. Fort), and (b) with the reconstructed surface-mass density determined from the shear field (courtesy of C. Seitz *et al.*). The reconstruction was done with a nonlinear, finite-field algorithm.

10.4.2.3 *Eliminating the mass-sheet degeneracy by measuring the convergence*
By equation (10.60),

$$\mu = \frac{1}{(1-\kappa)^2 - \gamma^2}, \tag{10.93}$$

and so the magnification scales with λ as $\mu \propto \lambda^{-2}$. Therefore, the mass-sheet degeneracy can be broken by measuring the magnification μ of the images in addition to the shear (Broadhurst, Taylor & Peacock 1995). Two methods have been proposed to measure μ. The first relies on comparing the galaxy counts in the cluster field with those in an unlensed "empty" field (Broadhurst et al. 1995). The observed counts of galaxies brighter than some limiting magnitude m are related to the intrinsic counts through

$$N'(m) = N_0(m)\, \mu^{2.5s-1}, \tag{10.94}$$

where s is the logarithmic slope of the intrinsic number count function,

$$s = \frac{d\log N(m)}{dm}. \tag{10.95}$$

In blue light, $s \sim 0.4$, and thus $N'(m) \sim N(m)$ independent of the magnification, but in red light $s \sim 0.15$, and the magnification leads to a dilution of galaxies behind clusters. The reduction of red galaxy counts behind the cluster A1689 has been detected by Broadhurst (1995).

The other method is to compare the *sizes* of galaxies in the cluster field to those of similar galaxies in empty fields. Since lensing conserves surface brightness, it is most convenient to match galaxies with equal surface brightness while making this comparison (Bartelmann & Narayan 1995a). The magnification is then simply the ratio between the sizes of lensed and unlensed galaxies. Labeling galaxies by their surface brightness has the further advantage that the surface brightness is a steep function of galaxy redshift, which allows the user to probe the change of lens efficiency with source redshift (see below).

10.4.2.4 Determining source redshifts

For a given cluster, the strength of distortion and magnification due to lensing increases with increasing source redshift z_s. The mean redshift $\bar{z}_s$ of sources as a function of apparent magnitude m can thus be inferred by studying the mean strength of the lensing signal vs. m (Kaiser 1995; Kneib et al. 1996).

The surface brightness S probably provides a better label for galaxies than the apparent magnitude because it depends steeply on redshift and is unchanged by lensing. Bartelmann & Narayan (1995a) have developed an algorithm, which they named the lens-parallax method, to reconstruct the cluster mass distributions and to infer simultaneously $\bar{z}_s$ as a function of the surface brightness. In simulations, data from ~ 10 cluster fields and an equal number of empty comparison fields were sufficient to determine the cluster masses to $\sim \pm 5\%$ and the galaxy redshifts to $\sim \pm 10\%$ accuracy. The inclusion of galaxy sizes in the iterative lens-parallax algorithm breaks the mass-sheet degeneracy, thereby removing the ambiguities in shear-based cluster reconstruction techniques arising from the transformation (10.92) and from the unknown redshift distribution of the sources.

10.4.2.5 Finite-field methods

As emphasized previously, the inversion equation (10.83) requires a convolution to be performed over the entire real plane. The fact that data are always restricted to a finite field thus introduces a severe bias in the reconstruction. Modified reconstruction kernels have been suggested to overcome this limitation.

Gravitational lensing

Consider the relation (Kaiser 1995)

$$\nabla \kappa = \begin{pmatrix} \gamma_{1,1} + \gamma_{2,2} \\ \gamma_{2,1} - \gamma_{1,2} \end{pmatrix}. \tag{10.96}$$

This shows that the convergence at any point θ in the data field is related by a line integral to the convergence at another point θ_0,

$$\kappa(\theta) = \kappa(\theta_0) + \int_{\theta_0}^{\theta} dl \cdot \nabla \kappa[\theta(l)] . \tag{10.97}$$

If the starting point θ_0 is far from the cluster center, $\kappa(\theta_0)$ may be expected to be small and can be neglected. For each starting point θ_0, equation (10.97) yields an estimate for $\kappa(\theta)-\kappa(\theta_0)$, and by averaging over all chosen θ_0 modified reconstruction kernels can be constructed (Schneider 1995; Kaiser et al. 1995; Bartelmann 1995c; Seitz & Schneider 1996). Various choices for the set of starting positions θ_0 have been suggested. For instance, one can divide the observed field into an inner region centered on the cluster and take as θ_0 all points in the rest of the field. Another possibility is to take θ_0 from the entire field. In both cases, the result is $\kappa(\theta) - \bar{\kappa}$, where $\bar{\kappa}$ is the average convergence in the region from which the points θ_0 were chosen. The average $\bar{\kappa}$ is unknown, of course, and thus a reconstruction based on equation (10.97) yields κ only up to a constant. Equation (10.97) therefore explicitly displays the mass sheet degeneracy since the final answer depends on the choice of the unknown $\kappa(\theta_0)$.

A different approach (Bartelmann et al. 1996) employs the fact that κ and γ are linear combinations of second derivatives of the same effective lensing potential ψ. In this method one reconstructs ψ rather than κ. If both κ and γ can be measured through image distortions and magnifications (with different accuracies), then a straightforward finite-field Maximum-Likelihood can be developed to construct $\psi(\theta)$ on a finite grid such that it optimally reproduces the observed magnifications and distortions. It is easy in this approach to incorporate measurement accuracies, correlations in the data, selection effects etc. to achieve an optimal result.

10.4.2.6 Results from weak lensing

The cluster reconstruction technique of Kaiser & Squires and variants thereof have been applied to a number of clusters and several more are being analyzed (see Fig. 10.25). We summarize some results in Table 10.5, focusing on the mass-to-light ratios of clusters and the degree of agreement between weak lensing and other independent studies of the same clusters.

Mass-to-light ratios inferred from weak lensing are generally quite high, $\sim 400\,h$ in solar units (cf. Table 10.5 and, e.g., Smail et al. 1997). The recent detection of a significant shear signal in the cluster MS 1054−03 at redshift 0.83 (Luppino & Kaiser 1997) indicates that the source galaxies either are at very high redshifts, $z \gtrsim (2-3)$, or that the mass-to-light ratio in this cluster is exceptionally high; if the galaxy redshifts are $z \lesssim 1$, the mass-to-light ratio needs to be $\gtrsim 1600\,h$.

The measurement of a coherent weak shear pattern out to a distance of almost

Table 10.5. *Mass-to-light ratios of several clusters derived from weak lensing*

Cluster	M/L	Remark	Reference
MS 1224	$800\,h$	virial mass ~ 3 times smaller ($\sigma_v = 770$ km s^{-1}) reconstruction out to $\sim 3'$	Fahlman et al. (1994)
A1689	$(400 \pm 60)\,h$	mass smoother than light near center; mass steeper than isothermal from $(200\text{–}1000)\,h^{-1}$ kpc	Tyson & Fischer (1995) Kaiser (1995)
Cl1455	$520\,h$	dark matter more concentrated than galaxies	Smail et al. (1995)
Cl0016	$740\,h$	dark matter more concentrated than galaxies	Smail et al. (1995)
A2218	$440\,h$	gas mass fraction $< 4\%\,h^{-3/2}$	Squires et al. (1996a)
A851	$200\,h$	mass distribution agrees with galaxies and X-rays	Seitz et al. (1996)
A2163	$(300 \pm 100)\,h$	gas mass fraction $\sim 7\%\,h^{-3/2}$	Squires et al. (1997)

1.5 Mpc from the center of the cluster Cl0024+1654 by Bonnet, Mellier & Fort (1994) demonstrates a promising method of constraining cluster mass profiles. These observations show that the density decreases rapidly outward, though the data are compatible both with an isothermal profile and a steeper de Vaucouleurs profile. Tyson & Fischer (1995) find the mass profile in A1689 to be steeper than isothermal. Squires et al. (1996b) derived the mass profile in A2390 and showed that it is compatible with both an isothermal profile and steeper profiles. Quite generally, the weak-lensing results on clusters indicate that the smoothed light distribution follows the mass well. Moreover, mass estimates from weak lensing and from the X-ray emission interpreted on the basis of hydrostatic equilibrium are consistent with each other (Squires et al. 1996a,b).

The epoch of formation of galaxy clusters depends on cosmological parameters, especially Ω_0 (Richstone, Loeb & Turner 1992; Bartelmann, Ehlers & Schneider 1993; Lacey & Cole 1993, 1994). Clusters in the local universe tend to be younger if Ω_0 is large. Such young clusters should be less relaxed and more structured than clusters in a low-density universe (Mohr et al. 1995; Crone, Evrard & Richstone 1996). Weak lensing offers straightforward ways to quantify cluster morphology (Wilson, Cole & Frenk 1996; Schneider & Bartelmann 1997), and therefore may be used to estimate the cosmic density Ω_0.

The dependence of cluster evolution on cosmological parameters also has a pronounced effect on the statistics of giant arcs. Numerical cluster simulations in different cosmological models indicate that the observed abundance of arcs can only be reproduced in low-density universes, $\Omega_0 \sim 0.3$, with vanishing cosmological constant, $\Lambda_0 \sim 0$ (Bartelmann et al. 1997). Low-density, flat models with $\Omega_0 + \Lambda_0 = 1$, or Einstein–de Sitter models, produce one or two orders of magnitude fewer arcs than observed.

10.4.3 Weak lensing by large-scale structure

10.4.3.1 Magnification and shear in 'empty' fields

Lensing by even larger scale structures than galaxy clusters has been discussed in various contexts. Kristian & Sachs (1966) and Gunn (1967) discussed the possibility of looking for distortions in images of background galaxies due to weak lensing by large-scale foreground mass distributions. The idea has been revived and studied in greater detail by Babul & Lee (1991); Jaroszyński et al. (1990); Miralda-Escudé (1991b); Blandford et al. (1991); Bartelmann & Schneider (1991); Kaiser (1992); Seljak (1994); Villumsen (1996); Bernardeau, van Waerbeke & Mellier (1997); Kaiser (1996); and Jain & Seljak (1997). The effect is weak–magnification and shear are typically on the order of a few percent–and a huge number of galaxies has to be imaged with great care before a coherent signal can be observed.

Despite the obvious practical difficulties, the rewards are potentially great since the two-point correlation function of the image distortions gives direct information on the power spectrum of density perturbations $P(k)$ in the universe. The correlation function of image shear, or *polarization* as it is sometimes referred to (Blandford et al. 1991), has been calculated for the standard CDM model and other popular models of the universe. Weak lensing probes mass concentrations on large scales where the density perturbations are still in the linear regime. Therefore, there are fewer uncertainties in the theoretical interpretation of the phenomenon. The problems are expected to be entirely observational.

Using a deep image of a blank field, Mould et al. (1994) set a limit of $\bar{p} < 4$ percent for the average polarization of galaxy images within a 4.8 arcmin field. This is consistent with most standard models of the universe. Fahlman et al. (1994) claimed a tighter bound, $\bar{p} < 0.9$ percent in a 2.8 arcmin field. On the other hand, Villumsen (1995a), using the Mould et al. (1994) data, claimed a detection at a level of $\bar{p} = (2.4 \pm 1.2)$ percent (95% confidence limit). There is clearly no consensus yet, but the field is still in its infancy.

Villumsen (1995b) has discussed how the two-point angular correlation function of faint galaxies is changed by weak lensing and how intrinsic clustering can be distinguished from clustering induced by lensing. The random magnification by large-scale structures introduces additional scatter in the magnitudes of cosmologically interesting standard candles such as supernovae of type Ia. For sources at redshifts

$z \sim 1$, the scatter was found to be negligibly small, of order $\Delta m \sim 0.05$ magnitudes (Frieman 1996; Wambsganss et al. 1997).

10.4.3.2 Large-scale QSO–galaxy correlations

Fugmann (1990) noticed an excess of Lick galaxies in the vicinity of high-redshift, radio-loud QSOs and showed that the excess reaches out to $\sim 10'$ from the QSOs. If real, this excess is most likely caused by magnification bias due to gravitational lensing. Further, the scale of the lens must be very large. Galaxy-sized lenses have Einstein radii of a few arcseconds and are clearly irrelevant. The effect has to be produced by structure on scales much larger than galaxy clusters.

Following Fugmann's work, various other correlations of a similar nature have been found. Bartelmann & Schneider (1993b, 1994; see also Bartsch, Schneider & Bartelmann 1997) discovered correlations between high-redshift, radio-loud, optically bright QSOs and optical and infrared galaxies, while Bartelmann, Schneider & Hasinger (1994) found correlations with diffuse X-ray emission in the 0.2–2.4 keV ROSAT band. Benítez & Martínez-González (1995, 1997) found an excess of red galaxies from the APM catalog around radio-loud QSOs with redshift $z \sim 1$ on scales $\lesssim 10'$. Seitz & Schneider (1995b) found correlations between the Bartelmann & Schneider (1993b) sample of QSOs and foreground Zwicky clusters. They followed in part an earlier study by Rodrigues-Williams & Hogan (1994), who found a highly significant correlation between optically selected, high-redshift QSOs and Zwicky clusters. Later, Rodrigues-Williams & Hawkins (1995) detected similar correlations between QSOs selected for their optical variability and Zwicky clusters. Wu & Han (1995) searched for associations between distant radio-loud QSOs and foreground Abell clusters and found a marginally significant correlation with a subsample of QSOs.

All these results indicate that there are correlations between background QSOs and foreground "light" in the optical, infrared and soft X-ray wavebands. The angular scale of the correlations is compatible with that expected from lensing by large-scale structures. Bartelmann & Schneider (1993a, see also Bartelmann 1995a for an analytical treatment of the problem) showed that current models of large-scale structure formation can explain the observed large-scale QSO–galaxy associations, provided a double magnification bias (Borgeest, von Linde & Refsdal 1991) is assumed. It is generally agreed that lensing by individual clusters of galaxies is insufficient to produce the observed effects if cluster velocity dispersions are of order 10^3 km s^{-1} (e.g., Rodrigues-Williams & Hogan 1994; Rodrigues-Williams & Hawkins 1995; Wu & Han 1995; Wu & Fang 1996). It appears, therefore, that lensing by large-scale structures has to be invoked to explain the observations. Bartelmann (1995a) has shown that constraints on the density perturbation spectrum and the bias factor of galaxy formation can be obtained from the angular cross-correlation function between QSOs and galaxies. This calculation was recently refined by including the nonlinear growth of density fluctuations (Sanz, Martínez-González & Benítez 1997; Dolag & Bartelmann 1997). The nonlinear effects are strong, and provide a good fit to the observational results by Benítez & Martínez-González (1995, 1997).

10.4.3.3 Lensing of the cosmic microwave background

The random deflection of light due to large-scale structures also affects the anisotropy of the cosmic microwave background (CMB) radiation. The angular autocorrelation function of the CMB temperature is only negligibly changed (Cole & Efstathiou 1989). However, high-order peaks in the CMB power spectrum are somewhat broadened by lensing. This effect is weak, of order $\sim 5\%$ on angular scales of $\lesssim 10'$ (Seljak 1996; Seljak & Zaldarriaga 1996; Martínez-González, Sanz & Cayón 1997), but it could be detected by future CMB observations, e.g., by the Planck Microwave Satellite.

10.4.3.4 Outlook: detecting dark-matter concentrations

If lensing is indeed responsible for the correlations discussed above, other signatures of lensing should be found. Fort et al. (1996) searched for shear due to weak lensing in the fields of five luminous QSOs and found coherent signals in all five fields. In addition, they detected foreground galaxy groups for three of the sources. Earlier, Bonnet et al. (1993) had found evidence for a coherent shear pattern in the field of the lens candidate QSO 2345+007. The shear was later identified with a distant cluster (Mellier et al. 1994; Fischer et al. 1994).

In general, it appears that looking for weak coherent image distortions provides an excellent way of searching for otherwise invisible dark-matter concentrations. A systematic technique for this purpose has been developed by Schneider (1996a). Weak lensing outside cluster fields may, in the near future, allow observers to obtain samples of mass concentrations which are selected purely on the basis of their lensing effect. Such a selection would be independent of the mass-to-light ratio, and would permit the identification and study of nonlinear structures in the universe with unusually large mass-to-light ratios. This would be complementary to the limits on compact masses discussed in §10.2.3.2.

Acknowledgments

The authors thank Rosanne Di Stefano, Andreas Huss, Chris Kochanek, Tsafrir Kolatt, Shude Mao, Peter Schneider and Uroš Seljak for helpful comments on the manuscript. This work was supported in part by NSF grant AST 9423209 and by the Sonderforschungsbereich SFB 375-95 of the Deutsche Forschungsgemeinschaft.

References

Alard, C. 1995, in *Astrophysical Applications of Gravitational Lensing*, Proc. IAU Symp. 173, eds. C. S. Kochanek & J. N. Hewitt (Boston: Kluwer)

Alard, C. & Guibert, J. 1997, *A&A*, **326**, 1

Alard, C., Mao, S. & Guibert, J. 1995, *A&A*, **300**, L17

Albrow, M., et al. 1996, in *Variable Stars and the Astrophysical Returns of Microlensing Surveys*, Proc. 12th IAP Conf., eds. R. Ferlet, J.-P. Maillard & B. Raban (Gif-Sur-Yvette: Editions Frontieres)

Alcock, C., et al. 1993, *Nature*, **365**, 621

Alcock, C., et al. 1995a, *PRL*, **74**, 2867

Alcock, C., et al. 1995b, *ApJ*, **454**, L125

Alcock, C., et al. 1996, *ApJ*, **461**, 84

Alcock, C., et al. 1997, *ApJ*, **479**, 119

Ansari, R., et al. 1996, *A&A*, **314**, 94

Aubourg, E., et al. 1993, *Nature*, **365**, 623

Aubourg, E., et al. 1995, *A&A*, **301**, 1

Babul, A. & Lee, M. H. 1991, *MNRAS*, **250**, 407

Baillon, P., Bouquet, A., Giraud-Héraud, Y. & Kaplan, J. 1993, *A&A*, **227**, 1

Bar-Kana, R. 1996, *ApJ*, **468**, 17

Bar-Kana, R. 1997, *ApJ*, **489**, 21

Bartelmann, M. 1995a, *A&A*, **298**, 661

Bartelmann, M. 1995b, *A&A*, **299**, 11

Bartelmann, M. 1995c, *A&A*, **303**, 643

Bartelmann, M. 1996, *A&A*, **313**, 697

Bartelmann, M. & Narayan, R. 1995a, *ApJ*, **451**, 60

Bartelmann, M. & Narayan, R. 1995b, in: *Dark Matter*, AIP Conf. Proc. 336, eds. S. S. Holt & C. L. Bennett (New York: AIP Press)

Bartelmann, M. & Schneider, P. 1991, *A&A*, **248**, 349

Bartelmann, M. & Schneider, P. 1993a, *A&A*, **268**, 1

Bartelmann, M. & Schneider, P. 1993b, *A&A*, **271**, 421

Bartelmann, M. & Schneider, P. 1994, *A&A*, **284**, 1

Bartelmann, M. & Steinmetz, M. 1996, *MNRAS*, **283**, 431

Bartelmann, M. & Weiss, A. 1994, *A&A*, **287**, 1

Bartelmann, M., Ehlers, J. & Schneider, P. 1993, *A&A*, **280**, 351

Bartelmann, M., Schneider, P. & Hasinger, G. 1994, *A&A*, **290**, 399

Bartelmann, M., Steinmetz, M. & Weiss, A. 1995, *A&A*, **297**, 1

Bartelmann, M., Narayan, R., Seitz, S. & Schneider, P. 1996, *ApJ*, **464**, L115

Bartelmann, M., Huss, A., Colberg, J. M., Jenkins, A. & Pearce, F. R. 1998, *A&A*, **330**, 1

Bartsch, A., Schneider, P. & Bartelmann, M. 1997, *A&A*, **319**, 375

Benítez, N. & Martínez-González, E. 1995, *ApJ*, **448**, L89

Benítez, N. & Martínez-González, E. 1997, *ApJ*, **477**, 27

Bennett, D. P., et al. 1995, in *Dark Matter*, AIP Conf. Proc. 336, eds. S. S. Holt & C. L. Bennett (New York: AIP)

Bergmann, A. G. & Petrosian, V. 1993, *ApJ*,

413, 18

Bergmann, A. G., Petrosian, V. & Lynds, R. 1990, ApJ, **350**, 23

Bernardeau, F., Van Waerbeke, L. & Mellier, Y. 1997, A&A, **322**, 1

Bernstein, G., Fischer, P., Tyson, J. A. & Rhee, G. 1997, ApJ, **483**, L79

Bissantz, N., Englmaier, P., Binney, J. & Gerhard, O. 1997, MNRAS, **289**, 651

Blaes, O. M. & Webster, R. L. 1992, ApJ, **391**, L63

Blandford, R. D. & Kochanek, C. S. 1987, ApJ, **321**, 658

Blandford, R. D., Kochanek, C. S., Kovner, I. & Narayan, R. 1989, Science, **245**, 824

Blandford, R. D. & Narayan, R. 1986, ApJ, **310**, 568

Blandford, R. D. & Narayan, R. 1992, ARA&A, **30**, 311

Blandford, R. D., Saust, A. B., Brainerd, T. G. & Villumsen, J. V. 1991, MNRAS, **251**, 600

Bonnet, H. & Mellier, Y. 1995, A&A, **303**, 331

Bonnet, H., Fort, B., Kneib, J.-P., Soucail, G. & Mellier, Y. 1993, A&A, **289**, L7

Bonnet, H., Mellier, Y. & Fort, B. 1994, ApJ, **427**, L83

Borgeest, U., von Linde, J. & Refsdal, S. 1991, A&A, **251**, L35

Brainerd, T. G., Blandford, R. D. & Smail, I. 1996, ApJ, **466**, 623

Broadhurst, T. 1995, in Dark Matter, AIP Conf. Proc. 336, eds. S. S. Holt & C. L. Bennett (New York: AIP)

Broadhurst, T., Taylor, A. & Peacock, J. 1995, ApJ, **438**, 49

Buchalter, A., Kamionkowski, M. & Rich, R. M. 1996, ApJ, **469**, 676

Burke, W. L. 1981, ApJ, **244**, L1

Canizares, C. R. 1982, ApJ, **263**, 508

Carlberg, R. G. 1994, ApJ, **433**, 468

Carlberg, R. G., Yee, H. K. C. & Ellingson, E. 1994, ApJ, **437**, 63

Carr, B. 1994, ARA&A, 32, 531

Cen, R. Y., Gott III, J. R., Ostriker, J. P. & Turner, E. L. 1994, ApJ, **423**, 1

Chang, K., Refsdal, S. 1979, Nature, **282**, 561

Chwolson, O. 1924, Astron. Nachr., 221, 329

Cole, S. & Efstahiou, G. 1989, MNRAS, **239**, 195

Colley, W. N. 1995, AJ, **109**, 440

Corbett, E. A., Browne, I. W. A., Wilkinson, P. N. & Patnaik, A. R. 1996, in Astrophysical Applications of Gravitational Lensing, Proc. IAU Symp. 173, eds. C. S. Kochanek & J. N. Hewitt (Boston: Kluwer)

Corrigan, R. T., et al. 1991, AJ, **102**, 34

Courbin, F., Magain, P., Keeton, C. R., Kochanek, C. S., Vanderriest, C., Jaunsen, A. O. & Hjorth, J. 1997, A&A, **324**, L1

Crone, M. M., Evrard, A. E. & Richstone, D. O. 1996, ApJ, **467**, 489

Crotts, A. P. S. 1992, ApJ, **399**, L43

Crotts, A. P. S. & Tomaney, A. B. 1996, ApJ, **473**, L87

Daines, S., Jones, C., Forman, W. & Tyson, A. 1996, preprint

Dalcanton, J. J., Canizares, C. R., Granados, A., Steidel, C. C. & Stocke, J.T. 1994, ApJ, **424**, 550

Dell'Antonio, I. P. & Tyson, J. A. 1996, ApJ, **473**, L17

De Rújula, A., Jetzer, P. & Massó, E. 1991, MNRAS, **250**, 348

Di Stefano, R. & Esin, A. A. 1995, ApJ, **448**, L1

Di Stefano, R. & Mao, S. 1996, ApJ, **457**, 93

Dolag, K. & Bartelmann, M. 1997, MNRAS, 291, 446

Durret, F., Gerbal, D., Lachièze-Rey, M., Lima-Neto, G. & Sadat, R. 1994, A&A, **287**, 733

Dyson, F. W., Eddington, A. S. & Davidson, C. R. 1920, Mem. R. Astron. Soc., **62**, 291

Eddington, A. S. 1920, Space, Time and Gravitation (Cambridge: University Press)

Einstein, A. 1911, Annalen der Physik, **35**, 898

Einstein, A. 1915, Sitzungsber. Preuß. Akad.

Wissensch., erster Halbband, p. 831

Einstein, A. 1936, Science, **84**, 506

Erdl, H. & Schneider, P. 1992, A&A, **268**, 453

Fahlman, G., Kaiser, N., Squires, G. & Woods, D. 1994, ApJ, **437**, 56

Falco, E. E., Gorenstein, M. V. & Shapiro, I. I. 1985, ApJ, **289**, L1

Falco, E. E., Gorenstein, M. V. & Shapiro, I. I. 1991, ApJ, **372**, 364

Falco, E. E., Kochanek, C. S. & Muñoz, J. A. 1998, ApJ, **494**, 47

Falco, E. E., Shapiro, I. I., Moustakas, L. A. & Davis, M. 1997, ApJ, **484**, 70

Fischer, P., Tyson, J. A., Bernstein, G. M. & Guhathakurta, P. 1994, ApJ, **431**, L71

Fischer, P., Bernstein, G., Rhee, G. & Tyson, J. A. 1997, AJ, **113**, 521

Flores, R. & Primack, J. 1996, ApJ, **457**, L5

Fort, B. & Mellier, Y. 1994, A&AR, **5**, 239

Fort, B., Prieur, J. L., Mathez, G., Mellier, Y. & Soucail, G. 1988, A&A, **200**, L17

Fort, B., Le Fèvre, O., Hammer, F. & Cailloux, M. 1992, ApJ, **399**, L125

Fort, B., Mellier, Y., Dantel-Fort, M., Bonnet, H. & Kneib, J.-P. 1996, A&A, **310**, 705

Frieman, J. 1996, Comments on Astrophysics, in press; preprint astro-ph/9608068

Fugmann, W. 1990, A&A, **240**, 11

Fukugita, M. & Turner, E. L. 1991, MNRAS, **253**, 99

Fukugita, M., Futamase, T., Kasai, M. & Turner, E. L. 1992, ApJ, **393**, 3

Garrett, M. A., Calder, R. J., Porcas, R. W., King, L. J., Walsh, D. & Wilkinson, P. N. 1994, MNRAS, **270**, 457

Gerbal, D., Durret, F., Lima-Neto, G. & Lachièze-Rey, M. 1992, A&A, **253**, 77

Gondolo, P., et al. 1997, in Dark and Visible Matter in Galaxies, Proc. Conf. Sesto Pusteria, Italy, eds. M. Persic & P. Salucci (San Francisco: ASP)

Gorenstein, M. V., et al. 1988, ApJ, **334**, 42

Gott, J. R. 1981, ApJ, **243**, 140

Gould, A. 1992a, ApJ, **386**, L5

Gould, A. 1992b, ApJ, **392**, 442

Gould, A. 1994a, ApJ, **421**, L71

Gould, A. 1994b, ApJ, **421**, L75

Gould, A. 1996, ApJ, **470**, 201

Gould, A. & Loeb, A. 1992, ApJ, **396**, 104

Griest, K. 1991, ApJ, **366**, 412

Griest, K. & Hu, W. 1992, ApJ, **397**, 362

Griest, K., Alcock, C. & Axelrod, T. S. 1991, ApJ, **372**, L79

Griffiths, R. E., Casertano, S., Im, M. & Ratnatunga, K. U. 1996, MNRAS, **282**, 1159

Grogin, N. & Narayan, R. 1996, ApJ, **464**, 92

Grossman, S. A. & Narayan, R. 1988, ApJ, **324**, L37

Grossman, S. A. & Narayan, R. 1989, ApJ, **344**, 637

Grossman, S. A. & Saha, P. 1994, ApJ, **431**, 74

Gunn, J. E. 1967, ApJ, **150**, 737

Hammer, F. 1991, ApJ, **383**, 66

Hammer, F. & Rigaut, F. 1989, A&A, **226**, 45

Henstock, D. R., et al. 1995, ApJS, **100**, 1

Haarsma, D. B., Hewitt, J. N., Burke, B. F. & Lehár, J. 1996, in Astrophysical Applications of Gravitational Lensing, IAU Symp. 173, eds. C. S. Kochanek & J. N. Hewitt (Boston: Kluwer)

Haarsma, D. B., Hewitt, J. N., Lehár, J. & Burke, B. F. 1997, ApJ, **479**, 102

Hewitt, J. N., et al. 1987, ApJ, **321**, 706

Huchra, J., Gorenstein, M., Kent, S., Shapiro, I. & Smith, G. et al. 1985, AJ, **90**, 691

Im, M., Griffiths, R. E. & Ratnatunga, K. 1997, ApJ, **475**, 457

Irwin, M. J., Webster, R. L., Hewett, P. C., Corrigan, R. T. & Jedrzejewski, R. I. 1989, AJ, **98**, 1989

Jain, B. & Seljak, U. 1997, ApJ, **484**, 560

Jaroszyński, M., Park, C., Paczyński, B. & Gott, J. R. 1990, ApJ, **365**, 22

Jetzer, P. 1994, A&A, **286**, 426

Kaiser, N. 1992, ApJ, **388**, 272

Kaiser, N. 1995, ApJ, **439**, L1

Kaiser, N. 1996, ApJ, submitted; preprint astro-ph/9610120

Kaiser, N. & Squires, G. 1993, *ApJ*, **404**, 441

Kaiser, N., Squires, G. & Broadhurst, T. 1995, *ApJ*, **449**, 460

Kaiser, N., Squires, G., Fahlman, G., Woods, D. & Broadhurst, T. 1995, in *Proc. 35th Herstmonceux Conf.*, ed. S. Maddox (Singapore: World Scientific)

Kamionkowski, M. 1995, *ApJ*, **442**, L9

Kassiola, A. & Kovner, I. 1993, *ApJ*, **417**, 450

Kassiola, A., Kovner, I. & Blandford, R. D. 1991, *ApJ*, **381**, 6

Kassiola, A., Kovner, I. & Blandford, R. D. 1992, *ApJ*, **396**, 10

Keeton II, C. R. & Kochanek, C. S. 1996, in *Astrophysical Applications of Gravitational Lensing*, IAU Symp. 173, eds. C. S. Kochanek & J. N. Hewitt (Boston: Kluwer)

Keeton II, C. R. & Kochanek, C. S. 1997, *ApJ*, **487**, 42

Keeton II, C. R. & Kochanek, C. S. 1998, *ApJ*, **495**, 157

Keeton II, C. R., Kochanek, C. S. & Seljak, U. 1997, *ApJ*, **482**, 604

Kiraga, M. & Paczyński, B. 1994, *ApJ*, **430**, L101

Kneib, J.-P., Ellis, R. S., Smail, I., Couch, W. T. & Sharples, R. M. 1996, *ApJ*, **471**, 643

Kneib, J.-P., Mellier, Y., Fort, B. & Mathez, G. 1993, *A&A*, **273**, 367

Kochanek, C. S. 1990, *MNRAS*, **247**, 135

Kochanek, C. S. 1991, *ApJ*, **373**, 354

Kochanek, C. S. 1992, *ApJ*, **384**, 1

Kochanek, C. S. 1993a, *ApJ*, **417**, 438

Kochanek, C. S. 1993b, *ApJ*, **419**, 12

Kochanek, C. S. 1995a, *ApJ*, **445**, 559

Kochanek, C. S. 1995b, *ApJ*, **453**, 545

Kochanek, C. S. 1996a, *ApJ*, **466**, 638

Kochanek, C. S. 1996b, *ApJ*, **473**, 595

Kochanek, C. S. & Narayan, R. 1992, *ApJ*, **401**, 461

Kochanek, C. S., Blandford, R. D., Lawrence, C. R. & Narayan, R. 1989, *MNRAS*, **238**, 43

Kovner, I. 1987, *ApJ*, **312**, 22

Kovner, I. 1989, *ApJ*, **337**, 621

Kristian, J. & Sachs, R. K. 1966, ApJ, 143, 379

Kundić, T., et al. 1996, *ApJ*, **482**, 75

Lacey, C. & Cole, S. 1993, *MNRAS*, **262**, 627

Lacey, C. & Cole, S. 1994, *MNRAS*, **271**, 676

Le Fèvre, O., Hammer, F., Angonin-Willaime, M. -C., Gioia, I. M. & Luppino, G. A. 1994, *ApJ*, **422**, L5

Lehár, J., Hewitt, J. N., Roberts, D. H. & Burke, B. F. 1992, *ApJ*, **384**, 453

Loeb, A. & Mao, S. 1994, *ApJ*, **435**, L109

Luppino, G. A., Gioia, I. M., Annis, J., Le Fèvre, O. & Hammer, F. 1993, *ApJ*, **416**, 444

Luppino, G. A. & Kaiser, N. 1997, *ApJ*, **475**, 20

Lynds, R. & Petrosian, V. 1986, *BAAS*, **18**, 1014

Malhotra, S., Rhoads, J. E. & Turner, E. L. 1997, *MNRAS*, **288**, 138

Maller, A. H., Flores, R. A. & Primack, J. R. 1997, *ApJ*, **486**, 681

Mao, S. & Paczyński, B. 1991, *ApJ*, **374**, L37

Mao, S. & Kochanek, C. S. 1994, *MNRAS*, **268**, 569

Mao, S. & Paczyński, B. 1996, *ApJ*, **473**, 57

Maoz, D. & Rix, H.-W. 1993, *ApJ*, **416**, 425

Maoz, D., Rix, H.-W., Gal-Yam, A. & Gould, A. 1997, *ApJ*, **486**, 75

Martínez-González, E., Sanz, J. L. & Cayón, L. 1997, *ApJ*, **484**, 1

Mathez, G., Fort, B., Mellier, Y., Picat, J.-P. & Soucail, G. 1992, *A&A*, **256**, 343

Mellier, Y., Fort, B. & Kneib, J.-P. 1993, *ApJ*, **407**, 33

Mellier, Y., Dantel-Fort, M., Fort, B. & Bonnet, H. 1994, *A&A*, **289**, L15

Miralda-Escudé, J. 1991a, *ApJ*, **370**, 1

Miralda-Escudé, J. 1991b, *ApJ*, **380**, 1

Miralda-Escudé, J. 1992, *ApJ*, **390**, L65

Miralda-Escudé, J. 1993, *ApJ*, **403**, 497

Miralda-Escudé, J. 1995, *ApJ*, **438**, 514

Miralda-Escudé, J. & Babul, A. 1995, *ApJ*, **449**, 18

Mohr, J. J., Evrard, A. E., Fabricant, D. G. & Geller, M. J. 1995, *ApJ*, **447**, 8

Mould, J., et al. 1994, *MNRAS*, **271**, 31

Myers, S. T., et al. 1995, *ApJ*, **447**, L5

Narayan, R. 1991, *ApJ*, **378**, L5

Narayan, R. & Grossman, S. 1989, in *Gravitational Lenses*, eds. J. M. Moran, J. N. Hewitt, K. Y. Lo (Berlin: Springer-Verlag)

Narayan, R. & White, S. D. M. 1988, *MNRAS*, **231**, 97p

Narayan, R. & Wallington, S. 1993, in *Gravitational Lenses in the universe*, Proc. 31st Liège Astroph. Coll., eds. J. Surdej, et al.

Narayan, R., Blandford, R. D. & Nityananda, R. 1984, *Nature*, **310**, 112

Natarajan, P. & Kneib, J.-P. 1996, *MNRAS*, **283**, 1031

Navarro, J. F., Frenk, C. S. & White, S. D. M. 1996, *ApJ*, **462**, 563

Nemiroff, R. J. & Wickramasinghe, W. A. D. T. 1994, *ApJ*, 424, L21

Nemiroff, R. J. et al. 1993, *ApJ*, **414**, 36

Nemiroff, R. J., Wickramasinghe, W. A. D. T., Norris, J. P., Kouveliotou, C. & Fishman, G. J., et al. 1994, *ApJ*, **432**, 478

Nityananda, R. & Samuel, J. 1992, *Phys. Rev. D.*, **45**, 3862

Østensen, R., et al. 1996, *A&A*, **309**, 59

Paczyński, B. 1986a, *ApJ*, **301**, 503

Paczyński, B. 1986b, *ApJ*, **304**, 1

Paczyński, B. 1987, *Nature*, **325**, 572

Paczyński, B. 1991, *ApJ*, **371**, L63

Paczyński, B. 1996, *ARA&A*, **34**, 419

Paczyński, B., et al. 1994, *ApJ*, **435**, L113

Pelló-Descayre, R., Soucail, G., Sanahuja, B., Mathez, G. & Ojero, E. 1988, *A&A*, **190**, L11

Pelló, R., Le Borgne, J. F., Soucail, G., Mellier, Y. & Sanahuja, B. 1991, *ApJ*, **366**, 405

Pelt, J., Hoff, W., Kayser, R., Refsdal, S. & Schramm, T. 1994, *A&A*, **286**, 775

Pelt, J., Kayser, R., Refsdal, S. & Schramm, T. 1996, *A&A*, **305**, 97

Pierre, M., Le Borgne, J. F., Soucail, G. &

Kneib, J. P. 1996, *A&A*, **311**, 413

Pratt, M. 1996, in *Microlensing*, Proc. 2nd International Workshop on Gravitational Microlensing Surveys, Orsay, France; ed. M. Moniez

Press, W. H. & Gunn, J. E. 1973, *ApJ*, **185**, 397

Press, W. H., Rybicki, G. B. & Hewitt, J. N. 1992a, *ApJ*, **385**, 404

Press, W. H., Rybicki, G. B. & Hewitt, J. N. 1992b, *ApJ*, **385**, 416

Rauch, K. P. & Blandford, R. D. 1991, *ApJ*, **381**, L39

Refsdal, S. 1964, *MNRAS*, **128**, 307

Refsdal, S. 1966a, *MNRAS*, **132**, 101

Refsdal, S. 1966b, *MNRAS*, **134**, 315

Refsdal, S. & Surdej, J. 1994, *Rep. Progr. Phys.*, **57**, 117

Renault, C., et al. 1997, *A&A*, **324**, L69

Richstone, D. O., Loeb, A. & Turner, E. L. 1992, *ApJ*, **393**, 477

Rix, H.-W., Maoz, D., Turner, E. L. & Fukugita, M. 1994, *ApJ*, **435**, 49

Rodrigues-Williams, L. L. & Hogan, C. J. 1994, *AJ*, **107**, 451

Rodrigues-Williams, L. L. & Hawkins, M. R. S. 1995, in *Dark Matter*, AIP Conf. Proc. 336, eds. S. S. Holt & C. L. Bennett (New York: AIP)

Roulet, E. & Mollerach, S. 1997, *Phys. Rep.*, 279, 67

Sackett, P. D. 1996, in *Microlensing*, Proc. 2nd International Workshop on Gravitational Microlensing Surveys, Orsay, France; ed. M. Moniez

Sahu, K. C. 1994, *Nature*, **370**, 275

Sahu, K. C. 1996, in *Microlensing*, Proc. 2nd International Workshop on Gravitational Microlensing Surveys, Orsay, France; ed. M. Moniez

Sanz, J. L., Martínez-González, E. & Benítez, N. 1997, *MNRAS*, **291**, 418

Schechter, P. L., et al. 1997, *ApJ*, **475**, L85

Schild, R. E. & Smith, R. C. 1991, *AJ*, **101**, 813

Schild, R. E. & Thomson, D. J. 1993, in *Gravitational Lenses in the Universe*, Proc. 31st Liège Int. Astroph. Coll., eds. J. Surdej, et al.

Schmidt, M. 1963, *Nature*, **197**, 1040

Schneider, P. 1985, *A&A*, **143**, 413

Schneider, P. 1993, *A&A*, **279**, 1

Schneider, P. 1995, *A&A*, **302**, 639

Schneider, P. 1996a, *MNRAS*, **283**, 837

Schneider, P. 1996b, in: *Cosmological Applications of Gravitational Lensing*, in *The universe at high-z, large-scale structure and the cosmic microwave background*, Lecture Notes in Physics, eds. E. Martínez-González & J. L. Sanz (Berlin: Springer Verlag)

Schneider, P. & Bartelmann, M. 1997, *MNRAS*, **286**, 696

Schneider, P. & Seitz, C. 1995, *A&A*, **294**, 411

Schneider, P. & Weiss, A. 1987, *A&A*, **171**, 49

Schneider, P., Ehlers, J. & Falco, E. E. 1992, *Gravitational Lenses* (Berlin: Springer Verlag)

Seitz, C. & Schneider, P. 1995a, *A&A*, **297**, 287

Seitz, S. & Schneider, P. 1995b, *A&A*, **302**, 9

Seitz, S. & Schneider, P. 1996, *A&A*, **305**, 383

Seitz, C., Wambsganss, J. & Schneider, P. 1994, *A&A*, **288**, 19

Seitz, C., Kneib, J.-P., Schneider, P. & Seitz, S. 1996, *A&A*, **314**, 707

Seljak, U. 1994, *ApJ*, **436**, 509

Seljak, U. 1996, *ApJ*, **463**, 1

Seljak, U. & Zaldarriaga, M. 1996, *ApJ*, **469**, 437

Shapiro, I. I. 1964, *PRL*, **13**, 789

Smail, I., Dickinson, M. 1995, *ApJ*, **455**, L99

Smail, I., Ellis, R.S., Fitchett, M. J., Norgaard-Nielsen, H. U., Hansen, L. & Jorgensen, H. J. 1991, *MNRAS*, **252**, 19

Smail, I., Ellis, R. S., Fitchett, M. J. & Edge, A. C. 1995, *MNRAS*, **273**, 277

Smail, I., et al. 1996, *ApJ*, **469**, 508

Smail, I., et al. 1997, *ApJ*, **479**, 70

Soldner, J. 1804, *Berliner Astron. Jahrb.* **1804**, 161

Soucail, G., Fort, B., Mellier, Y. & Picat, J. P. 1987a, *A&A*, **172**, L14

Soucail, G., Mellier, Y., Fort, B., Hammer, F. & Mathez, G. 1987b, *A&A*, **184**, L7

Squires, G., et al. 1996a, *ApJ*, **461**, 572

Squires, G., Kaiser, N., Fahlman, G., Babul, A. & Woods, D., 1996b, *ApJ*, **469**, 73

Squires, G., et al. 1997, *ApJ*, **482**, 648

Stanek, K. Z., Paczyński, B. & Goodman, J. 1993, *ApJ*, **413**, L7

Subramanian, K. & Cowling, S. A. 1986, *MNRAS*, **219**, 333

Surdej, J. et al. 1993, *AJ*, **105**, 2064

Surpi, G. C., Harari, D. D. & Frieman, J. A. 1996, *ApJ*, **464**, 54

Sutherland, W., et al. 1996, in *Proc. Workshop on Identification of Dark Matter*, Sheffield, 1996, in press; preprint astro-ph/9611059

Turner, E. L. 1980, *ApJ*, **242**, L135

Turner, E. L. 1990, *ApJ*, **365**, L43

Turner, E. L., Ostriker, J. P. & Gott, J. R. 1984, *ApJ*, **284**, 1

Tyson, J. A. 1988, *AJ*, **96**, 1

Tyson, J. A. 1990, *Gravitational Lensing*, Proc. Workshop Toulouse 1989 (Berlin: Springer)

Tyson, J. A. 1995, in *Dark Matter*, AIP Conf. Proc 336, eds. S. S. Holt & C. L. Bennett (New York: AIP)

Tyson, J. A. & Fischer, P. 1995, *ApJ*, **446**, L55

Tyson, J. A., Valdes, F. & Wenk, R. A. 1990, *ApJ*, **349**, L1

Udalski, A., Szymański, M., Kałuzny, J., Kubiak, M. & Mateo, M. 1992, *Acta Astron.*, **42**, 253

Udalski, A., et al. 1993, *Acta Astron.* **43**, 289

Udalski, A., et al. 1994, *ApJ*, **436**, L103

Vanderriest, C., Schneider, J., Herpe, G., Chèvreton, M., Moles, M. & Wlérick, G. 1989, *A&A*, **215**, 1

Villumsen, J. V. 1995a, *MNRAS*, submitted; preprint astro-ph/9507007

Villumsen, J. V. 1995b, *MNRAS*, submitted; preprint astro-ph/9512001

Villumsen, J. V. 1996, *MNRAS*, **281**, 369

Walsh, D., Carswell, R. F. & Weymann, R. J. 1979, *Nature*, **279**, 381

Wallington, S. & Narayan, R. 1993, *ApJ*, **403**, 517

Wallington, S., Narayan, R. & Kochanek, C. S. 1994, *ApJ*, **426**, 60

Wallington, S., Kochanek, C. S. & Koo, D. C. 1995, *ApJ*, **441**, 58

Wallington, S., Kochanek, C. S. & Narayan, R. 1996, *ApJ*, **465**, 64

Wambsganss, J. 1990, Ph.D. Thesis, MPA report #550

Wambsganss, J., Paczyński, B. & Schneider, P. 1990, *ApJ*, **358**, L33

Wambsganss, J., Cen, R. Y., Ostriker, J. P. & Turner, E. L. 1995, *Science*, **268**, 274

Wambsganss, J., Cen, R. Y. & Ostriker, J. P. 1998, *ApJ*, **494**, 29

Wambsganss, J., Cen, R., Xu, G. & Ostriker, J. P. 1997, *ApJ*, **475**, L81

Wang, Y. & Turner, E. L. 1997, *MNRAS*, **292**, 863

Webster, R. L., Ferguson, A. M. N., Corrigan, R. T. & Irwin, M. J. 1991, *AJ*, **102**, 1939

Wilson, G., Cole, S. & Frenk, C. S. 1996, *MNRAS*, **282**, 501

Witt, H. J. 1993, *ApJ*, **403**, 530

Witt, H. J. 1995, *ApJ*, **449**, 42

Witt, H. J. & Mao, S. 1994, *ApJ*, **430**, 505

Wu, X.-P. 1996, *Fund. Cosmic Phys.*, **17**, 1

Wu, X.-P. & Hammer, F. 1993, *MNRAS*, **262**, 187

Wu, X.-P. & Han, J. 1995, *MNRAS*, **272**, 705

Wu, X.-P. & Fang, L.-Z. 1996, *ApJ*, **461**, L5

Zhao, H. S. & Mao, S. 1996, *MNRAS*, **283**, 1197

Zhao, H. S., Spergel, D. N. & Rich, R. M. 1995, *ApJ*, **440**, L13

Zhao, H. S., Rich, R. M. & Spergel, D. N. 1996, *MNRAS*, **282**, 175

Zwicky, F. 1937a, *PRL*, **51**, 290

Zwicky, F. 1937b, *PRL*, **51**, 679

Part four

A conclusion

11 The mass of the universe

P. James E. Peebles

Abstract

The value of the cosmic mean mass density is constrained by a variety of independent measures within the relativistic Friedmann–Lemaître cosmological model. If we can establish consistency of the measures we will have a test of the cosmological model and a guide to the deeper theory needed to understand its initial conditions. The rapidly growing fund of cross-checks may develop into a web of evidence tight enough to be a believable measure of the mean mass density.

11.1 Introduction

We are witnessing a fascinating interaction between the older style of research in cosmology, that places heavy emphasis on arguments from first principles, and a newer phenomenological approach. It is impressive to see the high rate of success of ideas that originated when the observational basis for cosmology was a good deal sparser, but we have to expect that the considerable recent advances in the observations will require some reconsideration of the details, and there always is the chance Nature will teach us that we have to learn to love something new.

The example to be considered here is the mass problem. Most who have given the matter thought agree that the Einstein–de Sitter critical density, at which the universe in effect is expanding at escape velocity, is particularly elegant and natural; indeed many have added the Einstein–de Sitter case to the set of elements of the "standard model." This is not just an issue for theorists and philosophers, of course; the mass density can be measured. The indications still are mixed, but the methods of measurement are getting so good that we may soon find that the results fit together in one reasonable way, in a pattern tightly enough constrained to be believable. Out of this may come a demonstration that our universe is well described by the Einstein–de Sitter model, or perhaps we will have to learn to live with a lower mass density, as many of the preliminary results suggest.

This section starts with an overview of the issues, followed by a proposal for the list of elements of the standard cosmological model. The next section presents some details of the dynamical measurements of the mean mass density. Section 11.3 discusses some aspects of the cosmological tests, and §11.4 presents a table summarizing the state of the evidence and the issues.

11.1.1 Issues

The dynamical evidence is that the mass that clusters with the galaxies on scales of 200 kpc to 10 Mpc[1] is significantly less than the Einstein–de Sitter value. This led people to ask why we should assume the galaxies show us where the mass is. White, Tully & Davis (1988) state the issue in a direct way: "no theory for galaxy formation has ever demonstrated a physical basis for the conjecture" that galaxies trace mass; "it thus remains an assumption of convenience." In short, these authors make the point that until we have a believable theory we certainly are not competent to say where the physics has put the galaxies relative to the mass.

There is another side to the physics, however. The mass distribution in our universe certainly is not homogeneous: we are gravitationally bound to the Milky Way galaxy, and there is compelling evidence from lensing and intracluster plasma that gravity binds clusters of galaxies. The physics of the Friedmann–Lemaître model implies the mass distribution is gravitationally unstable: underdense regions tend to empty onto high-density patches. This draws together mass and galaxies, so recently added large-scale levels of the clustering hierarchy might be expected to offer a fairer measure of the global ratio of galaxies to mass. How fair a measure depends in part on when the positions of the galaxies were assigned: if it happened at low redshift the gravitational clustering effect could be weak; if the galaxies were present at the high redshift it could be quite important.

Since the relative motion of a pair of test particles is not much affected by mass fluctuations on scales large compared to the separation of the particles, an expected result of the gravitational erasing of initial biasing is that the apparent mass per galaxy increases with increasing scale of the dynamical system within which it is measured. The effect certainly is observed: the typical galaxy mass derived from the relative motions of gas and stars within a galaxy is an order of magnitude less than the mass derived from the relative motions of galaxies. The issue, first discussed by Ostriker et al. (1974), is whether the dynamical mass per galaxy approaches an asymptotic value that we can take to be a fair measure of the total. The debate on whether an asymptotic value has been reached runs through several chapters in this volume (e.g., Chs. 1, 4, 7, 8).

The evidence summarized in §11.2 of this chapter is that if our universe is Einstein–de Sitter the bulk of the mass has to be several megaparsecs away from the concentrations of large galaxies. This is what led Davis et al. (1985) to the astrophysical biasing picture within the cold dark matter (CDM) model for structure formation. Here, and

---

[1] Most current discussions put Hubble's constant in the range $H_0 = 70 \pm 15$ km s^{-1} Mpc^{-1}. For definiteness, the central value is used, except where indicated by the parameter $h = H_0/100$ in these units.

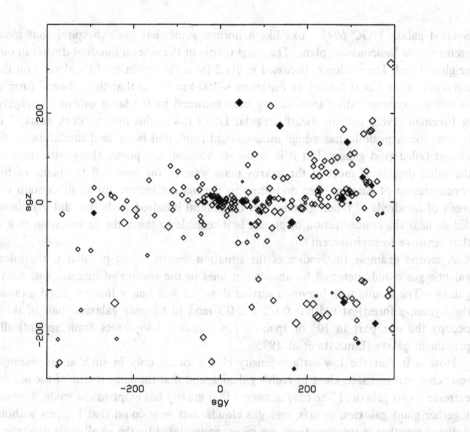

Figure 11.1. Map of the nearby galaxies (from J. Huchra's ZCAT). The Milky Way is at the center of the box, and the symbol size varies in proportion to the position in the box normal to the plane of the map. The more luminous galaxies are shown as filled symbols, the dwarfs as open symbols. The distance scale is the redshift in km s^{-1}. The ordinate is the distance normal to the plane of the de Vaucouleurs Local Supercluster; the abscissa points roughly toward the Virgo cluster at distance $cz \sim 1200$ km s^{-1}.

in related models (such as the mixed dark-matter or cold + hot dark-matter picture discussed in Ch. 1), the dominant mass component outside the concentrations of large galaxies has been seeded for galaxy formation by the Gaussian random fluctuations that gravitationally grew into the visible galaxies, but in the voids the seeds have germinated into at most inconspicuous dwarfs or irregulars. But if this were so why is there so little evidence for large-scale segregation of dwarfs from giant galaxies? Two examples are mentioned.

Figure 11.1 shows a map of the very nearby galaxies. These galaxies tend to lie on a plane, the local part of the de Vaucouleurs Local Supercluster; the map is oriented to show the plane in projection. The symbol size is meant to indicate relative position normal to the plane of this map. The more luminous, named galaxies are shown as filled symbols. Some of the giants are well away from the main concentration – the isolated galaxy NGC 6946 looks like a normal somewhat gas-rich spiral–but most prefer the de Vaucouleurs plane. The same is true of the several hundred dwarfs in our neighborhood. The evidence discussed in §11.2 from the redshifts of the dwarfs on the outskirts of the Local Group, at distances ~ 100 km s^{-1}, is that these dwarfs formed as isolated systems rather than having been spawned by the large galaxies. Carignan & Freeman (1988) call the dwarf irregular DDO 154 in this map a "dark galaxy." It is quite inconspicuous, has a high mass-to-light ratio, and is an ideal candidate for an almost failed void galaxy. But it is in the de Vaucouleurs plane, along with most of the other dwarfs. If most of the nearby mass were in the voids off the plane of the concentration of galaxies then according to the biasing picture the voids contain the seeds of hundreds of failed galaxies, yet the several hundred seeds that did germinate did so near the concentration of giants. Is it sensible to think the germination rate is that sensitive to environment?

A second example that indicates the situation nearby is not peculiar is the intergalactic gas clouds detected by absorption lines in the spectra of quasars and active galaxies. The clouds with baryonic surface densities well below that of galaxies (as in the Lyman-α forest) at redshift $0.1 \lesssim z \lesssim 0.5$ tend to be near galaxies: half of them occupy the one part in 10^5 of space a few hundred kiloparsecs from an optically prominent galaxy (Lanzetta et al. 1995).

How is it that the low-surface-density clouds occur only in such an exceedingly restricted part of space, close to visible galaxies, and that the same is true of the nearby extreme dwarf galaxies? The easy answer is that gravity has emptied the voids, drawing together giant galaxies, dwarfs, and gas clouds. But how could that happen without drawing together more mass than can be accommodated by the small-scale dynamical tests? Easy answers are that Ω is well below unity, or that the dominant contribution to Ω is incapable of clustering on scales of a few megaparsecs.

As mentioned, a related issue is the epoch of galaxy formation. In a commonly discussed scenario for the $\Omega = 1$ CDM model, galaxy assembly is recent, the giant galaxies forming at rare extreme peaks of the mass distribution. In this picture it is easy to imagine the bulk of the mass remains between the galaxies, much less clustered and so not very effective at gravitationally driving peculiar motions, consistent with the small relative velocities of galaxies outside the great clusters. If galaxies were assembled at high redshift it would allow time for intergalactic matter to gravitationally drape itself around the galaxies, in massive halos that one might expect to detect by the dynamics of relative motions of the galaxies. The spheroid components of the large galaxies contain about 10% of the baryon mass required for the successful theory of the origin of the light elements in thermal reactions at $z \sim 10^{10}$ (Copi et al. 1995). If $\Omega = 1$ and the spheroids were assembled at redshift z_f then gravitational accretion would tend to increase the baryon fraction clustered around the spheroids at the present epoch by the factor $\sim 1 + z_f$, bringing pressureless dark mass with it. Thus if $z_f \sim 10$ there is time for accretion of the bulk of the dark matter, but if $z_f \sim 1$ accretion could be quite inefficient. Perhaps it is not surprising that opinions on z_f correlate with opinions on the mass density. In §11.4, the evidence is briefly indicated that the large galaxies had been assembled by $z_f = 3$, an intermediate time.

In the standard cosmological model the mean mass density helps determine the rate of gravitational growth of clustering. An application is the extrapolation back from the present mass distribution to the density fluctuations present at high redshift that would have perturbed the cosmic thermal background radiation (the CMB). If the large-scale fluctuations in the galaxy distribution trace the mass, the computed mass density fluctuations at high redshift are close to what is required to produce the observed anisotropy of the CMB, $\delta T/T \sim 2 \times 10^{-5}$ at angular scales $\sim 1°$. We should pause to admire the remarkable success of the fundamental theory in making this connection. The search for more detailed checks of the CMB anisotropy and a tighter constraint on the mean mass density is described in Ch. 3.

We also have the constraints from the classical cosmological tests. It has been known since the 1930s that observations of objects at high redshift, and measurements of radioactive decay ages and stellar evolution ages, can in principle determine the mean mass density within the relativistic Friedmann–Lemaître cosmological model. A practical application is quite another matter, of course, but recent observational advances look promising. An estimate of the situation is presented in §§11.3 and 11.4.

11.1.2 The standard model

If future observations showed that the mean mass density is less than the critical Einstein–de Sitter value, and we could convince ourselves that $\Lambda \geq 0$, then the standard relativistic cosmology would predict that our world expands into the indefinitely remote future. I expect the general reaction would be to ask why we should trust an indefinite extraplolation of a theory we know can only be an approximation to reality. If the evidence within the standard model showed that the universe will collapse back to a Big Crunch we would be entitled to ask why we should trust a theory that ends in a singularity. More immediate and interesting goals are the test of General Relativity on about the largest scale we can hope to reach, and the search for clues to a deeper cosmology unblemished by the singularity theorems. For this purpose we should consider ahead of completion of the tests which elements of cosmology are so well motivated and tested we should be surprised to lose them. I expect most people would include the following.

(1) The universe is close to homogeneous and isotropic in the large-scale average. Precision tests are the isotropy of the X-ray background from active galaxies, and the deep counts of radio galaxies. A radial density gradient is less tightly constrained – we could be very close to the center of a universe that is spherically symmetric about only one point – but that is bizarre enough to be considered only if driven to it.

(2) The spectra of distant galaxies are shifted to the red. The redshift is defined by the ratio of the wavelength of a feature in the spectrum of the object to the measured (or computed) laboratory wavelength,

$$1 + z = \frac{\lambda_{obs}}{\lambda_{lab}}. \tag{11.1}$$

To the accuracy of the measurements z is independent of wavelength, from the 21-cm hydrogen line through the optical to the X-ray lines of highly ionized iron in intracluster plasma, a range of ten orders of magnitude. The linear Hubble relation between distance defined by the inverse-square law and z is a good representation for galaxies to $z \sim 3$. At $z \lesssim 0.5$ the scatter around the Hubble relation is $\delta z \lesssim 3 \times 10^{-3}$.

(3) Space is filled with the cosmic-background radiation, the CMB, with a near thermal spectrum at $T = 2.73$ K, uniform to $\delta T/T \sim 2 \times 10^{-5}$.

(4) A tired light model in a static universe could account for the redshift–distance relation but would quite mess up the spectrum of the CMB. A single metric tensor model for spacetime satisfies available constraints. The rotational and inferred translational symmetry imply the line element may be written in the Robertson–Walker form,

$$ds^2 = dt^2 - a(t)^2 \left[\frac{dr^2}{1 + r^2/R^2} + r^2(d\theta^2 + \sin^2\theta d\phi^2) \right]. \tag{11.2}$$

In this model, radiation emitted at time t and observed at time t_0 has redshift

$$1 + z = \frac{a(t_0)}{a(t)}, \tag{11.3}$$

independent of wavelength. It is an interesting exercise to check this, and to show that free, initially thermal radiation remains thermal, the temperature varying as $T \propto a(t)^{-1}$; the expansion or contraction of this model universe preserves a thermal radiation spectrum without the need for the traditional thermalizing grain of dust.

(5) The expansion rate equation is the relativistic expression

$$H^2 = \left(\frac{\dot{a}}{a}\right)^2 = \frac{8}{3}\pi G\rho + \frac{1}{a^2 R^2} + \frac{1}{3}\Lambda, \qquad \Omega H_0^2 = \frac{8}{3}\pi G\rho_0. \tag{11.4}$$

The present rate is Hubble's constant, H_0, and the density parameter Ω is a measure of the present mean mass density ρ_0. The cosmological constant term Λ has become socially acceptable only recently, under pressure of the evidence that Ω may be less than unity. We might bear in mind the possibility that fundamental physics allows a term in the stress-energy tensor that acts like a time-variable Λ (Freese et al. 1987; Ratra & Peebles 1988), or maybe something even more unusual. Equation (11.4) is essential to the successful theory of light-element production at $z \sim 10^{10}$, which is an important test, but at large z this expression is dominated by the radiation mass density, so nucleosynthesis does not test the roles of curvature and Λ in the expansion rate history and the spacetime geometry. This part of the theory is probed by the classical cosmological tests, including many new variations on the old themes. As discussed in §§11.3 and 11.4 a key to this enterprise is that the tests may be applied in independent ways that check consistency of the measurements as well as the theory.

(6) The theory of light-element production at high redshift meets demanding tests, so we can add to the standard model the predicted density parameter in baryons (Copi et al. 1995; Schramm & Turner 1996),

$$\Omega_b = 0.03 \pm 0.02 . \tag{11.5}$$

Since the dynamical tests indicate the net mass density is at least $\Omega = 0.05$, we are not quite forced to add nonbaryonic matter to the standard model.

In the Einstein–de Sitter model the mass density dominates the curvature and Λ terms, so $\Omega = 1$. The strongest case for this model is the coincidences argument. The initial conditions for the standard model set two characteristic epochs, the time t_1 at which the mass density ceases being the dominant term in the expansion rate equation (11.4), and the time t_2 at which we have come on the scene. We know t_1 is not much less than t_2, because dynamical measurements show $\Omega \gtrsim 0.1$. If $t_1 \sim t_2$ it would be a remarkable coincidence. More likely is $t_1 \gg t_2$, meaning the Einstein–de Sitter case is a good approximation.[2] If Ω proves to be significantly different from unity a widely discussed rationalization is that our universe has been selected from an ensemble with a distribution of t_1 that decreases rapidly with increasing t_1, with the result that we are likely to appear in a universe with the smallest value of t_1 compatible with the observations, $t_1 \sim t_2$. The concept is elegant and maybe deep, but hard to test. I feel comfortable with the philosophy that accidental coincidences do occur.

I expect most of us admire the concept of Inflation and are hoping it is right, if only because we lack any credible alternative picture of what the universe was like before it was expanding. But in my opinion that does not make the case for including it in the standard model.

11.1.3 The gravitational-instability picture

In the standard model the expanding universe is gravitationally unstable (see also Ch. 2 and §3.2). In the Einstein–de Sitter model, and on scales large enough that nongravitational stresses may be neglected and the smoothed mass-density fluctuations

---

[2] I learned this argument from R. H. Dicke, in about 1960. The earliest comment along roughly similar lines I have found in the literature is by Bondi (1960). Dicke finally published the coincidences argument in 1970.

may be analyzed in linear perturbation theory, the predicted evolution of the departure from homogeneity is

$$\delta\rho/\rho = At^{2/3} + B/t .\qquad(11.6)$$

Here t is the world time measured from high redshift. Because there is a decaying term we can choose initial conditions such that a chaotic universe grows smooth – this is the time-reversed solution for the growth of irregularities in a contracting universe – but the balance of initial values is not credible. That is, in the standard model our universe had to have been very close to homogeneous at high redshift. Equation (11.6) had to contain powers of time rather than exponential functions, because the Einstein–de Sitter model has no characteristic time. In the case of water, we know that the exponential growth of perturbations to an unstable flow means turbulence develops whatever the initial conditions. The slower power-law behaviour means the details of the initial conditions in cosmology are observationally significant; an example is the relation between the present-mass distribution and the angular fluctuations in the CMB that originated at high redshift when the density fluctuations were much smaller. This is why we can hope to use observations to test theories for the origin of the initial conditions for the standard model.

Nongravitational physics is essential – it turns gas into stars – but is it important in assembling the masses in galaxies? Ostriker, Thompson & Witten (1986) show that energetic blast waves could pile material into shells that fragment to form galaxies, but the model does not agree with the remarkably small relative velocities of galaxies outside the great clusters (Peebles 1988). The gravitational instability picture for the formation of structure accordingly assumes gravity dominated the early evolution of the mass distribution on the scale of galaxies and larger.

The cold dark-matter (CDM) model adds to the gravitational instability picture the assumption that the dominant mass component is nonbaryonic (which certainly eases interpretation of equation (11.5)), interacts significantly only through gravity, and has negligible pressure apart from the effect of crossed streams. The CDM model also assumes that the primeval departures from homogeneity are adiabatic (the primeval ratios of number densities of photons, massless neutrinos, baryons, and CDM particles are homogeneous), Gaussian, and with a mass fluctuation power spectrum such that the perturbations to space curvature are independent, or close to independent, of scale. With easy adjustments of the parameters the CDM model can be fitted to the large-scale fluctuations in the galaxy distribution and to the anisotropy of the CMB, which is impressive. The same can be said of other models for structure formation, however. Improved measurements of the anisotropy of the CMB will give a stringent test of models for structure formation. Pending the outcome it is premature to include the CDM picture (as opposed to the likely importance of nonbaryonic matter) in the standard cosmological model.

11.2 Dynamical mass measurements

It is standard procedure in physical science to judge reliability by consistency from independent methods of measurement, but perhaps this is nowhere more important than in extragalactic astronomy where analyses usually are indirect and easily corrupted. This is the first reason for my emphasis on measures of Ω in mass that clusters with the galaxies on relatively small scales, even though biasing is more likely to be important on smaller scales. In assessing results from large-scale flows it behooves us to consider how well we do at the easier problem of small-scale dynamics. This section reviews results from four well-based methods that offer a consistent story for the mass that clusters on scales less than about 10 Mpc.

The second reason for my emphasis is that the measures from small-scale dynamics constrain pictures for how the mass is distributed, and our understanding of the likely nature of the mass distribution informs our assessment of the measurements. To solve the mass problem we will need all the help we can get.

11.2.1 Galaxy relative velocity dispersion

11.2.1.1 Measuring the velocity dispersion

A pair of galaxies at separation $r = r_2 - r_1$ has relative peculiar velocity $v = v_2 - v_1$; the mean across all the pairs at this separation drawn from a fair sample of space is

$$\langle v^\alpha \rangle = v(r) r^\alpha / r \,. \tag{11.7}$$

The departure from the mean is

$$u^\alpha = v^\alpha - v(r) r^\alpha / r \,, \tag{11.8}$$

and the second moment of the scatter is

$$\langle u^\alpha u^\beta \rangle = \delta^{\alpha\beta} \sigma^2(r) + [\sigma_\pi^2(r) - \sigma^2(r)] r^\alpha r^\beta / r^2 \,. \tag{11.9}$$

The dispersion parallel to the separation is σ_π. If the scatter is close to isotropic the line-of-sight rms relative velocity dispersion is $\sigma(r)$. Confused discussions of the statistic $\sigma(r)$ to be found in the current literature likely result from the confused history. Here is my understanding of the situation.

The first estimates of $\sigma(r)$ from a reasonably large and complete redshift catalog yielded (Davis & Peebles 1983a)

$$\sigma(r \sim 1 \text{ Mpc}) = 340 \pm 40 \text{ km s}^{-1}, \tag{11.10}$$

the measured dispersion varying with separation at $0.2 \lesssim r \lesssim 1$ Mpc as expected if galaxies trace mass, in which case $\Omega \sim 0.2$. I was expecting to find a considerably larger dispersion; the result led me to lose faith in the Einstein–de Sitter model. This and the Virgocentric flow to be discussed below led Davis et al. (1985) to astrophysical biasing as a way to reconcile the galaxy relative velocity dispersion with the mass relative velocity dispersion σ_m computed in numerical CDM simulations. Rediscussions show equation (11.10) is biased low, however, because it does not take account of the large peculiar velocities of galaxies in the great clusters.[3] That is, equation (11.10) is not a valid constraint on σ_m in N-body simulations. It is a measure of the galaxy relative velocity dispersion σ_f outside the great clusters, in the field for short, and so may be used to estimate the contribution to Ω by the mass that clusters with field galaxies on the scale of the measurements, $r \lesssim 1$ Mpc.

Equation (11.10) for the field relative velocity dispersion σ_f is consistent with recent estimates from larger samples. The IRAS sample, which is biased against cluster members, indicates $\sigma_f = 317^{+40}_{-49}$ km s^{-1} (Fisher et al. 1994). Marzke et al. (1995) find that in the combined CfA and SSRS2 samples, and outside the $R \geq 1$ Abell clusters, $\sigma_f = 295 \pm 99$ km s^{-1}.

The density of sampling of galaxies in newer catalogs is too low for good measures of $\sigma(r)$ at separations less than one megaparsec. The Davis & Peebles (1983a) analysis and samples selected for the purpose of measuring small-scale relative motions (Peebles 1984a, Fig. 1; Zaritsky & White 1994) consistently indicate

$$\sigma_f(10 \lesssim r \lesssim 100 \text{ kpc}) = 225 \pm 50 \text{ km s}^{-1} . \tag{11.11}$$

At separations $r \lesssim 1$ Mpc the correction for the mean relative velocity $v(r)$ (equation [11.7]) is small and not an important factor in the estimates of the dispersion $\sigma(r)$. At larger separations $v(r)$ dominates the uncertainty in σ. Here the more useful statistic for dynamical mass estimates is $v(r)$. From a fit to $v(r)$ at $r \sim 10$ Mpc from the IRAS catalog Fisher et al. (1994) find in linear perturbation theory (following the method leading to equation (11.23))

$$\Omega \sim 0.25 , \tag{11.12}$$

if galaxies trace mass. Other aspects of this approach are discussed by Kaiser (1987). A number similar to equation (11.12) follows from the relative velocity dispersion on smaller scales, as discussed next.

11.2.1.2 Estimating the density parameter
The evidence is that the galaxy relative velocity dispersion $\sigma(r)$ is an unstable statistic, but that this is easily remedied by excising the regions of the great clusters of galaxies. The value of $\sigma(r)$ in the field at 10 kpc $\lesssim r \lesssim 1$ Mpc seems to be reliably known; it is related to the mean mass density through low-order correlation functions as follows.

The standard model assumes our universe is statistically homogeneous and isotropic, so the correlation functions depend only on relative positions. We can model the mass distribution by particles with mass m and mean number density n. The galaxy number

---

[3] The sample contained a significant number of galaxies from only one cluster, Virgo, with a relatively low velocity dispersion, and even those near the core were inadvertently under-represented, as shown by Somerville, Davis & Primack (1997). Other demonstrations that our result is biased low were found by Mo, Jing & Börner (1993) and Zurek et al. (1994).

density is n_g. The probability that galaxies are found centered in the volume elements dV_1 and dV_2 at separation $r = r_2 - r_1$ is

$$dP = n_g^2[1 + \xi_{gg}(r)]dV_1 dV_2 , \qquad (11.13)$$

where ξ_{gg} is the galaxy two-point correlation function. The probability that galaxies are found centered in dV_1 and dV_2 and a mass particle is in the element dV_3 at position z relative to dV_1 is

$$dP = n_g^2 n F \, dV_1 dV_2 dV_3 , \qquad (11.14)$$

where

$$F = 1 + \xi_{gg}(r) + \xi_{gm}(z) + \xi_{gm}(|z - r|) + \zeta_{ggm}(r, z, |z - r|) . \qquad (11.15)$$

The galaxy–mass cross correlation function $\xi_{gm}(z)$ is just the ensemble average mass density at distance z from a galaxy in excess of and relative to the mean density $\rho_b = mn$. Equation (11.15) defines the reduced galaxy–galaxy–mass three-point function. The contribution to the relative peculiar gravitational acceleration of a pair of galaxies at separation r by a mass particle at position z relative to one of the galaxies is

$$g(r, z) = Gm \left(\frac{z}{z^3} - \frac{z - r}{|z - r|^3} \right) . \qquad (11.16)$$

The ensemble average relative peculiar acceleration is the result of averaging this expression across the probability distribution in equation (11.14) and integrating over z:

$$\langle g \rangle = \frac{n}{1 + \xi_{gg}(r)} \int d^3z \, g(r, z) F . \qquad (11.17)$$

If at small galaxy separations the galaxy clustering is statistically stable, $\langle g \rangle$ is balanced by the relative motions of the ensemble of galaxy pairs:

$$n_g(1 + \xi_{gg})g^\alpha = \frac{\partial}{\partial r^\beta} n_g(1 + \xi_{gg})\langle v^\alpha v^\beta \rangle = \frac{\partial}{\partial r^\alpha} n_g(1 + \xi_{gg})\sigma(r)^2 , \qquad (11.18)$$

where $\alpha, \beta = 1, 2, 3$ label Cartesian coordinates. One can think of the first term as the mean gravitational force per unit volume, for galaxies with unit mass, and the second as the divergence of the Reynolds stress. The third term assumes the relative velocities are isotropic.

This computation is spelled out in more detail in Peebles (1980; §75). The assumption of statistical stability is reasonable at the separations 10 kpc $\lesssim r \lesssim$ 1 Mpc where we have good measures of σ, because the characteristic crossing times are small. The galaxy density n_g drops out of the stability equation, leaving the mean mass density $\rho_b = mn$. The assumption that the galaxies trace the mass leads us to try approximating the galaxy–galaxy–mass function as the galaxy three-point function, $\zeta_{ggm} = \zeta_{ggg}$. A positive check is that this predicts the observed variation of $\sigma(r)$ with r at 10 kpc $\lesssim r \lesssim$ Mpc. The easy way to deal with ζ_{ggm} at small scales is to cut it off at the typical optical size of a galaxy, because we know the bright parts of the galaxies contain relatively little mass. It makes sense to use the old estimates of ζ_{ggg} that under-represent rich clusters, and indicate (Peebles 1980, Fig. 75.1)

$$\Omega(r \sim 1 \text{ Mpc}) = (\sigma_f/900 \text{ km s}^{-1})^2 . \tag{11.19}$$

The density parameter from equations (11.10) and (11.11) is

$$\Omega(10 \text{ kpc} \lesssim r \lesssim 1 \text{ Mpc}) = 0.15 \pm 0.1 . \tag{11.20}$$

We have to add the mass within the Abell radii $r_A = 1.5\,h^{-1}$ Mpc of clusters. The rms line-of-sight velocity dispersion in rich clusters is $\sigma_{cl} = 750$ km s^{-1}, so in the limiting isothermal model the mean mass of a cluster within the Abell radius is $M_{cl} = 2\sigma_{cl}^2 r_a/G = 4 \times 10^{14} h^{-1} M_\odot$, and the product with the cluster number density, $n_{cl} = 4 \times 10^{-6} h^3$ Mpc^{-3}, yields $\Omega_{cl} \sim 0.01$. That is, most of the mass is outside the great clusters.

The error indicated in equation (11.20) represents my estimate of the uncertainty in σ_f. The uncertainty in the model for the three-point function is harder to assess, but that is why we are so interested in having many independent ways to estimate Ω.

11.2.2 Clusters of galaxies

The dark-mass problem was discovered in the first studies of the dynamics of motions of galaxies in the Coma and Virgo clusters (Zwicky 1933; Smith 1936). The mass per galaxy in clusters is important for our purpose as a possible measure of the global value. A systematic survey of cluster galaxy velocity dispersions by Carlberg et al. (1996) yields the mean ratio of cluster mass to R-band luminosity $M/L = 295 \pm 53h\,M_\odot/L_\odot$. If the same number applies to field galaxies, then the field luminosity density, also determined by Carlberg et al., gives

$$\Omega(\lesssim 10 \text{ Mpc}) = 0.24 \pm 0.05 , \tag{11.21}$$

at one standard deviation. The argument approximates the comoving radius of a sphere that contains the mean mass of a cluster.

One has to treat with caution the mass derived from an estimate of the galaxy velocity dispersion in a cluster: the dispersion can be biased high by accidental background and foreground galaxies or low by rejection of tails of the true distribution as accidentals, and the mass depends on a model for stability and the three-dimensional distribution of galaxy motions. Cluster masses from galaxy motions pass two non-trivial tests, however, from the gravitational field required to contain intracluster plasma and to account for gravitational lensing. Indeed, the Carlberg et al. value for M/L is quite familiar from gravitational lensing and intracluster plasma studies (as discussed in Ch. 4).

Many cluster members are ellipticals, spheroids that have lost or never acquired disk components with the attendant luminosity. This has the effect of making the cluster M/L larger than in the field, biasing equation (11.21) high. The effect can be overbalanced by other factors, of course: in the astrophysical biasing picture visible galaxies large and small originate in tight clumps in a smoother distribution of mass. Clusters are observed at $z \sim 1$, and it is a reasonable bet that the mass concentration within the Abell radius of a typical cluster has been present long enough to have accreted substantial amounts of matter from the field. If $\Omega = 1$, M/L in the field has to be considerably larger than in the clusters, so the accretion would tend to make M/L increase with increasing radius in clusters. The effect has not been found; further searches for it are of great interest.

To summarize, we have the wanted checks of consistency from independent measurements of the mass in clusters, and a convincing case also for the field luminosity density, leading to a number close to the value in equation (11.21) if the field mass-to-light ratio M/L is equal that of clusters. The mass-to-light ratios may be different, of course, but the straightforward reading from the astronomy is that the disk luminosity lowers the field M/L, making equation (11.21) an upper bound on the mass that clusters with field galaxies on similar scales.

11.2.3 Virgocentric flow

The concentration of galaxies in and around the Virgo cluster, at distance ~ 20 Mpc, is prominent in maps of the bright galaxies. The perturbation to the local Hubble flow offers a measure of the mass concentrated with the galaxies. Progress in mapping the peculiar-velocity field has made it clear that the local Virgocentric flow may be confused by shear from density fluctuations on larger scales (Faber & Burstein 1988). Let us begin by considering the theorist's approach to separating the effects.

In linear perturbation theory the mass conservation law relating the peculiar velocity $v(r,t)$ and the density contrast $\delta(r,t) = \delta\rho/\rho$ is

$$\nabla \cdot v = -a\frac{\partial \delta}{\partial t} = -a\frac{\dot{D}}{D}\delta . \tag{11.22}$$

The last expression assumes the density fluctuations grew by gravity out of small primeval irregularities, so $\delta \propto D(t)$, where $D(t)$ is the growing solution to the density contrast in linear perturbation theory. The result of integrating equation (11.22) over a sphere of radius R and applying Gauss's theorem is

$$\bar{v} = \frac{1}{3}fH_0R\bar{\delta} . \tag{11.23}$$

The radial inward peculiar velocity averaged over the surface of the sphere is $\bar{v}$, and the mass contrast averaged within the sphere is $\bar{\delta} = \delta M/M$. The dimensionless measure of the rate of growth of clustering is

$$f = \frac{\dot{D}}{D}\frac{a}{\dot{a}} \simeq \Omega^{0.6} . \tag{11.24}$$

The last expression is a useful approximation if $\Lambda = 0$ or space curvature vanishes.

In the literature equation (11.23) sometimes is called a spherical model. There is no condition in principle on how the mass is distributed, though unless the mass is close to centrally concentrated there is not much hope of measuring $\bar{v}$. When $\bar{\delta} \sim 2$ to 3, as for the galaxy concentration within our distance from the Virgo cluster, one might feel more comfortable using the nonlinear spherical model. The results are not much different from equation (11.23) (Davis & Peebles 1983b, Fig. 1), and I doubt they are much more believable unless the mass distribution around the Virgo cluster is more spherical than the galaxy distribution.

A preliminary result from a survey of precision measurements of peculiar velocities of galaxies on the far side of the Virgo cluster by Tonry et al. (1992) is $\bar{v} = 340 \pm 80$ km s^{-1} at our distance, $H_0 R = 1200$ km s^{-1}. Faber & Burstein (1988) find a considerably smaller upper bound. The density contrast in counts of IRAS galaxies within our distance from the Virgo cluster is $\bar{\delta} = 1.4$ (Strauss et al. 1992). IRAS galaxies avoid dense regions (likely because collisions and the ram pressure of intracluster gas have stripped the galaxies of the interstellar gas that fuels bursts of star formation); the standard correction factor of 1.4 would bring the contrast for optical galaxy counts to $\delta_g = 2$. A preliminary analysis of the Optical Redshift Survey (Santiago et al. 1996) by Strauss (1996) gives $\delta_g = 3$. Conservative bounds on the parameters appear to be

$$\bar{v} < 400 \text{ km s}^{-1}, \qquad \bar{\delta} \geq 2, \qquad H_0 R > 1200 \text{ km s}^{-1}. \tag{11.25}$$

In the spherical model this gives

$$\Omega(\lesssim 20 \text{ Mpc}) \lesssim 0.4 . \tag{11.26}$$

It is not easy to measure the mean radial infall, but it would be very surprising if Faber & Burstein (1988) missed an effect as large as $\bar{v} = 400$ km s^{-1}. More central values are $\bar{v} = 250$ km s^{-1} and $\bar{\delta} = 3$, which give

$$\Omega(\lesssim 20 \text{ Mpc}) \sim 0.1 . \tag{11.27}$$

If Ω were unity, the bound $\bar{v} < 400$ km s^{-1} says the mass contrast around the Virgo cluster and within a radius $r \sim 20$ Mpc would have to be $\delta\rho/\rho \lesssim 1$, no more than half that of the galaxy contrast. That is, if $\Omega = 1$ the Local Supercluster presents a strong case for biasing on the scale ~ 20 Mpc.

11.2.4 Dynamics in the neighborhood of the Local Group

We have particularly good information on the relative positions of the galaxies in and near the Local Group, so we can hope to find a particularly well-detailed picture of how they got to where they are from the nearly homogeneous primeval mass distribution required by the standard model. The method in the approach reviewed here uses a numerical version of the action principle of mechanics[4] that produces solutions to the

---

[4] To avoid confusion one should state that the method seeks the stationary points of the time integral of the Lagrangian. This condition traditionally is called Hamilton's Principle, but the habit of calling it the action principle has become so common that only particularly knowledgable people realise there is potential for confusion. This trend is followed in calling the time integral of the Lagrangian the action.

equation of motion consistent with these mixed boundary conditions. I begin with an explanation of the technique, and then comment on results. More detailed discussions may be traced through Peebles (1995).

11.2.4.1 The numerical action method

Consider one particle moving in one dimension. The Hamiltonian is $\mathcal{H}$ and the action is

$$A = \int_0^{t_0} dt\, [p\dot{x} - \mathcal{H}(x,p,t)] \,. \tag{11.28}$$

At a solution to the equation of motion, A is stationary under small variations $\delta p(t)$ of the momentum. That gives the first of Hamilton's equations,

$$\frac{dx}{dt} = \frac{\partial \mathcal{H}}{\partial p} \,. \tag{11.29}$$

The action also is stationary under small variations $\delta x(t)$ of the position. Before applying this condition let us rewrite the expression for A by integration by parts:

$$A = p(t_0)x(t_0) - p(0)x(0) - \int_0^{t_0} dt\, (x\dot{p} + \mathcal{H}) \,. \tag{11.30}$$

The result of varying this expression with respect to $x(t)$ while holding fixed the terms at $t = 0$ and $t = t_0$ is the second of Hamilton's equations,

$$\frac{dp}{dt} = -\frac{\partial \mathcal{H}}{\partial x} \,. \tag{11.31}$$

The terms from the integration by parts usually are eliminated by fixing the positions at the initial and final times. In the application in cosmology we can only fix the present position $x(t_0)$, because the momentum $p(0)$ at high redshift has to vanish. To see this note that the proper peculiar velocity is $v = a\dot{x}$, where $a(t)$ is the expansion parameter (equation (11.2)), so the kinetic energy is $m a^2 \dot{x}^2/2$, and the canonical momentum is $p = ma^2\dot{x} = mav$. In the gravitational instability picture structure grew out of small departures from homogeneity with bounded peculiar motions, so $p \to 0$ at $a \to 0$. The lesson is that in cosmology the usual boundary conditions of the action principle are replaced with the conditions that the mass tracers arrive at given present positions from the near homogeneous initial expansion.

For a numerical expression of this result write the positions at n time steps between high redshift and the present as x_i, $1 \le i \le n$, let the given present position be x_{n+1}, and let the momentum half-way between the times of evaluation of x_i and x_{i+1} be $p_{i+1/2}$. The initial momentum is $p_{1/2} = 0$ at $a_{1/2} = 0$. A numerical approximation to equation (11.28) is

$$A = \sum (x_{i+1} - x_i) p_{i+1/2} - \frac{p_{i+1/2}^2}{2ma_{i+1/2}^2}(t_{i+1} - t_i) - V_i(t_{i+1/2} - t_{i-1/2}) \,. \tag{11.32}$$

The derivatives with respect to the momenta are

$$\frac{\partial A}{\partial p_{i+1/2}} = x_{i+1} - x_i - \frac{p_{i+1/2}}{ma_{i+1/2}^2}(t_{i+1} - t_i),\tag{11.33}$$

and the derivatives with respect to position are

$$-\frac{\partial A}{\partial x_i} = p_{i+1/2} - p_{i-1/2} + \frac{\partial V_i}{\partial x_i}(t_{i+1/2} - t_{i-1/2}).\tag{11.34}$$

When the positions and momenta are adjusted to a stationary point of the action, so the first derivatives of A vanish, equations (11.33) and (11.34) are the usual leapfrog approximation to Hamilton's equations of motion. It is easy to choose the momenta such that they make the right-hand side of equation (11.33) vanish, leaving the x_i as the free variables. That brings equation (11.32) to a familiar numerical approximation to the action as the time integral of the Lagrangian.

For the computation of a stationary point write the first and second derivatives of the action with respect to the free variables x_i as A_i and A_{ij}. Then the iteratively applied shifts of the positions are

$$\delta x_i = -\epsilon \sum_j A_{ij}^{-1} A_j,\tag{11.35}$$

where the constant ϵ is positive. If A is a quadratic function of distance from a stationary point this equation with $\epsilon = 1$ brings the parameters to the solution. More generally, if ϵ is sufficiently small, iterative application reduces the sum of the squares of the first derivatives, $U = \sum(A_i)^2$, and in practice usually brings the positions to a stationary point where U vanishes to machine accuracy.

It is an easy exercise to generalize these results to the motions of N particles in three dimensions, and to refer the positions to one of the particles, the Milky Way. The results predict redshifts of the mass tracers given present positions. It is easy to make a canonical transformation to a numerical action principle that predicts positions given redshifts in the center of mass frame for the system of particles, awkward when the redshifts are referred to one of the particles.

11.2.4.2 The mass model

The particles in a numerical action solution are mass tracers; they are meant as a coarse-grain representation of the evolution of the mass distribution. If the density we are trying to represent does not change very much between neighboring mass tracers the representation is direct. We can go beyond this: a mass tracer can represent the position of the mass concentrated around a galaxy at low redshift, and at high redshift it can represent the position of the center of mass of the matter that is going to end up in and around the galaxy. This is a reasonable approximation if the mass draped around each galaxy comes from a set of initial positions that is not very far from spherical. I comment below on a test of the initial conditions.

Table 11.1. *Mass model*

Object	α	δ	d (Mpc)	cz (km s^{-1})	L (10^{10} L$_\odot$)	M (10^{11} M$_\odot$)	Cutoff (Mpc)
MW	0.0	0.0	0.0	0	2.0	21	0.1
Fnx	39.4	−34.7	0.14	−34	0.002	0.06	0.1
LI	151.4	12.6	0.22	177	0.0003	0.008	0.1
LII	167.7	22.4	0.22	16	0.0001	0.003	0.1
N6822	295.5	−14.9	0.52	33	0.02	0.57	0.1
M31	10.0	41.0	0.75	−123	3.0	28	0.1
I1613	15.6	1.8	0.77	−155	0.01	0.28	0.1
WLM	359.8	−15.8	0.95	−64	0.005	0.14	0.1
N3109	150.2	−25.9	1.26	203	0.02	0.56	0.1
Sex AB	150.6	0.6	1.43	172	0.01	0.28	0.1
I342	55.5	67.9	1.80	176	1.0	14	0.1
N300	13.1	−38.0	2.10	110	0.3	5.6	0.1
N55	3.1	−39.5	2.10	115	0.7	20	0.1
Sculpt	7.0	−26.0	3.20	225	1.0	21	0.5
Maffei I	38.2	59.4	3.50	168	1.0	14	0.1
M81	148.0	69.0	3.50	200	2.0	28	0.5
Cen	201.0	−43.0	3.50	380	2.0	42	0.5
CV I	190.0	40.0	4.50	300	1.0	14	0.5

11.2.4.3 Lack of uniqueness of solutions
The action generally has many stationary points. The lack of uniqueness is most troubling when the relative motions of particles have completed an orbit or more, because this can happen in many very different ways, but there is not a single stationary point even when the density fluctuations are small. It should be understood that the lack of uniqueness of the solution is a property of the equation of motion and the physical boundary conditions, not something peculiar to the numerical action method. For example, the ambiguity in the interpretation of galaxy redshifts near the Virgo cluster is called the triple-value problem (Davis & Peebles 1983b); when spherical symmetry is removed it becomes a multiple-value problem. A procedure for dealing with the problem is described next.

11.2.4.4 Analysis of the nearby galaxy distribution
The mass model in Table 11.1 is meant to represent the visible material within distance ~ 3.5 Mpc from the Milky Way, with a coarser lumping of objects at greater distances. The masses listed in column (7) are based on the assumed local mass-to-light ratio

$$M/L = 150 M_\odot/L_\odot , \qquad (11.36)$$

with an allowance for an increase in M/L with decreasing luminosity. The provenance of this mass model may be traced through Peebles (1995). The choice of M/L is discussed below.

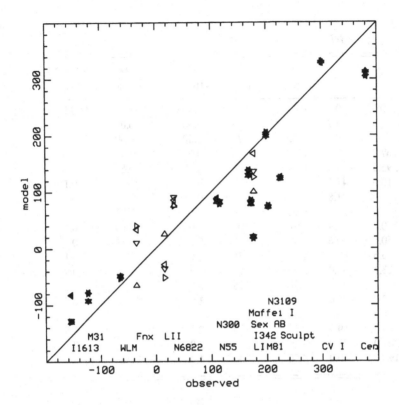

Figure 11.2. Comparison of observed and computed redshifts in four solutions for the motions of the nearby galaxies in a cosmological model with $t_0 = 12$ Gyr and $\Lambda = 0$. Orientations of the triangles are the same for all objects in a solution. Open symbols represent the four galaxies with constrained redshifts.

The four nearest dwarf galaxies in Table 11.1 are close enough to the Milky Way that solutions allow quite a broad range of redshifts, so the redshifts for these galaxies are used as constraints. The search for solutions starts by placing the positions of each of the mass tracers at each of the time steps uniformly at random within a sphere that contains the mass of all the tracers in Table 11.1. The positions are relaxed to a solution by iterated application of equation (11.35). If any of the redshifts of the four dwarfs in the solution differ from the measured values by more than 75 km s^{-1} the solution is discarded. Also, the solution is discarded if any two particles pass closer than proper separation 30 kpc, on the grounds that they likely would have merged.

Figure 11.2 compares observed and predicted redshifts relative to the center of the Milky Way in the first four acceptable solutions in a model with no cosmological constant and expansion time $t_0 = 12$ Gyr. These solutions apply for any combination of space curvature and Hubble parameter that produces the assumed t_0, because local physics is the same. Local physics is different if $\Lambda \neq 0$. In a cosmologically flat universe with $\Omega = 0.1$ and $h = 0.75$ the present peculiar velocities in the solutions are quite similar to the results in Fig. 11.2.

The filled symbols in Fig. 11.2, which are not constrained, show reasonably good agreement between observed and model redshifts. The results would be even better if the model redshifts for the outlying particles were larger. That would happen if t_0 were smaller, but it already is on the small side of what is thought to be acceptable, or if M/L were lower, so as to lower the gravitational deceleration. That is, we get a reasonable picture for the motions of the nearby galaxies with a local mass-to-light ratio not much larger than in equation (11.36), which is about 10% of the global M/L in an Einstein–de Sitter universe. If the local M/L were a fair measure, the density parameter would be

$$\Omega(0.2 \lesssim r \lesssim 3 \text{ Mpc}) \sim 0.1 \, . \tag{11.37}$$

The open symbols in Fig. 11.2 are significant and important as a test of consistency, showing that the mass model can account for the observed velocities of these dwarf galaxies. Zaritsky et al. (1989) note that the large radial velocity of Leo I suggests the mass of the Milky Way within radius ~ 200 kpc is comparable to the mass that, with a like mass around the Andromeda Nebula M31, accounts for the relative motion of these two large galaxies. There is not much room for more mass distributed through the Local Group. Figure 11.2 shows that a similar mass-to-light ratio accounts for the motions of the neighbors of the Local Group. That is, equation (11.37) is the local measure of the mass clustered with the galaxies on scales $0.2 \lesssim r \lesssim 3$ Mpc.

The computation assumes the galaxies in Table 11.1 trace the nearby mass. Branchini & Carlberg (1994) and Dunn & Laflamme (1995) point out that this could not be so in an $\Omega = 1$ CDM or mixed dark-matter cosmogony: the galaxies would have to be more strongly clustered than the mass. The effect is clearly illustrated in Fig. 5 of Branchini & Carlberg (1994), which shows that with their choice of CDM model parameters the mass per galaxy derived from relative motions of galaxies at a separation of 1 Mpc is only about 10% of the mass within 5 Mpc of a typical galaxy. The predicted sharp increase in the apparent mass per galaxy with increasing separation r on scales of 1 to 5 Mpc is not observed in the nearby galaxies: as we have just noted, a fixed value of the mass-to-light ratio, $M/L \sim 150$, gives a satisfactory account of the radial velocities of the Leo I dwarf spheroidal galaxy at 200 kpc distance, the Andromeda Nebula at 700 kpc, and the neighbors of the Local Group at 3 Mpc distance.

It is reasonable to ask whether the discrete mass model in Table 11.1 gives an adequate description of the small primeval departures from homogeneity. A test is to fit the mass tracer peculiar velocities at high redshift from a numerical action solution to the velocity covariance matrix computed in linear perturbation theory from a model for the primeval mass fluctuation power spectrum $P(k)$. The solutions in Fig. 11.2 constrain P at $k \sim 0.4$ Mpc in a model with $\Omega \sim 0.1$ to a value consistent with an extrapolation $P \propto k^{-1}$ from the spectrum of the large-scale galaxy distribution (Peebles 1996). That is, the initial conditions seem reasonable.

11.2.5 Velocity bias

The above analyses treat galaxies as test particles that have been responding to the gravitational field for a Hubble time. Carlberg & Dubinski (1991) point out that dynamical friction slows the galaxies relative to true test particles; Cen & Ostriker (1992) note that the velocity of a protogalaxy newly assembled out of merging gas clouds naturally tends to be lower than the cloud velocities. Evrard, Summers & Davis (1994) remark that the velocities of galaxies are biased low if the galaxies tend to form at minima of the gravitational potential, but this is a spatial bias in the distribution of galaxies relative to the dark matter.

Let us consider a few aspects of the physics and astronomy of velocity biasing in the dynamical analyses reviewed here. The galaxies in and around the Local Group are approaching each other for the first time, according to the solutions in Fig. 11.2. If the dark matter is falling together with the galaxies the Carlberg–Dubinski effect would not apply. Since the Milky Way and M31 are old galaxies and the Local Group is just being assembled the Cen–Ostriker effect is not relevant either. The same comments apply to the dark matter and galaxies flowing toward the Virgo cluster. The dwarf companion Leo I of the Milky Way is of interest for its high velocity. If dynamical drag were important in Leo I the direction of the effect would be to increase the mass of the Milky Way, but it already is close to what is wanted to pull together the Milky Way and M31. If the relative velocity dispersion at separation $r \sim 100$ kpc in equation (11.11) were biased low by dynamical drag one might have thought the large number of galaxy pairs at this separation would merge in a few crossing times, which is less than a Hubble time, contrary to the indications that the typical field galaxy has not suffered recent massive mergers. The evidence from gravitational lensing and confinement of intracluster plasma is that the motions of galaxies in clusters, as used by Carlberg et al. (1996), offers a fair measure of the mass.

In short, it it seems likely that velocity bias has not been a serious factor in the dynamical analyses summarized here.

11.2.6 Concluding remarks

Table 11.2 lists four measures of the mean mass density from dynamical analyses on relatively small scales. The error flags have been set to my estimate of three standard deviations within the uncertainty of the data. Each has been widely discussed and (again in my opinion) has been shown to give a reasonable and sensible measure of the mass that clusters with the galaxies on the indicated scale. Each also has problems. Among the most serious are the model for the galaxy–galaxy–mass correlation function for the relative velocity dispersion; the possibility that the Local Group is atypical, perhaps because we are in a local hole in the dark mass; the possibility the cluster galaxies are born with atypical mass-to-light ratios, and the corruption of $\bar{v}$ for Virgocentric flow by tidal shear. It is hard to believe Nature has fooled us four times, however, so it is concluded that the density parameter $\Omega(< r)$ in mass that clusters with galaxies on scales less than r is reliably bounded at

Table 11.2. *Dynamics on small scales*

Test	Scale	Ω
Relative velocity dispersion	10 kpc to 1 Mpc	0.15 ± 0.1
Local Group & neighbors	200 kpc to 3 Mpc	0.15 ± 0.15
Clusters of galaxies	10 Mpc	0.24 ± 0.15
Virgocentric flow	20 Mpc	0.2 ± 0.2

$$\Omega(\lesssim 10 \text{ Mpc}) = 0.2 \pm 0.15 \,. \tag{11.38}$$

The survey of dynamical mass measurements continues in §§7.5 and 7.6 on the analysis of the deeper Mark III catalog. Dekel obtains constraints that marginally overlap equation (11.38) at $\Omega \sim 0.35$, but he presents a strong case for $\Omega = 1$. Preliminary results from the analysis of the Tully (1988) *Nearby Galaxies Catalog*, at depth $cz = 3000$ km s^{-1}, by Shaya, Tully & Pierce (1992) and Shaya, Peebles & Tully (1995) are in the range of equation (11.38). One interpretation is that the mass of an Einstein–de Sitter universe is dominated by a component with coherence length broad enough to be missed in the measures in Table 11.2 and the deeper Tully sample, but narrow enough to figure in the still-deeper Mark III analysis. We also have to consider the possibility that one or both of the analyses of the Mark III and Tully catalogs have been corrupted by error or bias. The two samples have a significant overlap, and work in discussion on a detailed comparison of peculiar velocities in the overlap sample may shed light on the issue.

A detailed and authoritative survey of dynamical measures of Ω is given by Strauss & Willick (1995). They concentrate on what is to be learned from the large-scale components of the galaxy peculiar-velocity field, a reasonable approach considering this is where there is the best chance that the galaxy distribution is an unbiased measure of the mass distribution. It does skip two key points, however. First, measurements on smaller scales are less subject to systematic errors, and we ought to consider whether we have a convincing solution to the easier problem. It has been argued that the solution in equation (11.38) is reliable. Second, we have a valuable constraint on the mass distribution. If Ω is close to unity and equation (11.38) is correct the missing mass has to have a coherence length larger than about 10 Mpc or otherwise avoid the galaxies on this scale. The Branchini & Carlberg (1994) analysis suggests the $\Omega = 1$ CDM model cannot account for this broad coherence length. The search for a possible remedy will be conditioned by developments in the constraints on Ω from the cosmological tests, as discussed next.

11.3 The cosmological tests

Soon after the discovery of the standard relativistic Friedmann–Lemaître cosmological model, people had worked out how one could in principle test the model by measurements of the expansion time and observations of distant galaxies. A more recent addition to these classical cosmological tests is the constraint from the predicted gravitational evolution of mass clustering. This section reviews the physics of the tests and presents the history of galaxy counts as an illustration of progress in the application of the tests.

11.3.1 *The physics*

To help keep the units straight let us write the expansion rate as a function of redshift as

$$\frac{\dot{a}}{a} = H_0 E(z), \qquad 1 + z = \frac{a_0}{a(t)} . \tag{11.39}$$

The present value is Hubble's constant, H_0, and the dimensionless function $E(z)$ satisfies $E(0) = 1$. In the approximation of the line element by the Robertson–Walker form in equation (11.2) the cosmological tests may be expressed in terms of two functions. The rate of proper displacement of a light ray with respect to redshift is

$$\frac{dl}{dz} = \frac{dt}{dz} = \frac{H_0^{-1}}{(1+z)E(z)} . \tag{11.40}$$

The velocity of light is set to unity, and H_0^{-1} is the Hubble length. The angular size distance is

$$y(z) = H_0 a_0 r(z) = \frac{1}{\Omega_R^{1/2}} \sinh \int_0^z \frac{\Omega_R^{1/2} dz}{E(z)} , \tag{11.41}$$

where the curvature parameter is

$$\Omega_R = \frac{1}{(H_0 a_0 R)^2} . \tag{11.42}$$

Equation (11.41) assumes space has negative curvature. If the curvature is positive the hyperbolic function becomes a trigonometric sine. In a cosmologically flat model $\Omega_R \to 0$.

The integral of equation (11.40) gives the time of expansion from the very early universe to redshift z, to be compared to radioactive decay ages and stellar evolution ages. This expression also fixes the mean rate of appearance of absorption lines in the spectrum of a quasar produced by the intersection of the line-of-sight with clouds with proper number density $n(z)$ and cross section $\sigma(z)$ at redshift z:

$$\frac{dP}{dz} = \sigma n(z) \frac{dl}{dz} = \frac{\sigma n(z) H_0^{-1}}{(1+z)E(z)} . \tag{11.43}$$

Equation (11.41) determines the angular size θ of an object with proper diameter d observed at redshift z:

$$\theta = \frac{dH_0(1+z)}{y(z)} . \tag{11.44}$$

In this metric theory for the redshift, the observed surface brightness integrated over frequency varies as $I \propto (1+z)^{-4}$. The observed energy flux density f from an object at redshift z is the integral of its surface brightness over its angular size, from which it is an easy exercise to check that the energy flux density from an object with luminosity L at redshift z is

$$f = \frac{H_0^2 L}{4\pi y(z)^2 (1+z)^2} . \tag{11.45}$$

The count dN/dz of objects per steradian depends on both functions: the proper volume bounded by a given solid angle and redshift interval is the product of the subtended proper area and the rate of increase of proper radial displacement with increasing redshift. The product with the proper number density $n(z)$ is the predicted count,

$$\frac{dN}{dz} = F_n(z) \frac{n(z) H_0^{-3}}{(1+z)^3}, \quad F_n(z) = \frac{y(z)^2}{E(z)} . \tag{11.46}$$

Yet another combination of the two functions enters the probability of gravitational lensing. In the approximation of the mass distribution in a gravitational lens as an isothermal sphere with vanishing core radius the density run is $\rho = \sigma^2/2\pi G r^2$, where σ is the line-of-sight velocity dispersion. The gravitational deflection angle in this model is $\alpha = 4\pi\sigma^2$, independent of impact parameter. If the number density of lensing objects is $n(z)$ the probability the line-of-sight to a source passes close enough to a lens at redshift z in the range dz to produce a double image satisfies (Turner, Ostriker and Gott 1984)

$$\frac{dP}{dz} = \frac{16\pi^3}{H_0^3} \frac{n(z)\sigma^4}{(1+z)^3 E(z)} \left(\frac{y_{ol} y_{ls}}{y_{os}} \right)^2 , \tag{11.47}$$

where y_{ol} is the angular size distance from observer to lens, y_{ls} the angular size distance from lens to source, and y_{os} is the distance from observer to source.[5]

The lesson is that in a metric theory for a nearly homogeneous and isotropic spacetime the evolution $a(t)$ is constrained by a considerable variety of observations. In the Friedmann–Lemaître model this translates to a test of consistency of a parametric form for the expansion rate equation (equations (11.4) and (11.39)),

$$E(z) = \left[\Omega(1+z)^3 + \Omega_R(1+z)^2 + \Omega_\Lambda \right]^{1/2} . \tag{11.48}$$

The density parameter Ω is defined in equation (11.4), the curvature parameter is defined in equation (11.42), and the similar definition of Ω_Λ represents the cosmological constant Λ (which could generalize to a time-variable component).

---

[5] All these expressions assume the spatially homogeneous Robertson–Walker line element in equation (11.2). Space curvature fluctuations due to local irregularities in the mass distribution do not significantly affect the rate of intersection of objects along a line-of-sight (equation (11.43)), or mean counts as a function of redshift, but can perturb the energy flux density from an object (equation (11.45)). The smooth shapes of the luminous arcs produced by lensing in clusters of galaxies indicate the effect of density fluctuations along the line-of-sight is not large, but the issue will have to be reconsidered as the applications of the cosmological tests improve.

The model is tested also by the evolution of departures from homogeneity. Examples are the shape of the galaxy two-point correlation function (Peebles 1974), the rate of growth of the mass concentrations in clusters of galaxies (Richstone, Loeb & Turner 1992), and the angular fluctuations in the CMB from the interaction at high redshift with the primeval mass-density fluctuations that evolve into the present large-scale galaxy distribution. When nongravitational interactions may be neglected and the smoothed density contrast $\delta = \delta\rho/\rho$ is small enough to be described by linear perturbation theory the time evolution of the contrast satisfies

$$\frac{d^2\delta}{dt^2} + 2\frac{\dot a}{a}\frac{d\delta}{dt} = 4\pi G\rho\delta, \qquad (11.49)$$

where $\rho(t)$ is the mean mass density. If Ω_Λ is constant it is an easy exercise to check that the growing solution is

$$\delta(z) = \frac{5\Omega E(z)}{2} \int_z^\infty \frac{1+z}{E(z)^3} dz, \qquad (11.50)$$

where the normalization has been chosen so that at high redshift

$$D(z) \to \frac{1}{1+z} = \frac{a(t)}{a_0}. \qquad (11.51)$$

In a low-density open model with $\Lambda = 0$ this is a good approximation at $1+z \gtrsim \Omega^{-1}$, after which the growth of D slows. In a low-density cosmologically flat model the approximation holds to $1+z \sim \Omega^{-1/3}$. The more rapid transition from the early matter-dominated expansion rate in equation (11.48) makes the gravitational growth factor larger in a flat model than in an open model with the same Ω (Peebles, 1984b; Carroll, Press & Turner 1992).

11.3.2 Galaxy counts

The first large-scale cosmological test was Hubble's (1936) counts of galaxies as a function of apparent magnitude[6] to the depth reached by the 100-in telescope at Mount Wilson. Comparison with more recent measurements of the galaxy count–redshift relation indicates Hubble reached $z \sim 0.4$, an impressive feat. When the 200-in telescope on Mount Palomar was designed in the early 1930s the further application of the cosmological tests was a key project, to use the current term. When Sandage (1961) wrote his classic study of the ability of the 200-in telescope to discriminate among cosmological models the largest measured redshifts were well below $z = 0.5$, and Sandage concluded that at this depth there is too much noise in counts of galaxies to allow a useful test. Now galaxies are observed at redshifts well above unity, where the predicted count as a function of redshift can be quite sensitive to the cosmological parameters. This is illustrated in Fig. 11.3, which shows the factor $F_n(z)$ in equation (11.46). The correction for galaxy evolution can be large too, of course, but

---

[6] The apparent magnitude is $m = 2.5 \log_{10} f + \text{constant}$, where f is the flux density, as in equation (11.45)

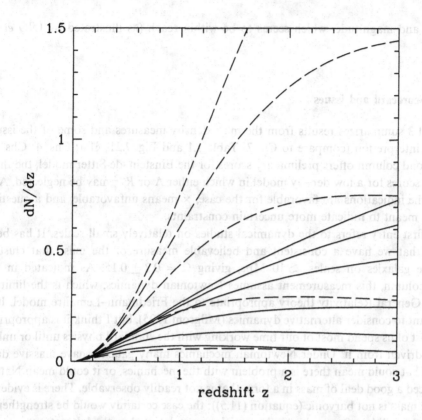

Figure 11.3. Galaxy counts as a function of redshift. Solid curves assume $\Lambda = 0$, dashed curves a cosmologically flat model. In each set of curves the density parameter from bottom to top is $\Omega = 2, 1, 0.5, 0.2, 0.1$, and 0.05.

there are measures of what happened. For example, it is known that the large elliptical galaxies in clusters have quite uniform spectra back to $z \sim 1$, suggesting the bulk of the stars formed well before that (Oke 1984). If so, the evolution of a galaxy can be modelled as the evolution of a population of stars with similar ages.

In an elegant approach Driver et al. (1996) use Hubble Space Telescope images to identify distant elliptical galaxies. They find that their counts of elliptical galaxies as a function of magnitude are inconsistent with the predicted rapid increase of $F_n(z)$ in a cosmologically flat model with $\Omega \lesssim 0.2$. The deeper K-band Keck counts by Djorgovski et al. (1995) are consistent with what would be expected in a low-density open universe with $\Lambda = 0$, and larger than predicted in an Einstein–de Sitter universe. But the analysis by Yoshii & Peterson (1995) favors a cosmologically flat model with $\Omega \sim 0.2$. In short, the counts have reached the precision adequate for a significant constraint on Ω, but the accuracy still is limited by uncertainties in completeness of the counts as a function of redshift and in the correction for galaxy evolution. Both will be better constrained by the deep measurement of the joint distribution of galaxy counts as a function of redshift and magnitude, which seems to be within reach (as illustrated by Lilly et al. 1996).

11.4 Scorecard and issues

Table 11.3 summarizes results from the mass density measures and some of the issues in their interpretation (compare to Ch. 7 Table 7.1 and Fig. 7.21; §1.4; Chs. 4; Chs. 8). The second column offers preliminary scores for the Einstein–de Sitter model, the third column scores for a low density model in which either Λ or R^{-2} may be neglected. A $\sqrt{}$ means the indications are favorable for the case, × means unfavorable, and a question mark is meant to indicate more uncertain constraints.

The first entry refers to the dynamical studies on relatively small scales. It has been argued that we have a consistent and believable measure of the mass that clusters with the galaxies on scales $\lesssim 10$ Mpc, giving $\Omega = 0.2 \pm 0.15$. As indicated in the fourth column, this measurement assumes Newtonian dynamics, which is the limiting case of General Relativity theory appropriate to the Friedmann–Lemaître model. It is important to consider alternative dynamics (Milgrom 1995), but I think it is appropriate that most of us spend most of our time working with the standard physics until or unless we are driven from it. Under Newtonian mechanics many galaxies have massive dark halos. That could mean there is a problem with the mechanics, or it could mean Nature has placed a good deal of mass in a form that is not readily observable. There is evidence that the mass is not baryonic (equation (11.5)); the case certainly would be strengthened if a good candidate for nonbaryonic particles were detected in the laboratory.

In the $\Omega = 1$ CDM family of models the dark mass outside the galaxy concentrations is seeded for galaxy formation but apparently has been quite inefficient at producing compact giant or dwarf galaxies. An issue for astronomy is whether this matter might be traced by low-surface-density galaxies (e.g. Roennback & Bergvall 1994; de Blok, van der Hulst & Bothun 1995; Ferguson & McGaugh 1996). If so, it will be an interesting challenge to account for the apparent lack of clear transition cases, objects intermediate between the high-surface-density galaxies that cluster and the low-surface-density ones that avoid them.

The second entry refers to dynamical analyses on larger scales, where the density fluctuations are closer to linear. Since this is more difficult perhaps it is not surprising that the results are less concordant. An issue for theoretical cosmology is whether some variant of the CDM model could account for the large coherence length of the missing mass of an Einstein–de Sitter universe. Theoretical physicists might be led to consider dark matter with an effective Jeans length large enough to have been missed in samples shallower than Mark III.

The expansion-time test in the third entry is appealingly direct. A recent discussion of the evolution of stars in globular clusters bounds the age at $t_0 > 12.1$ Gyr at 95% confidence within the astrophysics as now understood (Chaboyer et al. 1996). In the Einstein–de Sitter model this translates to the bound $H_0 = 2/3t_0 < 54$ km s^{-1} Mpc^{-1}.

Table 11.3. *Scorecard and issues*

Test	Einstein–de Sitter	$\Omega = 0.2 \pm 0.15$	Issues for physics	Issues for astronomy & astrophysics
1. Dynamics and biasing on scales $\lesssim 10$ Mpc	×	√	Newtonian dynamics; detecting dark matter	What's in the voids? Low surface density objects?
2. Dynamics on scales $\gtrsim 10$ Mpc	√?	√?	Could the matter Jeans length be $\gtrsim 10$ Mpc?	Get a consistent story for Ω
3. Expansion times & ages of stars & elements	?	√	Rate of change of radial distance dl/dz	It's the distance scale, stupid
4. Angular size–redshift relation	?	?	Angular size distance; $I \propto (1+z)^{-4}$	Find some standard objects
5. Galaxy count–magnitude–redshift relation	√?	√?	Radial & angular size distances	Galaxy merging & star formation histories
6. Rate of lensing of quasars by galaxies	√	√?	Gravitational deflection of light	Properties of early-type L_* galaxies
7. Plasma mass fraction in clusters of galaxies	×	√	Initial conditions	Measuring structure & evolution of clusters
8. Large-scale spectrum of the galaxy distribution	×?	√	Inflation & the CDM model	Measuring large-scale peculiar motions
9. Clusters of galaxies at high & low redshift	√?	√?	Evolution, gravitational & nongravitational	Cluster mass function as a function of z
10. The redshift of assembly of galaxies	√?	√?	The primeval seeds of galaxies	Finding young galaxies
11. 3 K background spectrum & anisotropy	√	√	Relativity; nature of seeds of structure	Mass biasing; early structure formation

Two recent measurements yield a Hubble parameter larger than this by 1.5 standard deviations (Freedman *et al.* 1994) and 1.9 standard deviations (Tanvir *et al.* 1995), not quite far enough to be a compelling problem for the model. The well-planned programs for improving the accuracy and reliability of the measurements of H_0 certainly will be followed with interest.

Sandage (1961) chose the magnitude–redshift relation as a promising cosmological test. The measurements now reach $z \sim 3$ for radio galaxies, but the errors grow with redshift about as fast as the predicted difference of predictions for large and small Ω in the standard model (McCarthy 1993). In a metric theory the surface brightness varies as $I \propto (1+z)^{-4}$, such that the redshift–magnitude test is equivalent to the redshift–angular size test (equations (11.44) and (11.45)). Observational studies of the latter relation are discussed by Daly (1994).

In the standard model the galaxy count–redshift relation and the gravitational lensing rate depend on combinations of the functions probed in entries three and four. I have entered a marginally less favorable grade for the Einstein–de Sitter model from galaxy counts because the large counts found in deep K-band surveys are easier to understand in low-density models (Fig. 11.3). Galaxy evolution has significantly affected optical counts. The effect is thought to be easily understood: the UV luminosity of this radiation at emission is readily affected by a relatively small star formation rate. The issue is whether there is a better chance of understanding the effect of evolving star populations in the I-band and K-band observations of radiation emitted in the optical (e.g., Lilly et al. 1996).

Fukugita and Turner (1991) show that the rate of lensing of high-redshift quasars by foreground galaxies is quite sensitive to the cosmological parameters. Kochanek (1996) concludes that the observed lensing rate requires $\Omega > 0.34$ if the universe is cosmologically flat; the limit on Ω is lower if $\Lambda = 0$. This and other analyses of the lensing rate use the natural assumption that elliptical galaxies have dark halos with density related to the star velocities in the same way as for spiral galaxies. Massive dark halos are needed to gravitationally contain the pools of plasma observed around giant ellipticals in clusters of galaxies, but it is much harder to establish the case for the less-luminous ellipticals that would be responsible for lensing at angular separations less than about one second of arc, which is the key to the constraint on Ω. Analyses of star velocity distributions indicate ellipticals contain dark matter, but are not clear on the amount (Carollo et al. 1995). There is one case, the L_* elliptical M105, where a detailed study of the motions of planetary nebulae (Ciardullo, Jacoby & Dejonghe 1993) and of a conveniently placed ring of atomic hydrogen (Schneider 1991) yield a well-motivated mass model that requires little dark matter. If this were a common situation among the less-luminous ellipticals, present estimates of Ω from strong lensing would be biased high.

We might pause to note that in the standard model the gravitational deflection of light contains the Einstein factor of two over the naive Newtonian computation. Perhaps the best chance to check this is from the gravitational lensing in clusters, where we have mass measures from galaxy velocities and the intracluster plasma.

In Ch. 8 the plasma mass fraction in clusters, the expansion time, the lensing rate, and the large-scale fluctuation spectrum of the galaxy distribution are used to constrain the density parameter in a cosmologically flat model to $\Omega \sim 0.35$ (see also Chs. 4, 1 and 7). The last of these constraints is powerful but depends on the CDM model. This and other models for structure formation will be very closely tested by the measurements of the angular fluctuations of the CMB, as indicated in entry eleven in Table 11.3 and discussed in detail in Ch. 3.

The epoch of assembly of the mass concentrations in galaxies is important for three issues. First, it constrains biasing scenarios: if galaxies are assembled early there is time for intergalactic matter to drape itself around galaxies and time for galaxies plus halos to respond to the gravitational attraction of neighboring galaxies. Second, for a given present mass fluctuation spectrum the density fluctuations that might trigger galaxy formation at a given redshift tend to be larger in a lower-density universe, hence capable of producing galaxies earlier. Third, the epoch of galaxy formation is a measure of the character of the primeval departures from homogeneity.

In hierarchical or bottom-up up scenarios galaxies form before groups and clusters of galaxies. This certainly agrees with the situation in our neighborhood: the Local Group is forming now out of old galaxies. The evidence is that the large galaxies were on place at $z \sim 3$: Steidel et al. (1996) identify objects at $z = 3$ to 3.5 with half-light radii, velocity dispersions, and comoving number density characteristic of ellipticals, and Wolfe et al. (1994) and Lu et al. (1996) find evidence for protodiscs at similar redshifts, from velocity structures in the high column density gas clouds that appear in quasar spectra. The presence of high column density clouds at $z = 3$ challenges some models for astrophysical biasing (Ma & Bertschinger 1994). If galaxies of stars were in place this early it would be a still more serious constraint.

The baryon mass in stars and gas in these candidate young galaxies is at least 10% of the total (based on light-element abundances; equation (11.5)), and the young galaxies presumably have captured a like fraction of the dark matter. At least as much matter is between the galaxies at $z = 3$, associated with the Lyman-α forest clouds. The baryonic mass in the forest clouds is lower at $z = 1$, and at $z \sim 0.5$ the clouds typically are much nearer a galaxy than the mean distance between galaxies (Lanzetta et al. 1995). This could mean the bulk of the intergalactic matter present at $z = 3$ has been collected by the galaxies. On the other hand, Stocke et al. (1995) find that very-low-surface-density clouds at the present epoch tend to avoid the galaxies. Perhaps this is a sign of a still significant intergalactic component?

11.5 Concluding remarks

Further progress in this very active field may reduce the number of question marks in Table 11.3 and reveal that the number of passing grades ($\sqrt{}$) peaks around a small range of values of Ω at which the number of failing grades ($\times$) is small and attributable to systematic errors. If that happens we will have cross-checks of consistency of interpretation of the observations and a beautiful test of the fundamental theory. It may prove impossible to find a consistent story, of course; that will teach us we have to reconsider the fundamental theory.

Here is my list of portents from Table 11.3. Dekel presents the strongest evidence for the Einstein–de Sitter case, from the analysis of the Mark III catalog. The weight of the rest of the evidence rather consistently favors $\Omega = 0.15 \pm 0.1$. If Ω should prove to be well below unity the elegant interpretation would be that there is only one significant contribution to the mass, the baryons we all believe exist. This idea is not excluded but it certainly is pressed. If $\Omega = 1$ then in my opinion the CDM family of models is not a promising way to account for the slow trend of the apparent value of Ω with the size of the dynamical system; a better bet is a term in the stress-energy tensor with a large present value of an effective Jeans length. Finally, the central lesson from Table 11.3 is that the observational tests have reached the precision needed for interesting constraints on the cosmological parameters. Systematic errors are not yet under adequate control, but improvements in the accuracy are within reach of work in progress, an exciting prospect.

Acknowledgments

This work was supported in part by the US National Science Foundation.

References

Bondi, H. 1960, *Cosmology* 2nd edn., (Cambridge: Cambridge University Press), p. 166

Branchini, E. & Carlberg, R. G. 1994, *ApJ*, **434**, 37

Carignan, C. & Freeman, K. C. 1988, *ApJ*, **332**, L33

Carlberg, R. G. & Dubinski, J. 1991, *ApJ*, **369**, 13

Carlberg, R. G., et al. 1996, *ApJ*, **462**, 32

Carollo, C. M., et al. 1995, *ApJ*, **441**, L25

Carroll, S. M., Press, W. H. & Turner, E. L. 1992, *ARA&A*, **30**, 499

Cen, R. & Ostriker, J. P. 1992, *ApJ*, **399**, L113

Chaboyer, B., Demarque, P., Kernan, P. J. & Krauss, L. M. 1996, *Science*, **271**, 957

Ciardullo, R., Jacoby, G. H. & Dejonghe, H. B. 1993, *ApJ*, **414**, 454

Copi, C. J., Schramm, D. N. & Turner, M. S. 1995, *Science*, **267**, 192

Daly, R. A. 1994, *ApJ*, **436**, 38

Davis, M., Efstathiou, G., Frenk, C. S. & White, S. D. M. 1985, *ApJ*, **292**, 371

Davis, M. & Peebles, P. J. E. 1983a, *ApJ*, **267**, 465

Davis, M. & Peebles, P. J. E. 1983b, *ARA&A*, **21**, 109

de Blok, W. J. G., van der Hulst, J. M. & Bothun, G. D. 1995, *MNRAS*, **274**, 235

Dicke, R. H. 1970, *Gravitation and the Universe* (Philadelphia: American Philosophical Society)

Djorgovski, S., et al. 1995, *ApJ*, **438**, L13

Driver, S. P., Windhorst, R. A., Phillipps, S. & Bristow, P. D. 1996, *ApJ*, **461**, 525

Dunn, A. M. & Laflamme, R. 1995, *ApJ*, **443**, L1

Evrard, A. E., Summers, F. J. & Davis, M. 1994, *ApJ*, **422**, 11

Faber, S. M. & Burstein, D. 1988, in *Large-Scale Motions in the Universe*, eds. V. C. Rubin & G. V. Coyne (Princeton: Princeton University Press)

Ferguson, H. C. & McGaugh. S. S. 1996, *ApJ*, **440**, 470

Fisher, K. B., et al. 1994, *MNRAS*, **267**, 927

Freedman, W. L., et al. 1994, *Nature*, **371**, 757

Freese, K., Adams, F. C., Frieman, J. A. & Mottola, E. 1987, *Nucl. Phys. B.*, **287**, 797

Fukugita, M. & Turner, E. L. 1991, *MNRAS*, **253**, 99

Hubble. E. 1936, *Realm of the Nebulae* (New Haven: Yale University Press)

Kaiser, N. 1987, *MNRAS*, **227**, 1

Kochanek, C. S. 1996, *ApJ*, **466**, 638

Lanzetta, K. M., Bowen, D. V., Tytler, D. & Webb, J. K. 1995, *ApJ*, **442**, 538

Lilly, S. J., Le Fèvre, O., Hammer, F. & Crampton, D. 1996, *ApJ*, **460**, L1

Lu, L., Sargent, W. L. W., Womble, D. S. & Barlow, T. A. 1996, *ApJ*, **457**, L1

Ma, C. -P. & Bertschinger, E. 1994, *ApJ*, **434**, L5

Marzke, R. O., Geller, M. J., da Costa, L. N. & Huchra, J. P. 1995, *AJ*, **110**, 477

McCarthy, P. J. 1993, *ARA&A*, **31**, 639

Milgrom, M. 1995, *ApJ*, **455**, 439

Mo, H. J., Jing, Y. P. & Börner, G. 1993, *MNRAS*, **264**, 825

Oke, J. B. 1984, in *Clusters and Groups of Galaxies*, eds. Mardirossian, et al. (Dordrecht: Reidel), p. 99

Ostriker, J. P., Peebles, P. J. E. & Yahil, A. 1974, *ApJ*, **193**, L1

Ostriker, J. P., Thompson, C. & Witten, E. 1986, *Phys. Lett. B.*, **180**, 231

Peebles, P. J. E. 1974, *ApJ*, **189**, L51

Peebles, P. J. E. 1980, *The Large-Scale Structure of the Universe* (Princeton: Princeton University Press)

Peebles, P. J. E. 1984a, *Ann. N. Y. Acad. Sci.*, **422**, 118

Peebles, P. J. E. 1984b, *ApJ*, **284**, 439

Peebles, P. J. E. 1988, *ApJ*, **332**, 17

Peebles, P. J. E. 1995, *ApJ*, **449**, 52

Peebles, P. J. E. 1996, *ApJ*, **473**, 42

Ratra, B. & Peebles, P. J. E. 1988, *Phys. Rev. D.*, **37**, 3406

Richstone, D., Loeb, A. & Turner, E. L. 1992, *ApJ*, **393**, 477

Roennback, J. & Bergvall, N. 1994, *A&A*, **108**, 193

Sandage, A. 1961, *ApJ*, **133**, 355

Santiago, B. X., et al. 1996, *ApJ*, **461**, 38

Schneider, S. E. 1991, in *Warped Discs and Inclined Rings Around Galaxies*, eds. S. Casertano, P. D. Sackett & F. H. Briggs (Cambridge: Cambridge Univ. Press), p. 25

Schramm, D. N. & Turner, M. S. 1996, *Nature*, **381**, 193

Shaya, E. J., Peebles, P. J. E. & Tully, R. B. 1995, *ApJ*, **454**, 15

Shaya, E. J., Tully, R. B. & Pierce, M. J. 1992, *ApJ*, **391**, 16

Smith, S. 1936, *ApJ*, **83**, 23

Somerville. R., Davis, M. & Primack, J. R. 1997, *ApJ*, **479**, 616

Steidel, C. C., et al. 1996, *ApJ*, **462**, L17

Stocke, J. T., Shull, J. M., Penton, S., Donahue, M. & Carilli, C. 1995, *ApJ*, **451**, 24

Strauss, M. A. 1996, private communication

Strauss, M. A., Davis, M., Yahil, A. & Huchra, J. P. 1992, *ApJ*, **385**, 421

Strauss, M. A. & Willick, J. A. 1995, *Phys. Rep.*, **261**, 271

Tanvir, N. R., Shanks, T., Ferguson, H. C. & Robinson, D. R. T. 1995, *Nature*, **377**, 27

Tonry, J. L., et al. 1992, *BAAS*, **23**, 1341

Tully, R. B. 1988, *Nearby Galaxies Catalog* (New York: Cambridge Univ. Press)

Turner, E. L., Ostriker, J. P. & Gott, J. R. 1984, *ApJ*, **284**, 1

White, S. D. M., Tully, R. B. & Davis, M. 1988, *ApJ*, **333**, L45

Wolfe, A. M., et al. 1994, *ApJ*, **435**, L101

Yoshii, Y. & Peterson, B. A. 1995, *ApJ*, **444**, 15

Zaritsky, D., Olszewski, E. W., Schommer, R. A., Peterson, R. C. & Aaronson, M. A. 1989, *ApJ*, **345**, 759

Zaritsky, D. & White, S. D. M. 1994, *ApJ*, **435**, 599

Zurek, W., et al. 1994, *ApJ*, **431**, 559

Zwicky, F. 1933, *Helv. Phys. Acta*, **6**, 110

Index

Abell clusters 136, 138, 162, 163
active galactic nuclei (AGN) 32, 337–8
age of the universe 12–16, 310–12
arclets 415–23
axions 35, 36

baryon clusters, measuring Ω_0 28–9, 151–2
baryonic mass, galaxies 5, 107–8
bias
   galaxy mass-density 274–8, 303–6, 312
   linear 183, 196–7
   velocity 454
Big Bang Nucleosynthesis (BBN) 27–9, 32
black holes (BHs) 337–57
   detection 345–7
   M87 352–3
   NGC3115 350–2
   NGC4258 353–4
   predictions from quasars 347–9
   searches for 349–54
black-body radiation 118
bosons 37–8
bottom-up formation models 110–11
brightest cluster galaxies, distance indicators 237–9
bulk velocity 281–3

Cepheid variables
   Hubble constant 16–17, 214, 216–19, 236, 245
   surface brightness fluctuations 232–3
CfA2 redshift survey 176
clusters *see* galaxy clusters
COBE-normalized models 56, 62–8, 72, 105, 123
cold dark matter (CDM)
   ΛCDM models 22–3, 54–75
   axions 35, 36
   cold + hot dark matter (CHDM) models 25, 30–1, 54–75
   density fluctuations 103–5
   gravitational lensing 402–3
   linear theory 62–5
   Massive Astrophysical Compact Halo Objects (MACHOs) 34, 36, 39, 374–9
   models 6, 33–9, 54–75, 285–6, 442
   nonrelativistic 5
   small-scale simulations 65–8
   standard cold dark matter (SCDM) 65, 189–90
   supersymmetry 36–9
   Weakly Interacting Massive Particles (WIMPs) 35, 36–9
comoving coordinates 8–9
constants, values 6
continuity equation 262
correlation function, galaxy clusters 160–4
COsmic Background Explorer *see* COBE
cosmic flows, estimates of Ω_0 and β 252–3, 304
cosmic inflation *see* inflation
cosmic microwave background (CMB)
   acoustic peaks 309–10
   anisotropies 53, 118–29, 439
   fluctuations 44, 118–23, 273–5, 286–7, 313
   gravitational lensing 425
   maps 126–9, 267–9, 275
   measuring β 300–1
cosmic string 42–4, 57
cosmic virial theorem (CVT) 25
cosmochronometry 14
cosmological constant (Λ) 9–10, 21–3, 32–3, 406, 440–1
cosmological density parameter (Ω_0)
   cluster baryons 27–9, 151–2
   cluster correlations 161–2
   cluster velocity 165–6
   constraints 306–12, 313–14
   cosmic flows 252–3, 304
   cosmological tests 455–60
   definition 321–2
   early structure formation 30–2
   galaxy halo scales 26–7
   Hubble diagrams 234–6
   inflation models 51–2
   initial probability distribution function 292–3
   large-scale measurements 24
   measurements 23–33, 151–2, 252–3, 290–3
   peculiar velocities 290–3
   point masses 381–2
   small-scale measurements 25–6
   value 5, 286–7, 463
   value in simulations 323–6
   very-large-scale measurements 23–4
cosmological models 7–8, 54–7, 439–41

Index

cosmological simulations 321–34
cosmology, physical constants 6

damped Lyman α systems 31, 33, 56, 68–9, 116–17
   see also Lyman α clouds
dark matter 3–75
   astrophysical properties 5
   classification 33–5
   distribution 195–8
   galaxies relationship 195–8
   galaxy clusters 155
   galaxy halos 26–7
   gravitational force 10–12
   gravitational lensing 425
   large-scale structure formation 280, 313
   nonbaryonic 5
   simulations 322–3
   see also cold dark matter; hot dark matter; warm dark matter
de Sitter cosmology 44–5
degenerate parameter β, measuring 204, 253, 293–306
density comparisons, measuring β 294–7
density fluctuations 102–5
deuterium, measurements 27–9, 32
distance indicators (DIs) 215–16
   brightest cluster galaxies 237–9
   Cepheid variables 216–19
   Fundamental Plane 225–9
   galaxies 213–46
   Hubble constant 215–16
   Malmquist bias 241–4
   redshift-distance catalogs 239–41
   supernovae 233–7
   surface brightness fluctuations 229–33
   Tully–Fisher relation 219–25

Einstein Cross 383–4
Einstein radius 371–2
Einstein rings 362
Einstein–de Sitter cosmology 22–3, 92, 308, 435, 441
elliptical galaxies
   boxy and disky types 344–5, 357
   formation 115–16
   Fundamental Plane relations 225–9, 245
   lens models 394–5
Equivalence Principle 8
ESO Slice Project (ESP) 178
Euler equation of motion 88, 262
extragalactic background infrared light (EBL) 32

Faber–Jackson relation 225–6
Fermat's principle, gravitational lensing 387–92
fermions 37
flatness problem 46
fluctuations
   cosmic microwave background 44, 118–23, 181, 273–5, 286–7
   density 102–5
   mass-density 261–4, 280–90
   origins 42–54, 47–8

surface brightness 229–33, 245–6
Fourier Transform 183, 184–5
Friedmann equation 10, 44–5, 46–7
Friedmann–Robertson–Walker universe 9–10, 87, 92, 251
Fundamental Plane relations, elliptical galaxies 225–9, 245

Galactic bulge, microlensing 377–8
Galactic Zone of Avoidance (ZOA) 269
galaxies
   baryonic mass 5, 107–8
   core formation 354–7
   counts 458–9
   dark-matter relationship 195–8
   density comparisons 294–7
   distances 213–46
   distribution 195–8
   elliptical 115–16, 278–80, 338, 344–5
   gravitational lensing 384–406
   irregular 338
   luminosity 107–8, 149–50, 178–81
   Mark III catalog 240–1, 254–5, 265–9
   mass-density biasing 274–8, 303–6, 312
   nearby galaxies, mass model 451–3
   nucleus 342
   redshift 116–18, 172–207, 451–3
   spiral 111–15, 278–80, 338, 400
   structure 400–1
   surface brightness 109–10, 339–45
   velocity analysis 253–80
   velocity bias 454
galaxy centers
   black holes 337–57
   central parameter relations 342–4
   surface brightness 339–42
galaxy clusters 135–68
   Abell clusters 136, 138, 162, 163
   abundance 306–7
   baryon fraction 28–9, 151–2
   cooling flows 150
   core radii 410–11, 414
   correlation function 160–4
   dark matter 155
   density parameter 444–6
   density profiles 140–1
   galactic content 141–2
   giant arcs 407–15
   gravitational lensing 362–3, 406–23
   intracluster gas 144–9
   luminosity 237–9
   mass function 155–6
   mass-to-luminosity ratio 142–3, 154–5, 156, 421–2, 446–7, 451–3
   masses 152–4, 408–15
   measuring Ω_0 29–30, 446–7
   number density 138–9
   optical properties 136–43
   peculiar motions 164–6
   power spectrum 182–95
   quasar association 156–8, 424

galaxy clusters (*cont.*)
   Sunyaev–Zel'dovich effect 151
   unsolved problems 167–8
   velocity dispersion 142, 443–6
   X-ray properties 143–51, 330–1, 332–3
galaxy formation 4, 436–9
   age 438–9
   backwards approach 111–18
   gravitational lensing 401–2
   simulations 328–9
galaxy halos, measuring Ω_0 26–7
galaxy mass function 105–7
galaxy superclusters 5–6, 158–60
   pencil-beam surveys 159–60
   unsolved problems 168
gas
   hot 331
   near galaxies 332
Gaussian fluctuations 288–90
general relativity (GR) theory 7–9, 360–1
globular clusters (GCs), age 12–16
Grand Unified Theories (GUTs) 36–7, 43, 46, 50–1
gravitational acceleration 10–12
gravitational instability 86–96
   galaxy mass function 105–7, 274–8, 441–2
   hypothesis testing 273–4
   large-scale structure 251, 312–13
   linear theories 88–90, 99–105
   mixed boundary conditions 90–1, 161
   nonlinear methods 90–6, 105–11
gravitational lensing 360–425
   applications 363
   arclets 415–23
   astrophysical results 400–6
   circularly symmetric lens models 391–4
   cluster masses 152–3
   convergence and shear 386–7, 396, 416–17
   cosmic microwave background 425
   cosmological constant 406
   effective potential 386–7
   Einstein rings 362
   Fermat's principle 387–91
   galaxies 384–406
   galaxy clusters 362–3, 406–23
   giant arcs 407–15
   Hubble constant measurement 18–19, 361, 403–6
   Kaiser & Squires algorithm 416–18
   large-scale structure 423–5
   microlensing 39, 362, 374–84
   models 397–400
   non-circularly symmetric lens models 395–7
   point masses 364–84
   principles of 360–3
   quasars 362, 399–400, 424
   simulations 333
   strong lensing 407–15
   time-delay function 387–92
   weak lensing 407, 415–25
gravitino 37, 50
gravity, simulations 326
gravity theory 7, 10–12

Great Attractor (GA) 20, 266–9

Hamilton's principle 94–6, 449
hierarchical formation models 110–11
Higgs field 42, 46
horizon problem 45–6
Hot Big Bang 3–4
hot dark matter (HDM) 5, 33–5, 40–2
hot gas, simulations 331
Hubble constant
   Cepheid variables 16–17, 214, 216–19, 236, 245
   cosmological model 54–5
   distance indicators 215–16
   gravitational lensing 18–19, 361, 403–6
   measurement 16–21, 213–19, 233–7, 403–6
   peculiar velocities 214, 232
   supernovae 16–18, 216, 219, 233–7, 246, 308–9
   value 5, 8
Hubble diagram 234–6, 246, 308
Hubble Space Telescope (HST)
   black hole research 350–4
   Cepheid distances 16–17, 19–20, 217–19
   galaxy center observations 337–45
hydrodynamics, simulations 326
hypotheses, testing 273–5

inflation 44–54
   $\Omega_0 < 1$ 51–2
   eternal 48–50
   fluctuations 47–8
   Linde's classification 49
   summary 52–4
   supersymmetric 50–1
inflaton 46, 50–1
Infrared Astronomical Satellite *see* IRAS
initial probability distribution 287–90
intracluster gas, galaxy clusters 144–9
IRAS
   measuring β 294–7, 299–303
   redshift surveys 174–6

Jacobian matrix 386–7
Jeans mass 102

Kaiser & Squires algorithm 416–18
Kamiokande data 40–1
Karhunen–Loève (K-L) Transform 186–7

Large Magellanic Cloud (LMC) 39
   Cepheids 217
   gravitational lensing 362
   microlensing 377–9
large-scale dynamics 250–314
large-scale structure (LSS)
   COBE-normalized models 123–6
   formation theory 251–2
   gravitational instability 251, 312–13
   gravitational lensing 423–5
   mass-density fluctuations 280–90
   peculiar velocities 253–80, 312
Las Campanas Redshift Survey (LCRS) 176–7, 187

Index

least action principle (LAP) 26
LensClean technique 399
lensing geometry 369–71
LensMEM 399
light, gravitational lensing 360–425
lightest supersymmetric partner (LSP) 38–9, 50
linear biasing 183, 196–7
linear theories
  cold dark matter (CDM) 62–5
  gravitational instability 88–90, 99–105
  perturbation 88–9, 447
  redshift 190–1
Liquid Scintillator Neutrino Detector (LSND) 40
Local Group
  dynamics 448–54
  redshift survey 175–6, 194
  velocity bias 454
luminosity
  brightest cluster galaxies 237–9
  galaxies 107–8, 149–50, 178–81, 446–7
Lyman-α clouds
  simulations 331, 333
  see also damped Lyman α systems

magnetic fields, simulations 327–8
Malmquist bias, velocity analysis 241–4, 256, 257–8, 271–3
maps
  cosmic microwave background (CMB) 126–9, 274
  velocity and density fields 267–9
masers 346, 353–5
mass density, measurement 435–63
mass distribution, Local Group 448–54
mass-to-luminosity ratio 142–3, 154–5, 156, 421–2, 446–7, 451–3
Massive Astrophysical Compact Halo Objects (MACHOs) 34, 36, 39, 374–9
massive dark objects (MDO), black holes 346–7
matter power spectrum 124
microlensing 39, 362, 374–84
  extragalactic 381–4
  ongoing searches 377, 380–1
  optical depths 377–9
  pixel lensing 380
  quasar QSO 2237+0305 383–4
Milky Way
  dark halo 36, 39
  formation 112–15
mixed boundary conditions 90–1, 262
mock catalogs, simulations 264–6

Near Infrared Camera/Multi-Object Spectrograph (NICMOS) 31–2
neutralino 38
neutrinos 5, 35
  mass 40–2, 74
  warm dark matter 57–8
nonbaryonic dark matter 5

observations, comparison with simulations 322, 330–4

Occam's Razor 308
OGLE, microlensing 377–8
Optical Redshift Survey (ORS) 174–6

Peccei–Quinn symmetry 36
peculiar motions, galaxy clusters 164–6
peculiar velocities
  catalogs 240–1
  Hubble constant measurement 214, 232
  large-scale structures 253–80
  measuring Ω_0 290–3
  redshift 183, 190–5, 239–41
  Tully–Fisher relation 219–25, 245
pencil-beam surveys, galaxy superclusters 159–60
perturbation equations 87–8, 92–4
Petrosian flux 202–3
photino 37–8
point masses, gravitational lensing 364–84
Poisson field equation 262, 300
POTENT analysis
  bulk velocity 281–3
  density comparisons 294–7
  measuring Ω_0 24, 32
  power spectrum 283–4
  velocity data 256–69
power spectrum 182–95, 283–6
Primordial Isocurvature Baryonic model (PIB) 280
probability distribution function (PDF) 24, 280, 288–9, 292–3

Quantum Chromo-Dynamics (QCD) confinement transition 34–5
quasars (QSOs)
  black hole predictions 347–9
  galaxy centers 337–8
  galaxy cluster association 156–8, 424
  gravitational lensing 362, 399–400, 424
  Hubble constant measurement 18
  measuring Ω_0 30–1, 33, 309
  QSO 0957+561 18–19, 362, 390, 401
  QSO 2237+0305 383–4

redshift
  biasing 196–7
  cosmological principle 8
  distance indicators 239–41, 245
  galaxy formation 116–18
  galaxy mass 451–3
  gravitational lensing 420
  luminosity function 178–81
  measuring Ω_0 23–4, 30–1
  measuring β 204, 299–300, 301–3
  power spectrum 182–95
  selection function 180–1
  surveys 173–8, 199–207
  Tully–Fisher relation 225
relative distance measurement 16–17
Ring Cycle algorithm 399
Robertson–Walker metric 8
rocket effect 301

Sachs–Wolfe effect 274

Sakharov oscillations 121
scalar fields 42–3
Schechter function 107
Schwarzschild radius 367
secondary radiation fields, simulations 329–30
selectron 37
Shapiro delay 365
simulations
  comparison with observations 322, 330–4
  cosmological 321–34
  methods 323–30
  parameters 323–4
singular isothermal sphere, gravitational lensing 384–5
slice surveys 187
Sloan Digital Sky Survey (SDSS) 33, 178, 199–205
Space Infrared Telescope Facility (SIRTF) 31
spiral galaxies
  dark-matter density 73–4
  formation 111–15
  gravitational lensing 400
  Tully–Fisher relation 219–25
standard model of the universe 439–41
stellar age estimates 12–16
Sunyaev–Zel'dovich effect
  distance indicator 216
  galaxy clusters 151
  Hubble constant measurement 18
superclusters see galaxy superclusters
supernovae, Hubble constant measurement 16–18, 216, 219, 233–7, 246, 308–9
supersymmetry 36–9, 50–1
surface brightness
  fluctuations 229–33, 245–6
  galaxies 109–10, 339–45

time machines 96, 287–90
topological defects 42–4
Tully–Fisher relation
  inverse 256–7, 271–3, 299–300
  law 12

scatter 224–5
spiral galaxies 17, 219–25, 245
velocity analysis 253–7, 271–3
Two-Degree Field Survey 178, 205–6

universe
  age 12–16, 310–12
  mass density 4–5, 435–63
  standard model 439–41

VELMOD, measuring β 298–9
velocity analysis
  bulk velocity 281–3
  comparisons 298–9
  dispersion measurement 142, 443–4
  large-scale structure 253–80
  Malmquist bias 241–4, 256, 257–8, 271–3
  maps 267–9
  mass-density fluctuations 280–90
  methods 255–7
  power spectrum 283–6
  Wiener reconstruction 269–71
velocity bias 454
Virgo cluster, Cepheid variables 218–19
Virgocentric flow 447–8
Virgocentric infall 19–20, 198
Void Probability Function (VPF) 65, 67
voids, measuring Ω_0 290–2

warm dark matter (WDM) 5, 33–5, 57–62
Weakly Interacting Massive Particles (WIMPs) 35, 36–9
white dwarfs 14–15, 233
Wiener Filter (WF), velocity analysis 269–71

X-ray properties, galaxy clusters 143–51, 330–1, 332–3

Zel'dovich approximation 89–90, 96, 262, 288
Zel'dovich spectrum 48, 55
Zel'dovich–Bernoulli equation 288